MISSOURI'S CONSERVATION ATLAS

A Guide To Exploring Your Conservation Lands

Missouri Department of Conservation

Charlotte Overby, Editor
Russ Mueller, Cartographer
Kevin Binkley, Designer

Missouri counties with Atlas page numbers

CONTENTS

WELCOME

Missouri's Conservation Atlas is your invitation to explore almost 900 Missouri Department of Conservation (MDC) areas that have been acquired for your enjoyment. More than 900,000 acres of land, thousands of miles of streams and rivers, and 610 lakes owned and leased by the Department await your discovery.

To help you on your way, this atlas contains maps of every county in the state. We've included complete directions to all MDC areas within the counties to help you in your search for some of the best places to fish, hike, hunt, photograph or study nature. But this atlas contains more than road maps! A detailed index in the back of this book will show you exactly what facilities you'll find at each area, which species of fish and wildlife you may come across, and which activities you may participate in while you're there. We also show state parks and federal land on the maps.

We've left plenty of room to write in the margins. As you wander the wild, natural areas of Missouri, we encourage you to jot down notes about your favorite scenic trail, directions to a good spot for a tree stand—even comments about your favorite local cafe. You may wish to record dates of when you visited a particular area or the flowers and wildlife you spied there. This atlas is yours. It is designed to help you enjoy public land in Missouri—land that is valued by and entrusted to all of us.

Acquiring land for all Missourians

The Missouri Department of Conservation was founded in 1937 by a determined, visionary group of conservationists and sportsmen. In its infancy, MDC relied heavily upon the generosity of Missourians who donated land, and any purchases of land were usually made possible with partial federal funding.

It wasn't until 1977, following passage of the conservation sales tax, that the Department began more actively to acquire land under the "Design for Conservation" program. A Land Acquisition Unit within the Department was established, and a major goal of the "Design" program was to purchase land for many purposes. Today, emphasis gradually is shifting from acquisition to development. For example, in the coming years, the Department will build more river accesses so more people can enjoy Missouri's abundant waterways.

All lands purchased by the Department only come from willing sellers, and purchase prices are based on appraisals by disinterested, outside appraisal firms. It is a tribute to the Department that lands often are offered for sale below appraised values, and many as part of a donation package from citizens interested in furthering the cause of conservation or wishing to memorialize a friend or relative.

If you are interested in donating or selling land, please call or write the Policy Coordination Division, Land Acquisition Unit, Missouri Department of Conservation, 2901 W. Truman Blvd., Jefferson City, MO 65102, 573/751–4115.

Remember

Everyone who uses public land in Missouri is responsible for knowing and following the rules and regulations which govern activities on MDC areas. The *Wildlife Code of Missouri* is a booklet of rules outlined by the Conservation Commission. It is available free from the Conservation Department or from hunting and fishing license vendors. In addition, visitors always should check area bulletin boards for any posted special regulations.

Everyone is welcome on Conservation Department land, state parks and federal lands. You are not welcome on private land without the owner's consent! Respect the rights of private landowners—never enter or cross private land without prior permission.

HOW TO USE THIS BOOK

Missouri's Conservation Atlas consists of two parts:

114 county maps

Travelers may turn to any county map–all arranged alphabetically–and find road directions and brief introductions to all Department areas within that county. State parks and federal lands also appear on each county map.

LEGEND

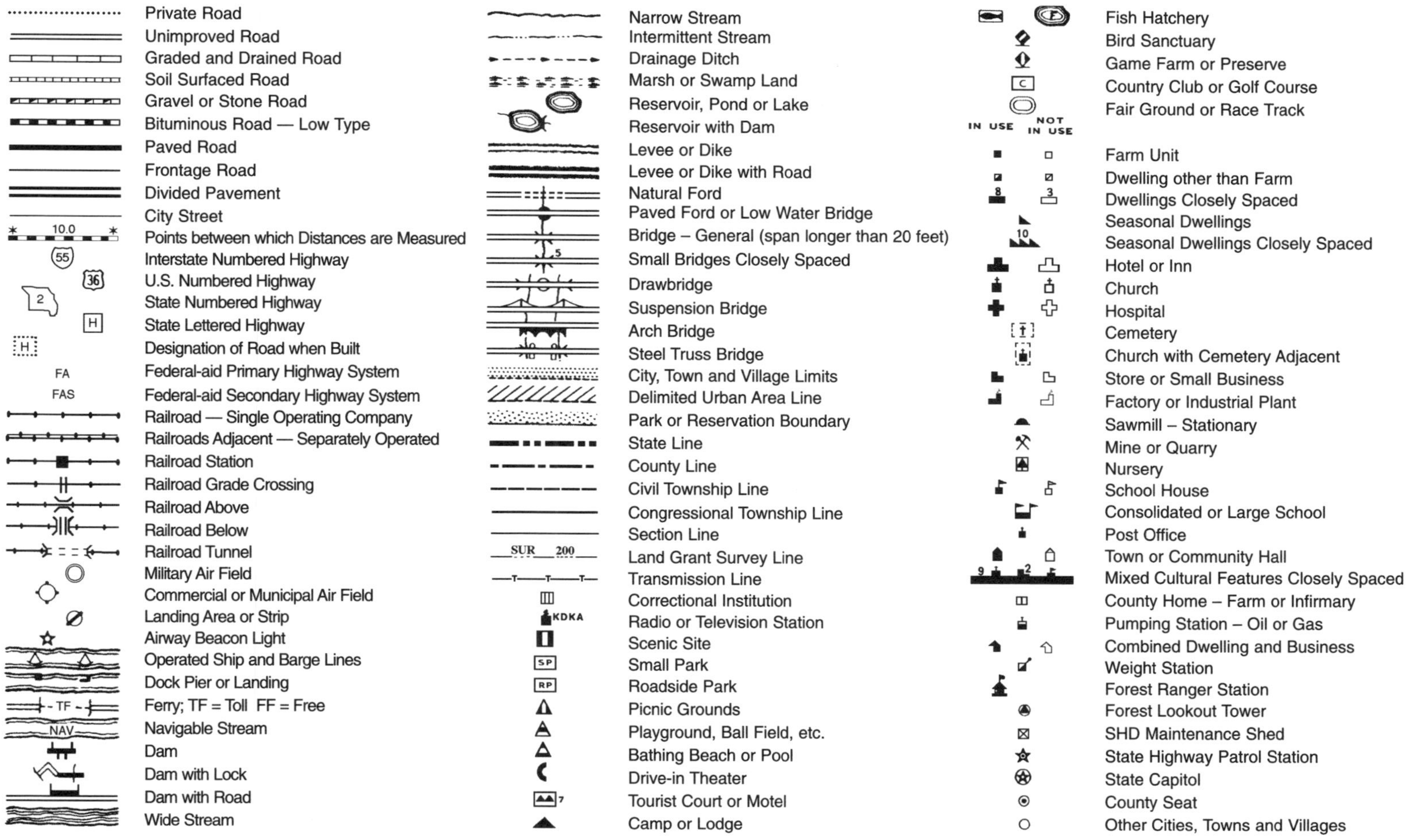

21-page index

By knowing the name of a particular Conservation Department area, users can turn to the alphabetical index to locate the appropriate county map. The index also provides users with a great deal more information about each area. For example, details about activities, types of facilities, wheelchair accessibility and the abundance of wildlife and fish for each area are listed in the index. Many areas are named for people, for example, the Ella Ewing Lake Conservation Area located in Scotland County. In the index, these areas are alphabetized by last name; in this example, the area appears under "Ewing."

The Community Assistance Program (CAP), sponsored by the Missouri Department of Conservation, provides fisheries management and assistance to local communities for developing or up–grading their lakes or ponds. Local communities retain ownership of these facilities. In return for fisheries assistance from the Department, local communities agree to maintain the areas and facilities, and open their lakes to the general public. Improved lakes and ponds, better fishing... Everyone benefits! Because some land managed by the Conservation Department may be owned by another agency or local government—a CAP lake for example—information may not always be complete or up to date. Effort was made to include information about facilities and activities on this land, but visitors should be prepared to read and follow regulations posted at each area.

ADAIR COUNTY

1 **Big Creek Conservation Area** is 1 mile southwest of Kirksville, just off State Road H. This 1,201-acre area contains 12 ponds (totaling about 10 acres) and one stream suitable for fishing. To the west, the area borders Thousand Hills State Park. The north end is managed for demonstration grazing in cooperation with Northeast Missouri State University.

2 **Elmer Cook Memorial Access** is 7 miles west of Kirksville on Highway 11, just before the Chariton River bridge. This is a 2.5-acre area.

3 **Hazel Creek Lake** (530 acres) is 2.5 miles north of Kirksville on Highway 63, then 1.5 miles west on a gravel county road (marked by a cantilever sign). This road leads to the south boat ramp. Activities on the surrounding land are regulated by the city of Kirksville.

4 **Montgomery Woods Conservation Area** is 18 miles west of Kirksville on Highway 11 and .75 mile south on County Road 417. This is a 322-acre forest.

5 **Shoemaker Conservation Area** is near Novinger. Take Highway O north about 7 miles, turn left on County Road 174, then make another quick left on County Road 171. The forest is .75 mile down 171. This 200-acre area is mainly forestland with about 27 acres of old pasture. There is one pond, and Spring Creek borders the area for .75 mile.

6 **Sugar Creek Conservation Area** is 4 miles west of Kirksville on Highway 11 and .25 mile south on Highway N. The area is mainly forestland (2,604 acres) but also has about 170 acres of old fields. There is a 1.8-mile natural hiking trail and an 8-mile trail open for horseback riding. A permit is required for groups of 10 or more riders.

7 **Henry Truitt Access** is 1 mile east of Novinger on State Highway 6. It is a public access (but no boat ramp) on the Chariton River. The 3.5 acres of surrounding land are primarily old pastures.

8 **Union Ridge Conservation Area** is on Highway D, 2 miles north of Greencastle. Of the 8,164 acres, about half is forestland and the rest is grass and old pastures. Thirty-two ponds, one lake and a stream provide 14 total acres for fishing. Horseback riding is allowed with a permit. Additional tracts are located in Putnam and Sullivan counties.

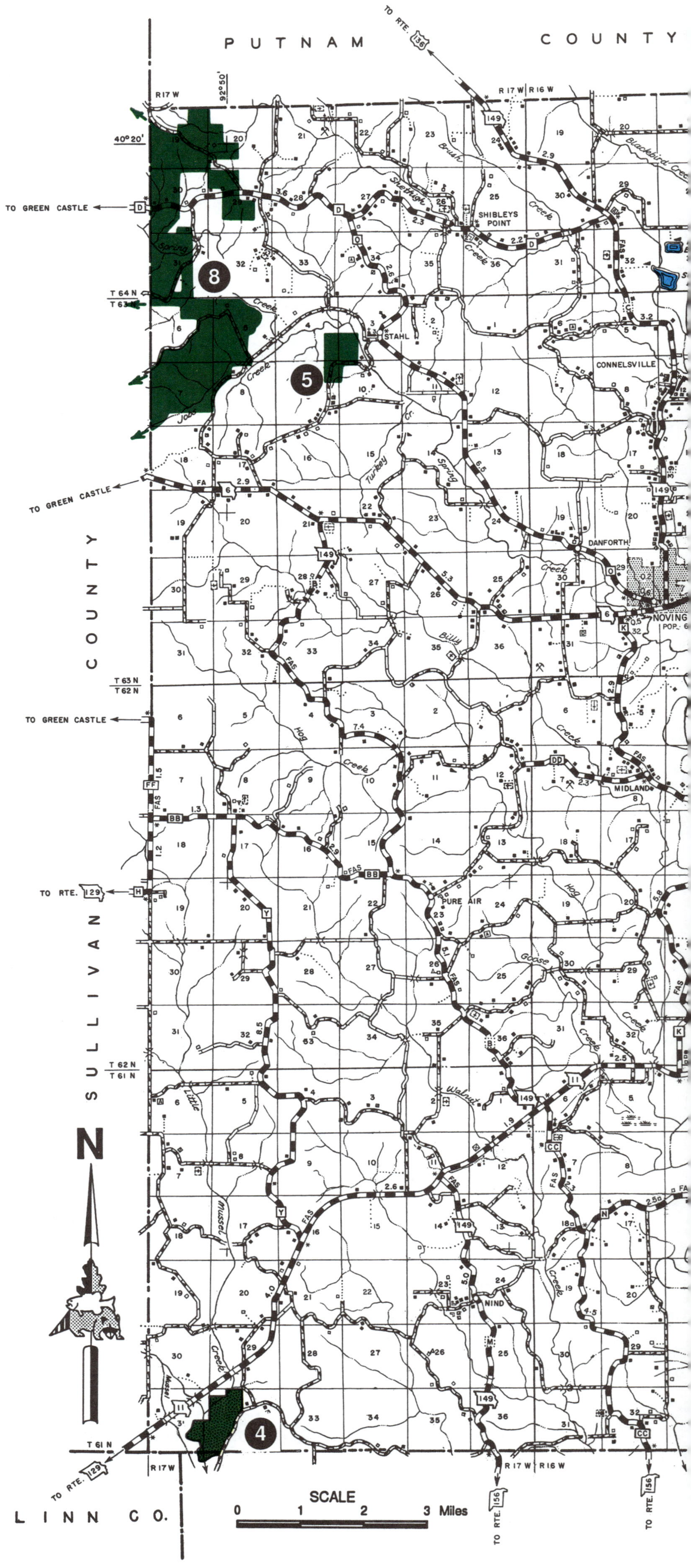

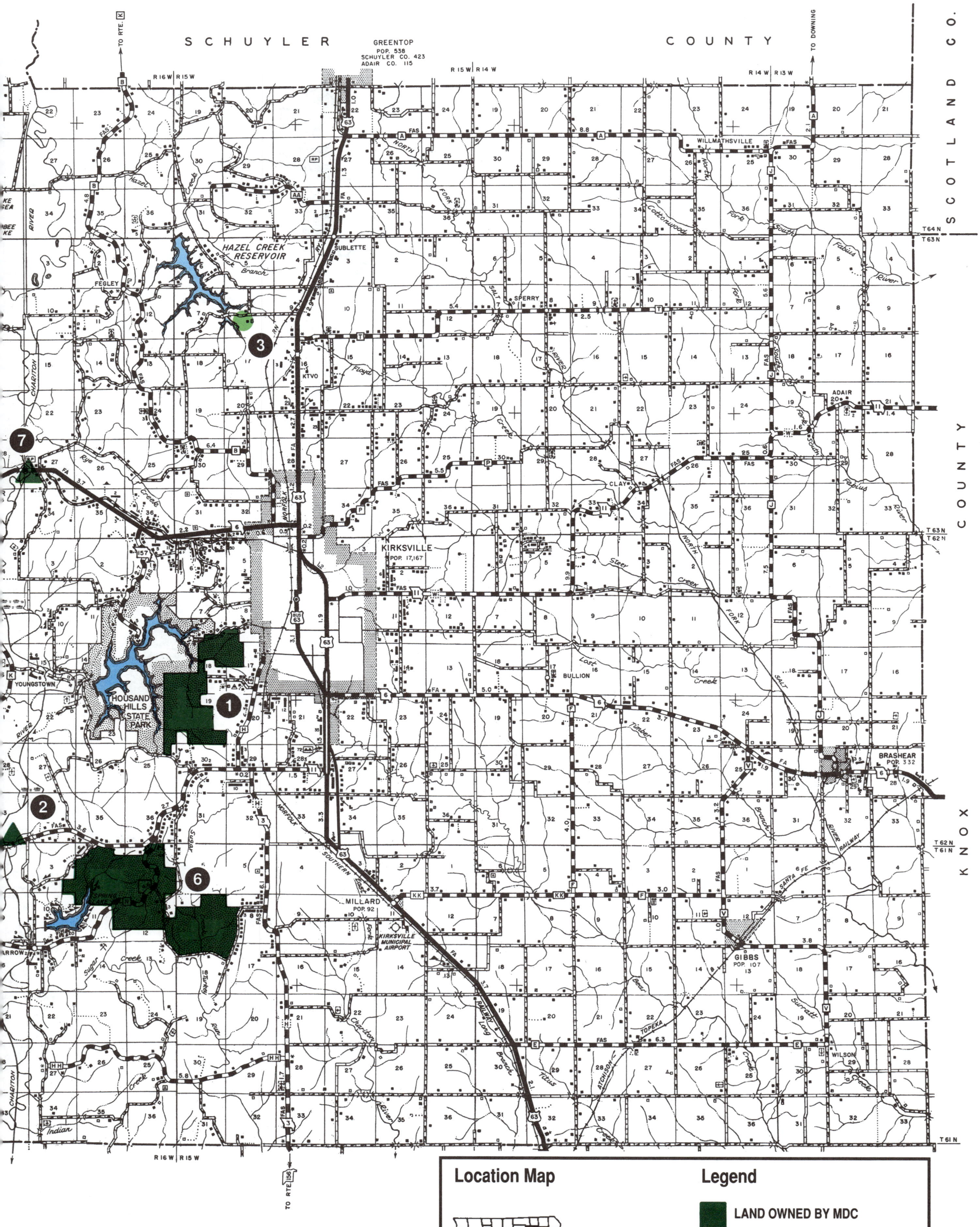

Location Map

Legend

- LAND OWNED BY MDC
- LAND LEASED BY MDC
- LAND OWNED BY MDC LESS THAN 160 ACRES
- LAND LEASED BY MDC LESS THAN 160 ACRES
- CAP ACCESS

ANDREW COUNTY

1 **Tom Brown Access** is 3 miles west of Amazonia off Highway T. This is a 7.6-acre area on the Missouri River with no boat ramp.

2 **James D. Christie Conservation Area** and **Eva Neely Davis Memorial Conservation Area** are off Highway 48, 1.5 miles northeast of Rosendale. The 204-acre areas are made up equally of forestland, cropland and old pastures. There are three ponds with 3 total acres suitable for fishing.

3 **Elrod Mill Access** on the Platte River is 2 miles west of Union Star, 3 miles north on Highway M, then 4 miles west on a gravel road. There are two accesses on the river in the 57-acre area, though there are no boat ramps.

4 **Hadorn Bridge Access** is 4 miles north of Savannah off Highway C. The 102 River extends for nearly a mile along the 93-acre area. There is no boat ramp.

5 **Happy Holler Conservation Area** is composed of three tracts. The main entrance is 4 miles east of Savannah on Highway E, then 6 miles north off Highway D. Within the 2,207-acre area, there are three accesses on the 102 River. There are three hiking trails totaling 5 miles in length.

6 To reach **Honey Creek Conservation Area,** take the Fillmore exit (exit 65) off I-29. The 1,448-acre area is then 1.5 miles south on Road RA. It is primarily forested and has three accesses on the Nodaway River and four ponds. There is one trail .75 mile long.

7 **Midway Access** is 1 mile west of Bolckow on Highway B, just where the highway crosses the 102 River. It consists of 1 acre and has no boat ramp.

8 **Nodaway Island Access** is south of I-29. From the Amazonia exit (exit 60), go 2 miles south on Highway K, then turn north on Highway T. The access extends 400 feet along the Missouri River beside Highway T. The 6 acres surrounding the river access is mainly pasture. There is a boat ramp.

9 The main entrance to **Nodaway Valley Conservation Area** is 8 miles north of Highway 59 on Highway B in Holt County. Three hundred acres of this 3,833-acre area is wetlands, and the rest mainly cropland and old fields. The Nodaway River and one pond provide 40 acres of fishable water.

10 **Rochester Falls Access** is about 20 miles northeast of St. Joseph on Highway 169. The 14-acre area is an access on the Platte River. The land is mainly pasture and extends .25 mile along the river.

11 **Rock Quarry Access** is 2 miles north of St. Joseph off Highway 71. From Highway 71, take the gravel road marked with an area sign 1.5 miles. When this road comes to a T-intersection, turn .25 mile left to reach the 8.6-acre access. This area has a boat ramp on the 102 River.

12 **Savannah City Lake** is reached by following signs from Highway Business 71 north of Savannah. This 22-acre reservoir has a boat ramp and offers good catfish, bass and bluegill fishing.

13 **Worthwine Island Conservation Area** is on Highway K, 5 miles south of I-29 (Amazonia exit). Two and three-quarter miles of the Missouri River flows through this 430-acre area. There is access to the river and a 2.5-mile hiking trail.

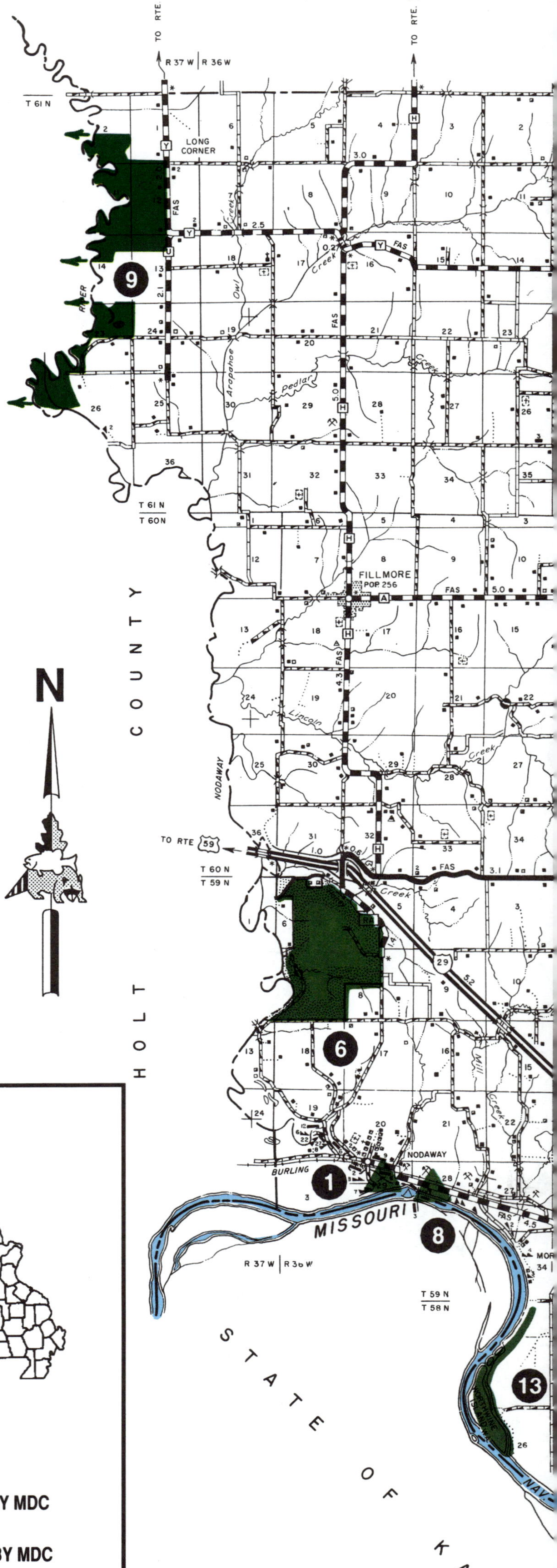

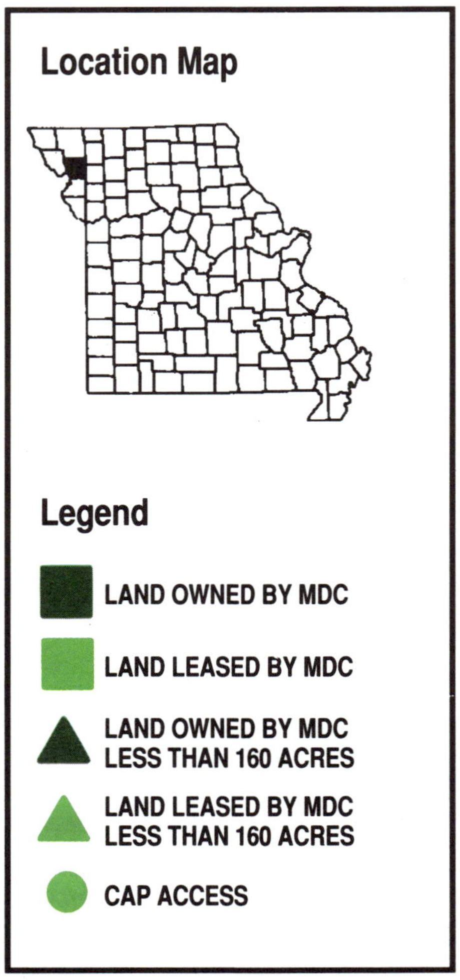

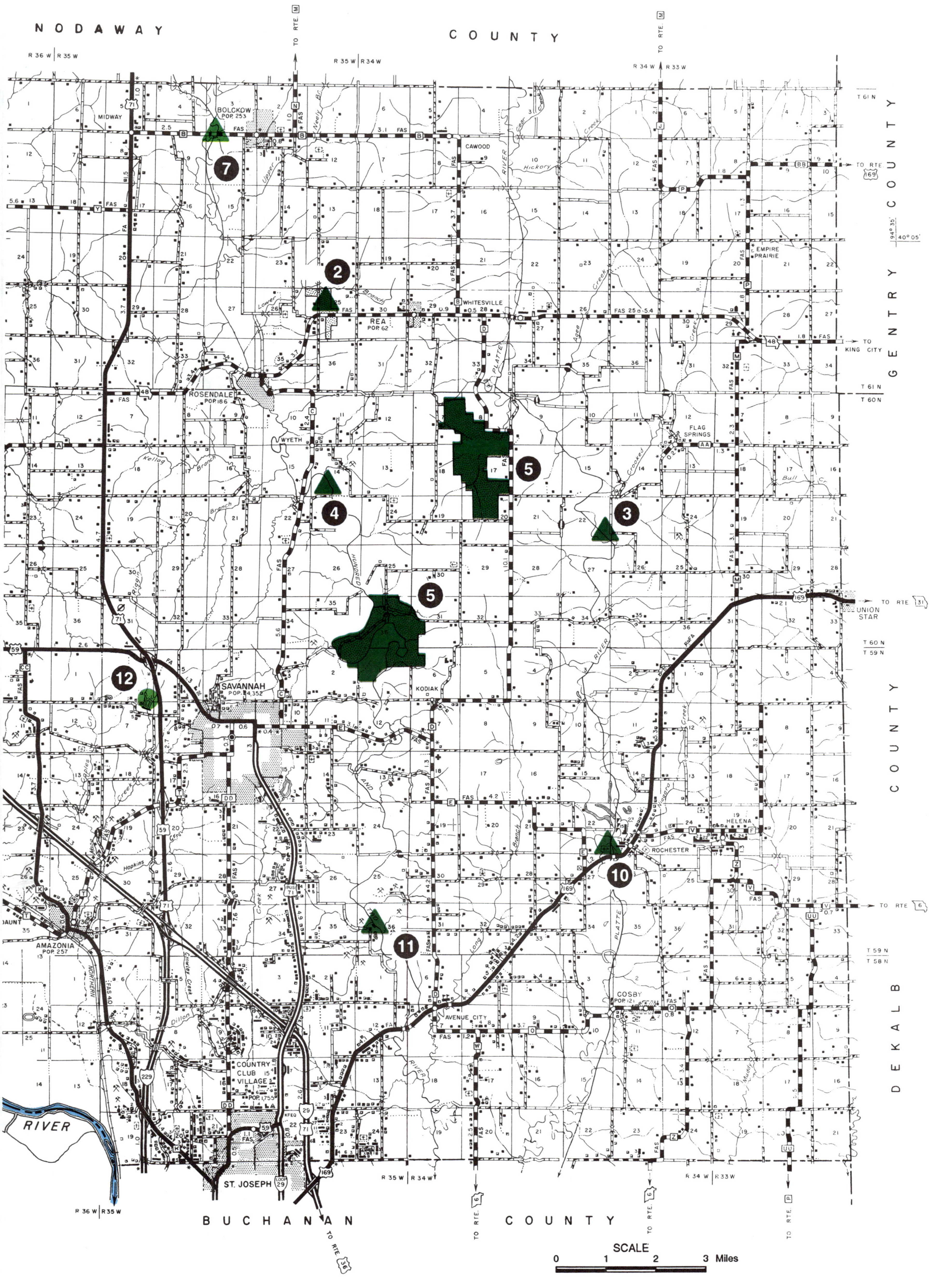
NODAWAY COUNTY
GENTRY COUNTY
DEKALB COUNTY
BUCHANAN COUNTY
MIDWAY
BOLCKOW POP. 253
CAWOOD
EMPIRE PRAIRIE
WHITESVILLE
REA POP. 62
ROSENDALE POP. 186
WYETH
FLAG SPRINGS
TO KING CITY
UNION STAR
SAVANNAH POP. 4,352
KODIAK
HELENA
ROCHESTER
AMAZONIA POP. 257
COSBY POP. 121
AVENUE CITY
COUNTRY CLUB VILLAGE POP. 1,755
ST. JOSEPH
RIVER
PLATTE RIVER
HUNDRED AND TWO RIVER
Hickory
Kellog
Bull Cr.
Crooked
Agee Creek
Hopkins
Dillon
Long Branch
Muddy Cr.
T 61 N
T 60 N
T 59 N
T 58 N
R 36 W | R 35 W
R 35 W | R 34 W
R 34 W | R 33 W
94°35′
40°05′
TO RTE. M
TO RTE. 169
TO RTE. 31
TO RTE. 6
TO RTE. 36
TO RTE. P
SCALE
0 1 2 3 Miles

ATCHISON COUNTY

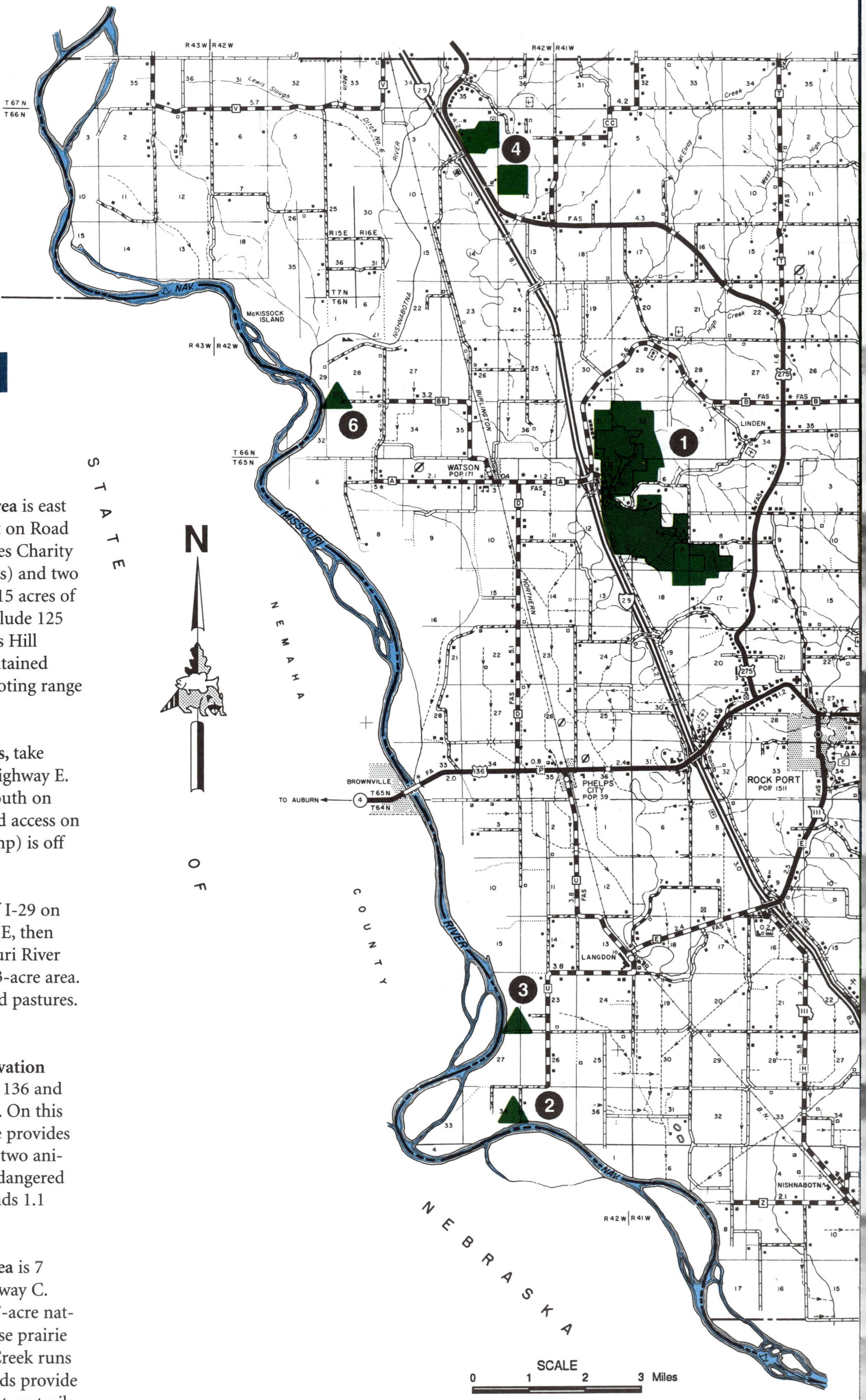

1 **Brickyard Hill Conservation Area** is east of I-29 on Highway A, then east on Road RA. This 2,262-acre area includes Charity Lake (equipped with boat ramps) and two ponds, which together provide 15 acres of fishable water. Natural areas include 125 acres of the Brickyard Hill Loess Hill Prairie. The area has three maintained trails extending 4.8 miles, a shooting range and an archery range.

2 To get to **Hoot Owl Bend Access,** take Highway 111 west off I-29 to Highway E. Continue west on E and turn south on Highway U. This 8-acre area and access on the Missouri River (no boat ramp) is off Highway U.

3 **Langdon Bend Access** is west of I-29 on Highway 111, west on Highway E, then west on Highway U. The Missouri River extends .25 mile through this 13-acre area. The area is mainly forest and old pastures. There is a boat ramp.

4 **Star School Hill Prairie Conservation Area** is east of I-29 on Highway 136 and 12 miles north on Highway 275. On this 289-acre area, a loess hill prairie provides habitat for 12 plant species and two animal species on the Rare and Endangered Species lists. A hiking trail extends 1.1 miles.

5 **Tarkio Prairie Conservation Area** is 7 miles east of Westboro off Highway C. This 560-acre area contains a 57-acre natural area, which features a diverse prairie flora on glacial till soil. Tarkio Creek runs through the area, and three ponds provide 3 acres of fishable water. Two nature trails extend 1.5 miles over natural and grass terrain.

6 **Watson Access** is west of I-29 on Highway A and north on Highway BB. This 4-acre area is an access on the Nishnabotna River. It is equipped with a boat ramp.

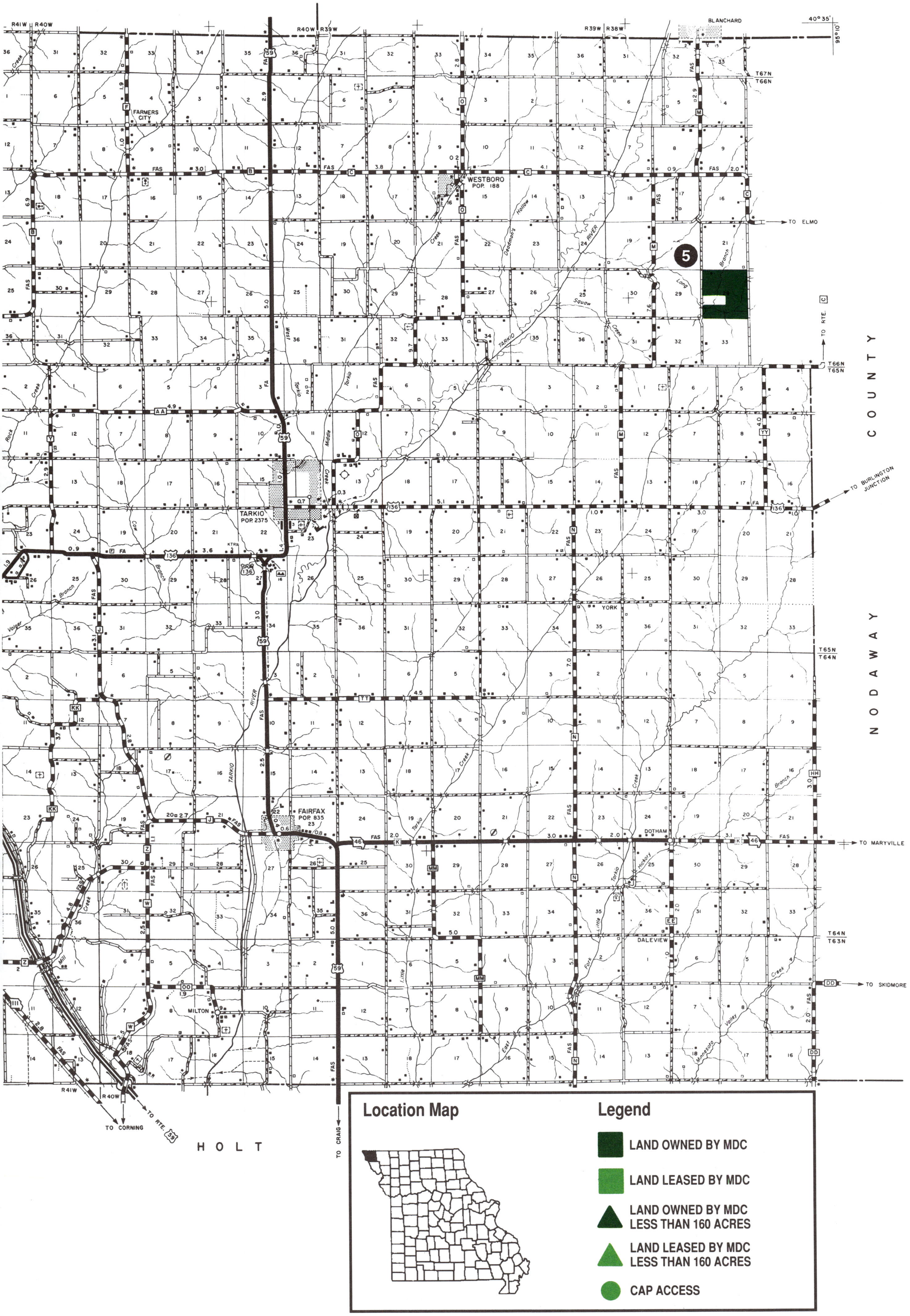
BLANCHARD
40° 35'
95° 10'
R41W R40W
R40W R39W
R39W R38W
T67N T66N
T66N T65N
T65N T64N
T64N T63N
FARMERS CITY
WESTBORO POP. 188
TARKIO POP. 2375
FAIRFAX POP. 835
YORK
DOTHAM
DALEVIEW
MILTON
KTRX
TO ELMO
TO RTE. C
TO BURLINGTON JUNCTION
TO MARYVILLE
TO SKIDMORE
TO CORNING
TO RTE. 59
TO CRAIG
NODAWAY COUNTY
HOLT
TARKIO RIVER
Location Map
Legend
LAND OWNED BY MDC
LAND LEASED BY MDC
LAND OWNED BY MDC LESS THAN 160 ACRES
LAND LEASED BY MDC LESS THAN 160 ACRES
CAP ACCESS

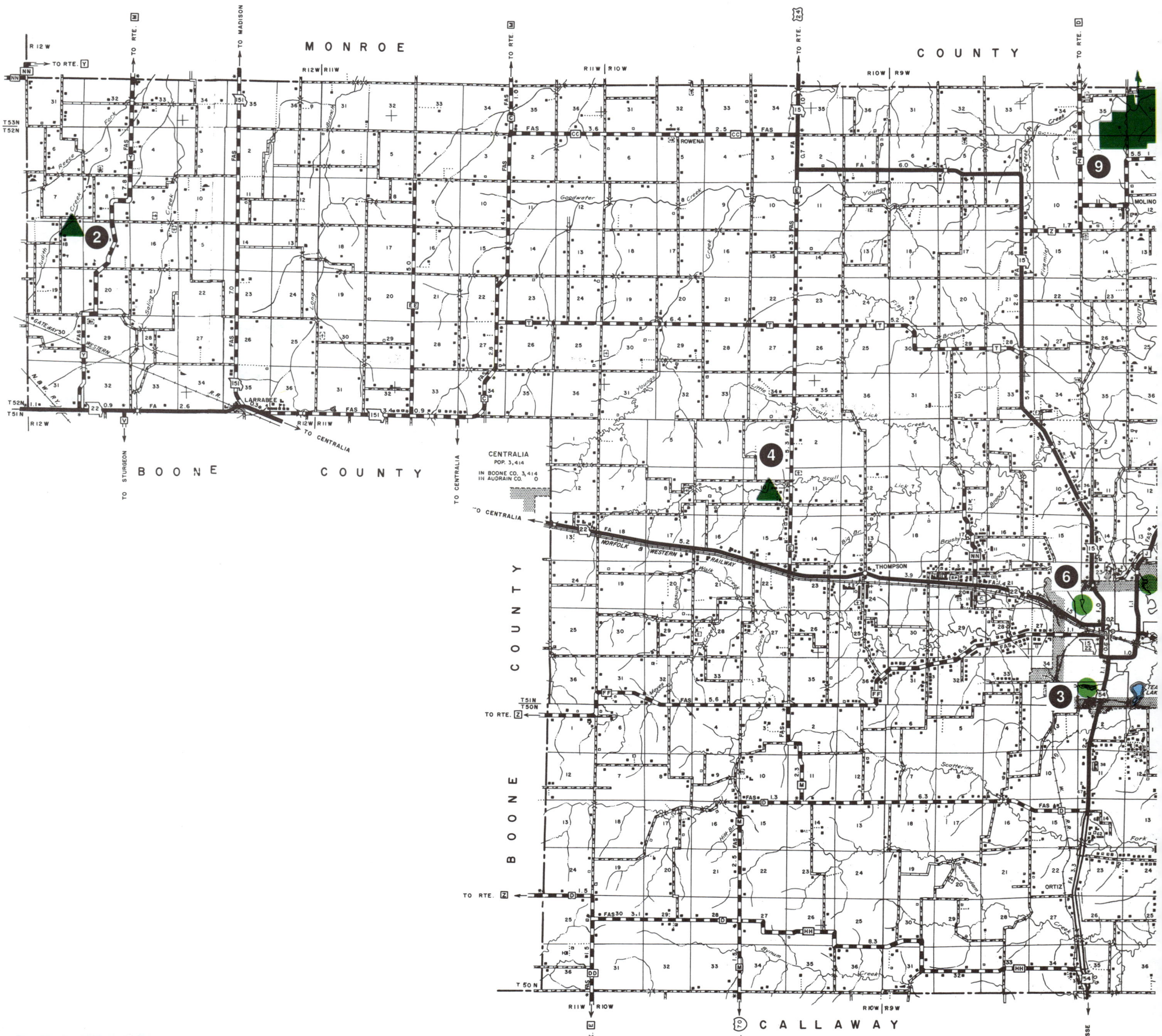

AUDRAIN COUNTY

1 **Marshall I. Diggs Conservation Area** is approximately 3 miles west of Wellsville on Montgomery County Route ZZ and .25 mile west on Route RA. Little Loutre Creek runs 3 miles through this 1,015-acre area and two lakes, Lake Walter and Lake Whitesell, provide 24 acres for fishing. There is a boat ramp and a shelter house.

2 **Maude Shores Jacks Conservation Area** is 8 miles west of Centralia on Highway 22, 4 miles north on Highway Y, then 3 miles west on County Gravel Road 367. The area is 23 acres.

3 To reach **Lakeview Park Lake,** take Highway 54 north to Mexico. Near the south side of town, take Lakeview Road west and follow the signs 1 block to the lake. This is a 45-acre lake with a trail and picnic areas.

4 **C. L. Northcutt Memorial Conservation Area** is 5.5 miles west of Mexico on Highway 22, 1.75 miles north on Route E, and .25 mile left on Gravel Road 324. This is an 80-acre forest.

5 **Northeast Park Lake** is north on Highway 54 to Route J. Take J north to Anderson Street. Go east on Anderson and follow the signs. This is a 25-acre area.

6 **Plunkett Park Lake** is near Kiwanis Lake in Mexico. From Highway 54, take Highway 15 north to Hendricks Drive. Take Hendricks Drive west to reach the 110-acre area.

7 **F. O. and Leda J. Sears Memorial Wildlife Area** is 3.5 miles north of Rush Hill on Route B. This 160-acre area is 40 percent native prairie.

8 **Vandalia Community Lake** is 6 miles east of Laddonia off Highway K. The 44-acre lake is equipped with a boat ramp. The land around the lake (102 acres) is mainly forestland and old fields. There is a shelter house, and some picnic areas have grills and tables.

9 **Robert M. White II Conservation Area** is 11 miles north of Mexico on Highway 15, 1.75 miles on Route Z, and 2 miles down Route ZZ to the entrance. Long Branch and Young's Creek flow through this 1,163-acre area. There also are two lakes and five ponds providing anglers with 30 acres of fishable water. The area is composed of forest (447 acres), cropland (430 acres), old pasture (175 acres) and prairie (80 acres).

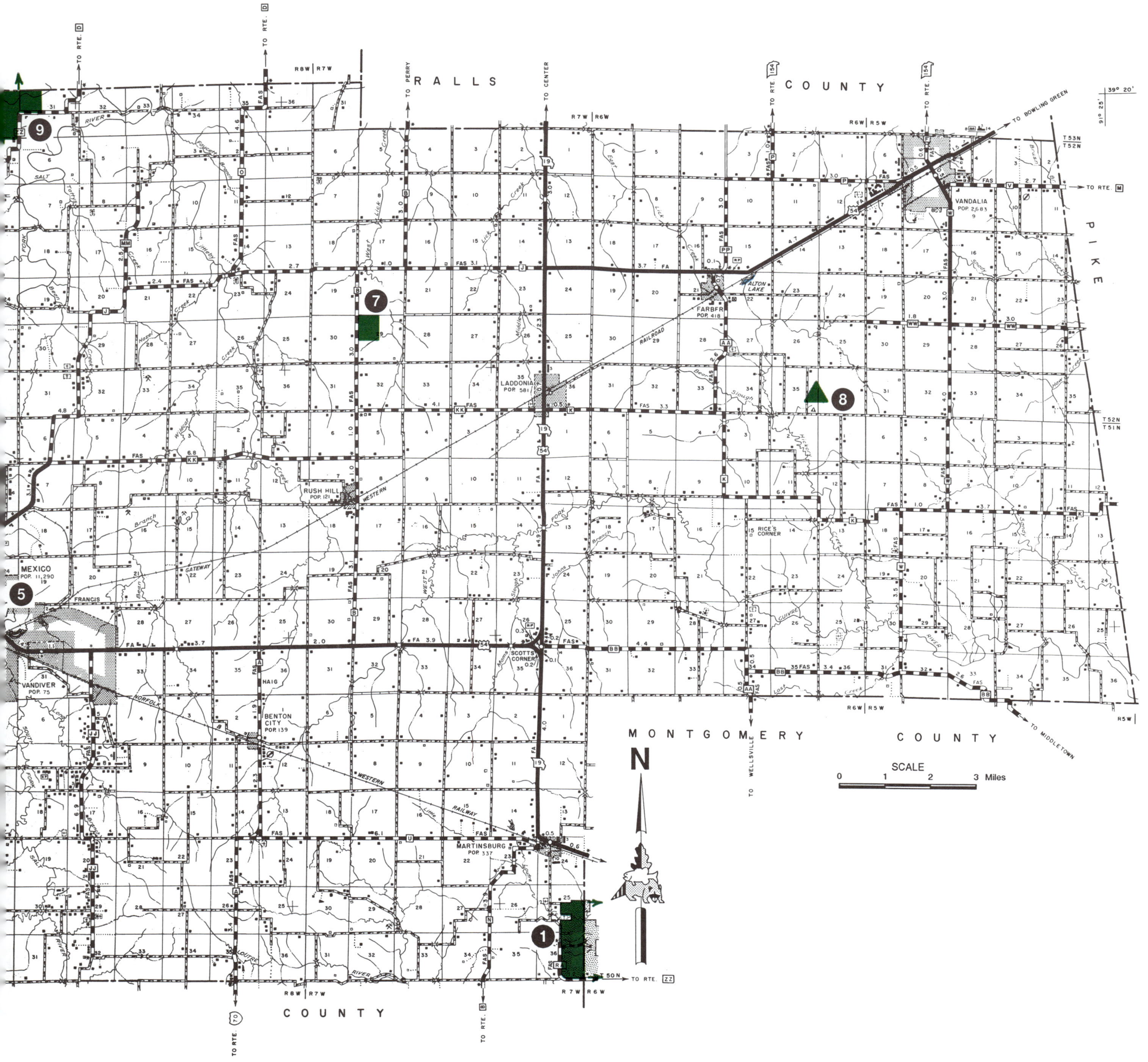

Location Map

Legend

- LAND OWNED BY MDC
- LAND LEASED BY MDC
- LAND OWNED BY MDC LESS THAN 160 ACRES
- LAND LEASED BY MDC LESS THAN 160 ACRES
- CAP ACCESS

BARRY COUNTY

1 **Cassville City Greenway** is east off Business 37 on 14th Street, then left on 13th Street. Parking is at Rocky Edmonson Park. The area is 1 acre.

2 **Flag Spring Conservation Area** is northwest of Washburn about 1.75 miles on Highway UU, then left 3 miles on the first public gravel road. This 3,955-acre area is mainly forestland and has 13 ponds (not suitable for fishing). Approximately 15 miles of interior, natural surface roads are closed to vehicle traffic but may be used for horseback riding. There also is a shooting range.

3 **Jenkins Towersite** (17.5 acres) is south of Aurora. From Highway 39, take Highway WW until the road turns to gravel. Continue on the gravel road 1 mile, then turn .5 mile left on the first gravel road.

4 **Lower Flat Creek Access** is 3 miles east of the Highway 39 and Highway EE junction. This area is 4 acres of gravel bar and lawn along Flat Creek. There is parking, but no boat ramp.

5 **Roaring River Conservation Area** consists of two tracts. The main tract is 2 miles east of Roaring River State Park on Highway F. The area is 439 acres of forestland and 20 acres of old pastureland. Roaring River flows through the area for about .5 mile.

6 **Roaring River Fish Hatchery** is inside Roaring River State Park, 7 miles south of Cassville on Highway 112. The hatchery grounds are 3 acres and home to a 7.5-mile stretch of the Roaring River and the Roaring River Spring. Natural areas include the spring, a 90-foot bluff and Devil's Icebox Spring, which flows in a waterfall from a bluff-top crevice. Hatchery tours are given daily at 10:30 a.m. and 2 p.m. from Memorial to Labor days, and daily at 2 p.m. in May and September. Appointments for tours are accepted during other months of the year.

7 To reach **Stubblefield Access**, go 2.5 miles west of the junction of highways 39 and 248 (on Highway 248), then turn 2.5 miles north (right) on an unnamed gravel road that lies across the highway from Shiloh Church. This is 56 acres on Flat Creek with no boat ramp.

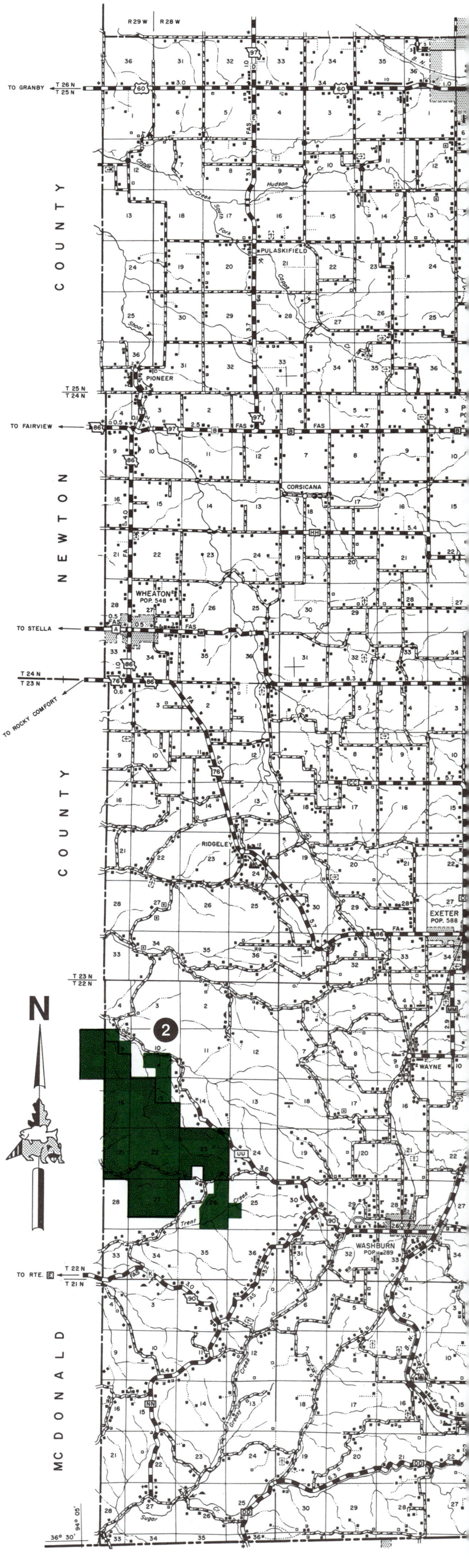

Location Map

Legend

- LAND OWNED BY MDC
- LAND LEASED BY MDC
- LAND OWNED BY MDC LESS THAN 160 ACRES
- LAND LEASED BY MDC LESS THAN 160 ACRES
- CAP ACCESS

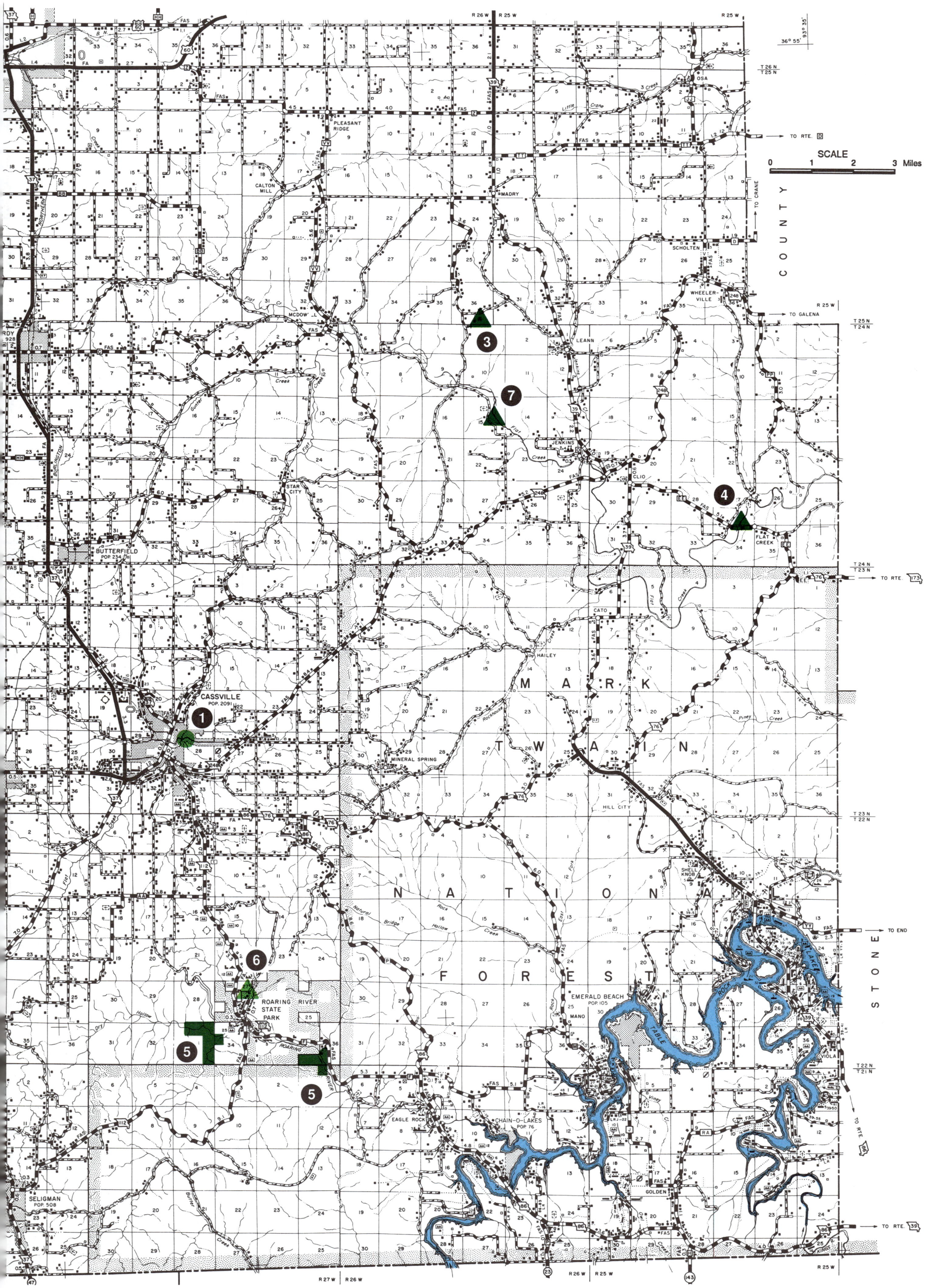
SCALE
0 1 2 3 Miles
R 26 W R 25 W
T 26 N
T 25 N
T 24 N
T 23 N
T 22 N
T 21 N
OSA
PLEASANT RIDGE
CALTON MILL
MADRY
SCHOLTEN
WHEELER-VILLE
TO RTE. D
TO CRANE
TO GALENA
COUNTY
LEANN
JENKINS
CLIO
STAR CITY
FLAT CREEK
BUTTERFIELD POP. 234
TO RTE. 173
CATO
HAILEY
CASSVILLE POP. 2091
MINERAL SPRING
M A R K
T W A I N
N A T I O N A L
F O R E S T
HILL CITY
SHELL KNOB
ROARING RIVER STATE PARK
EMERALD BEACH POP. 105
MANO
TO END
STONE
VIOLA
EAGLE ROCK
CHAIN-O-LAKES POP. 76
TO RTE. 86
GOLDEN
SELIGMAN POP. 508
TO RTE. 39
R 27 W R 26 W
R 26 W R 25 W
R 25 W

BARTON COUNTY

1 **Bethel Prairie Conservation Area** is 4 miles west of Highway 71 on Highway 126. This 260-acre area is mostly prairie.

2 **Buffalo Wallow Prairie Conservation Area** consists of two tracts. The main tract is 3 miles south of Sheldon on Highway 71. The access road is marked with a sign. Native prairie extends for 409 acres in this area and is home to prairie chickens. Forestland, old fields and cropland make up the rest of the 1,017 total acre area.

3 **Bushwhacker Lake Conservation Area** is 1 mile east of Highway 43 on the Barton-Vernon County Road. This 4,337-acre area has native prairie, cropland, forestland and old fields. Bushwhacker Lake and Willow Lake together provide 198 acres of fishable water. There is a boat ramp and jetty on the lake, and a stream also extends 8.7 miles through the area.

4 **Clear Creek Conservation Area** is 4 miles east of Highway 71 on NE 80 Road, and 1 mile north on NE 40 Lane. About a mile of Clear Creek flows through this 762-acre area and, together with five ponds, provide anglers with 5.2 acres of fishable water. A hiking trail winds 2.5 miles through the area, which is composed mainly of prairie (457 acres) and forestland (225 acres). Additional tracts lie in Vernon County.

5 **Comstock Prairie Conservation Area** is 1 mile west of Highway 43 on NW 100 Road. A .5-mile stretch of Comstock Creek flows through this 320 acres of prairie, providing anglers with 1.8 acres of fishable water.

6 **Davis Memorial State Forest** is 4 miles north of Mindenmines off Highway 160. This is an 85-acre forested area with 5 acres of strip pits.

7 **Dorris Creek Prairie Conservation Area** is 4 miles east of Highway 71 and 1 mile south on Gravel Road SE 40 Lane. This 160-acre native prairie has three ponds providing anglers with 9.5 acres of fishable water.

8 **Drywood Conservation Area** is north of Joplin on Highway 43 to Route V (9 miles north of Highway 160). Go 4 miles west on Route V, then .5 mile south on Route K to the area. This is 100 acres of mainly old pastureland with about 2.5 acres of fishable water.

9 **Lamar Access** is 2 miles north of Lamar on Northeast 15th Lane (off Truman). This is a 30-acre area with a boat ramp on the North Fork of Spring River.

10 **Mo-No-I Prairie Conservation Area** is 4 miles west of Highway 160 on Highway DD. This is a 302-acre prairie.

11 **Mon-Shon Prairie Conservation Area** (80 acres) is 2.5 miles south of Highway 126 on SW State Line Lane.

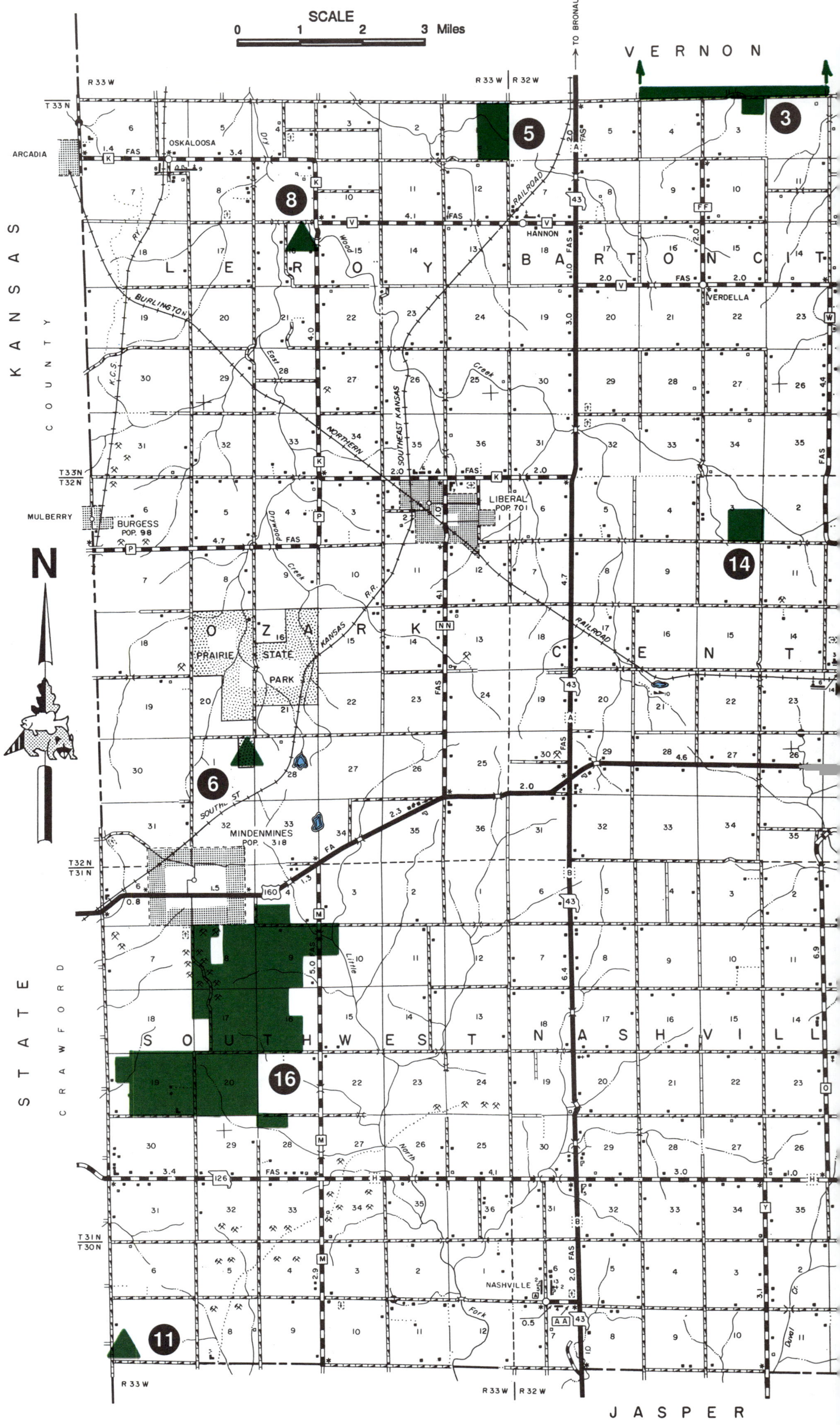

12 **Pa Sole Prairie Conservation Area** (240 acres) is 5 miles east of Highway 160 on Highway 126, then 1.25 miles north on SE 50 Lane.

13 **Pawhuska Prairie** (77 acres) is 3 miles north of Highway 160 on Highway HH, then .5 mile west on NE 30 Road. This is a Designated Natural Area.

14 **Redwing Prairie Conservation Area** (160 acres) is 2.5 miles east of Highway 43 on NW 30 Road.

15 **Edward B. and Marie O. Risch Conservation Area** is 4 miles east of Highway 71 on NE 80 Road, 1 mile north on NE 40 Lane, and 1 mile east on NE 90 Road. There are three ponds on this 162-acre area (composed mainly of prairie) providing anglers with 1 acre of fishable water.

16 **Shawnee Trail Conservation Area** is on County Road M, 1 mile south of Highway 160. This 3,635-acre area has 18 lakes, four ponds and numerous strip pits.

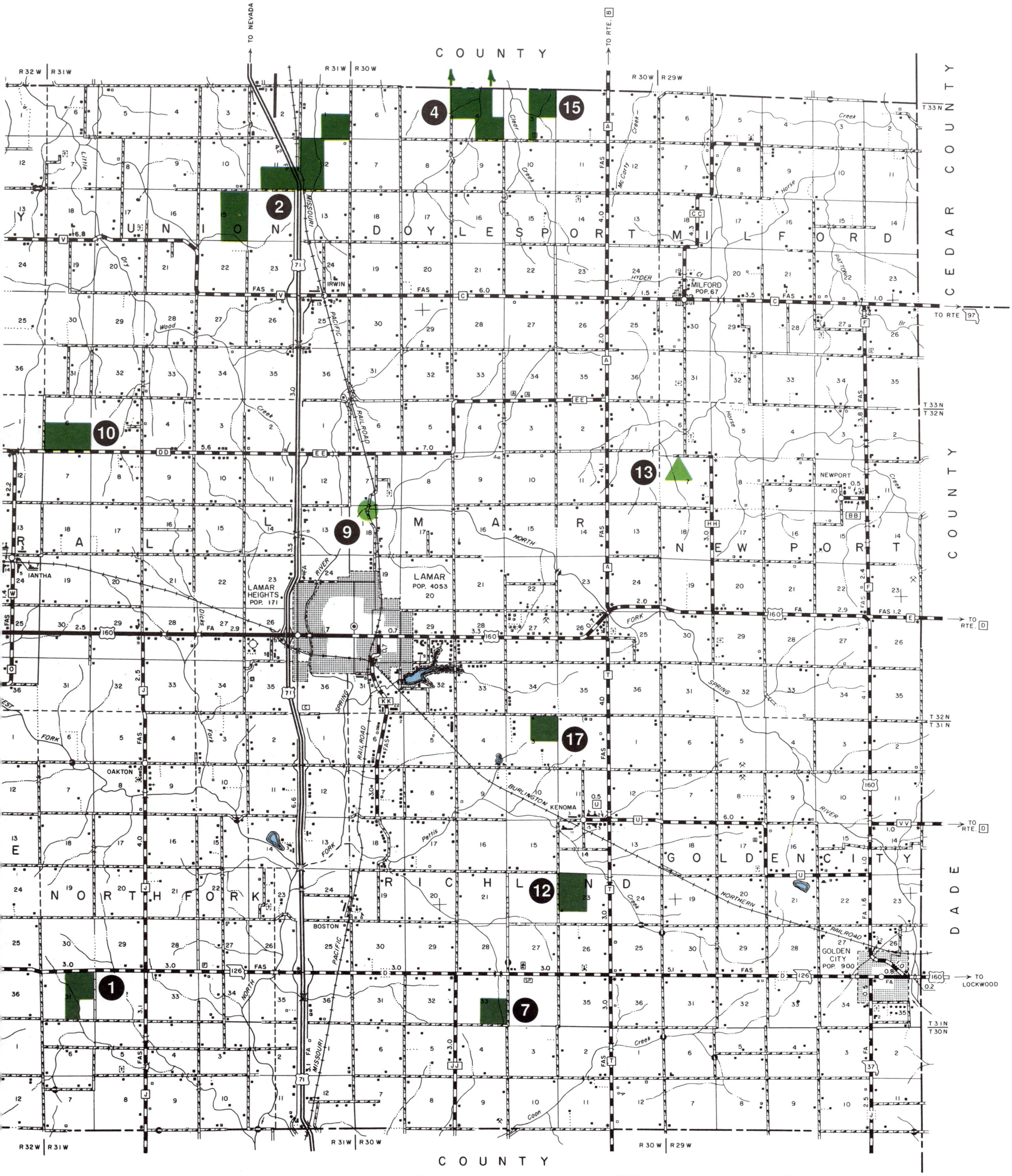

17 **Treaty Line Prairie Conservation Area** (168 acres) is 1.5 miles south of Highway 160 on SE 40 Lane and .5 mile east on SE 20 Road.

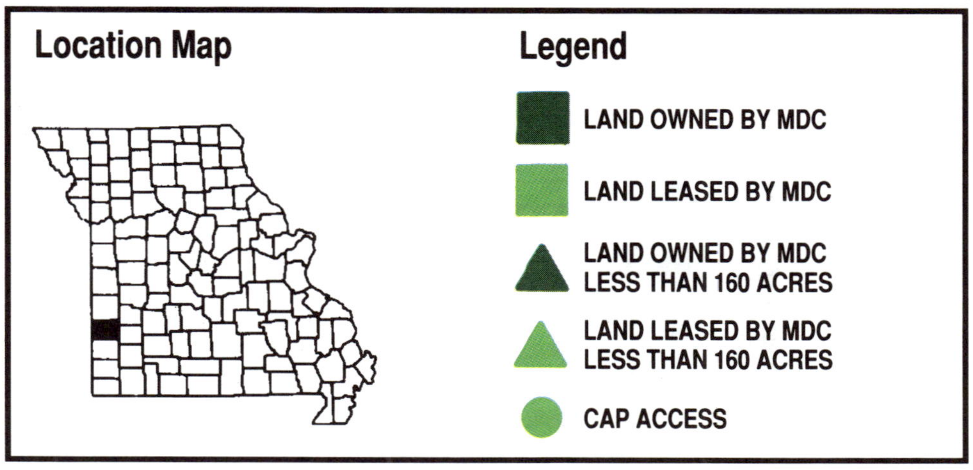

BATES COUNTY

1 **Butler City Lake** is 4 miles west of the city of Butler off Highway 52. This is a 71-acre lake with a boat ramp and wheelchair accessible fishing dock.

2 **Four Rivers Conservation Area** lies in Bates and Vernon Counties. The main entrance is in Vernon County, 15 miles north of Nevada on Highway 71, 1 mile east on Highway TT, then 2.5 miles south on a gravel road marked with a cantilever sign. The area has a public contact office and 59 acres of fishable water. Little Osage River, Marmaton River, Marais Des Cygnes River and Muddy Creek all run through it, and there are four ponds and five strip pits. Designated Natural Areas include Horton Bottoms Natural Area, 227 acres of wet prairie and wet savannah. The area also contains an Osage Native American village site and 2,029 acres of seasonal wetlands.

3 **Harmony Mission Lake Conservation Area** is .75 miles west of Rich Hill on Highway A and 4.25 miles southwest on Highway PP. The pavement on PP ends, and an unnamed county gravel road continues west about 1 mile to the area. This 1,080-acre area contains Harmony Mission Lake, which provides anglers with 96 acres of fishable water. There is a boat ramp and wheelchair accessible fishing jetty.

4 **Old Town Access** (310 acres) is 3 miles north of Rich Hill on Highway 71. Marais Des Cygnes River and the Bates County Drainage Ditch flow through the area, providing anglers with 10 acres of fishable water. There is parking and a boat ramp. Natural features of interest include an oxbow lake and several shallow wetlands.

5 **Peabody Conservation Area** is approximately 4 miles west of Rich Hill on Highway A. This 299-acre area is dominated by strip pits and spoil banks. There are 23 acres of lakes and ponds and two boat ramps.

6 **Ripgut Prairie Natural Area** is .75 mile east of Rich Hill on Highway B, 1.25 miles north and east on a gravel road (marked on Highway B with a sign), then .25 mile north on another unnamed gravel road (but also marked with a sign). This is a 280-acre area with 124 acres of natural, wet prairie.

7 The main entrance to **Settle's Ford Conservation Area** is 4 miles west of Creighton on Highway B and 1 mile south on Index Road (gravel) in Cass County. This is 6,611 acres composed mainly of wetlands (2,500 acres) and old pastures. About 822 acres are forestland. South Grand River flows through the area.

Location Map

Legend

- LAND OWNED BY MDC
- LAND LEASED BY MDC
- LAND OWNED BY MDC LESS THAN 160 ACRES
- LAND LEASED BY MDC LESS THAN 160 ACRES
- CAP ACCESS

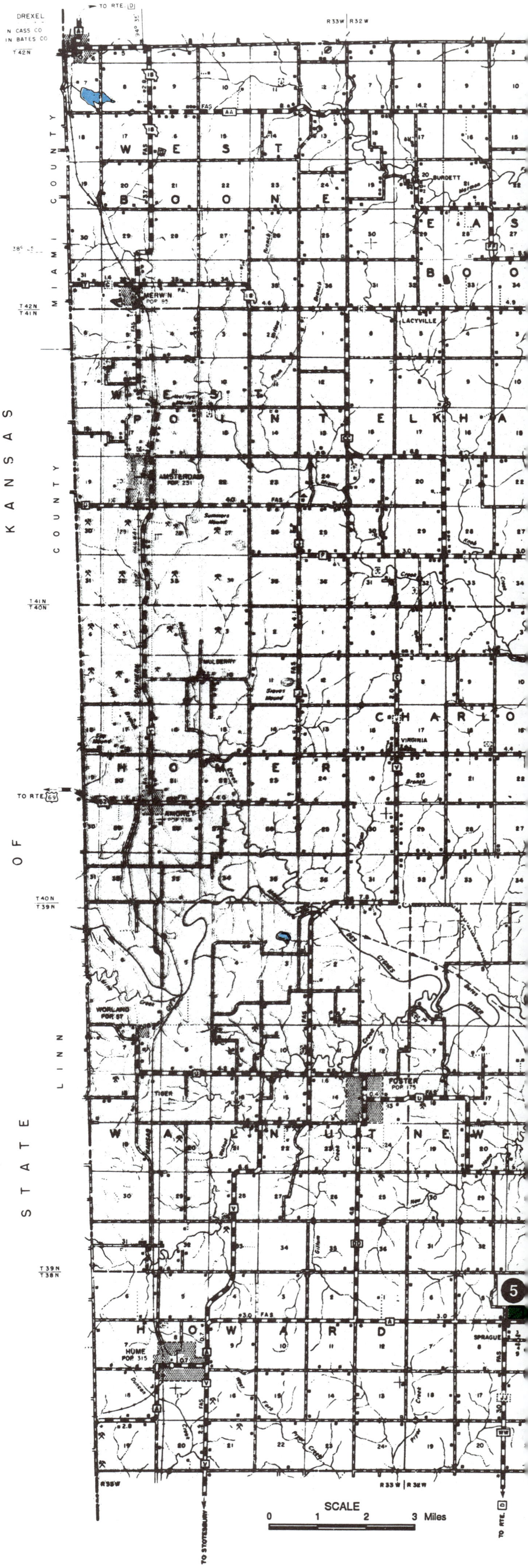

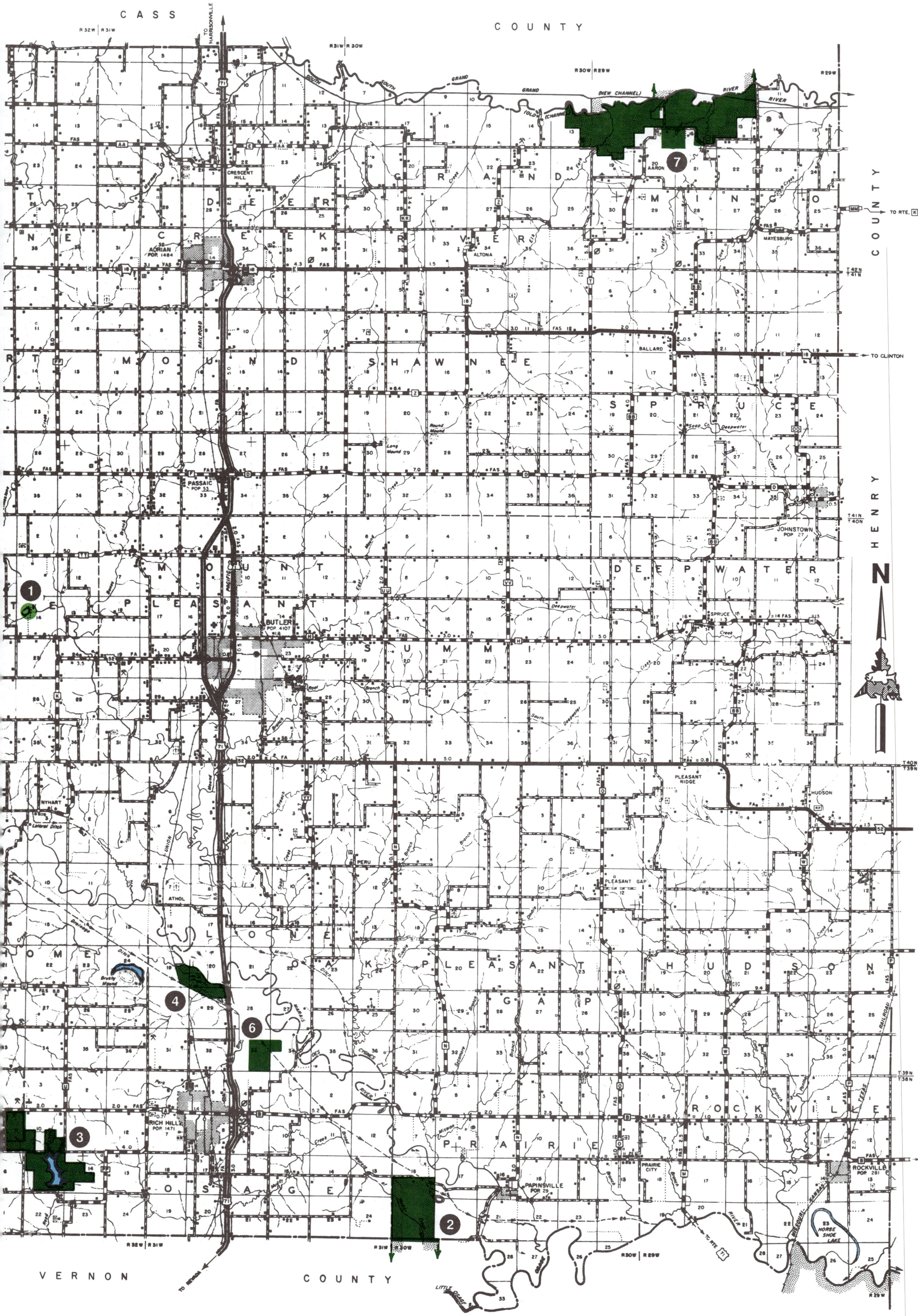
CASS COUNTY
HENRY COUNTY
ST. CLAIR COUNTY
VERNON COUNTY
TO HARRISONVILLE
TO RTE. K
TO CLINTON
TO RTE. H
TO NEVADA
GRAND RIVER
DEER CREEK
MINGO
MOUND
SHAWNEE
SPRUCE
DEEPWATER
MOUNT PLEASANT
SUMMIT
LONE OAK
PLEASANT GAP
HUDSON
ROCKVILLE
PRAIRIE
OSAGE
CRESCENT HILL
ADRIAN POP. 1484
ALTONA
MAYESBURG
AARON
BALLARD
PASSAIC POP 53
JOHNSTOWN POP 27
SPRUCE
BUTLER POP. 4107
PLEASANT RIDGE
HUDSON
NYHART
PERU
PLEASANT GAP
ATHOL
RICH HILL POP. 1471
PRAIRIE CITY
PAPINSVILLE POP 29
ROCKVILLE POP 281
HORSE SHOE LAKE
LITTLE OSAGE
1
2
3
4
6
7
N

BENTON COUNTY

1 To reach **Big Buffalo Creek Conservation Area,** go north of Edmonson on Highway W. Turn south on Highway B, then east on Highway WW. This road ends perpendicular to a gravel road. Turn left and this gravel road leads to the area. There are low water crossings on this road, and they may be impassable during heavy rain. This 1,555-acre forested area has one fishing pond, three springs and a 1.5-mile stretch of the Big Buffalo Creek. Natural areas of interest include the Big Buffalo Creek Fen (45 acres of calcerous wet meadow, deep muck fen and sedge meadow shrub fen) and Big Buffalo Bottomland Natural Area, a 5-acre mesic bottomland forest. There are several unmaintained trails and roads suitable for hiking.

2 **Brickley Hollow Access** is 1 mile south of Lincoln on Highway 65, and 8 miles east on Highway H to the H and Highway W junction. Continue 1.5 miles to an unmarked blacktop road and turn right. Turn right again just past the road to Henry's Acres. This is a 186-acre area on the Cole Camp Creek Arm of Lake of the Ozarks. Brickley Creek flows 2 miles through the area.

3 **Edmonson Access** (110 acres) is 1 mile south of Lincoln on Highway 65, then 7.5 miles east on Highway H. Turn north on an unnamed gravel road near the Cole Camp Creek bridge. Sixty-four acres are forestland consisting of new hardwood plantations and some riparian corridor. The rest is old pastures. Three-eighths mile of Cole Camp Creek flows through the area, and there is one access (no boat ramp).

4 **Hi Lonesome Prairie Conservation Area** is 1 mile west of Cole Camp on Highway 52, then .5 mile north on County Road NE 221. This 627-acre area is nearly all unplowed native prairie. Five ponds provide anglers with .5 acre of fishable water.

5 To reach **Lincoln Towersite,** take Highway 65 south of Lincoln and turn east on Highway H. The towersite is 3.5 miles down Highway H on the south side of the road. This is a 4-acre area.

6 **Mora Conservation Area** is 3.5 miles north of Cole Camp on Highway U. There is one pond in the area suitable for fishing (1.7 acres). Nearly all 320 acres of surrounding land is cropland, pasture or old fields.

7 **Mount Hulda Towersite** is south of Lincoln on Highway 65, east on Highway H, then 6.5 miles north on Highway W. This is a 40-acre forested area.

8 **Harry S Truman Reservoir Wildlife Management Lands** are scattered widely throughout Benton, Henry, Hickory and St. Clair counties. These areas total 54,511 acres and contain 130 ponds, 10 streams, 10 lakes and five rivers for a total of 250 acres of fishable water. There are three caves, several springs and many miles of unmaintained old roads and fields suitable for hiking. One natural feature of interest includes a transitional zone where tallgrass prairie and Ozark forest habitats meet. Features in this area include native prairies, glades, savannahs, bluffs and wetlands. Golden Valley Shooting Range lies near the reservoir in Henry County, 6 miles west of Clinton on Highway SW 100. It is wheelchair accessible.

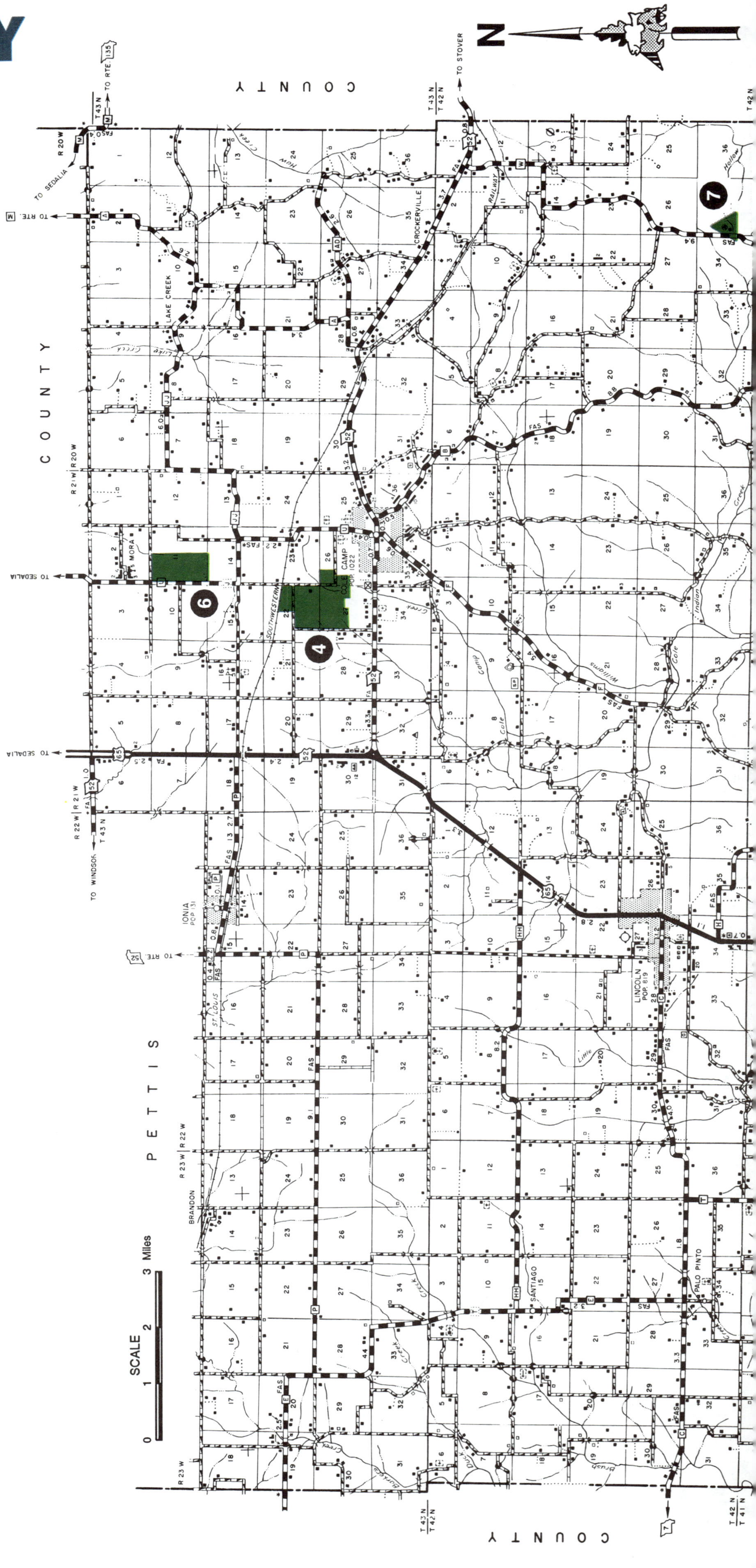

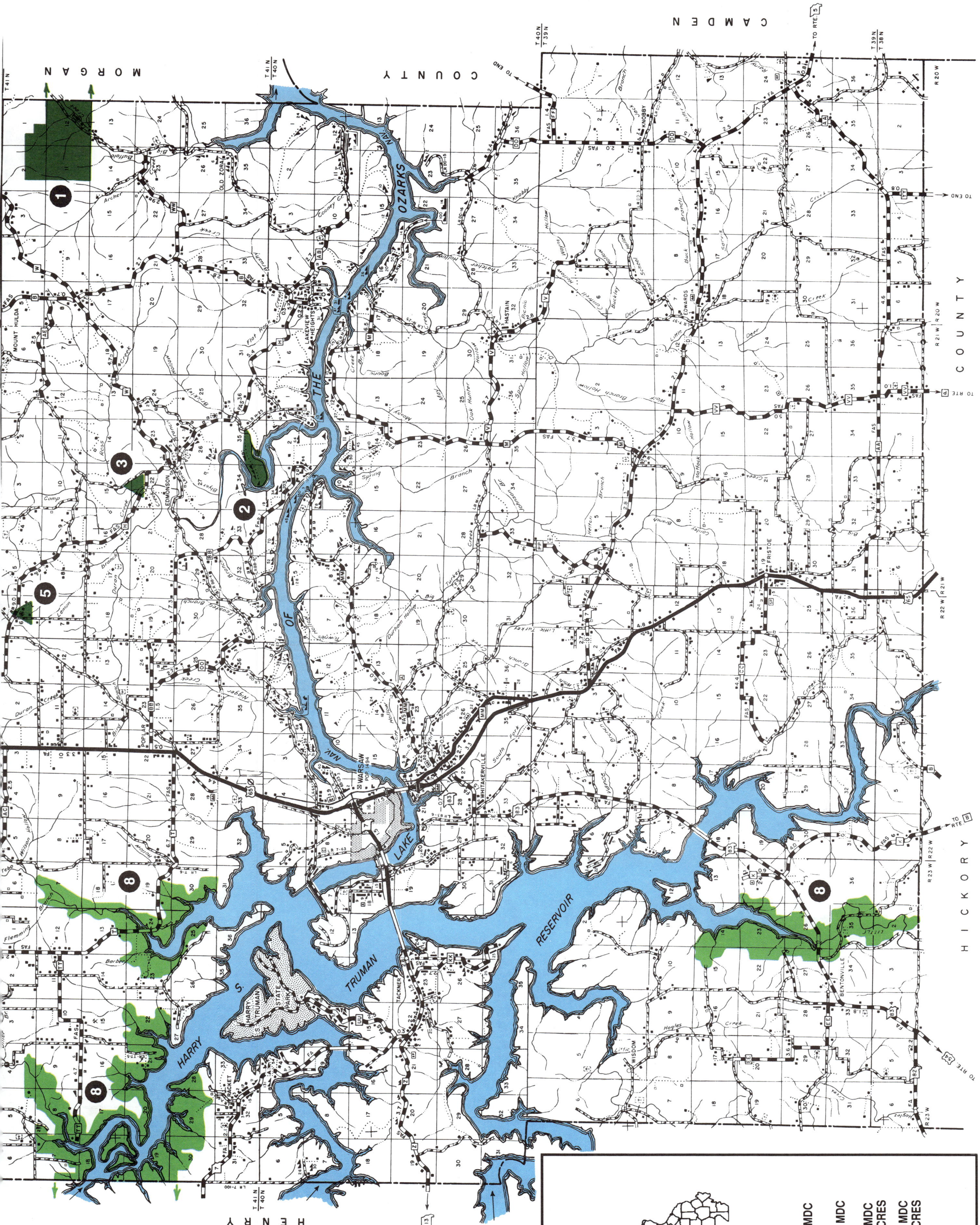

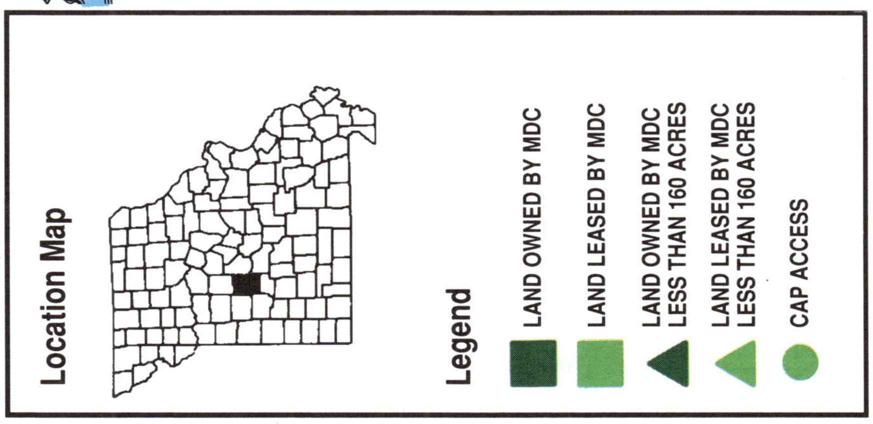

There is also an archery range, several accesses on the rivers and one primitive camping spot. For additional information, contact the Truman District Office, 2010 S. Second St., Clinton, MO 64735, 816/885-6981.

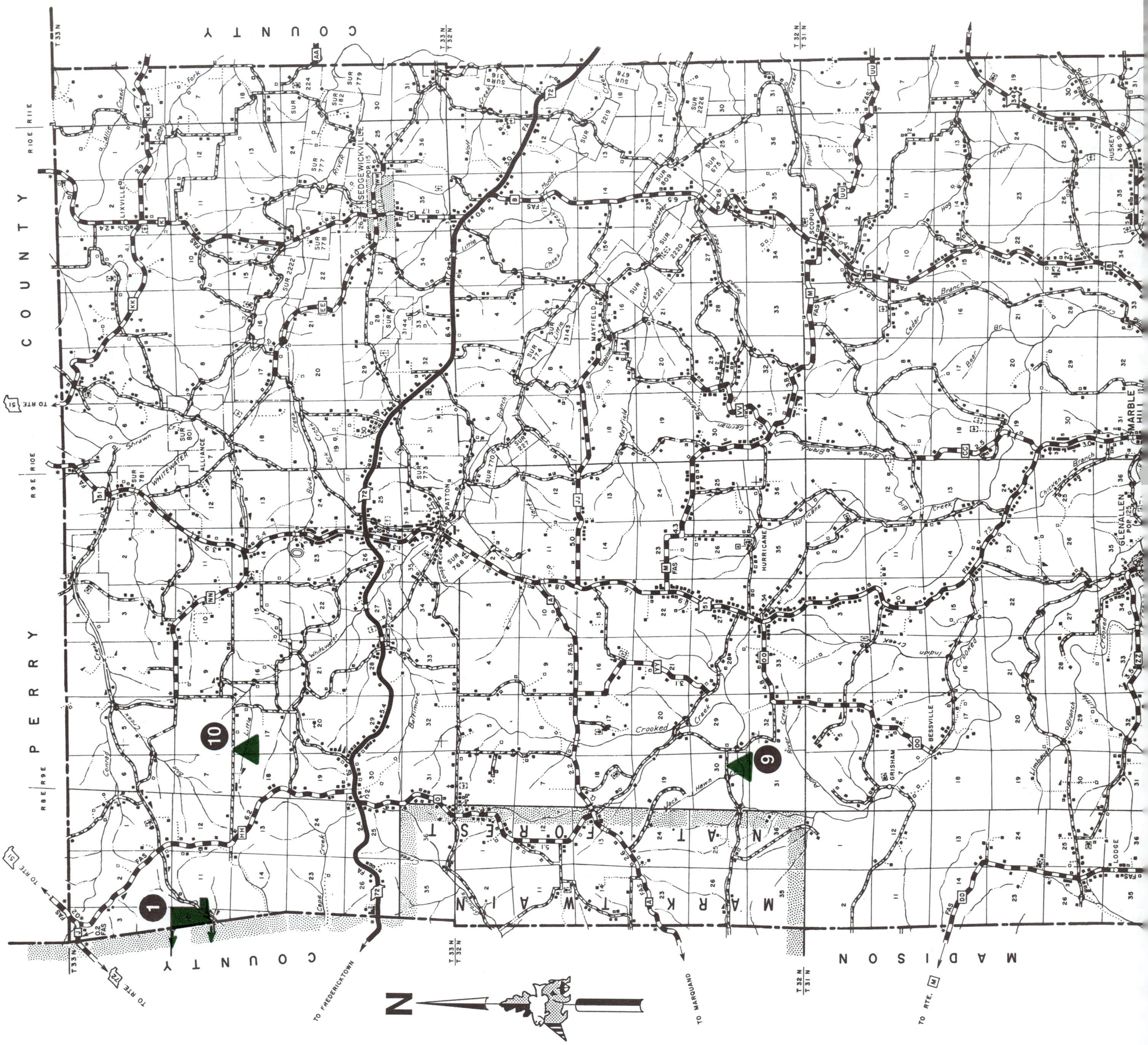

BOLLINGER COUNTY

1 **Amidon Memorial Conservation Area** lies in Bollinger and Madison Counties. The main entrance is in Madison County, east of Fredericktown on Highway J, south on Highway W, then east on County Road 208. The 1,632-acre area is nearly all forestland and contains a 2.5-mile stretch of the Castor River (5 acres for fishing). Duchesne Access lies within the area. Cedar Glades hiking trail is 1 mile long, and natural areas of interest include Castor River Shut-ins Natural Area and 65 acres of granite glades.

2 **Castor River Conservation Area** is composed of several tracts. The main tract is west of Marble Hill on Highway 34. The area lies on both sides of the highway. This 9,122-acre area is nearly all forestland and is home to Blue Pond Natural Area–the deepest natural pond in Missouri. Other natural features of interest include some dry sink holes and a 1.5-acre fen.

3 **Clubb Creek Conservation Area** totals 631 acres. The area is made up of eight tracts located southwest of Marble Hill, primarily along highways PP and 51. For specific information about each tract, contact the Perryville Forest District, 2206 W. St. Joseph, Perryville, MO 63775, 314/547-4537.

4 **Coldwater Conservation Area**–see the Wayne County map, page 232.

5 **Dark Cypress Swamp Conservation Area** is 5 miles west of Advance on Highway C. This 470-acre area extends along a 1.5-mile stretch of the Castor River (9 acres for fishing). Two hundred and twenty acres are forested, and the rest is cropland.

6 **Duck Creek Conservation Area** consists of two tracts. The main entrance is 6 miles south of Zalma on Highway 51. This 6,234-acre area is noted for its 1,800-acre lake. Suitable for fishing, the lake has four boat ramps and six fishing jetties, some of which are wheelchair accessible.

7 **Gipsy Towersite** is off Highway TT in the southwest corner of Bollinger County. The area is composed of 40 acres of mainly forestland.

8 **Grassy Towersite** is 5 miles west of Marble Hill on Highway 34, and 1.5 miles southeast on County Road 508. The tower is surrounded by 40 acres of forestland.

9 **Hawn Access** (79 acres) is 7 miles northwest of Marble Hill on Highway 51, 1.5 miles west on Highway OO, and 1.25 miles right on an unnamed county road.

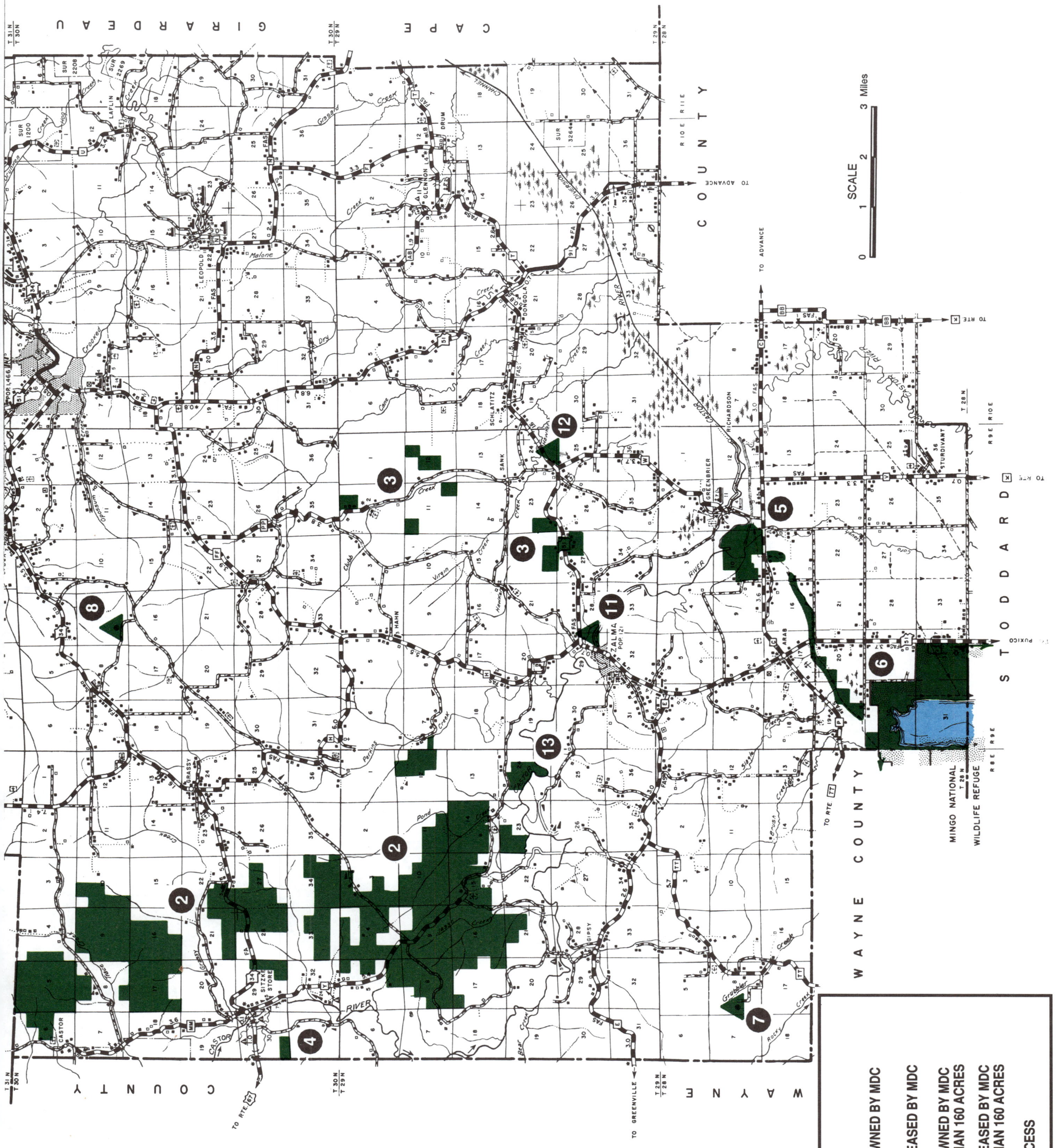

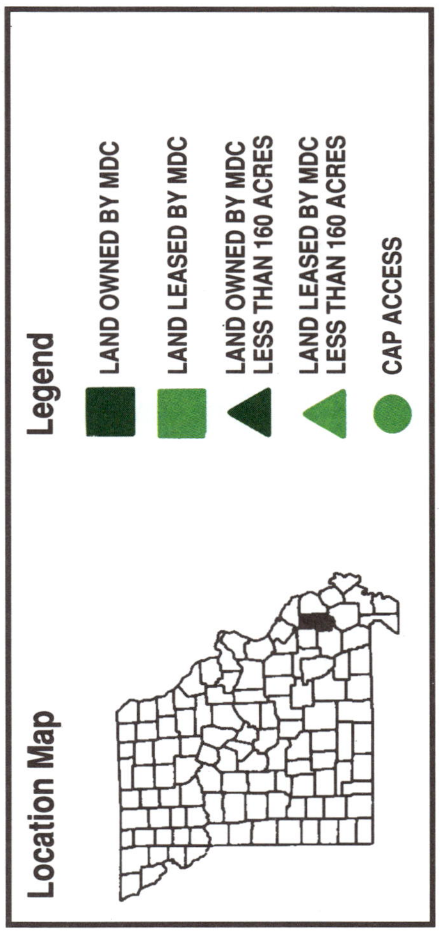

10 **Little Whitewater Conservation Area** is 5 miles west of Patton on Highway 72, 2.5 miles north on Highway HH, and 2 miles east on Gravel Road 916. This is an 80-acre forest.

11 **Maple Flats Access** is .5 mile north of Zalma on Highway 51. A .75-mile stretch of the Castor River runs along this 72-acre area. There is a boat ramp.

12 **Sank Conservation Area** is 8 miles north of Zalma on Highway 51. It is composed of 90 acres of forestland and about 28 acres of cropland and old pastures.

13 **Sweetgum Access** is north of Zalma on Highway 51, north on Highway H, then 1.5 miles west on County Road 708. This is 161 acres with a boat ramp on the Castor River.

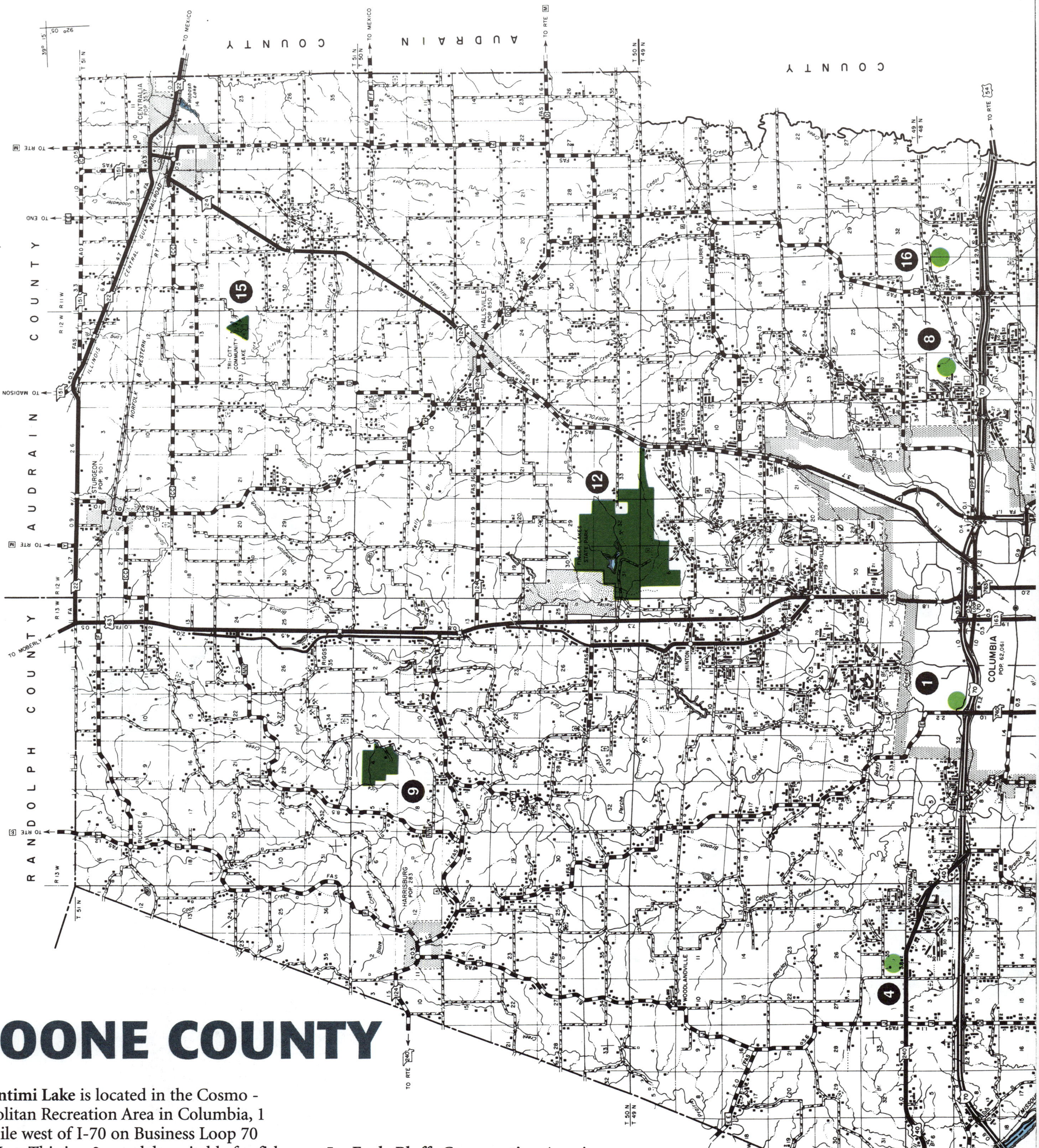

BOONE COUNTY

1 **Antimi Lake** is located in the Cosmo - politan Recreation Area in Columbia, 1 mile west of I-70 on Business Loop 70 West. This is a 2-acre lake suitable for fishing. There are six picnic areas and two trails totaling 3.75 miles in length.

2 **Ashland Lake** is 3 miles east of Ashland on Highway Y. This is a 12-acre area with a boat ramp.

3 **Cosmo-Bethel Lake** is south on Providence Road, 1 block west on Southampton, then north on Bethel Church Road. This is a 6-acre lake in a city park.

4 **Dairy Farm Lakes 1** and **3** are 6 miles west of Columbia on I-70, 1 mile north on Highway J, and 1 mile east on Highway 40. These are 15.5 and 5 acres in size.

5 **Eagle Bluffs Conservation Area** is near McBaine on Route K, 1 mile west on Burr Oak Road, then 2 miles south on Star School Road. This 4,269-acre area borders portions of the Missouri River and Perche Creek (11.5 total acres of fishable water). It is the largest wetland area in the country that uses treated municipal wastewater to supplement wetland water needs. The Katy Trail State Park parallels the area's eastern boundary for 3 miles.

6 **Hartsburg Access** is 17 miles south of Columbia on Highway 63, west on Highway AA, then 2 miles on River Road. This is a 30-acre area with .5-mile frontage on the Missouri River. There is a boat ramp.

7 **Hinkson Woods Conservation Area** is 1 mile southwest from the Forum Boulevard entrance to the MKT Nature Fitness Trail in Columbia. This 70-acre area is accessible by bike or on foot. Natural features include riparian timber and creek bluffs.

8 **Lake of the Woods Recreation Area** is north of I-70 on Lake of the Woods Road, then east on St. Charles Road. This is 3 acres of fishable water.

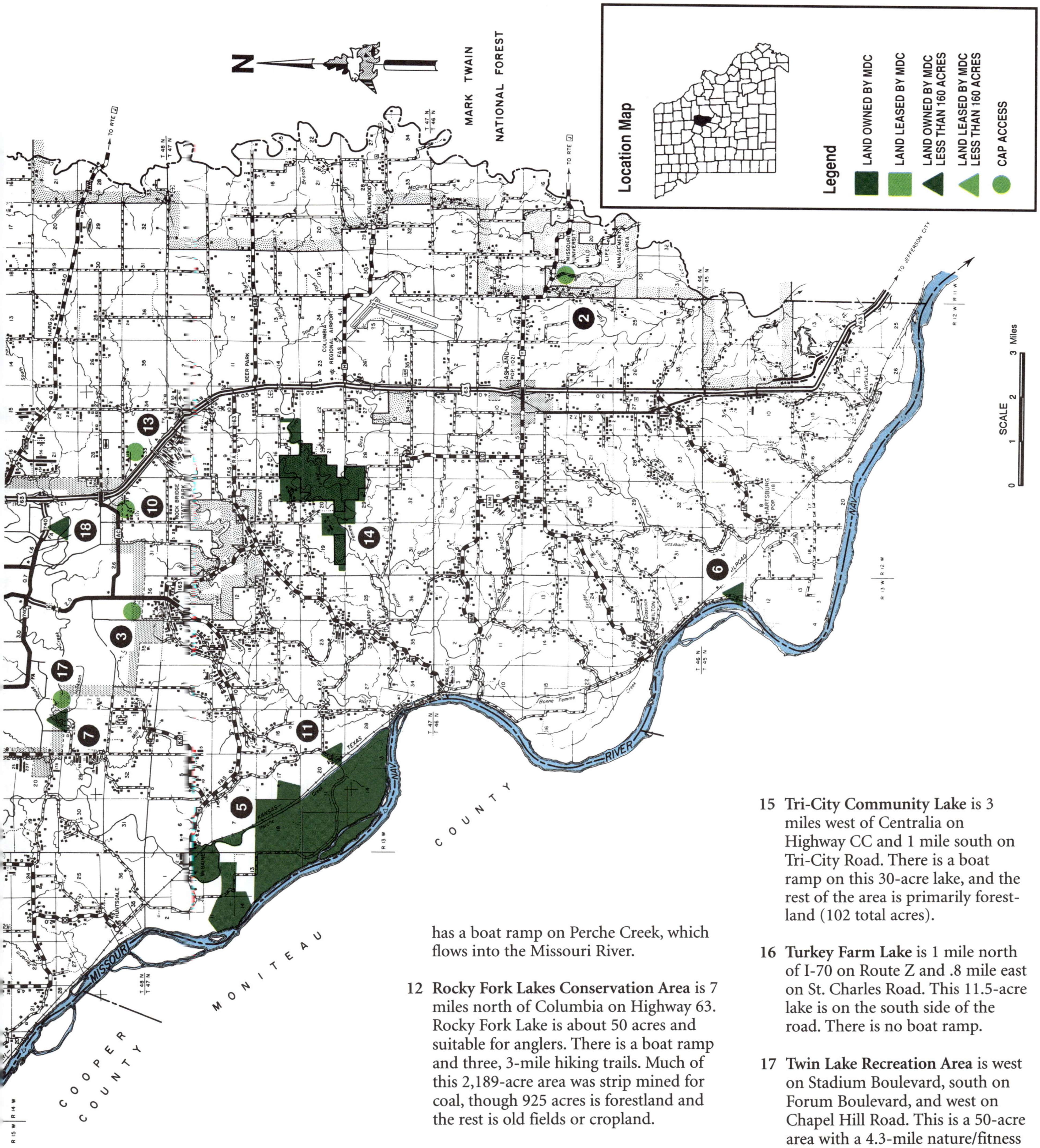

9 **Lick Creek Conservation Area** is northwest of Columbia on Highway 124, then 3.5 miles north on Oak Grove School Road. This is a 317-acre area.

10 **Nifong Park Pond** is .25 mile west of Highway 63 on Nifong Boulevard. This is a 15-acre area.

11 **Providence Access** is south of Columbia on Route K and 2.5 miles west to the dead end of Old Plank Road. This 5-acre area has a boat ramp on Perche Creek, which flows into the Missouri River.

12 **Rocky Fork Lakes Conservation Area** is 7 miles north of Columbia on Highway 63. Rocky Fork Lake is about 50 acres and suitable for anglers. There is a boat ramp and three, 3-mile hiking trails. Much of this 2,189-acre area was strip mined for coal, though 925 acres is forestland and the rest is old fields or cropland.

13 **South Farm R-1 Lake** is 1.5 miles east of Highway 63 on Route AC, then .5 mile south and .25 mile west on a gravel road. The 9-acre lake is on the south side of the road.

14 **Three Creeks Conservation Area** is 5 miles south of Columbia on Highway 63 and 1.75 miles west on Deer Park Road. This 1,277-acre area contains stretches of the Turkey, Bass and Bonne Femme creeks. A 2.7-mile natural trail extends through the area, and several test plots of Running Buffalo Clover are maintained. Natural features include Hunter's Cave and Tumbling Cave Spring.

15 **Tri-City Community Lake** is 3 miles west of Centralia on Highway CC and 1 mile south on Tri-City Road. There is a boat ramp on this 30-acre lake, and the rest of the area is primarily forestland (102 total acres).

16 **Turkey Farm Lake** is 1 mile north of I-70 on Route Z and .8 mile east on St. Charles Road. This 11.5-acre lake is on the south side of the road. There is no boat ramp.

17 **Twin Lake Recreation Area** is west on Stadium Boulevard, south on Forum Boulevard, and west on Chapel Hill Road. This is a 50-acre area with a 4.3-mile nature/fitness trail, fishing and swimming opportunities.

18 **H. J. Waters and C. B. Moss Memorial Wildlife Area** is in Columbia, .5 mile west of U.S. 63 on Stadium Boulevard, then .25 mile south on Old 63. This 102-acre area has a 1.5-mile, grass hiking trail and a public contact office on the grounds. There is also a bat house with public demonstrations. Natural features include 75-foot high limestone bluffs along Grindstone Creek, which flows .25 mile through the area.

BUCHANAN COUNTY

1 **Agency Access** is 1 mile east of the town of Agency on Highway H, just where the highway crosses the Platte River. This is a 1-acre area with no boat ramp.

2 **Agency Conservation Area** is 3 miles east on Highway DD off Highway 29, and .5 mile east on County Road 322. This is a 94-acre area on the Platte River.

3 **Belcher Branch Lake Conservation Area** is east on Highway DD off Highway 29, then south on County Road 330. This is a 57-acre lake on 405 acres of mainly cropland (262 acres).

4 **Bluffwoods Conservation Area** is 9 miles south of St. Joseph off Highway 59. Turn southeast at the Highway Department directional sign, cross the railroad tracks, and go about 1 mile to the area. This is a 2,281-acre area (mainly forestland) with a scenic overlook, 2.5-mile hiking trail, 2-mile horseback riding trail and picnic areas. **Goodell Memorial Annex** to Bluffwoods is about 7 miles south of St. Joseph on Highway 59. This is 63 acres of forestland.

5 **Burton Bridge Access** is 1.5 miles east on Highway MM off Highway DD near Faucett, then east on County Gravel Road 342. This is a 16-acre area on the Platte River with no boat ramp.

6 **Arthur Dupree Memorial Conservation Area** is 1 mile north of Rosecrans Airport on County Road 2, then 1 mile right on County Road 3. This is 236 acres on the Missouri River.

7 **French Bottom Access** is off Highway 229 at the junction of McArthur Drive and Waterworks Road. This is 1.5 acres on the Missouri River.

8 **Jentell Brees Access** is south of St. Joseph on Highway 59, 7 miles west on Highway U, and right on Contrary Road. This is a 31-acre access to the Missouri River.

9 **Anthony and Beatrice Kendzora Conservation Area** is 1.5 miles north of Edgarton on Union Mill Road. This is a 740-acre area with fishing opportunities on Kendzora Lake and the Platte River.

10 **Kneib Memorial Conservation Area** is 3 miles west of Highway 59 on Highway U, and .5 mile north on County Gravel Road 201. Watch for the area sign. This 40-acre area has a 1-mile hiking trail and native warm season grass demonstration plots. It is composed mainly of old pastures.

11 **Lewis and Clark Fish Hatchery** is .5 mile west of Highway 45 on Highway 138. This 44-acre area has 16 acres of hatchery ponds and tours are available. The hatchery is visited by a wide variety of waterfowl and songbirds depending on the season. Tours are available with prior scheduling.

12 **Pigeon Hill Conservation Area** is east on Highway 169 (off Highway 29) to Highway O. From Highway O, turn west on Highway FF, and watch for the cantilever sign. This is a 336-acre area and is home to the National Bowhunting Club's annual competition. There also is a firearms range and a 1-mile hiking trail.

13 **Saxton Access** is 4 miles east of St. Joseph on Highway 36, 2 miles south on Highway C, and .5 mile west on Saxton/Easton Road. This is 5 acres with no boat ramp on the Platte River.

14 To reach **Sunbridge Hills Conservation Area,** take the Highland exit off Highway 229 east to Huntoon Road. Go north on Huntoon and watch for the cantilever sign. This is a 225-acre area on the Missouri River. There is a 4-mile hiking trail. **Caroline Sheridan Logan Memorial Wildlife Area** is next to Sunbridge Hills. It is a 40-acre forested area.

15 The main entrance to **Mark Youngdahl Urban Conservation Area** is on 32nd Street between Faroan and Messanie avenues. This is an 85-acre area of old fields (65 acres) and forestland (20 acres) with 2.5 miles of hiking trails.

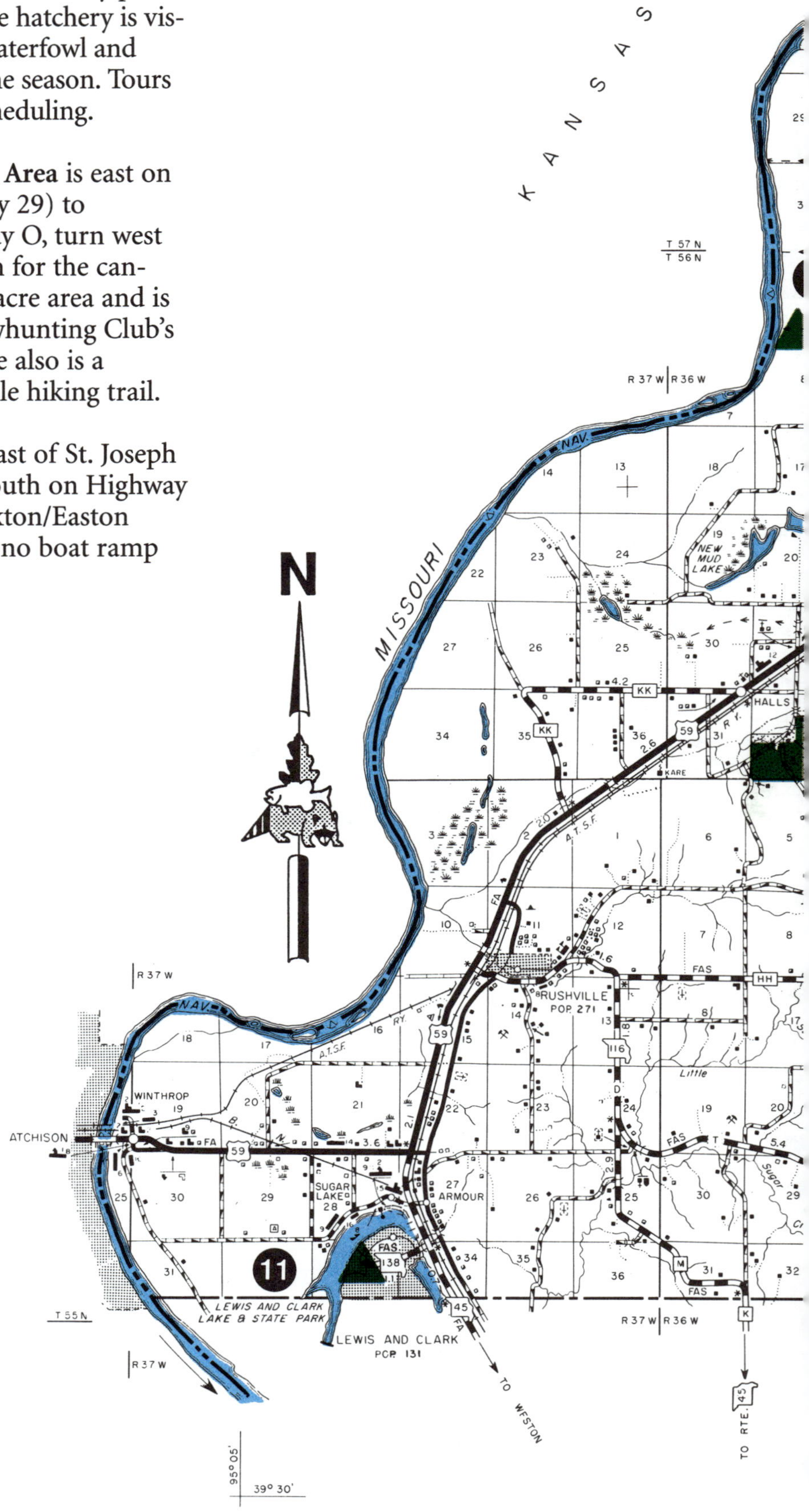

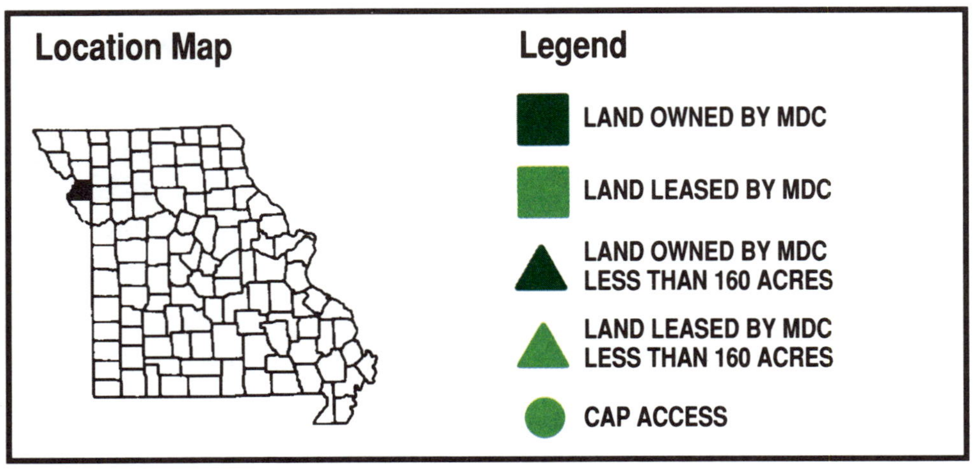

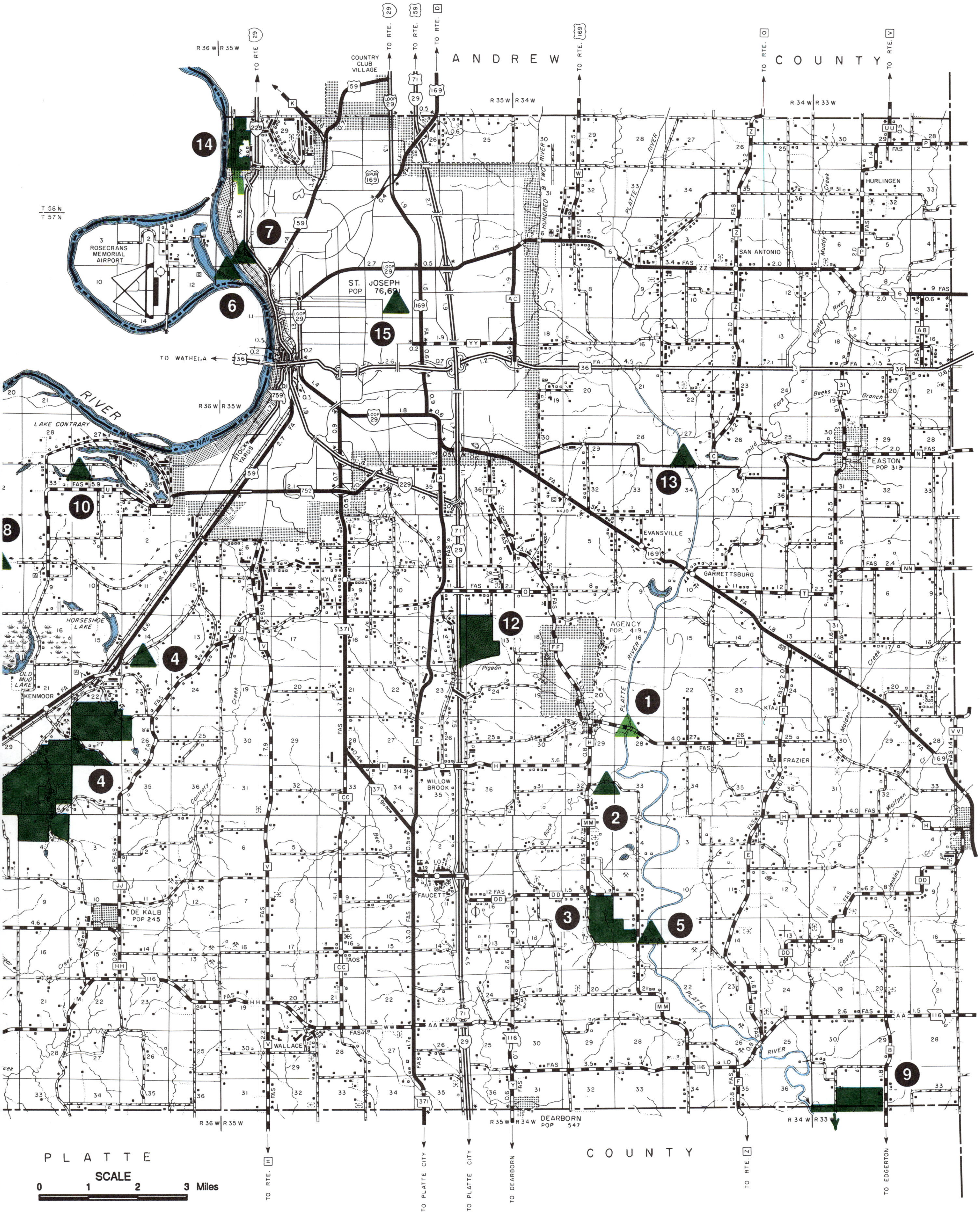
ANDREW COUNTY
PLATTE COUNTY
SCALE
0 1 2 3 Miles
COUNTRY CLUB VILLAGE
ROSECRANS MEMORIAL AIRPORT
ST. JOSEPH POP 76,691
TO WATHENA
LAKE CONTRARY
RIVER
STOCK YARDS
HORSESHOE LAKE
OLD MUD LAKE
KENMOOR
KYLE
AGENCY POP. 419
EVANSVILLE
GARRETTSBURG
SAN ANTONIO
HURLINGEN
EASTON POP 313
FRAZIER
WILLOW BROOK
FAUCETT
DE KALB POP 245
TAOS
WALLACE
DEARBORN POP 547
PLATTE RIVER
TO RTE. H
TO PLATTE CITY
TO PLATTE CITY
TO DEARBORN
TO RTE. Z
TO EDGERTON
T 56 N
T 57 N
R 36 W R 35 W
R 35 W R 34 W
R 34 W R 33 W

BUTLER COUNTY

1 **Allred Lake Natural Area** is 5 miles east of Neelyville on Highway 142, and 2.5 miles south on Highway H to a gravel road that heads south. The 160-acre area is .25 mile down this road. This is a Designated Natural Area, and fishing is not permitted in Allred Lake. There is a .25-mile boardwalk trail.

2 **Big Cane Conservation Area** consists of two tracts. The largest tract is east of Neelyville on Highway 142, 2 miles south on Highway EE, 1 mile right on County Road 269, and .5 mile left on County Road 276. Big Cane Creek flows 2 miles through this 2,154-acre forested area.

3 **Mac and Zelma Carmichael State Forest** is southeast of Neelyville off Highway EE. At the end of EE, turn 1 mile east on County Road 260, then south on County Road 263. This is a 40-acre forest.

4 **Coon Island Conservation Area and Access** is southeast of Poplar Bluff on Highway 53, then 8 miles south on Highway HH to County Road 224. This is a 3,153-acre area with two boat ramps on the Black River. Six miles of the river flows through the area.

5 **Corkwood Conservation Area** is .5 mile west of Neelyville on Highway 142. This 434-acre area contains a wetland habitat (3 acres) for the corkwood plant, and the rest of the area is forestland.

6 **Dan River Access** is southeast of Poplar Bluff on Highway 53 and 1 mile north on County Road 611. This is a 68-acre area on the Black River.

7 **Fisk Access** is east of the town of Fisk on Old Highway 53. This is a 4-acre area with a boat ramp on the St. Francis River.

8 **Harviell Access** is 3.5 miles south of Poplar Bluff on Highway 67, and 2.5 miles south on County Road 321. This is a 35-acre area with a boat ramp on Cane Creek.

9 **Hilliard Access** (1 acre) is 2 miles north of Poplar Bluff on Highway W on the Black River. There is a boat ramp.

10 **Lone Hill Towersite** is southwest of Poplar Bluff on Highway 67, 5 miles west on Highway M, 1 mile south on County Road 461, and .5 mile south on County Road 462. This is a 13-acre forest.

11 **Mussel Boat Landing Access** is 10 miles southeast of Poplar Bluff on Highway 53, then north on an access road that lies immediately north of the Black River bridge. This is a 2-acre area.

12 The main entrance to **Otter Slough Conservation Area** is west of Dexter on Highway 60, then 10 miles south on Highway ZZ in Stoddard County. This is a 4,863-acre area on the St. Francis River and also home to Otter Lake, a 250-acre body of water suitable for fishing. Natural areas include Bradyville Hardwoods (139 acres of bottomland hardwoods) and Otter Slough Natural Area (20 acres). There is a picnic area and a 1.5-mile hiking trail through the beaver pond area.

13 **Poplar Bluff/Stephen J. Sun Conservation Area** is 2 miles east of Poplar Bluff on U.S. Business 60, then 3.5 miles north on Highway NN. When this road becomes gravel, watch for the first gravel road on the right marked 546 and go .5 mile east to the area. This 1,172-acre area is composed of five tracts and includes Carpenter Lake (6 acres) and the 80-acre Poplar Bluff Forest Natural Area.

14 **Ringo Ford Access** is at the Highway 160 crossing of the Little Black River, 10 miles southwest of Poplar Bluff. The area is 1 acre and has no boat ramp.

15 **Sportsman's Park Access** is at the Highway 60 crossing of the Black River near Poplar Bluff. This is an 8-acre forested area with a boat ramp.

16 **University Forest Conservation Area** consists of five tracts. The main tract is 15 miles north of Poplar Bluff on Highway 67, 4.5 miles east on Highway 172, then 3 miles south on Highway W. This is a 7,149-acre area with a 21-mile horseback riding trail.

17 **Wilhelmina Conservation Area** lies in Butler and Dunklin Counties. The main tract is in Dunklin County north of Campbell on Highway 53, 4.5 miles left on Highway DD, then left again on County Road 203. This is 1,437 acres with a stretch of the old St. Francis River channel.

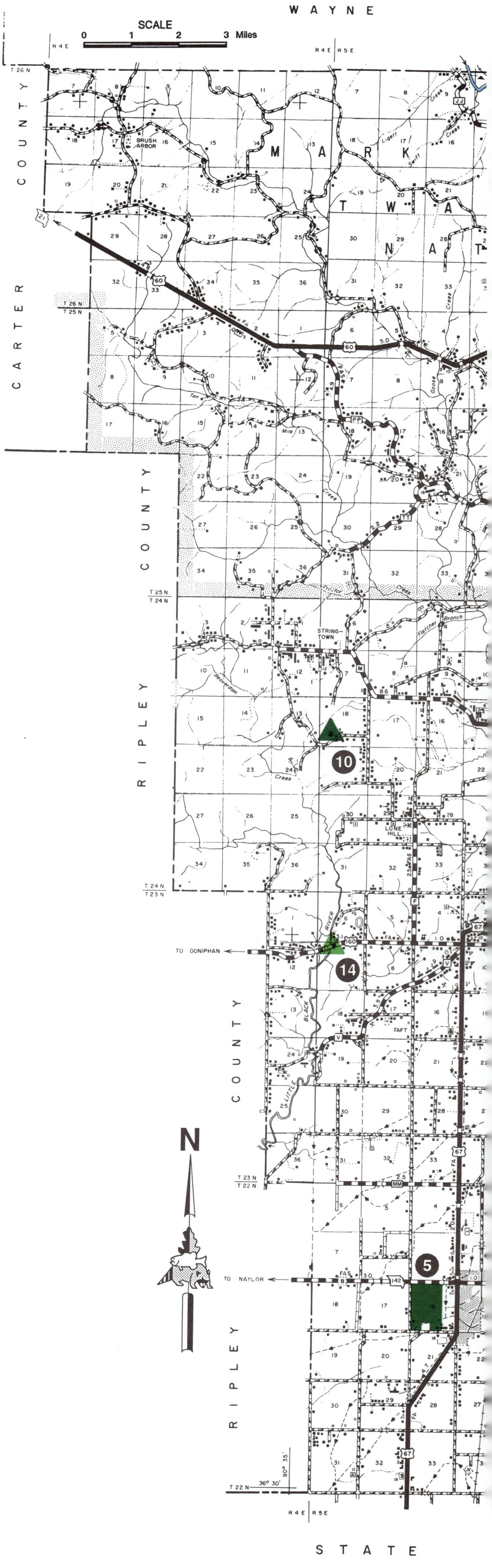

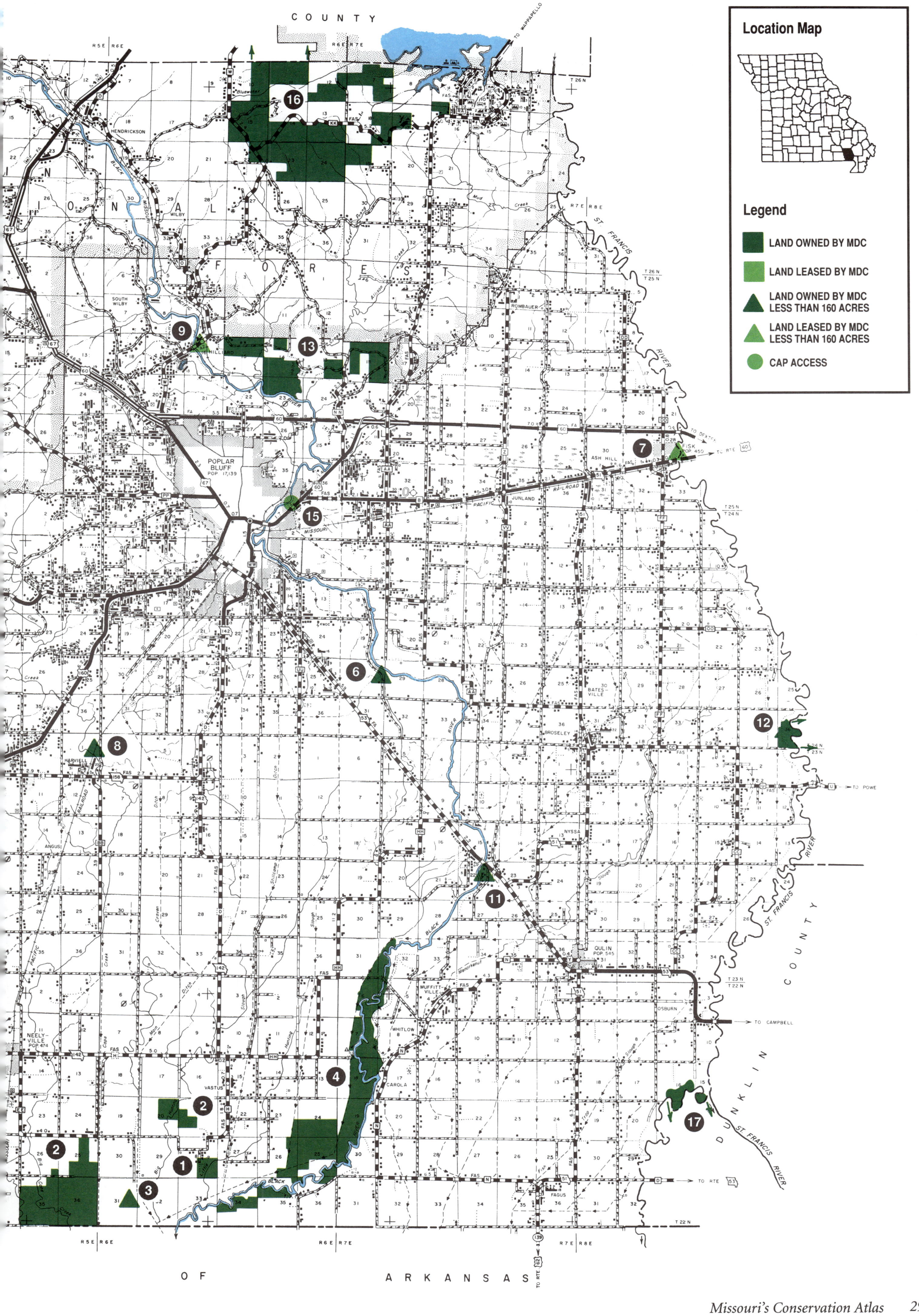
Location Map
Legend
LAND OWNED BY MDC
LAND LEASED BY MDC
LAND OWNED BY MDC LESS THAN 160 ACRES
LAND LEASED BY MDC LESS THAN 160 ACRES
CAP ACCESS
COUNTY
OF ARKANSAS
DUNKLIN COUNTY
FOREST
POPLAR BLUFF POP 17,139
QULIN POP 545
NEELYVILLE POP 474
HENDRICKSON
WILBY
SOUTH WILBY
HILLIARD
ROMBAUER
ASH HILL
FISK POP 450
JUNLAND
BATESVILLE
BROSELEY
HARVIELL
ANGUS
NYSSA
MUFFITTVILLE
WHITLOW
CAROLA
VASTUS
FAGUS
OSBURN
BLACK RIVER
ST. FRANCIS RIVER
TO WAPPAPELLO
TO DEXTER
TO RTE 60
TO POWE
TO CAMPBELL
TO RTE 53
TO RTE 67
MISSOURI PACIFIC RAILROAD

CALDWELL COUNTY

1 **Bonanza Conservation Area** (two tracts) is 2 miles south of Kingston on Highway 13, then 5 miles east on Route F. This is a 1,773-acre area containing stretches of Shoal Creek, Log Creek and Brushy Creek, as well as 30 ponds – providing anglers with about 15 acres of fishable water.

2 **Hamilton City Lake** is 1 mile west of the city of Hamilton on Business 36. This is an 85-acre reservoir (with a boat ramp) suitable for fishing. A golf course is adjacent to the lake.

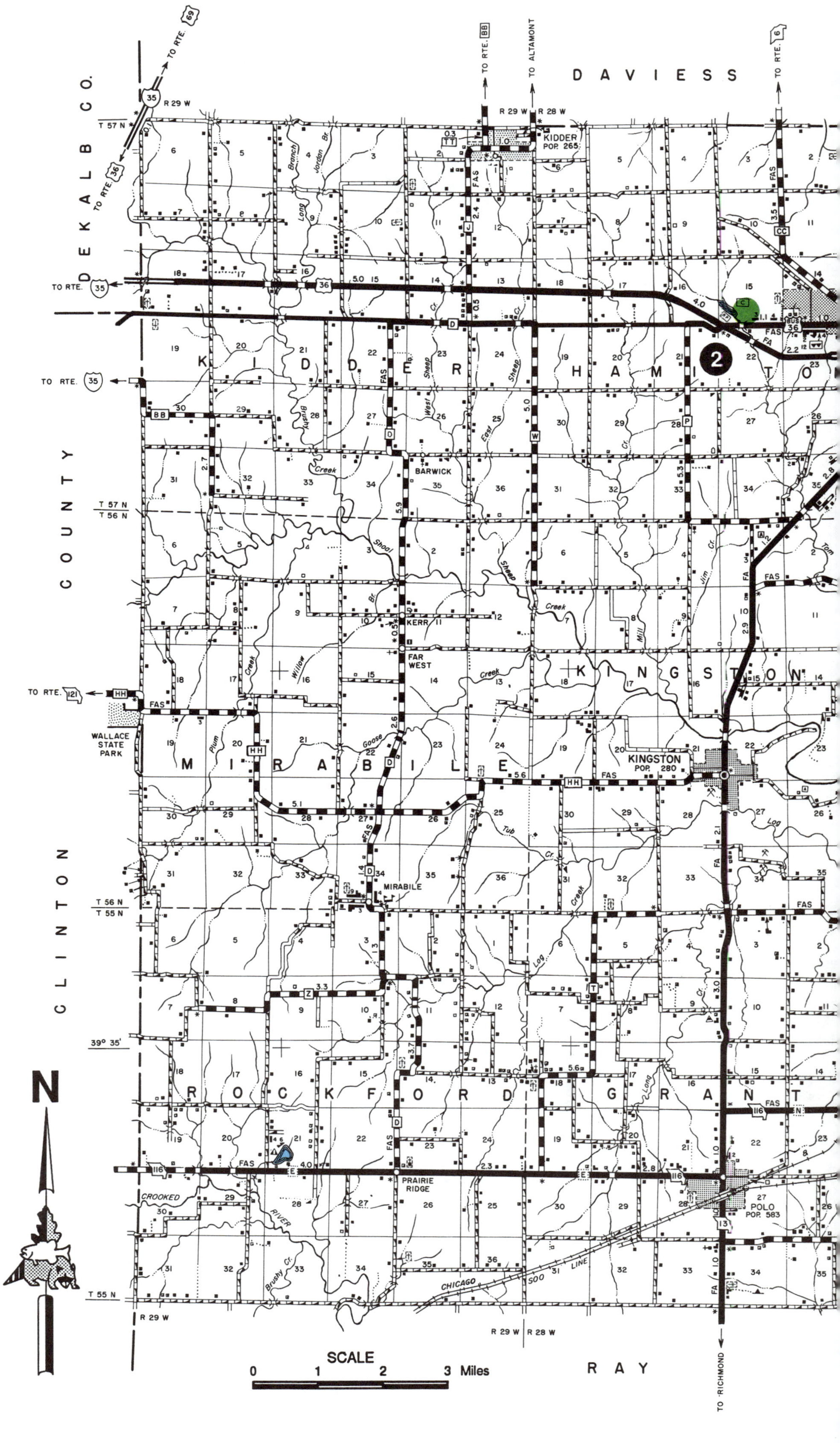

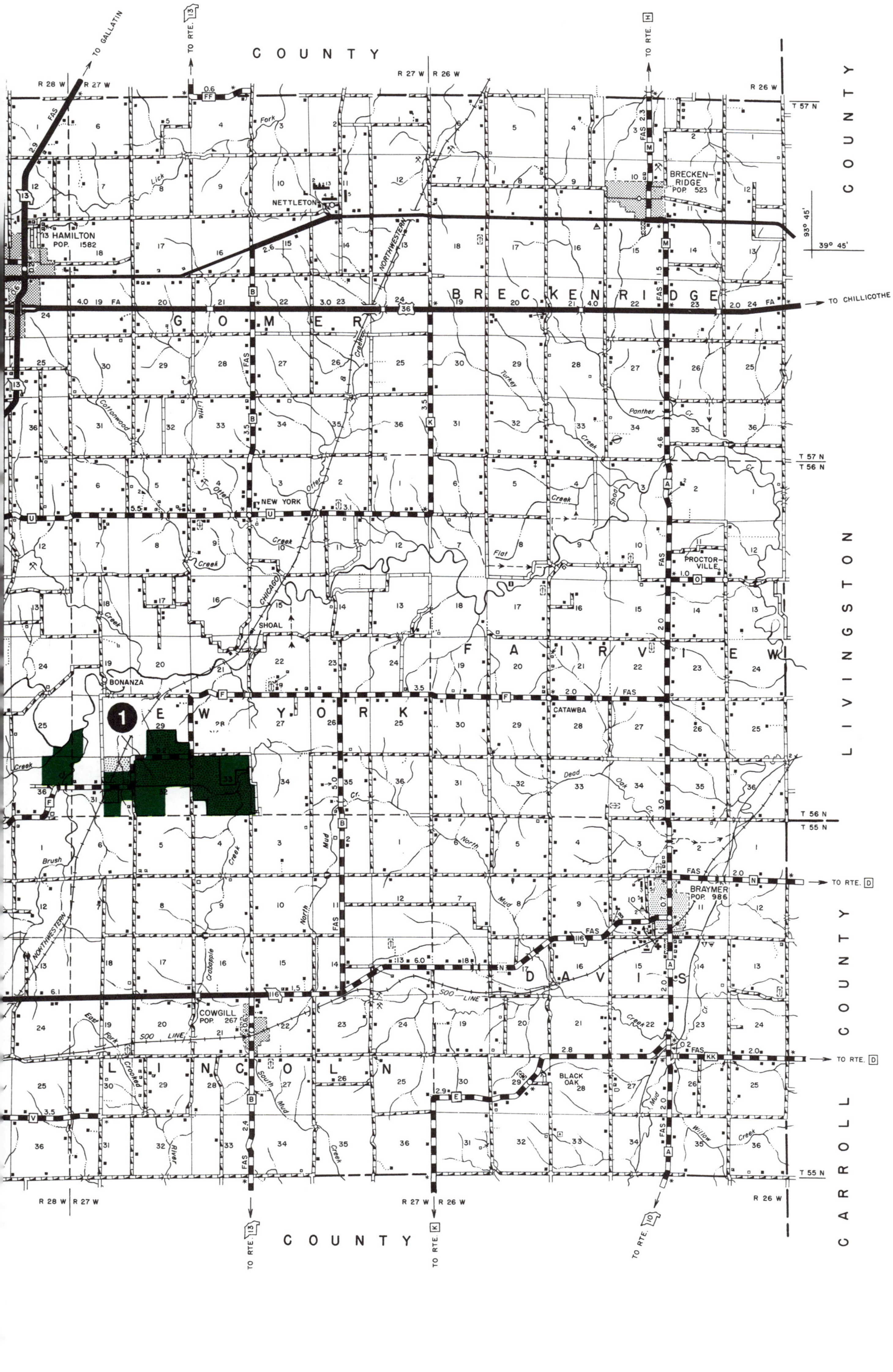

Location Map

Legend

- LAND OWNED BY MDC
- LAND LEASED BY MDC
- LAND OWNED BY MDC LESS THAN 160 ACRES
- LAND LEASED BY MDC LESS THAN 160 ACRES
- CAP ACCESS

CALLAWAY COUNTY

1 **Capitol View Access** is 5 miles north of Jefferson City on Highway 63. This is 13 acres on Cedar Creek.

2 **Earthquake Hollow Conservation Area** is 2 miles north of Holts Summit on Highway 54, 1 mile east on Highway TT, and 1 mile down County Road 452. This 87-acre area has a chert conglomerate cliff and many species of ferns. There is a .25-mile hiking trail.

3 **Hams Prairie Access** is 9 miles east of Fulton on Highway O, 1.5 miles south on Highway CC, then 2 miles west on County Road 428. This is a 28-acre area on Auxvasse Creek with no boat ramp.

4 **Little Dixie Lake Conservation Area** is west of Kingdom City on I-70. Take the Route J exit and head south to Route RA. The 733-acre area is marked with a sign on the left. Little Dixie Lake is 206 acres. There is a boat ramp and rental, nine fishing jetties (some are wheelchair accessible), three hiking trails and picnic areas.

5 **McCredie Farm Lake** is north of I-70 on Highway 54, then 2 miles east on the first outer road (that lies north of I-70). This is an 11.5-acre lake with one boat ramp.

6 **Mokane Access** is on the south side of Highway 94, 1 mile west of the town of Mokane. This is a 10-acre area with a boat ramp on the Missouri River.

7 **Moore's Mill Access** is south on Route Z off I-70 (the Calwood exit). In Calwood, take Highway JJ south to County Road 136. Go east on 136, then north on County Road 139 to the area entrance. This is 101 acres on Auxvasse Creek (1.5 miles for fishing).

8 **Reform Conservation Area** is 9 miles east of Fulton on Highway O. Logan Creek flows through this 7,044-acre area, and there are several ponds providing 30 acres of fishable water. There is a 2-mile hiking trail.

9 To reach **Whetstone Creek Conservation Area,** take Highway D (the Williamsburg exit off I-70) north to a stop sign. Turn west, and Highway D continues to County Road 1003. Go 2 miles north on County Road 1003 to the area entrance. The area has six lakes and a creek providing anglers with 60 acres of fishable water. The area is 5,147 acres of mainly forestland (2,700 acres) and old pastures.

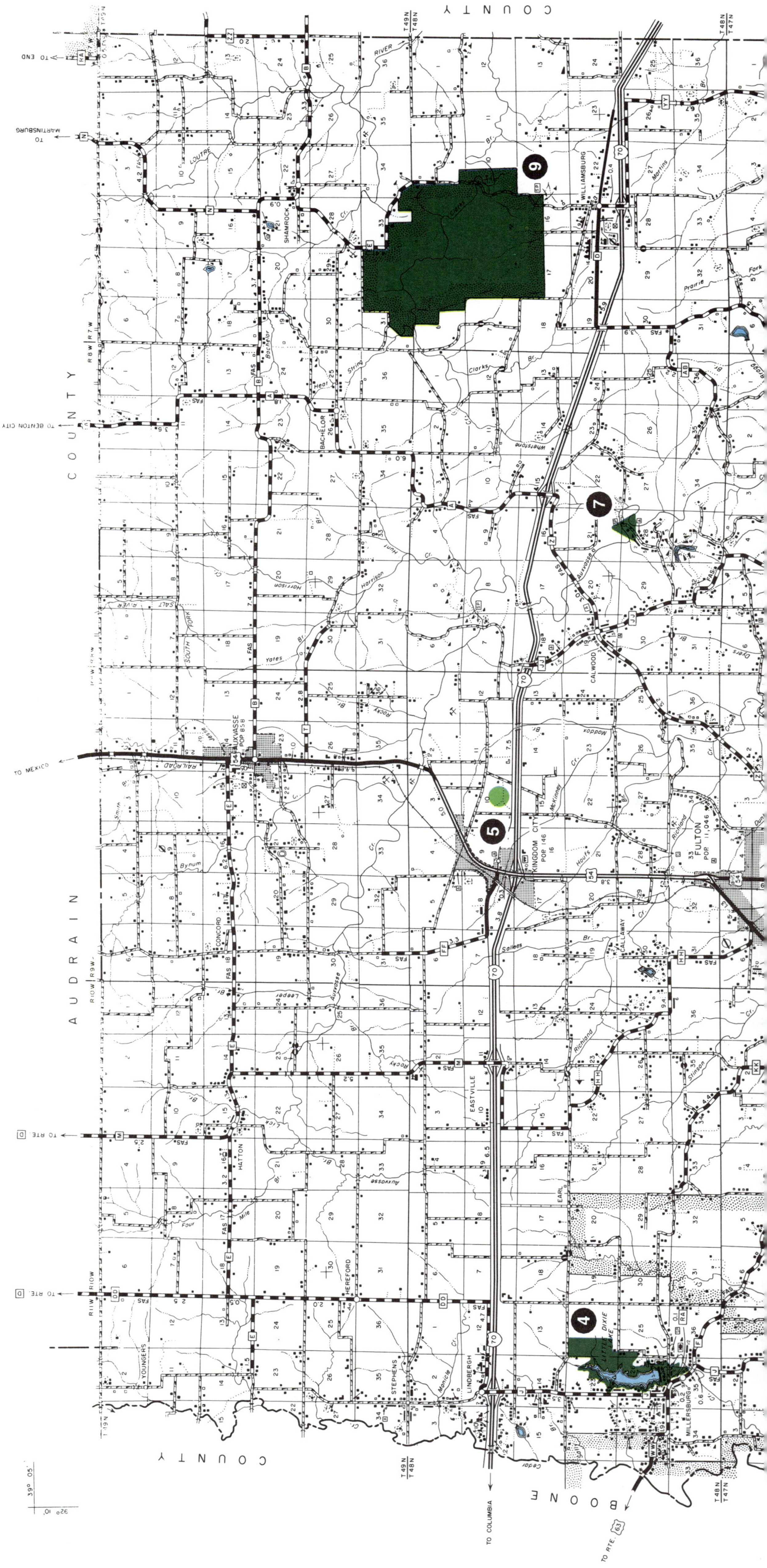

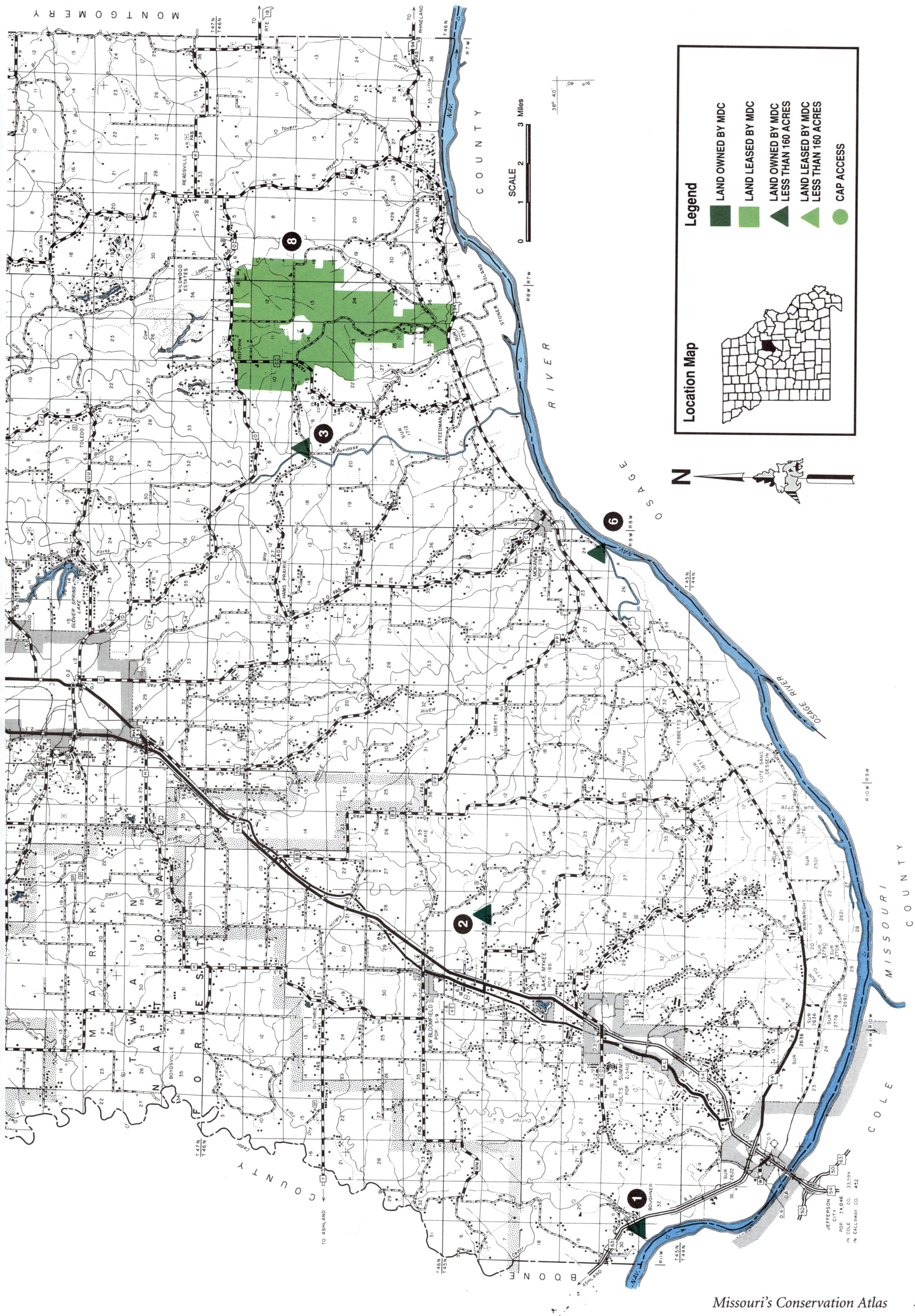
Legend
LAND OWNED BY MDC
LAND LEASED BY MDC
LAND OWNED BY MDC LESS THAN 160 ACRES
LAND LEASED BY MDC LESS THAN 160 ACRES
CAP ACCESS
Location Map
SCALE
0 1 2 3 Miles
N
MONTGOMERY
COUNTY
OSAGE
RIVER
MISSOURI
COUNTY
COLE
BOONE
COUNTY
OSAGE RIVER
JEFFERSON CITY
HOLTS SUMMIT
NEW BLOOMFIELD
LAKE MYKEE
MOKANE
PORTLAND
READSVILLE
TOLEDO
YUCATAN
WILDWOOD ESTATES
REFORM
STEEDMAN
HAMS PRAIRIE
LIBERTY
DIXIE
TEBBETTS
COTE SANS DESSEIN
WAINWRIGHT
GUTHRIE
BOYDSVILLE
CARRINGTON
FOREST
KINGDOM
BOUGHNER
TO ASHLAND
TO RHINELAND
TO ASHLAND
GLOVER SPRING LAKE

CAMDEN COUNTY

1 **Branch Towersite** is west of Camdenton on Highway 54 and 2 miles south on Highway 73. This is a 40-acre area with a .25-acre pond suitable for fishing.

2 **Brown Bend Access** is 3.5 miles west of Climax Springs on Highway 7, 4 miles north on Highway DD, 4.5 miles on Highway FF, then 2 miles on FF-14 (gravel). This is 2 acres on the Osage Arm of Lake of the Ozarks. There is a boat ramp and a fishing jetty and dock accessible to wheelchairs.

3 **Climax Springs Towersite** is north of Camdenton on Highway 5, then 20 miles west on Highway 7. This is a 4-acre forested area.

4 **Fiery Fork Conservation Area** is west of Camdenton on Highway 5, 10 miles west on Highway 7, then 2 miles down Gravel Road 7-17. Watch for signs. This 1,606-acre area includes stretches of Little Niangua River, Fiery Forks Creek and Toby Creek (10 acres for fishing). There are two hiking trails.

5 **Larry R. Gale Access** is 4 miles west of Camdenton on Highway 54, 2.5 miles down Highway AA, and 1 mile down County Road AA-101D. Watch for signs. This 194-acre access on Lake of the Ozarks has boat ramps and a fishing jetty accessible to wheelchairs.

6 **Hurricane Deck Towersite** (21 acres) is off Highway 5. At the city limits of Sunrise Beach, turn on Lake Road 5-32, and the tower is located 60 yards down 5-32 on the right.

7 **Alice Ahart Mansfield Conservation Area** is 7 miles east of Camdenton on Highway 54, 2 miles on Highway KK, and .25 mile down Knolls Road. This is an 80-acre forest with a .5-mile hiking trail.

8 **Moles Cave Conservation Area** is northwest of Camdenton. From Highway 5, travel 17 miles to Highway DD, then 3 miles to Highway FF, then 1 mile to a gravel road on the right. The 108-acre area is 1.25 miles down this gravel road. It is home to Moles Cave Spring and a large heron rookery.

9 **Shawnee Bend Access** is near Sunrise Beach. Take Highway F to Highway TT, then watch for signs near the end of TT (about 5 miles). This is a 6-acre access on Lake of the Ozarks with a boat ramp and wheelchair accessible fishing dock.

10 **Toronto Springs Conservation Area** is 3 miles east of Montreal on State Road E, then 1.5 miles east on State Road A. This 568-acre area contains a stretch of the Wet Glaize River (no boat ramp), five ponds, two caves and a spring.

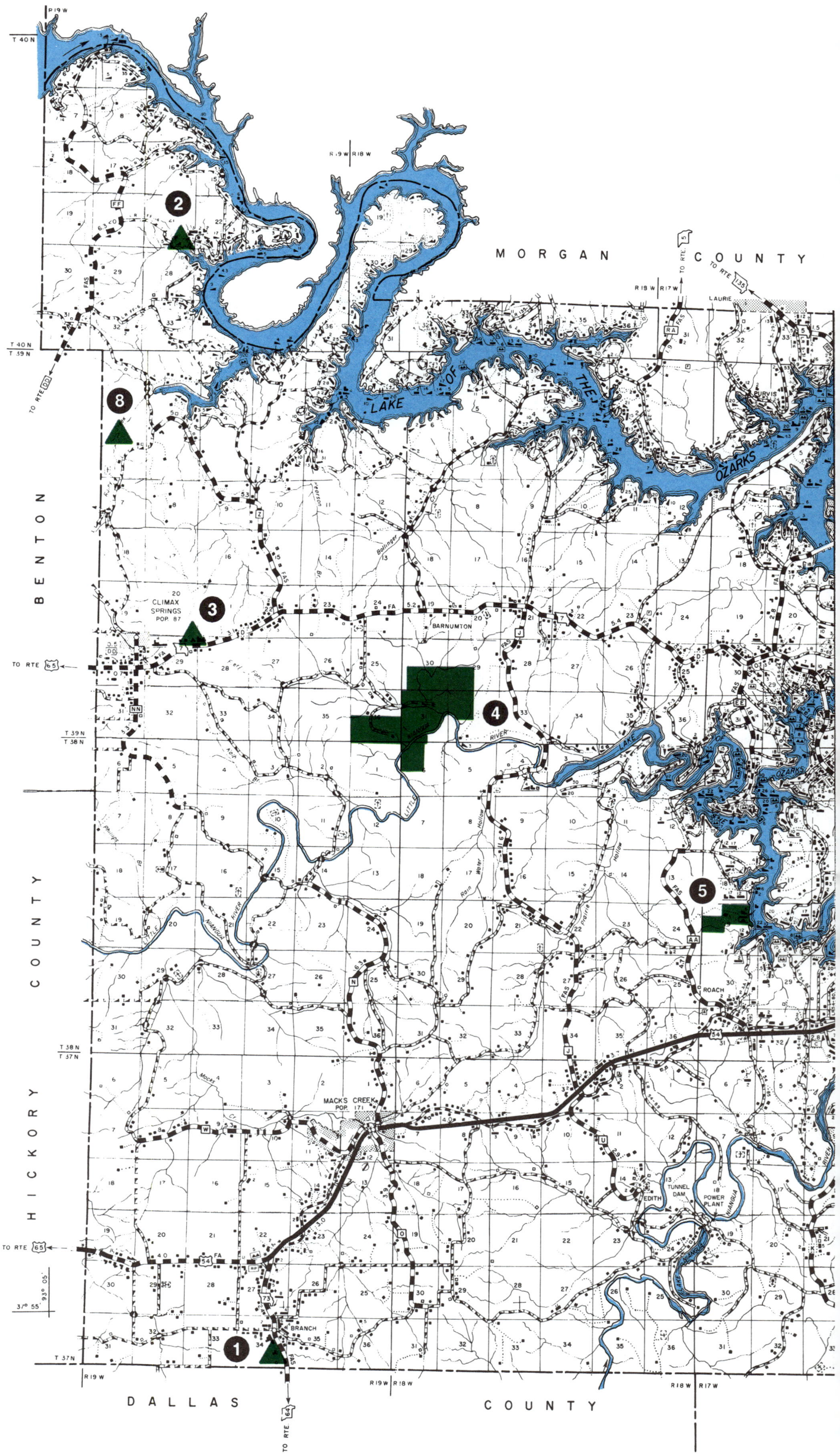

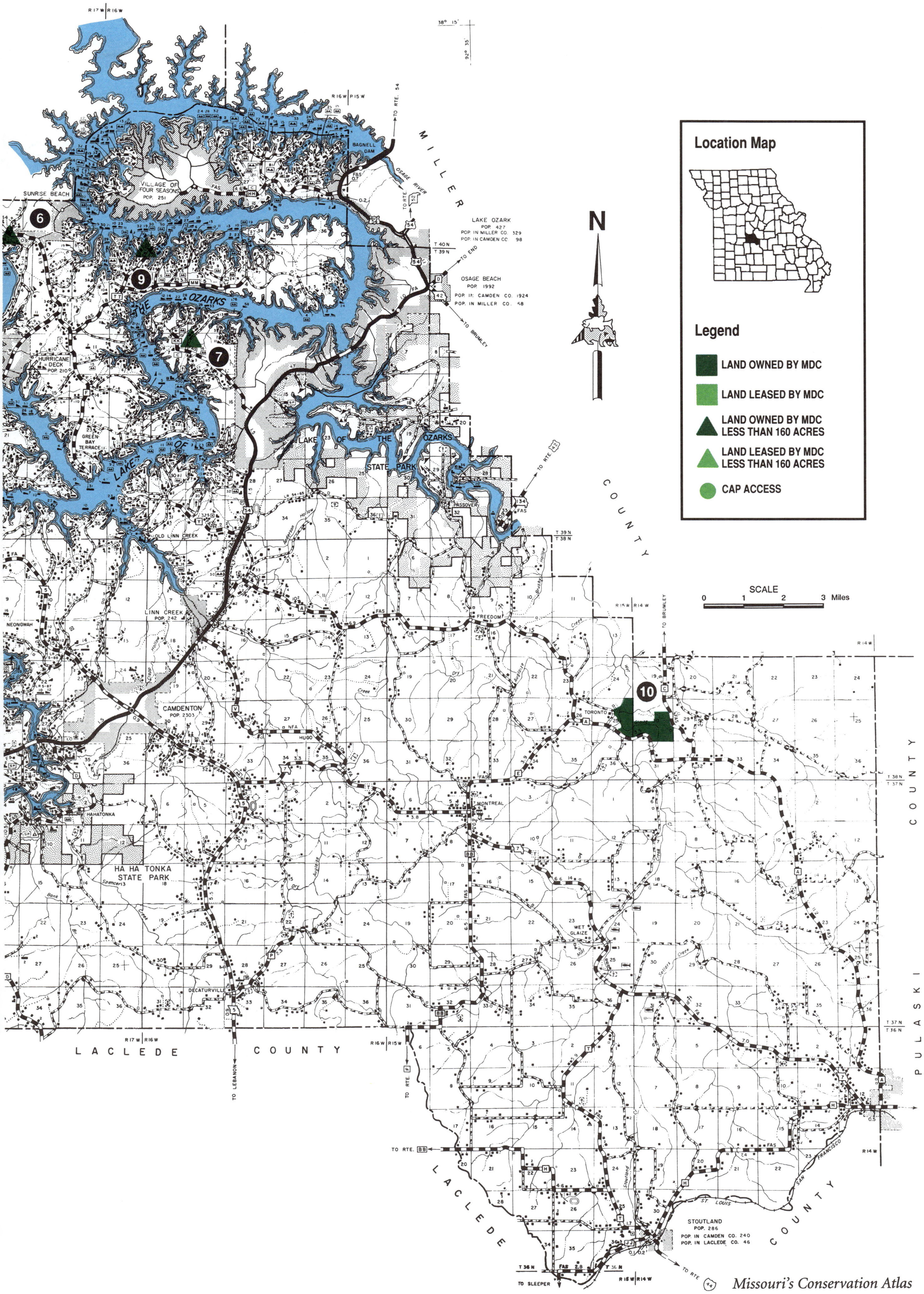
Location Map
Legend
LAND OWNED BY MDC
LAND LEASED BY MDC
LAND OWNED BY MDC LESS THAN 160 ACRES
LAND LEASED BY MDC LESS THAN 160 ACRES
CAP ACCESS
SCALE
0 1 2 3 Miles
N
MILLER COUNTY
LACLEDE COUNTY
PULASKI COUNTY
LAKE OZARK
POP. 427
POP. IN MILLER CO. 329
POP. IN CAMDEN CO. 98
OSAGE BEACH
POP. 1992
POP. IN CAMDEN CO. 1924
POP. IN MILLER CO. 58
VILLAGE OF FOUR SEASONS
POP. 251
SUNRISE BEACH
BAGNELL DAM
OSAGE RIVER
HURRICANE DECK
POP. 210
GREEN BAY TERRACE
LAKE OF THE OZARKS
LAKE OF THE OZARKS STATE PARK
PASSOVER
OLD LINN CREEK
LINN CREEK
POP. 242
NEONGWAH
FREEDOM
CAMDENTON
POP. 2303
HUGO
TORONTO
MONTREAL
HAHATONKA
HA HA TONKA STATE PARK
WET GLAIZE
DECATURVILLE
STOUTLAND
POP. 286
POP. IN CAMDEN CO. 240
POP. IN LACLEDE CO. 46
TO RTE. 54
TO RTE. 52
TO END
TO BRUMLEY
TO RTE. 42
TO LEBANON
TO RTE. F
TO RTE. BB
TO SLEEPER
TO RTE. 44

CAPE GIRARDEAU COUNTY

1 **Apple Creek Conservation Area** is about 20 miles north of Cape Girardeau. To reach this 2,100-acre area, take the Fruitland exit from I-55, go 7 miles north on Highway 61, then go 5 miles east on Highway CC. Five ponds and a 2-mile stretch of Apple Creek provide anglers with about 15 acres of fishable water. There are wheelchair accessible firearms and archery ranges, and a 5-mile horseback riding trail.

2 **Block Hole Access** is off Highway U in southwest Cape Girardeau County, .5 mile east of the junction of highways T, N and U. This is a 10-acre area.

3 To reach **Cape Woods Conservation Area,** take Kingshighway (Bus. 55-61-34) north through Cape Girardeau. Turn south on Mount Auburn Road, and the area lies on the east side. This is a 37-acre area with hiking trails.

4 **Headwaters Access** is east of I-55 on Highway 74. It is an access with a boat ramp on the Diversion Channel (Mississippi River) and totals about 10 acres.

5 To reach **Juden Creek Conservation Area** (8 acres), take Highway 177 (Big Bend Road) north through Cape Girardeau. At the north edge of the city limits, turn right on Old Missouri Route V and continue 125 yards to the area. The area is known for its wildflowers, and there is a short hiking trail that is reached from neighboring Kelso Bird Sanctuary or Twin Trees County Park.

6 **Lake Girardeau Conservation Area** is 7 miles west of Jackson on Highway 34, then 6 miles south on Highway U. Watch for signs. This 162-acre lake has a boat ramp and fishing jetty that is accessible to wheelchairs. The land around the lake (area is 351 acres total) is mainly forested and pastureland. There are picnic areas.

7 **Maintz Wildlife Preserve** is northwest of Jackson. Take Highway B north to Highway BB. Turn west on Highway BB to County Road 471. This 804-acre area is north on 471. Thirteen ponds provide 8 acres of fishable water, and the area is home to a 400-acre demonstration farm.

8 **Old Plantation Access** on the Whitewater River is west of Jackson on Highway 72 and right on County Road 385. This 70-acre area is mainly cropland with 1 mile of Whitewater River frontage.

9 **Rotary Park Lake** is in the Jackson City Park off Highway 61. It is a 3-acre fishable lake with wheelchair access.

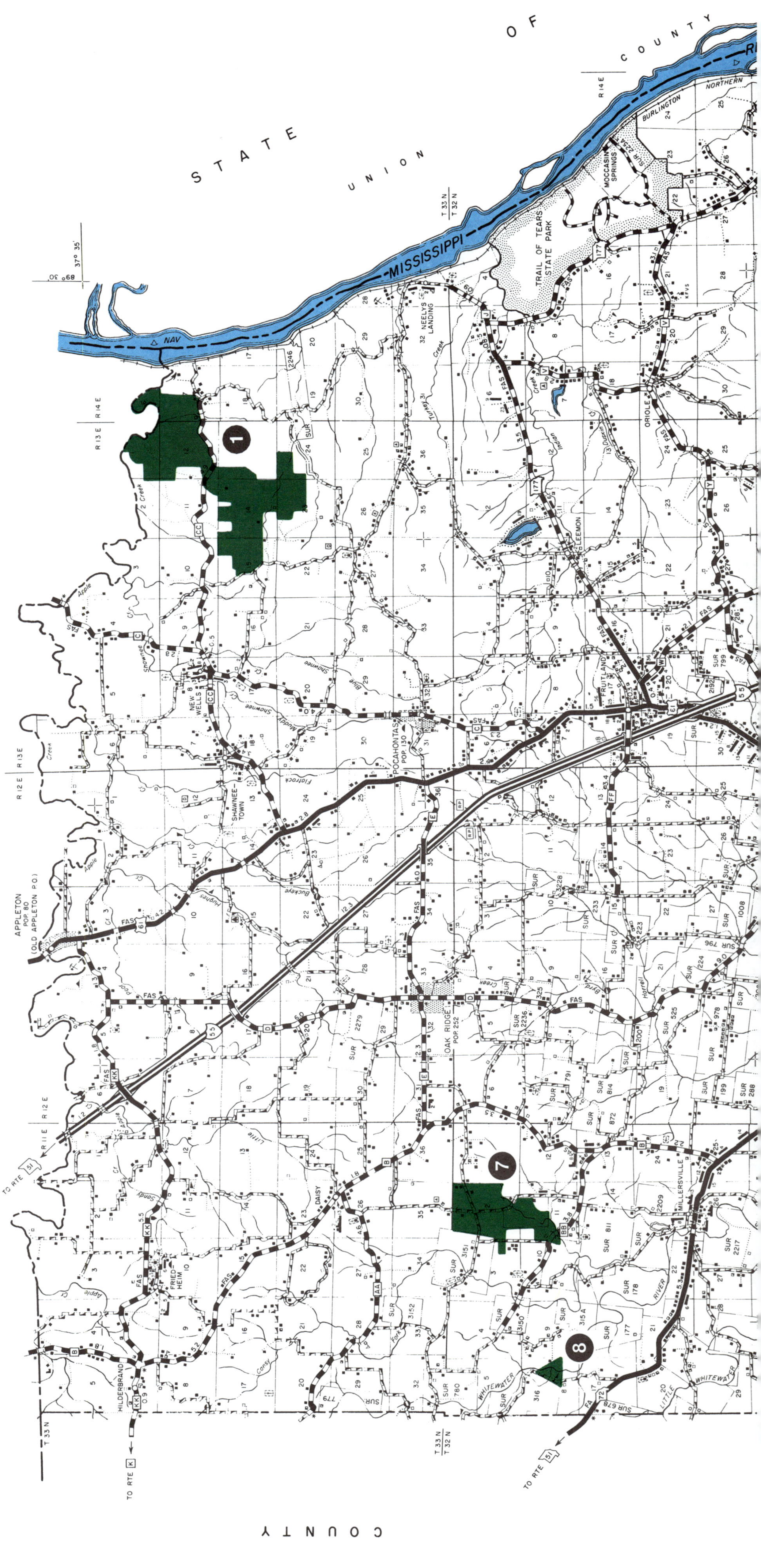

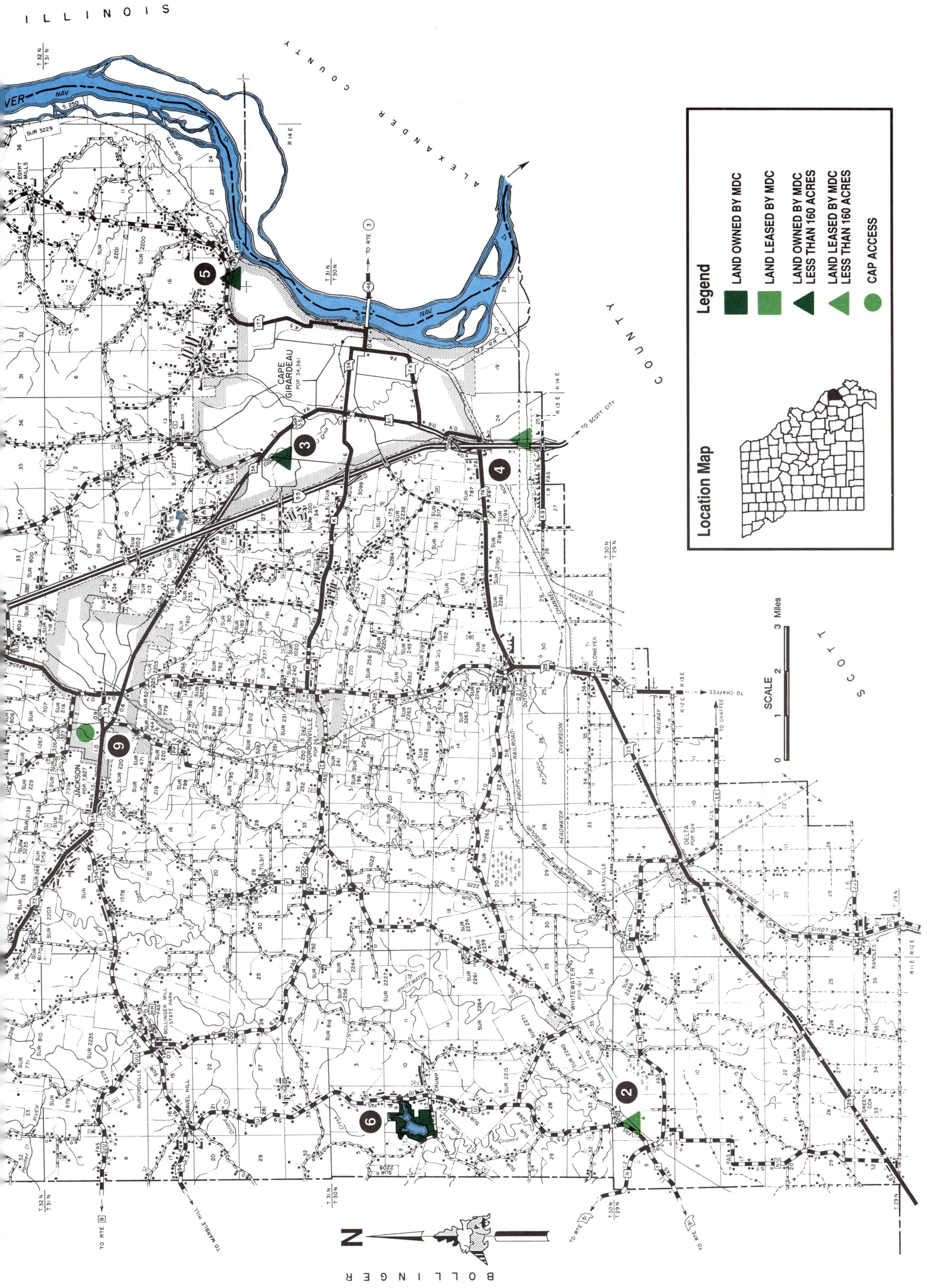
ILLINOIS
ALEXANDER COUNTY
SCOTT COUNTY
BOLLINGER
Legend
LAND OWNED BY MDC
LAND LEASED BY MDC
LAND OWNED BY MDC LESS THAN 160 ACRES
LAND LEASED BY MDC LESS THAN 160 ACRES
CAP ACCESS
Location Map
SCALE
0 1 2 3 Miles
CAPE GIRARDEAU POP 34,361
JACKSON POP 7,827
GORDONVILLE POP 267
DELTA POP 524
WHITEWATER POP 161
BOLLINGER MILL STATE PARK
BURFORDVILLE
GRAVEL HILL
DUTCHTOWN
BLOMEYER
ALLENVILLE
TILSIT
RANDLES
GREEN COX
EGYPT MILLS
HEADWATER DIVERSION CHANNEL
WHITEWATER RIVER
TO RTE 3
TO SCOTT CITY
TO CHAFFEE
TO RTE B
TO MARBLE HILL
TO RTE 51
TO RTE 91
1 2 3 4 5 6 9
N

CARROLL COUNTY

1 **Bosworth Access** is near the town of Bosworth on Highway M. It is a 7-acre access on the Grand River and has a boat ramp.

2 **Bunch Hollow Conservation Area** is 10 miles north of Carrollton on Highway 65, 7 miles north on Route Z, and 2 miles west on an unnamed gravel road. This area is 3,294 acres, has 17 ponds (6 acres for fishing) and a 3-mile horseback riding trail.

3 **Little Compton Lake Conservation Area** is 4 miles south of Hale on Highway 139, and 3 miles east on Gravel Road 214. Little Compton Lake is 40 acres, and the Grand River also flows through the area (345 total acres).

4 **J. J. McKinny Conservation Area** is on Highway 41, 1 mile south of the junction of highways 41 and 24. This 120-acre area is forestland (60 acres), cropland and old pastures.

5 **W. L. Schifferdecker Memorial Conservation Area** is 10 miles west of Carrollton at the junction of routes D and E. This 241-acre area has seven ponds, providing anglers with 6 acres of fishable water.

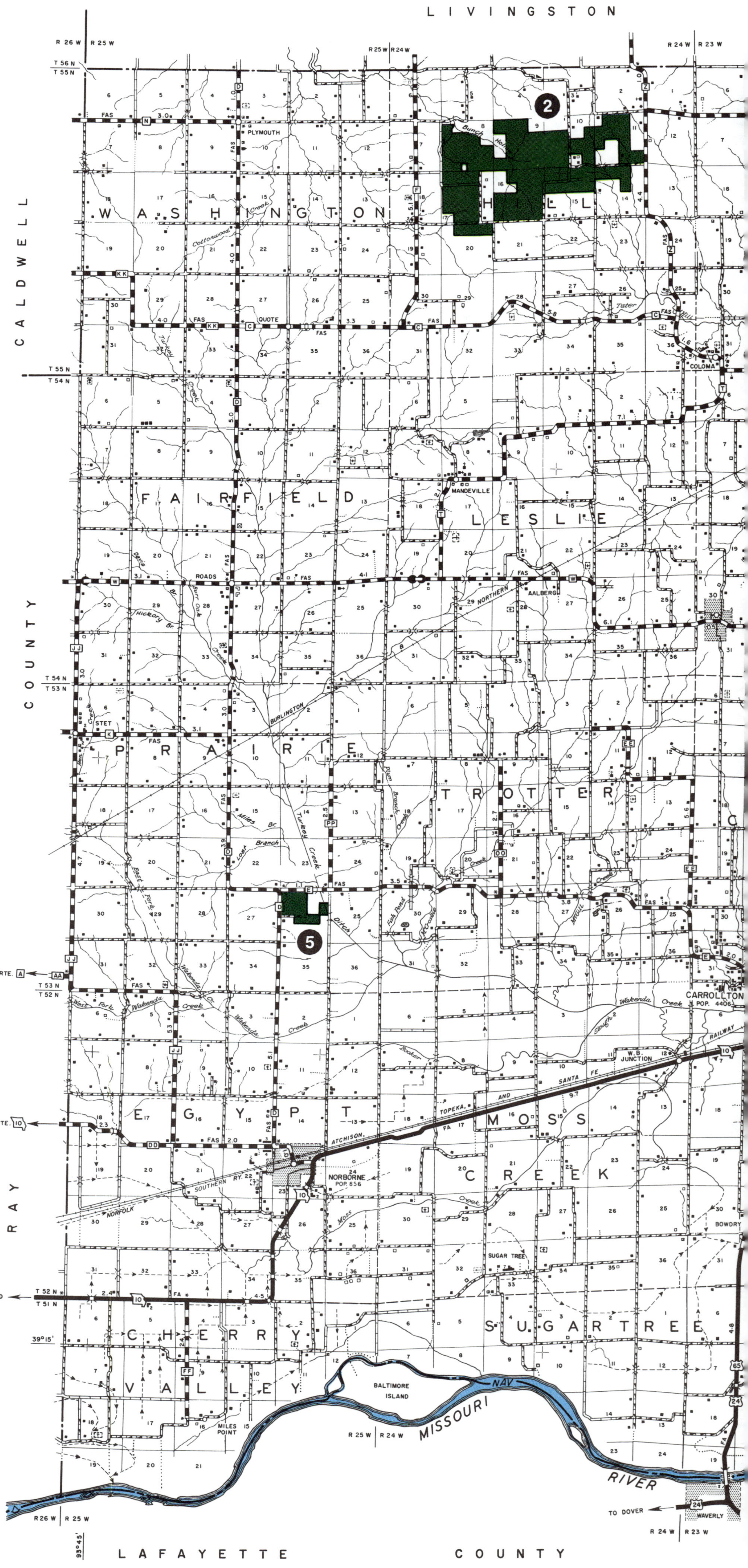

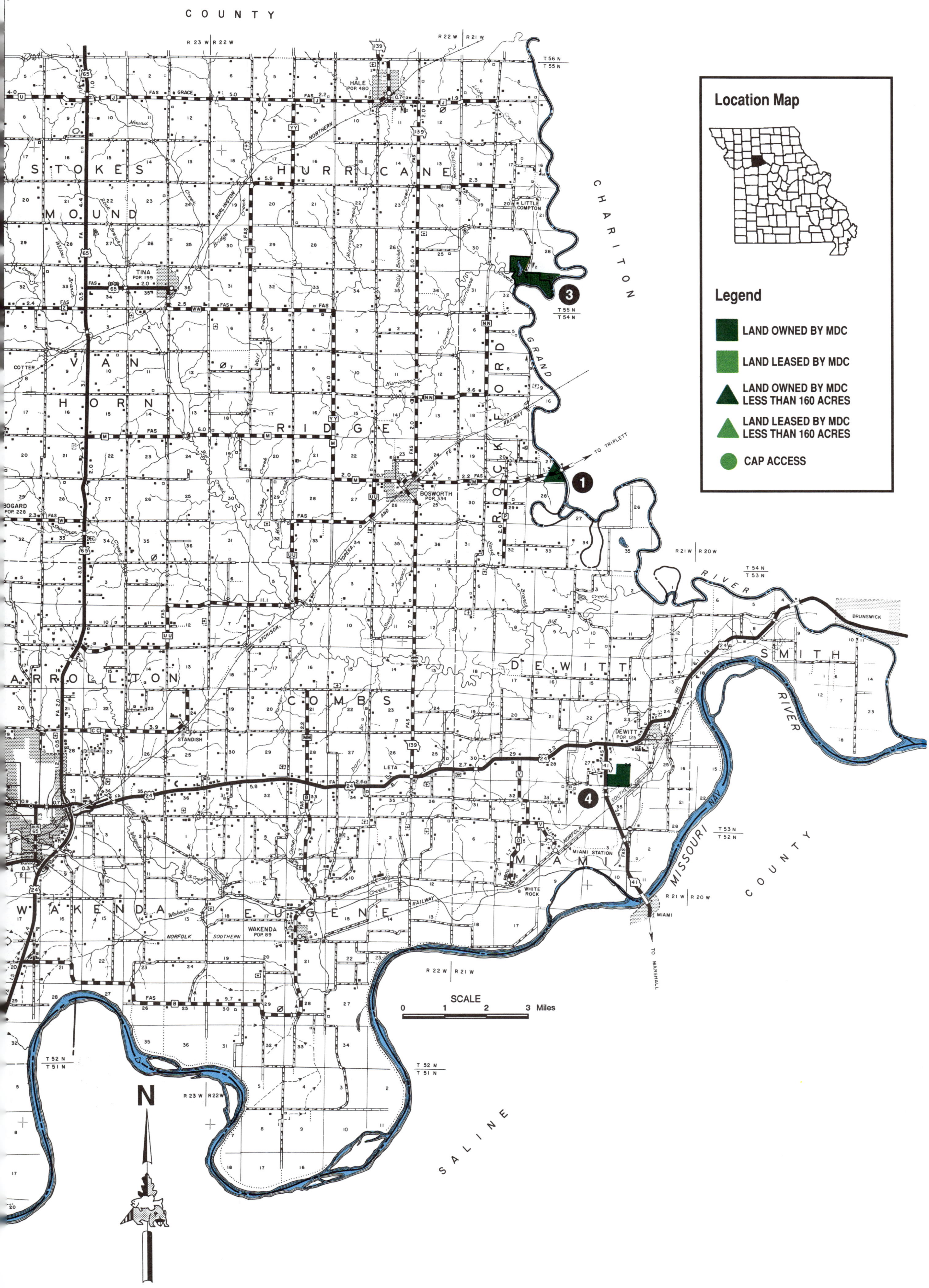
COUNTY
Location Map
Legend
LAND OWNED BY MDC
LAND LEASED BY MDC
LAND OWNED BY MDC LESS THAN 160 ACRES
LAND LEASED BY MDC LESS THAN 160 ACRES
CAP ACCESS
CHARITON
GRAND
RIVER
STOKES
HURRICANE
MOUND
VAN HORN
RIDGE
ROCKFORD
CARROLLTON
COMBS
DEWITT
SMITH
MIAMI
WAKENDA
EUGENE
HALE POP. 480
TINA POP. 199
LITTLE COMPTON
BOSWORTH POP. 334
BOGARD POP. 228
COTTER
STANDISH
LETA
DEWITT POP. 125
MIAMI STATION
WHITE ROCK
WAKENDA POP. 89
BRUNSWICK
TO TRIPLETT
TO MARSHALL
MISSOURI NAV.
MISSOURI RIVER
SALINE
COUNTY
SCALE
0 1 2 3 Miles
N

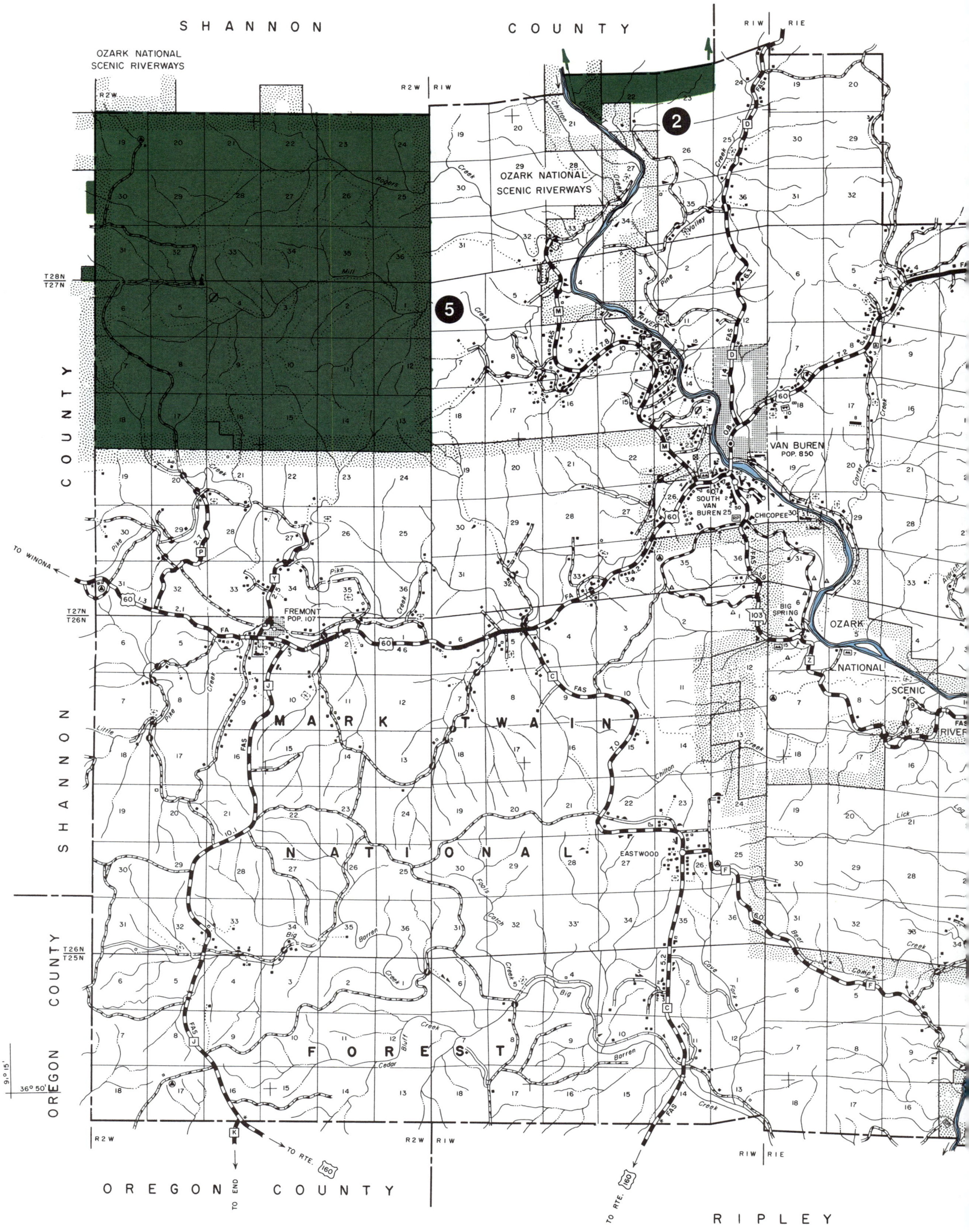

CARTER COUNTY

1 **Carter Creek Conservation Area** (440 acres) is 5 miles east of Van Buren along Highway 60. One tract is south of the highways 21 and 60 junction. The other tract is off Highway 21 on County Road 21-306.

2 **Current River Conservation Area** is west of Ellington on Highway 106, then 4 miles south on Highway HH to Cardareva Mountain. This is a 24,181-acre area. Parts of it border the Current River, Carrs Creek, Paint Rock Creek and Long Branch. There are two overlook platforms near Blue Springs – a Designated Natural Area.

3 **Hunter Towersite** is off Highway E near the town of Hunter, .25 mile down County Road 0-251. The tower overlooks 75 acres of forestland.

4 **Miller Community Lake** is 1 mile north of Highway 60 on Highway 21. This 27-acre lake is equipped with a boat ramp and picnic area. The 35 acres of land surrounding the lake is mainly forestland, and there is also a 1-mile dirt hiking trail.

5 The main entrance to **Peck Ranch Conservation Area** is 5 miles east of Winona on Highway H and another 7 miles east on a gravel road (marked with a sign) in Shannon County. This 22,948-acre area has 119 ponds (totaling 125 acres), seven caves and numerous springs.

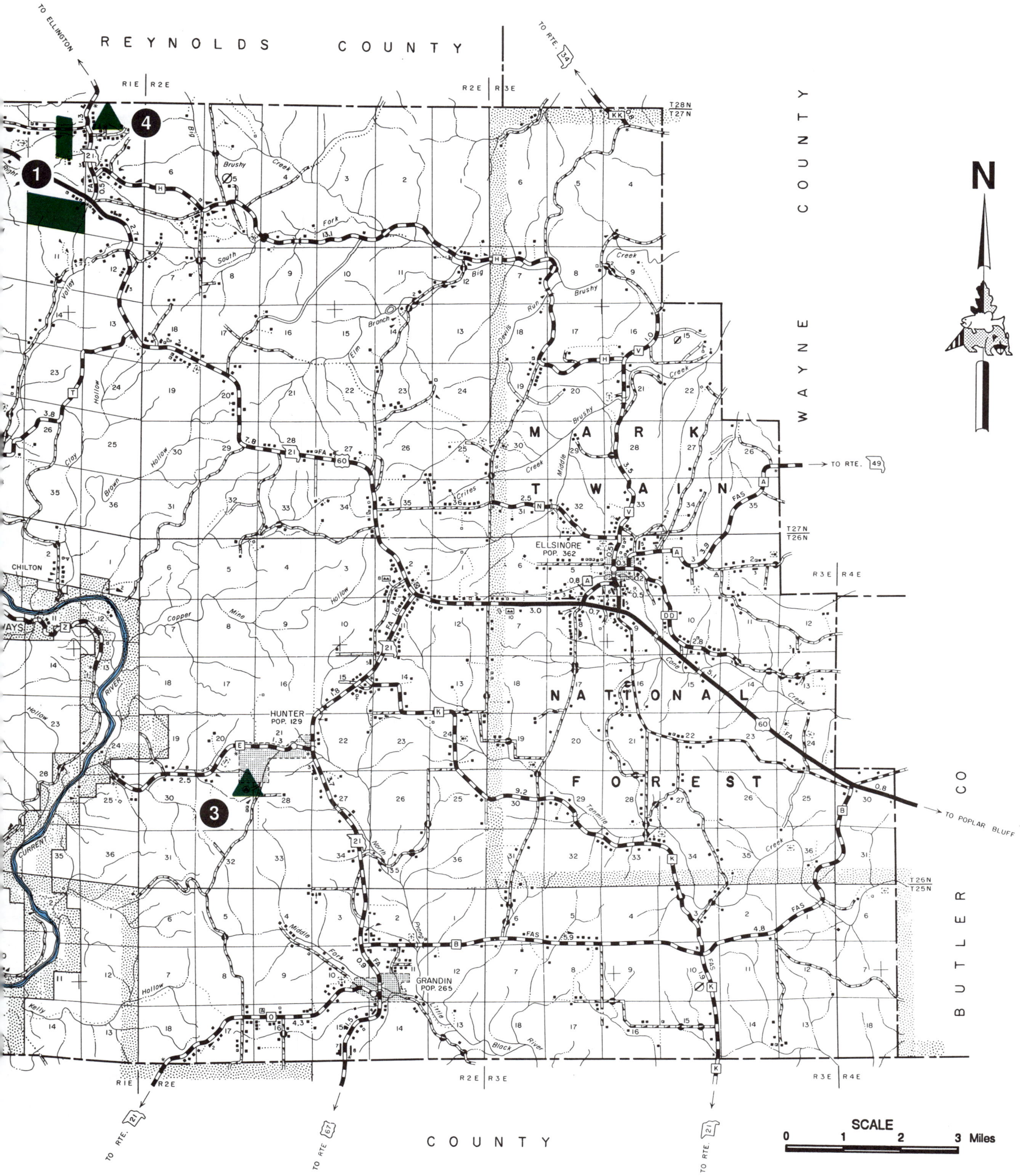

Designated Natural Areas include Stegall Mountain (3,872 acres), Golden Seal (100 acres), Mule Hollow Glade (44 acres) and Grassy Pond (1 acre). Most of the area is forested, and a 9.5-mile stretch of the Ozark Trail goes through the area. It is also equipped with a shooting range and public contact office.

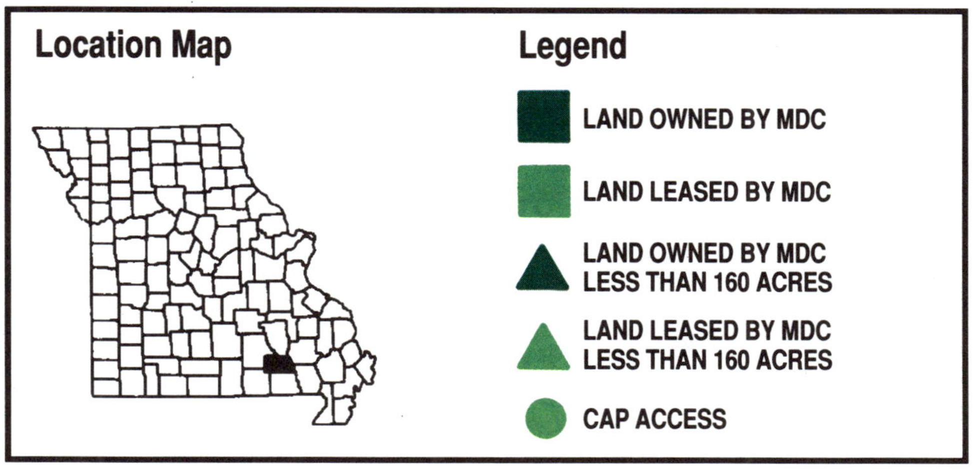

CASS COUNTY

1 **Amarugia Highlands Conservation Area** is 7 miles northwest of Archie on Route W. This 1,041-acre area is home to several ponds and Amarugia Lake. The lake is 60 acres and has a boat ramp and wheelchair accessible fishing dock.

2 **Archie Access** is on Highway 71 in the city of Archie, north of Route B and just east of 71. This is a 1-acre access with no boat ramp on the South Grand River.

3 **Bittern Bottoms Conservation Area** is 5 miles south of Harrisonville on Highway 71, and 2.5 miles west on Gravel Road 307. Thirty-three acres of the area are wetlands, and 27 acres are forestland. Natural features include a 20-acre marsh and a 13-acre open sedge- and grass-covered wet meadow. There is a .25-mile maintained trail.

4 **Dorsett Hill Prairie Conservation Area** is northeast of Everett on Route W, 1 mile east on 333rd Street, and .25 mile north (left) on Dorsett Hill Lane. The area is composed of 14 acres of native prairie inside 79 total acres of forestland and old pastures. There is a short walking trail (1/8 mile) from the parking lot to the prairie.

5 **James R. Harter Conservation Area** is at the intersection of Highway 7 and Kircher Road, 4 miles west of Garden City. This is a 456-acre area.

6 **North Lake-Harrisonville** is 11 miles south of Lee's Summit on Highway 291 and .5 mile east on County Road 235. This 34-acre lake is suitable for fishing. It also has trap shooting and skeet shooting facilities that are operated by the city of Harrisonville.

7 **Settle's Ford Conservation Area**–see the Bates County Map, page 18.

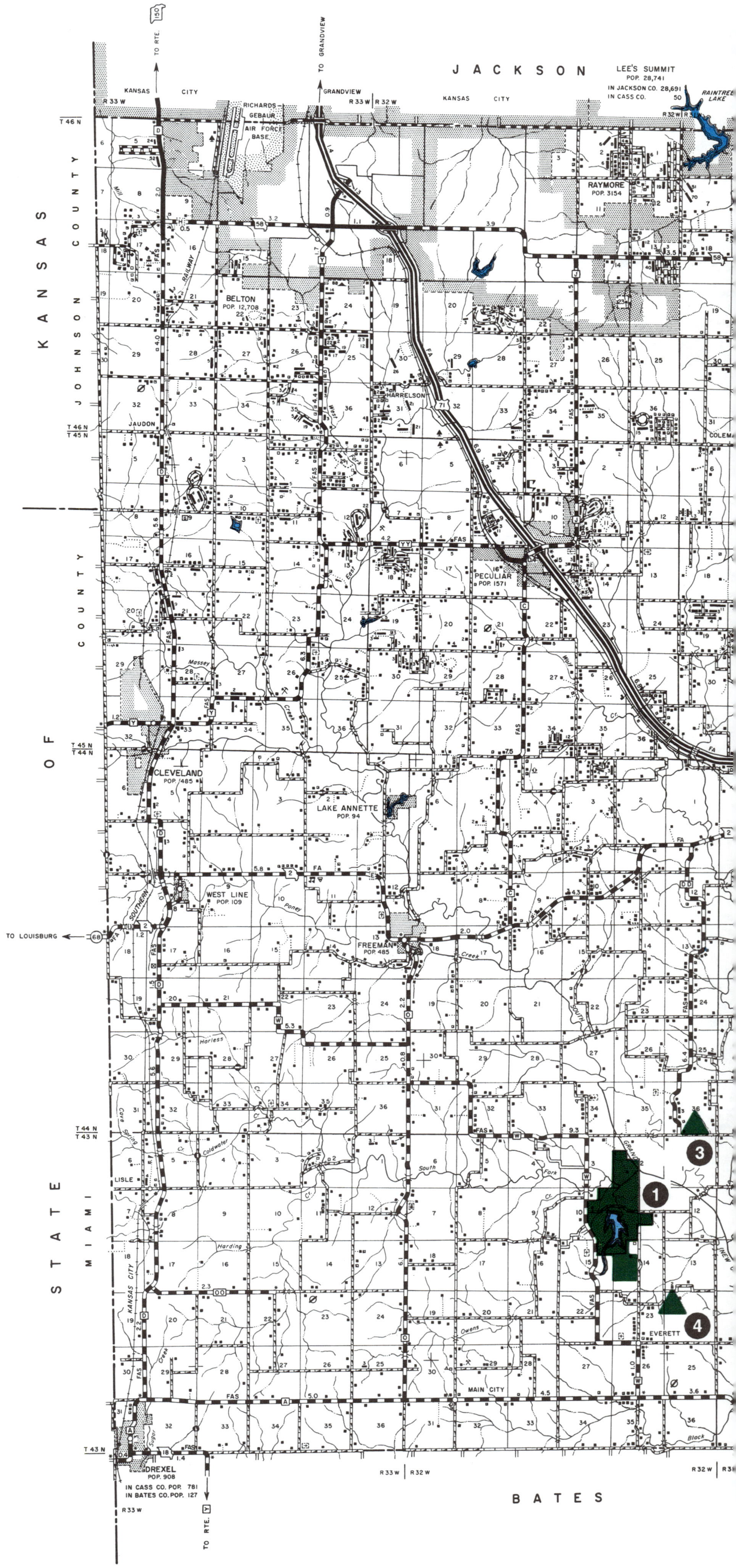

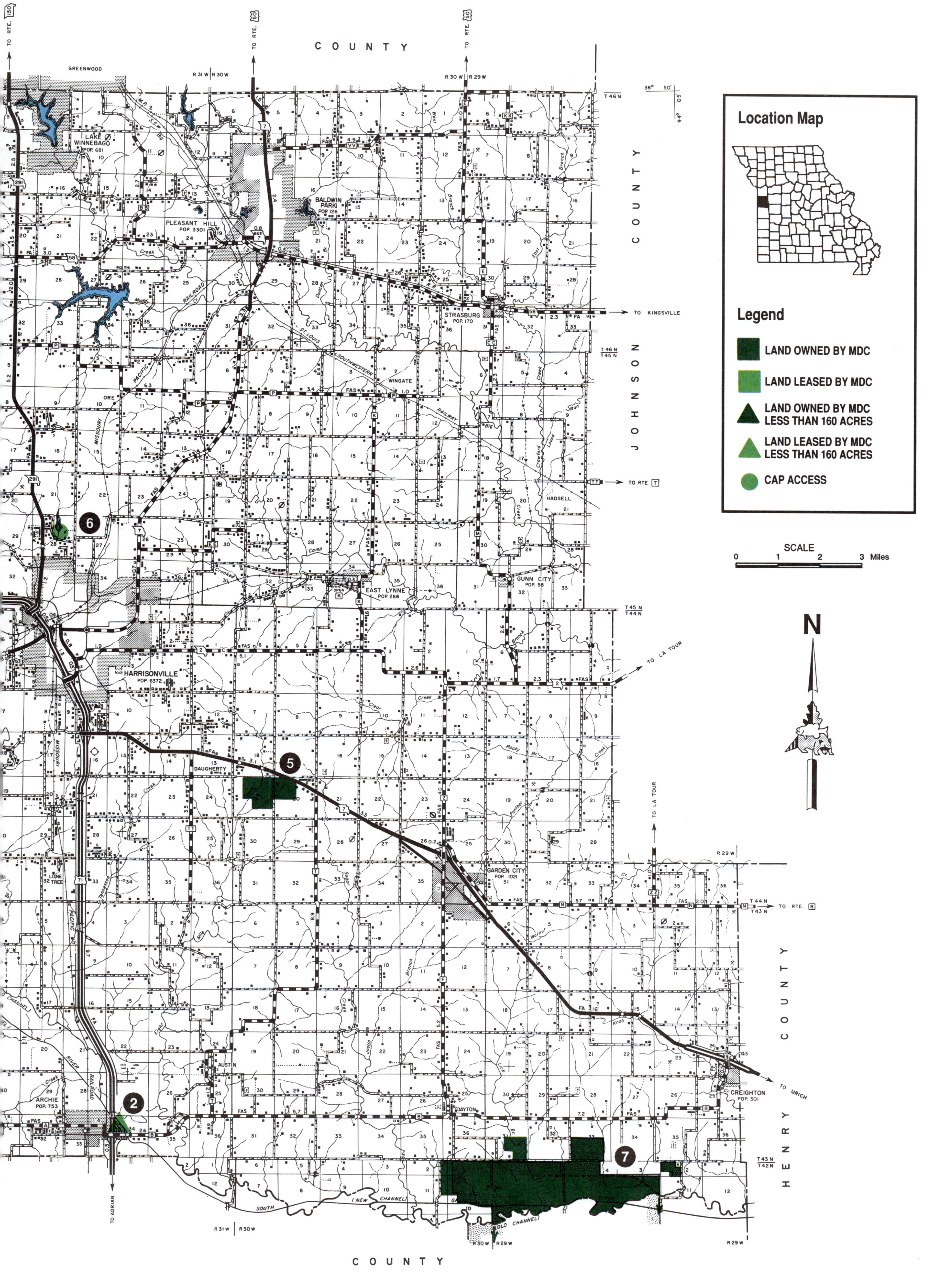

Location Map
Legend
LAND OWNED BY MDC
LAND LEASED BY MDC
LAND OWNED BY MDC LESS THAN 160 ACRES
LAND LEASED BY MDC LESS THAN 160 ACRES
CAP ACCESS
SCALE
0 1 2 3 Miles
N
COUNTY
JOHNSON COUNTY
HENRY COUNTY
COUNTY
GREENWOOD
LAKE WINNEBAGO POP. 681
PLEASANT HILL POP. 3301
BALDWIN PARK POP. 126
STRASBURG POP. 170
WINGATE
ORE
HADSELL
GUNN CITY POP. 58
EAST LYNNE POP. 286
HARRISONVILLE POP. 6372
DAUGHERTY
GARDEN CITY POP. 1021
LONE TREE
AUSTIN
ARCHIE POP. 753
DAYTON
CREIGHTON POP. 301
TO KINGSVILLE
TO RTE. T
TO LA TOUR
TO RTE. B
TO URICH
TO ADRIAN
TO RTE. 150
TO RTE. 50
R 31 W
R 30 W
R 29 W
T 46 N
T 45 N
T 44 N
T 43 N
T 42 N
SOUTH
NEW CHANNEL
OLD CHANNEL

CEDAR COUNTY

1 **Bluff Springs Conservation Area** is 8 miles north of Stockton on Highway J. This 415-acre area has three ponds (1.25 acres), four springs and a 135-acre sandstone savannah.

2 **Mo-Ko Prairie** is 1 mile east of El Dorado Springs on Lake Hill Road. This 420-acre area is dominated by native prairie (400 acres) and some forestland (20 acres).

3 **Monegaw Prairie Conservation Area** is on Highway 54, 2.5 miles east of El Dorado Springs. This 270-acre native prairie has two small ponds (1.25 acres for fishing) and a spring.

4 **Sky Prairie Conservation Area** is about 12.5 miles south of El Dorado Springs off Highway 32. At the junction of highways 32 and CC, take the unnamed county road 2.5 miles south to the second county road. Go right, or west, 1 mile. The prairie is 125 acres, and the rest (75 acres) is forestland. The area has one small pond, a spring and areas where running buffalo clover has been reestablished.

5 **Stockton Reservoir** is 15 miles north of Springfield. This 24,900-acre lake extends into Cedar, Dade and Polk Counties and has many accesses. For specific information, contact the Southwest Management District, 2630 N. Mayfair, Springfield, MO 65803, 417/895-6880.

6 **Turkey Creek Conservation Area** is west of Humansville on Highway N, 1.5 miles south on Highway AA, then west on an unnamed county gravel road. The southeast corner of the area is .5 mile down this road. It is 160 acres of forestland.

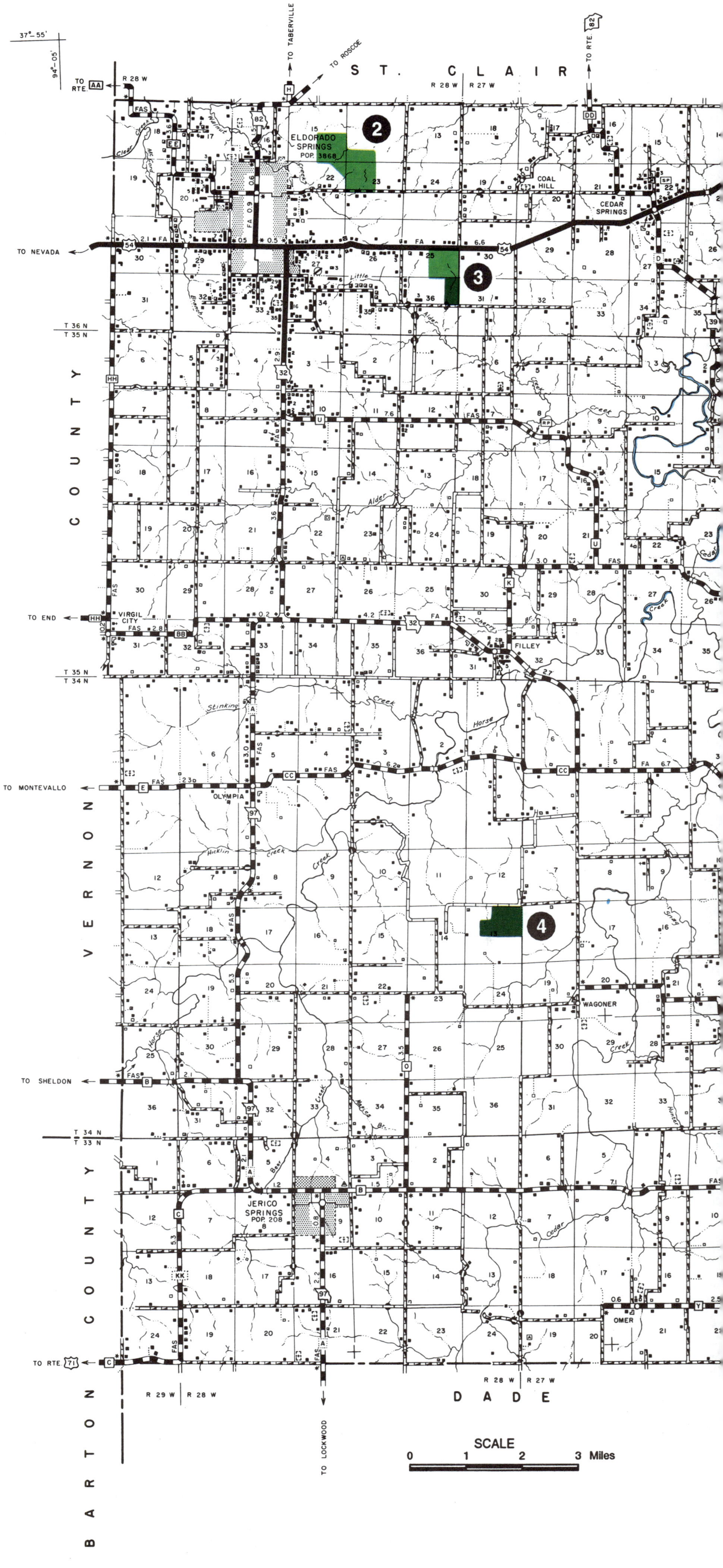

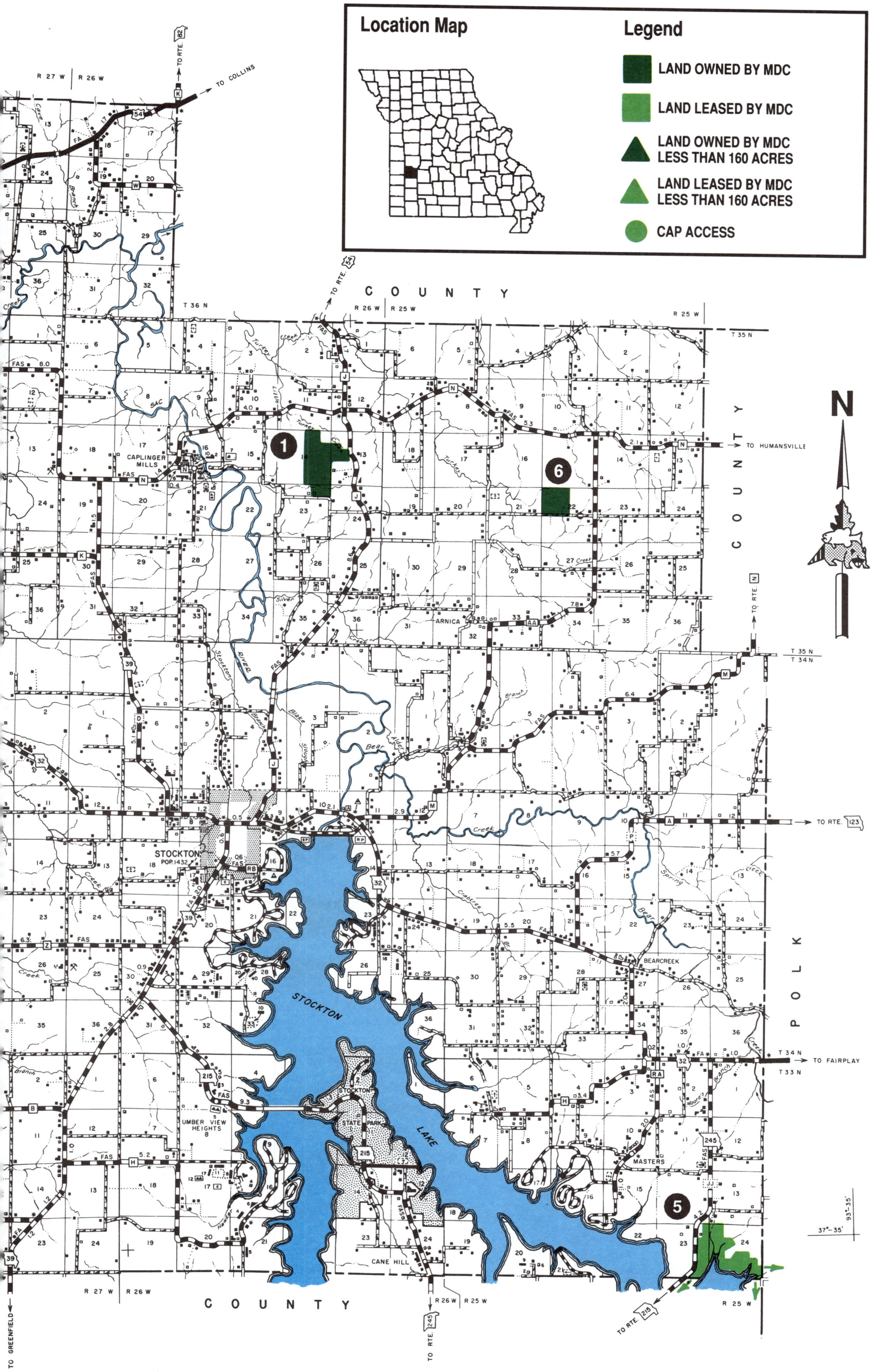
Location Map
Legend
LAND OWNED BY MDC
LAND LEASED BY MDC
LAND OWNED BY MDC LESS THAN 160 ACRES
LAND LEASED BY MDC LESS THAN 160 ACRES
CAP ACCESS
N
C O U N T Y
P O L K
C O U N T Y
C O U N T Y
TO COLLINS
TO HUMANSVILLE
TO RTE. 123
TO FAIRPLAY
TO GREENFIELD
TO RTE. 245
TO RTE. 215
R 27 W
R 26 W
R 25 W
T 36 N
T 35 N
T 34 N
T 33 N
CAPLINGER MILLS
ARNICA
STOCKTON POP. 1432
STOCKTON LAKE
STOCKTON STATE PARK
UMBER VIEW HEIGHTS
BEARCREEK
MASTERS
CANE HILL
SAC RIVER
37°-35'
93°-35'

CHARITON COUNTY

1 **Brunswick Access** (2 acres) is at the intersection of Highway 24 and Polk in the town of Brunswick. It is an access on the Grand River and is equipped with a boat ramp.

2 **Dalton Bottoms Access** is 5 miles south of the town of Dalton on Highway J, then 1.5 miles east and south on a gravel road. This is a 47-acre access to the Chariton and Missouri rivers.

3 **Nannie B. Floyd Memorial Conservation Area** is 9 miles south of Brookfield on Highway 11, 1.5 miles east on Highway E, then .25 mile north on a gravel road marked with a cantilever sign. This 20-acre area is primarily forestland and extends for .25 mile along West Yellow Creek.

4 **Lewis Mill Access** is north of Glascow, 1.5 miles west off Highway 5 on an unnamed county road. This is a 6-acre area with a boat ramp on the Little Chariton River.

5 **Marceline City Lake** is 3 miles south of the town of Marceline on Highway 5, then 3 miles west on Highway E to the second county road that heads north. The entrance to the 175-acre lake is less than 1 mile north on this county road. There is a boat ramp. Activities on the surrounding land are regulated by the city of Marceline.

6 **Pine Ridge Lake** is located in Potts Memorial Park near the city of Salisbury. The park entrance is two miles south of town on Highway 129. The lake is 25 acres and has a boat ramp. The city of Salisbury maintains picnic areas near the lake and manages other activities on the surrounding land.

7 **Price Bridge Access** is 3 miles south of Keytesville off Highway VV. This is a 56-acre access to the Chariton River.

8 **Sterling Price Community Lake** is 2 miles west of Salisbury on Highway 24 and 1 mile south (watch for the area sign) on County Road 427. This 35-acre lake is surrounded by 49 acres of forestland and old pastureland. It has a boat ramp, four fishing jetties and a picnic area.

9 **Sumner Access** is 2 miles southwest of the town of Sumner, just where Highway 139 crosses the Grand River. This is a small area (1.5 acres) with a boat ramp.

10 **Swan Lake Wildlife Area** is 2 miles northwest of Mendon on Highway CC. This 10,670-acre area is known for three lakes–Silver Lake, Swan Lake and South Pool–and has a total of 3,000 fishable acres. There are waterfowl blinds, an observation tower, hiking trail and some interpretive programs and exhibits.

11 **Yellow Creek Conservation Area** is 3 miles south of Sumner on RA, then right at the cantilever sign on a gravel road. The area is about 3 miles down this road. It consists of 617 acres of lowland hardwoods and is a Designated Natural Area. Yellow Creek and Elk Creek flow for about 4 miles through the area, providing anglers with approximately 20 acres of fishable streams. There are several oxbows and sloughs, and two hiking trails run through the area for a total of 4 miles.

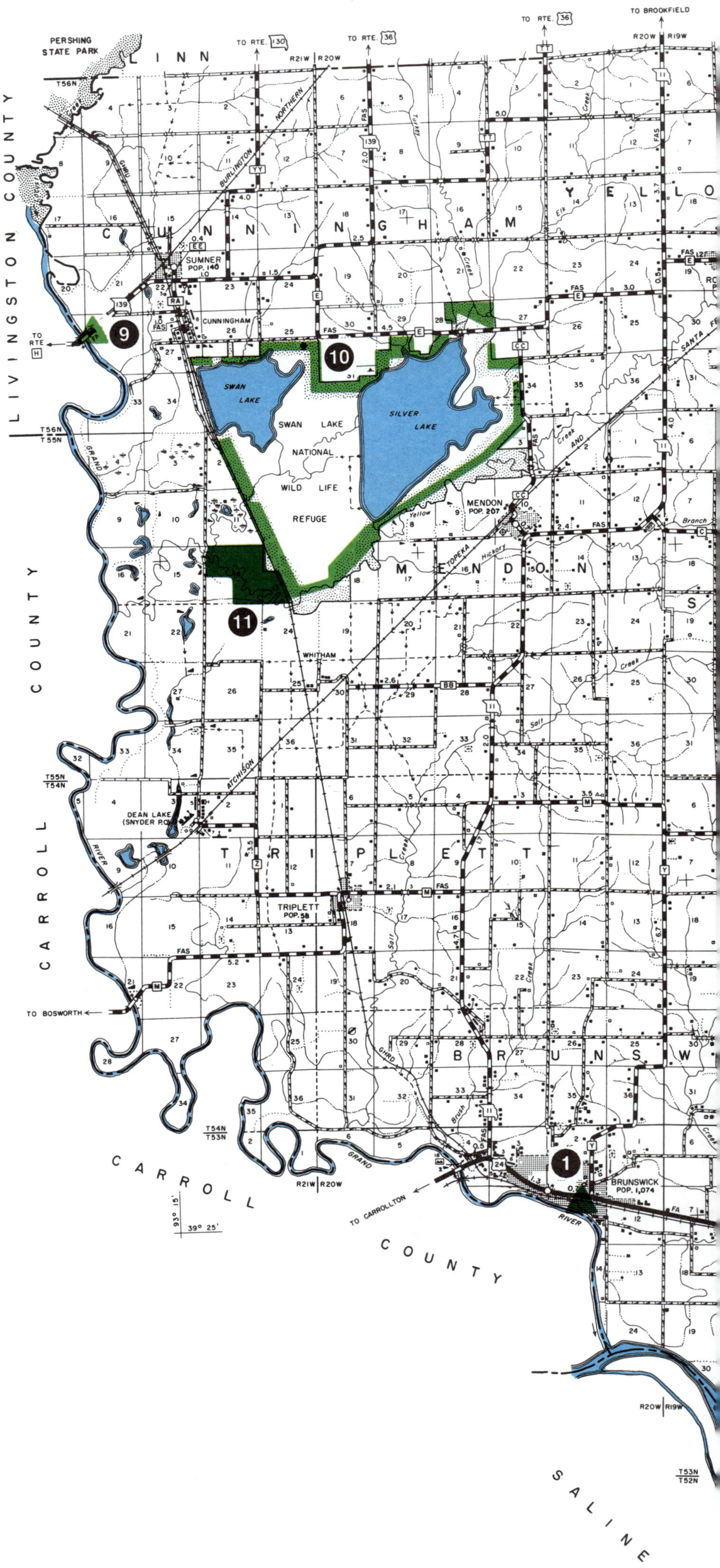

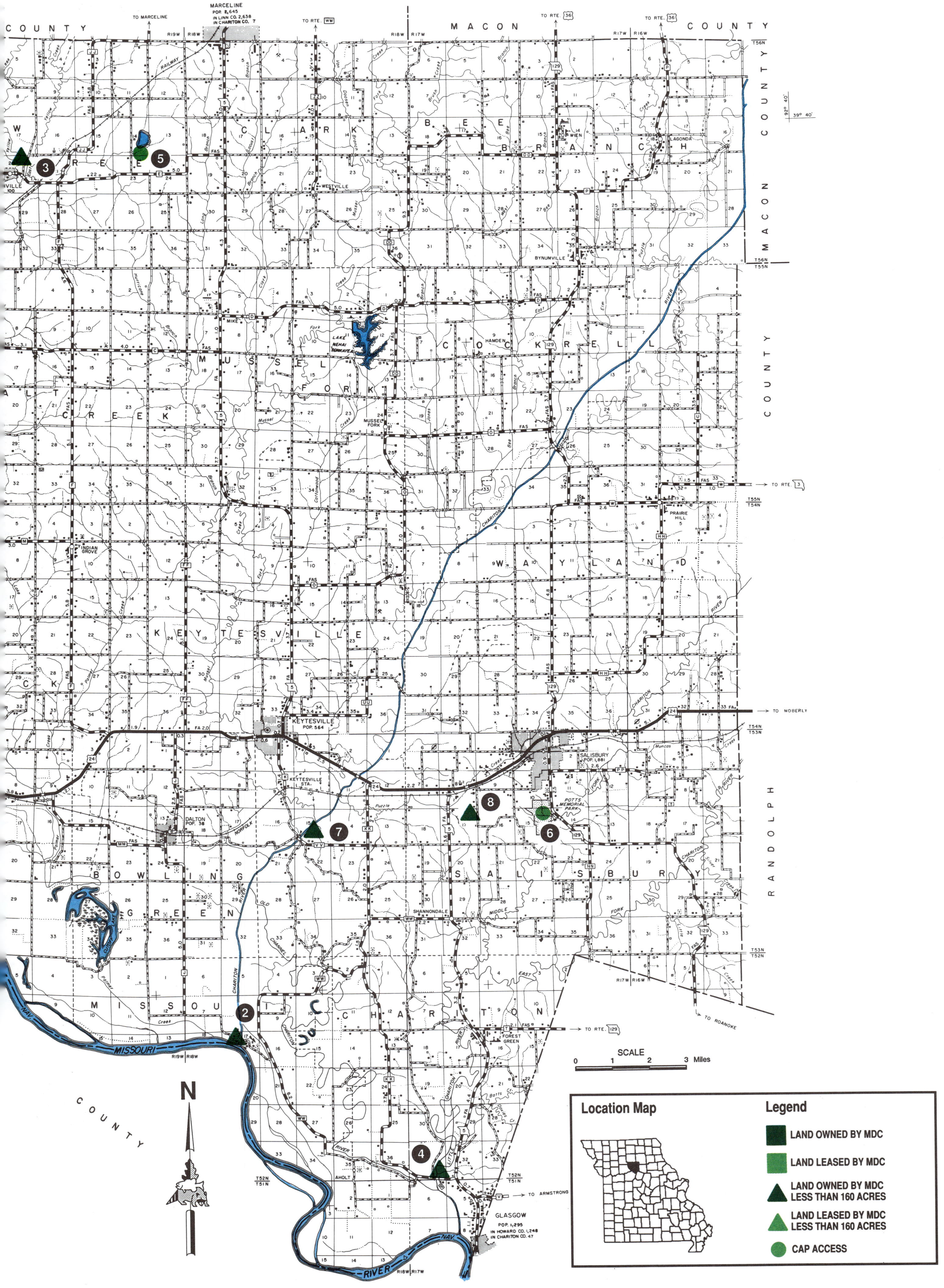

MACON COUNTY
RANDOLPH COUNTY
MARCELINE POP. 2,645 IN LINN CO. 2,638 IN CHARITON CO. 7
TO MARCELINE
TO RTE. WW
TO RTE. 36
TO RTE. 3
TO MOBERLY
TO ROANOKE
TO RTE. 129
TO ARMSTRONG
CLARK
BEE BRANCH
MUSSEL FORK
COCKRELL
SALT CREEK
KEYTESVILLE
WAYLAND
BOWLING GREEN
SALISBURY
MISSOURI
CHARITON
LAKE NEHAI TONKAYEA
KEYTESVILLE POP. 564
KEYTESVILLE STA.
SALISBURY POP. 1,881
POTTS MEMORIAL PARK
DALTON POP. 38
GLASGOW POP. 1,295 IN HOWARD CO. 1,248 IN CHARITON CO. 47
WESTVILLE
BYNUMVILLE
MIKE
HAMDEN
MUSSEL FORK
WIEN
LAGONDA
PRAIRIE HILL
INDIAN GROVE
SHANNONDALE
FOREST GREEN
AHOLT
MISSOURI RIVER
SCALE
0 1 2 3 Miles
Location Map
Legend
LAND OWNED BY MDC
LAND LEASED BY MDC
LAND OWNED BY MDC LESS THAN 160 ACRES
LAND LEASED BY MDC LESS THAN 160 ACRES
CAP ACCESS

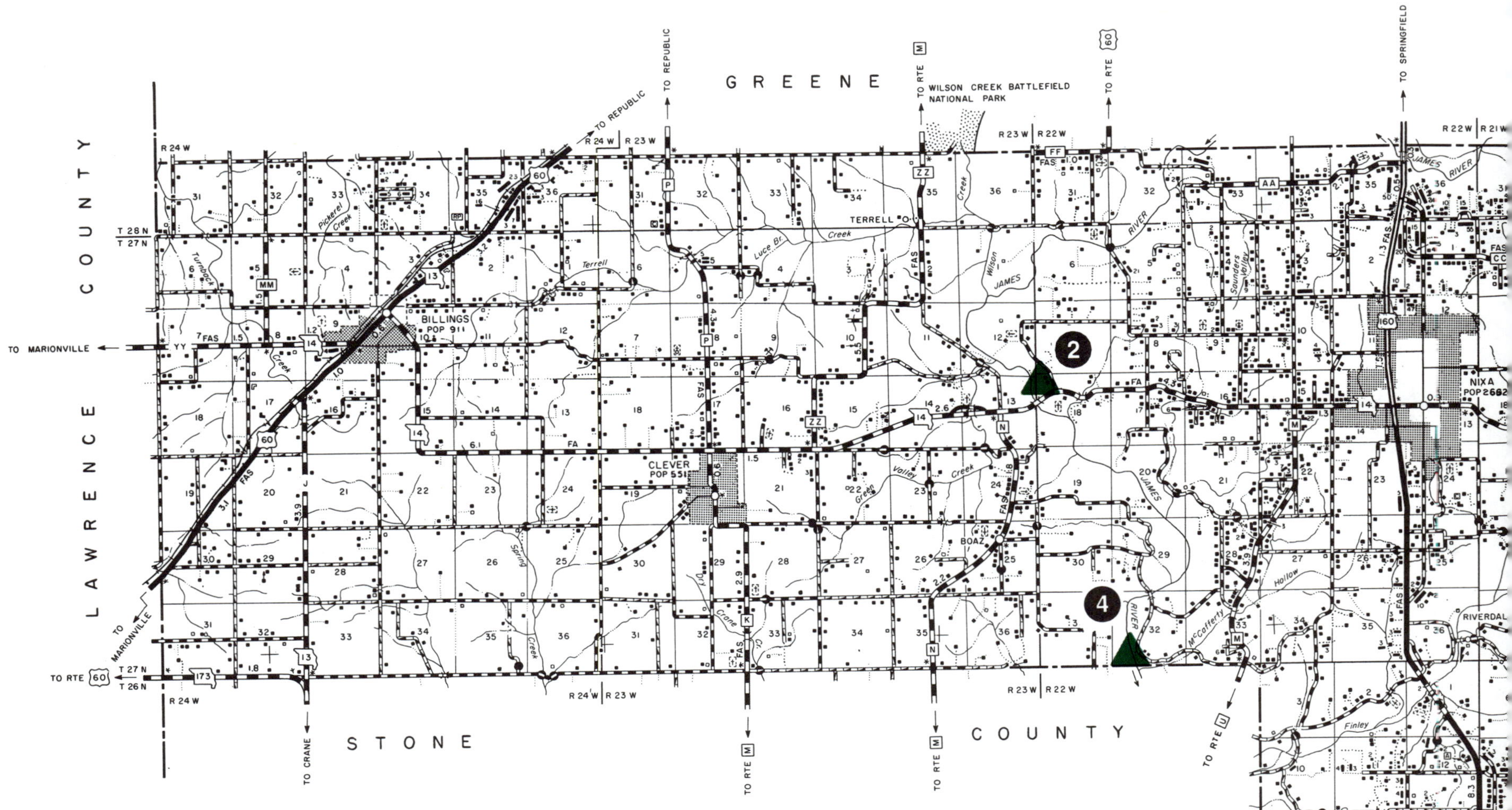

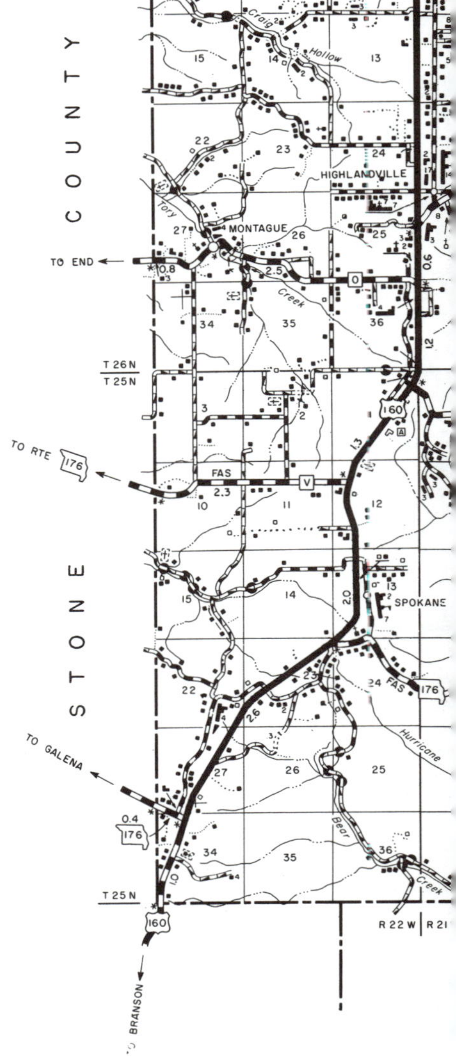

CHRISTIAN COUNTY

1 **Busiek State Forest and Wildlife Area** is on Highway 65, 18 miles south of Springfield. This 2,505-acre area is mainly forestland, though there are 270 acres of dolomite glades. There is a firearms range. Horseback riding and mountain biking are permitted on trails.

2 **Delaware Town Access** on the James River is 5 miles west of Nixa on Highway 14, then to the right on County Road 14-31. There is a picnic area at this 24-acre access and a boat ramp.

3 To reach **Finley Creek Park,** go south of Springfield on Highway 65. Take exit 14 at Ozark east to Riverside Road. This 1-acre area is north on Riverside. There is a boat ramp and wheelchair accessible fishing jetties.

4 **Shelvin Rock Access** is west of Nixa on Highway 14, 4 miles south on Highway M, then 2 miles west on an unnamed gravel road (look for an area cantilever sign on Highway M). There is a boat ramp on the James River.

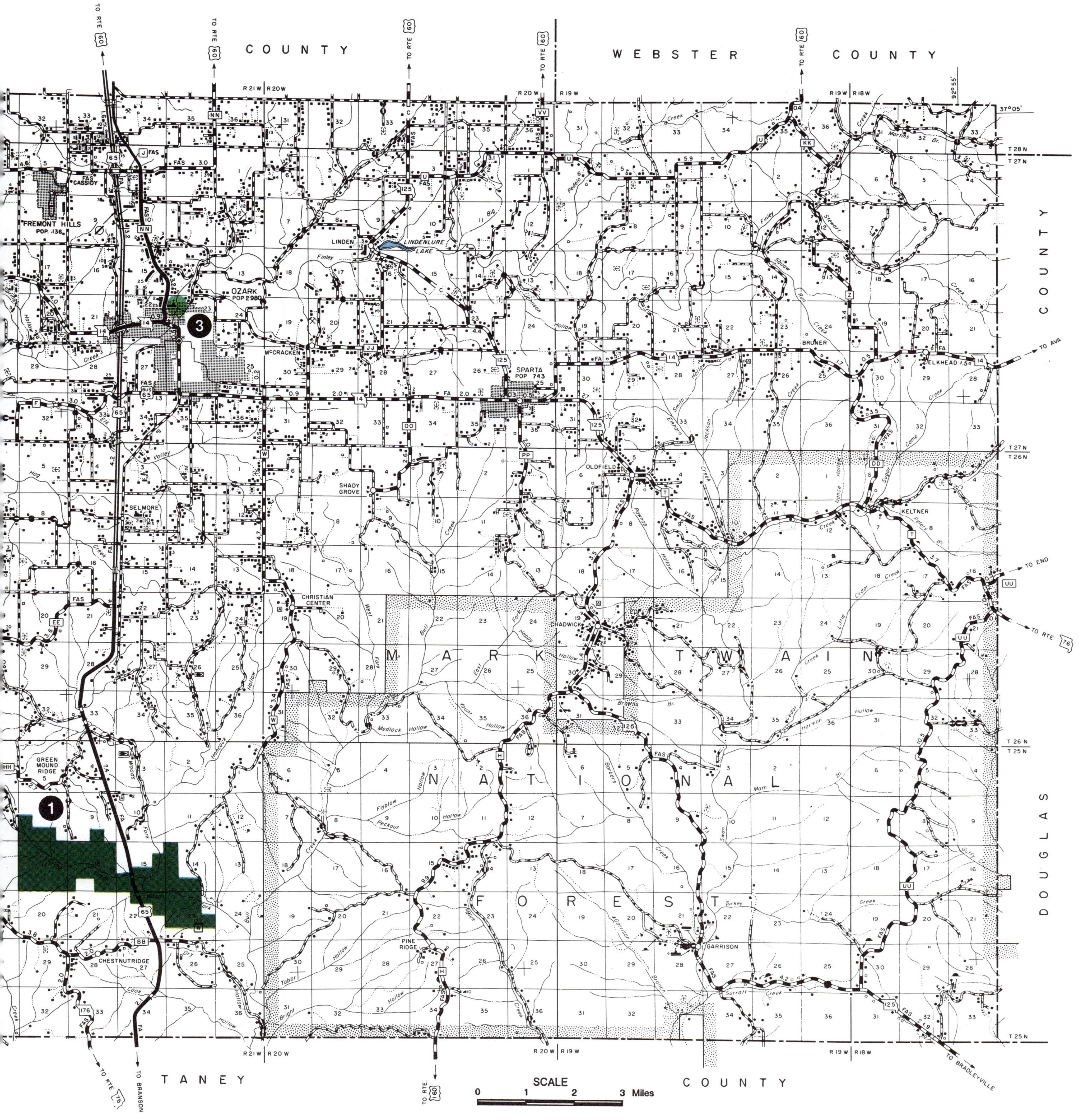
COUNTY
WEBSTER COUNTY
DOUGLAS COUNTY
TANEY COUNTY
MARK TWAIN NATIONAL FOREST
OZARK POP 2980
SPARTA POP 743
FREMONT HILLS POP. 136
CASSIDY
LINDEN
LINDENLURE LAKE
McCRACKEN
SHADY GROVE
SELMORE
CHRISTIAN CENTER
CHADWICK
OLDFIELD
BRUNER
ELKHEAD
KELTNER
GREEN MOUND RIDGE
CHESTNUTRIDGE
PINE RIDGE
GARRISON
TO RTE 60
TO AVA
TO END
TO RTE 76
TO BRADLEYVILLE
TO BRANSON
TO RTE 160
SCALE
0 1 2 3 Miles
37°05'
92°55'
T 28 N
T 27 N
T 26 N
T 25 N
R 21 W R 20 W
R 20 W R 19 W
R 19 W R 18 W

N

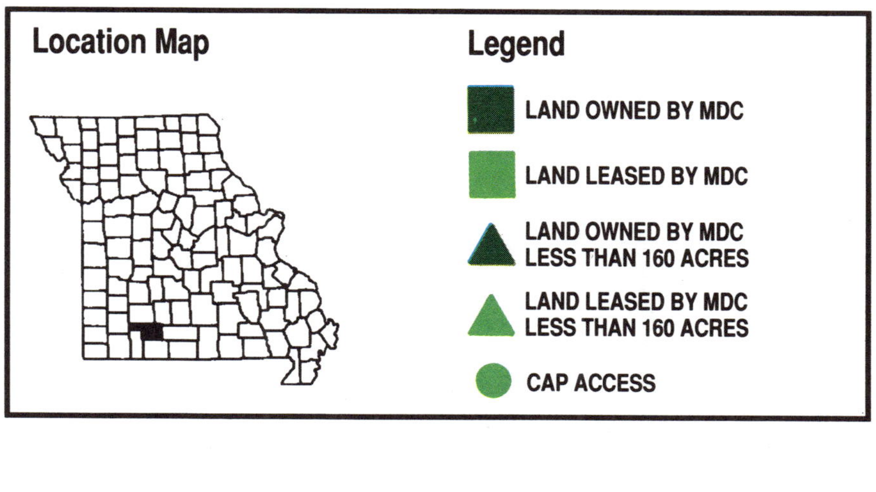
Location Map
Legend
LAND OWNED BY MDC
LAND LEASED BY MDC
LAND OWNED BY MDC LESS THAN 160 ACRES
LAND LEASED BY MDC LESS THAN 160 ACRES
CAP ACCESS

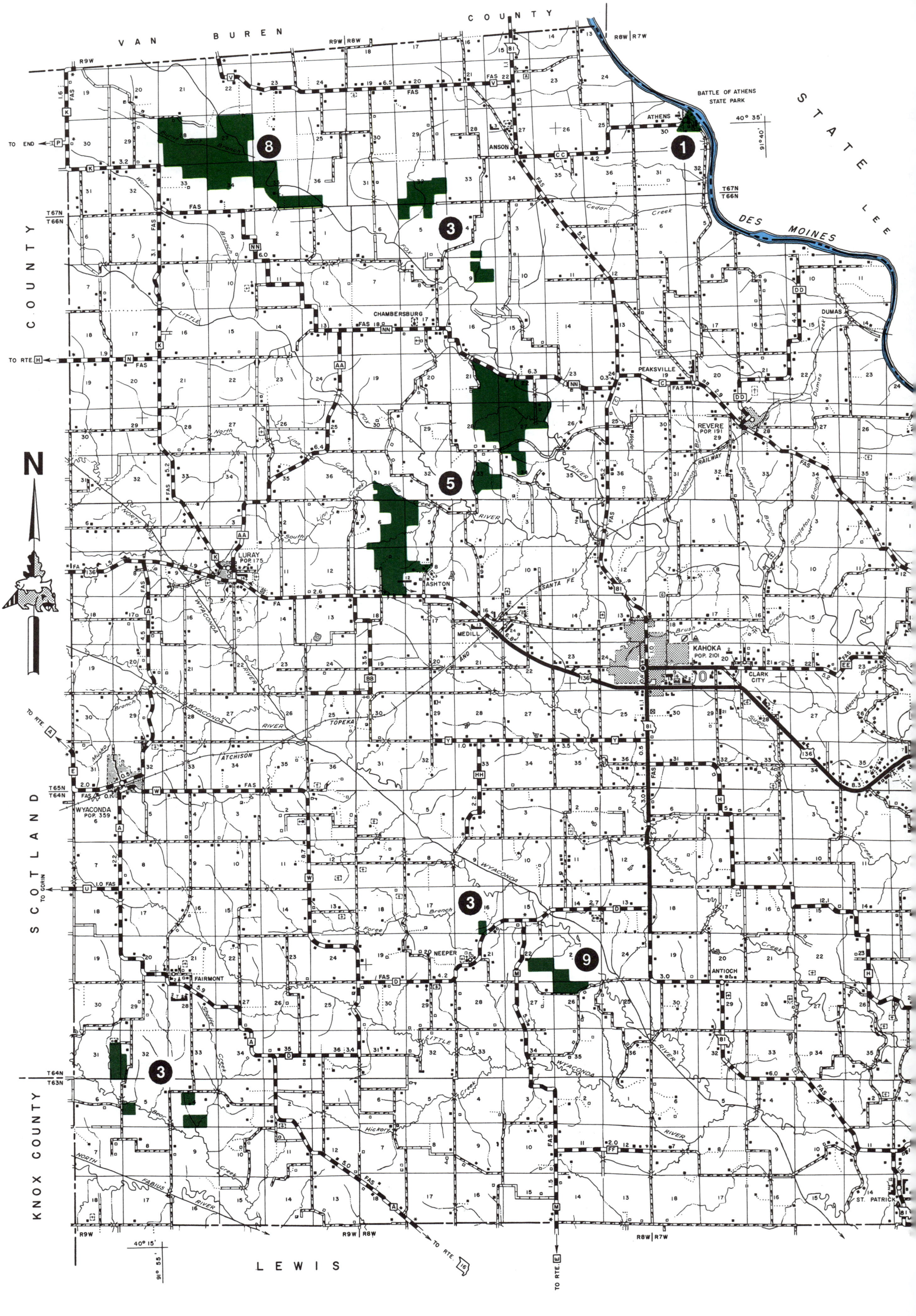
VAN BUREN COUNTY
STATE LEE
DES MOINES
SCOTLAND COUNTY
KNOX COUNTY
LEWIS
BATTLE OF ATHENS STATE PARK
ATHENS
ANSON
CHAMBERSBURG
PEAKSVILLE
REVERE POP. 191
DUMAS
LURAY POP. 175
ASHTON
MEDILL
KAHOKA POP. 2101
CLARK CITY
WYACONDA POP. 359
FAIRMONT
NEEPER
ANTIOCH
ST. PATRICK
TOPEKA
ATCHISON
SANTA FE
N
TO END
TO RTE H
TO RTE A
TO GORIN
TO RTE 15
TO RTE E
T67N
T66N
T65N
T64N
T63N
R9W
R8W
R7W
40° 35'
91° 40'
40° 15'
91° 55'

CLARK COUNTY

1 **Athens Access** is 10 miles north of Kahoka on Highway 81, and 4 miles east on Highway CC. This 2.5-acre area is on a small stretch of the Des Moines River. There is no boat ramp.

2 **Buck and Doe Run Conservation Area** is north of the Lewis County line on Highway 61, then 1 mile east on an unnamed gravel road (which lies .25 mile north of the Lewis County line). This 192-acre area is a wooded wetland, and the entire area floods regularly.

3 **Clark Conservation Area** is composed of many tracts. Nixon Branch Tract is 9 miles north of Kohoka on Highway 81, and 1.5 miles west on an unnamed gravel road. Bear Creek Tract is 3.5 miles south of Wyaconda on Highway A, then 1.75 miles south on an unnamed gravel road. This 736-acre area contains parts of Linn Creek and several ponds, providing anglers with 35 acres of fishable water.

4 **Fort Pike Access** is 4 miles north of Wayland on Route B. This is a 2-acre area on the Des Moines River with one boat ramp.

5 **Fox Valley Lake Conservation Area** consists of three tracts. The main tract is 4.75 miles north of Kahoka on Highway 81, then 2.5 miles west on Highway NN. Fox River and Fox Valley Lake (109 acres of fishable water) are on this 2,158-acre area. There is a boat ramp, wheelchair accessible fishing jetty and a shooting and archery range. A natural area of interest includes a native stand of eastern gamagrass.

6 **Frost Island Conservation Area** is 4 miles north of Wayland on Highway B, then 1 mile east on an unnamed gravel road. This is a 962-acre area on the Des Moines River.

7 **Geode Access** is 1 mile west of Wayland on Highway 136. This is a 1-acre area with no boat ramp on the Fox River.

8 **Charlie Heath Memorial Conservation Area** is on Highway K, 8 miles north of Highway 136 at Luray. The Fox River flows through this 1,636-acre area, and there are three ponds (12 acres of fishable water). The area is suitable for horseback riding, and trails consist mainly of mowed field roads. There is a 50-nest heron rookery.

9 **Neeper Conservation Area** is southwest of Kahoka near the junction of highways D and M. From the junction, the area entrance is 2 miles south on Highway M. This 227-acre area is mainly forestland, though there are two small ponds providing 1 acre of water suitable for fishing.

10 To reach **Rose Pond Conservation Area,** go north of the Lewis County line on Highway 61, turn east on Highway F, and go 2.9 miles to Highway P. Take Highway P north until the pavement ends (2.3 miles), then take the gravel road to the left for .2 mile. Turn right on the next gravel road, and the area is .5 mile down this road. This 379-acre area contains stretches of Big Branch and Honey Creek (5 acres suitable for fishing). It is home to three rare or endangered species, the Illinois mud turtle, Blanding's turtle and western fox snake.

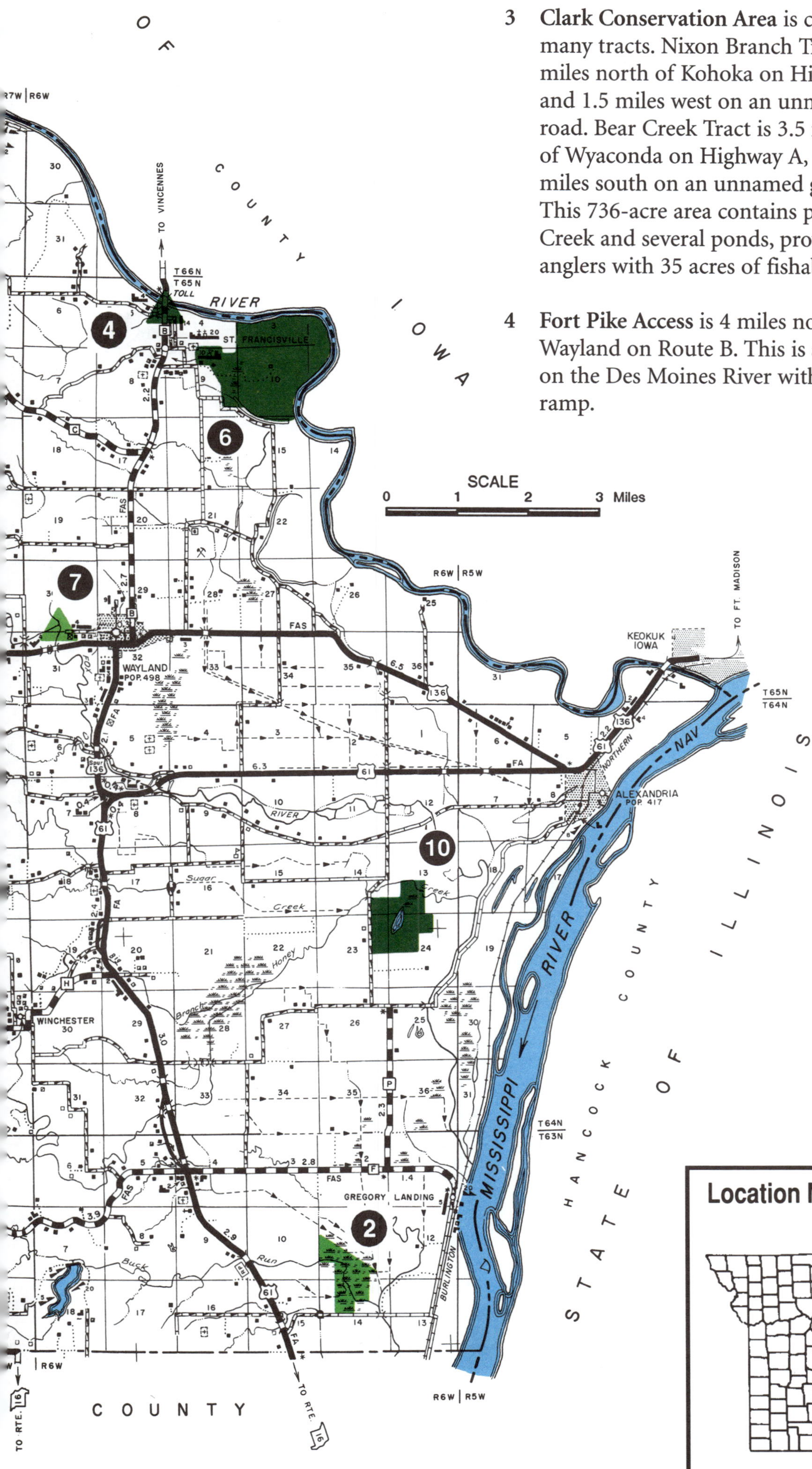

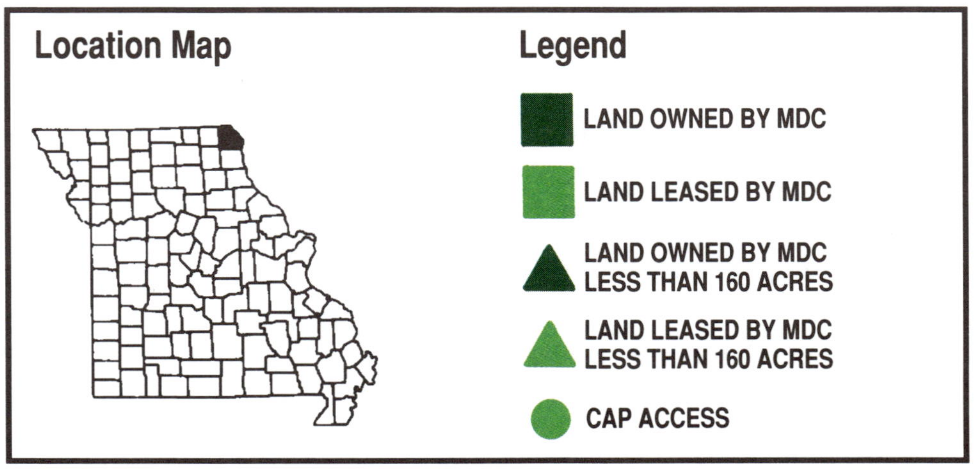

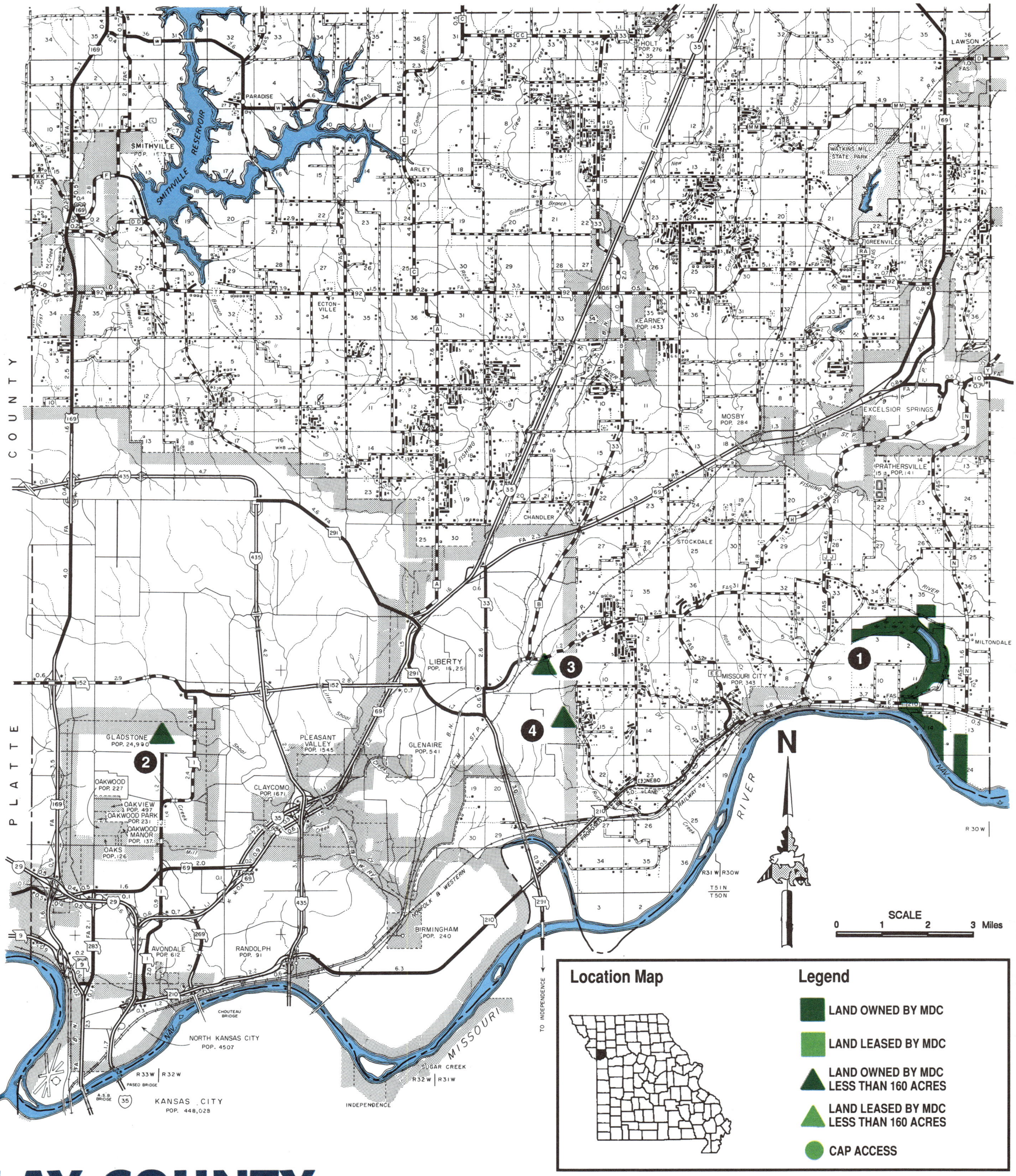

CLAY COUNTY

1 **Cooley Lake Conservation Area** is 6 miles west of Orrick on Highway 210. This 1,227-acre area contains Cooley Lake, which is 335 acres, and a 1-mile stretch of the Missouri River. A natural-surface hiking trail wanders .5 mile through the area, and a migratory waterfowl tour is held every fall. There is boat ramp access to the Missouri River.

2 **Maple Woods Natural Area** is 1.25 miles east of North Oak Trafficway on 76th Street. This 39-acre area has a mature stand of sugar maples and is a Designated Natural Area. There are three trails that extend 2.25 miles through the area.

3 To reach **Rush Creek Conservation Area,** take Highway 152 (Kansas Avenue) 2 miles east to Highway 33 (Lightburne Road). Go 1 block south to Mill Street, .25 mile to Richfield Road, .7 mile east to LaFrenz Road, then .25 mile north on La Frenz. There is a nature center operated by a non-profit agency adjacent to the area.

4 **Walnut Woods Conservation Area** is 2 miles east on Highway 152 (Kansas Avenue), 1 block south on Highway 33 (Lightburne Road), .25 mile east on Mill Street, .7 mile east on Richfield Road, then 2 miles south on La Frenz Road to Stocksdale Park. The area is on the east side of this park. The park has parking, restrooms, picnic areas and a disabled accessible trail.

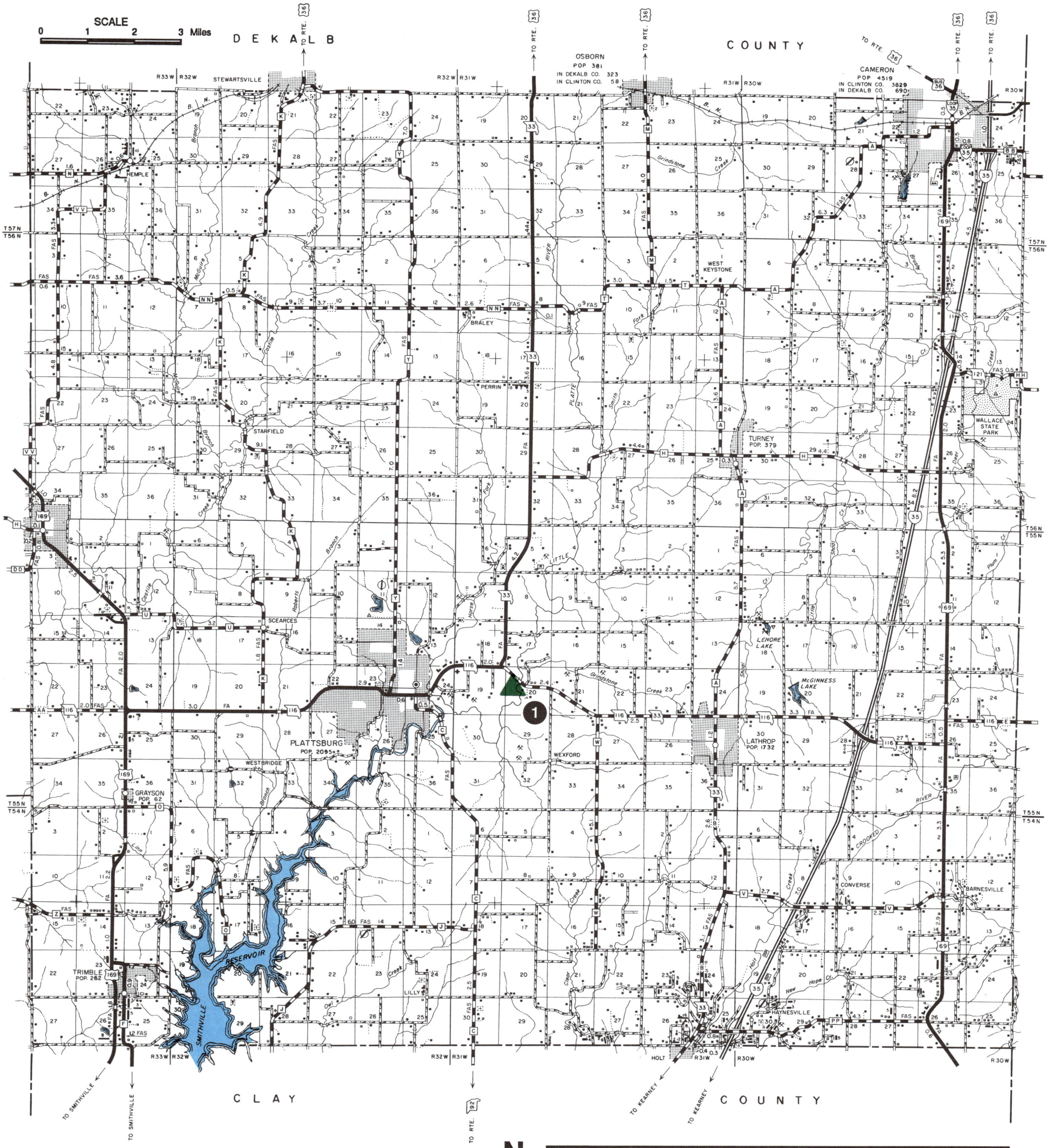

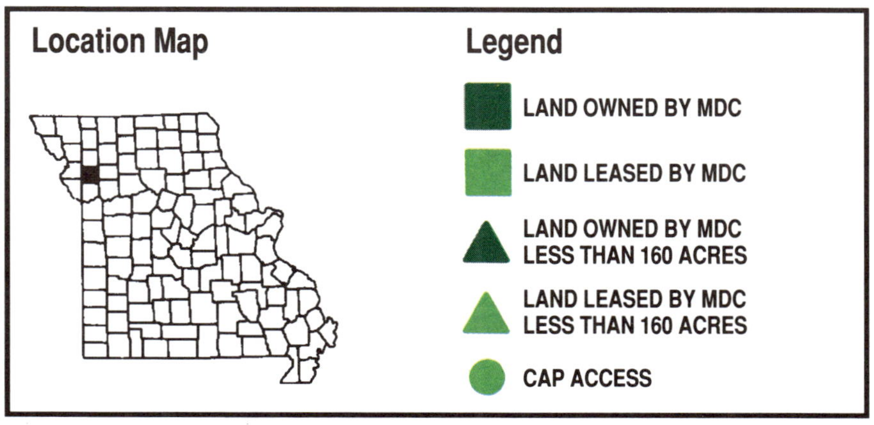

CLINTON COUNTY

1 **Lathrop Bridge Access** is on the Little Platte River 2 miles east of Plattsburg on Highway 116. This is a 25-acre area.

COLE COUNTY

1 **Binder Community Lake** is near Jefferson City. Take Binder Road off Highway 50, and go .5 mile to Rainbow Drive. Turn left, and the entrance gate is across the first bridge. Binder Lake is 150 acres and has a boat ramp, three fishing jetties accessible to wheelchairs and a viewing blind. There are several picnic areas, a concession stand and boat rental.

2 **Cole County Jaycee Park Lake** is off Highway 50 on South Country Club Drive (which becomes Fairgrounds Road). Take this road 1.5 miles, and turn .5 mile left on County Park Road. The lake is on the right. This is a 7-acre lake with picnic facilities.

3 **Honey Creek Access** is 3 miles south of Jefferson City on Highway 54, and 2.25 miles east on Buffalo Road. This 85-acre area contains stretches of Honey Creek and the Moreau River. There is a boat ramp and primitive camping.

4 **Hough Park Lake** is in Jefferson City, 1 mile south of Highway 54 on Ellis Boulevard in the Oak Hills Golf Course. This is a 7-acre lake suitable for fishing.

5 **McKay Park Lake** is in Jefferson City off Highway 54, 2 blocks north on Southwest Boulevard, then 2 blocks to the right on McKay Street. The lake is 6 acres.

6 **Mari-Osa Access** is 7 miles east of Jefferson City on Highway 50 along the Osage River. This 28-acre area has restrooms and a boat ramp.

7 **Marion Access** is 12 miles northwest of Jefferson City on Highway 179. This 3-acre area extends for .75 mile along the Missouri River. There is a boat ramp.

8 **Marion Bottoms Conservation Area** is on Highway 79 in Marion. This is a 2,999-acre area made up of old fields and forestland.

9 **Moreau 50 Access** is 1 mile east of Jefferson City on Highway 50. This is a 10-acre area along a .25-mile stretch of the Moreau River. There is a boat ramp.

10 **Pikes Camp Access** is 1 mile south of Wardsville on Highway W. This is a 170-acre area on the Osage River. There is a boat ramp.

11 **Runge Conservation Nature Center** is in Jefferson City, 1 mile north of Route 50 on Highway 179. This is a 90-acre area with both paved and wood-chipped trails, four ponds and a .5-mile stream. Facilities include an exhibit area with interactive computer displays, classrooms, auditorium, wildlife viewing area and library. See page 56 for additional information.

12 **St. Thomas Ferry Access** is on Highway B, 9 miles south of Jefferson City. This is a 2-acre area with a boat ramp on the Osage River.

13 To reach **Scrivner Road Conservation Area,** go south of Russellville on Cole County Route AA, and turn left on Scrivner Road. After traveling 1.25 miles, turn left again on Scott Road, and the area is 1 mile down this road. This 919-acre area has two major bodies of water on it, Winegar Lake (12 acres) and South Moreau Creek (2 miles). There is a 3-mile trail suitable for hiking and, in places, visitors can fish from the trail.

14 **Roger V. and Viola Wachal Smith Conservation Area** is 2 miles north of Centertown on Highway NN, left on Pea Ridge about .25 mile, then right on Bird Haven Road. This 517-acre area has several small dolomite glades, six small ponds (not suitable for fishing) and one stream.

15 **Smoky Waters Conservation Area** is 2 miles east of Osage City on Engineers Road. This is a 976-acre area near the Osage and Missouri rivers. There are two blue holes.

16 **Stringtown Bridge Access** is 5 miles west of Jefferson City on Highway C, and 1 mile west on Hemstreet Road. This 50-acre area is nearly all forestland and contains 1 mile of the Moreau River.

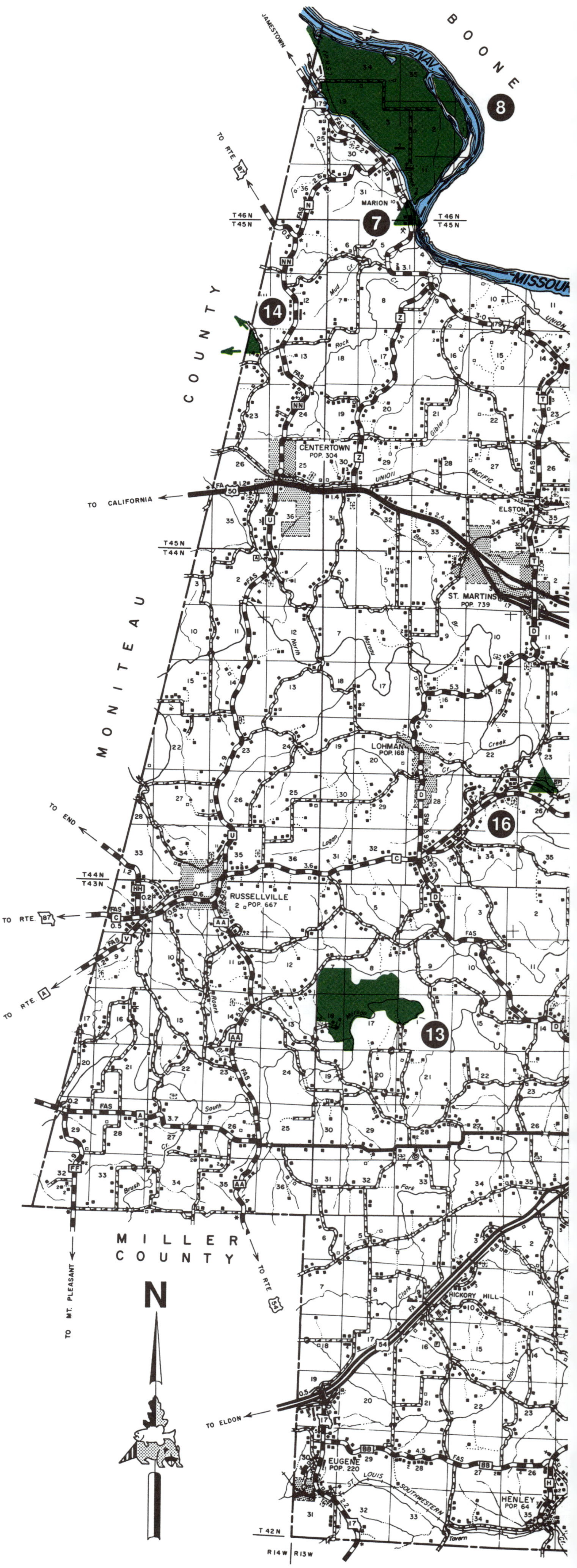

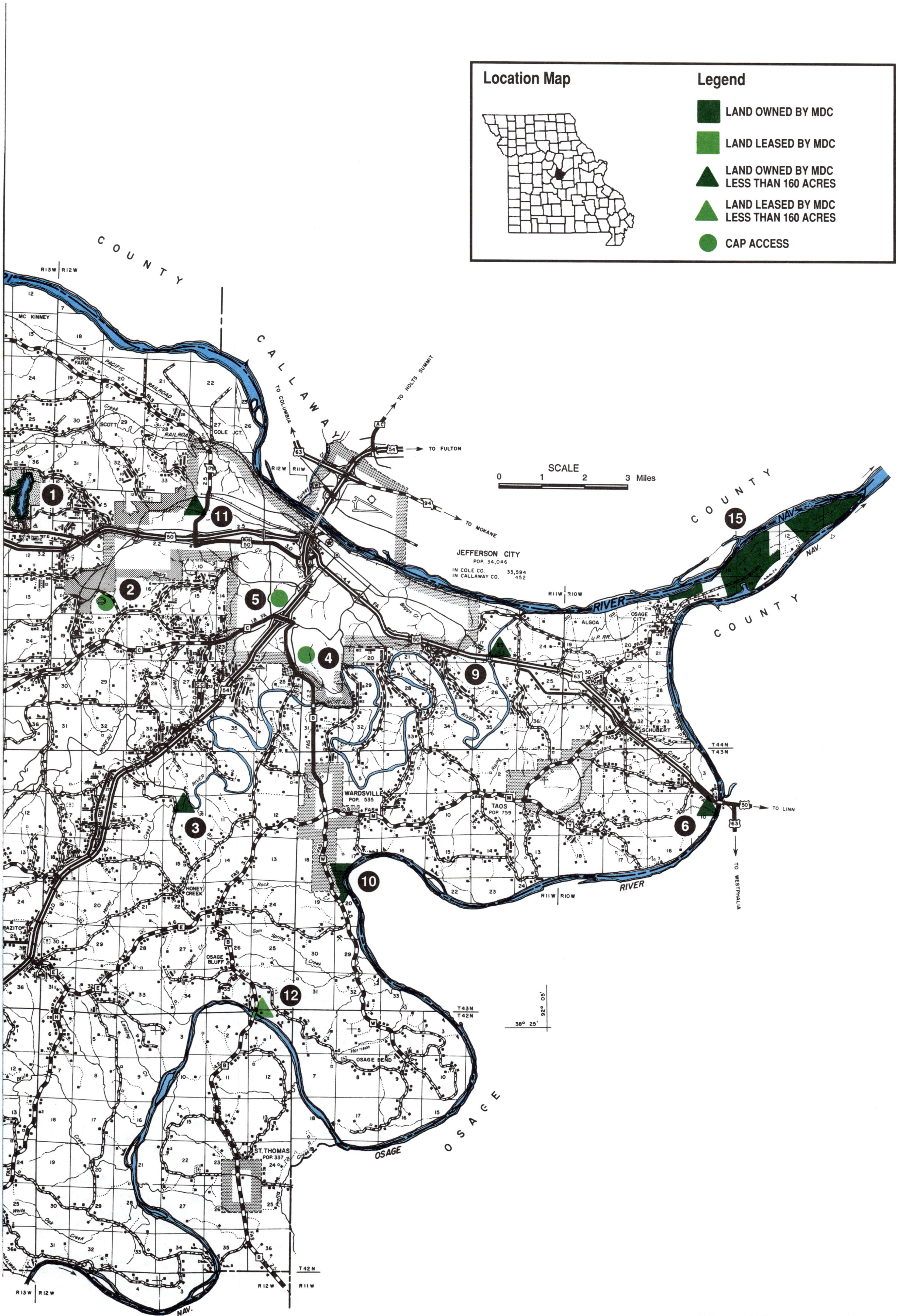
Location Map
Legend
LAND OWNED BY MDC
LAND LEASED BY MDC
LAND OWNED BY MDC LESS THAN 160 ACRES
LAND LEASED BY MDC LESS THAN 160 ACRES
CAP ACCESS
SCALE
0 1 2 3 Miles
COUNTY
CALLAWAY
COUNTY
COUNTY
COUNTY
OSAGE
TO COLUMBIA
TO HOLTS SUMMIT
TO FULTON
TO MOKANE
TO LINN
TO WESTPHALIA
JEFFERSON CITY
POP. 34,046
IN COLE CO. 33,594
IN CALLAWAY CO. 452
MC KINNEY
PRISON FARM
PACIFIC
RAILROAD
SCOTT
COLE JCT.
RIVER
NAV.
ALGOA
OSAGE CITY
U.P. RR.
SCHUBERT
WARDSVILLE
POP. 535
TAOS
POP. 759
HONEY CREEK
OSAGE BLUFF
OSAGE BEND
ST. THOMAS
POP. 337
OSAGE
RIVER
R13W R12W
R12W R11W
R11W R10W
T44N T43N
T43N T42N
T42N
38° 25'
92° 05'

While in Cole County, visit

Runge Conservation Nature Center

Runge Conservation Nature Center is .5 mile north of U.S. 50 on Highway 179 in Jefferson City. At the Runge Conservation Nature Center, visitors can explore Missouri's diverse and changing landscape. Trails, exhibits and nature programs are just a few of the attractions available to drop-by visitors.

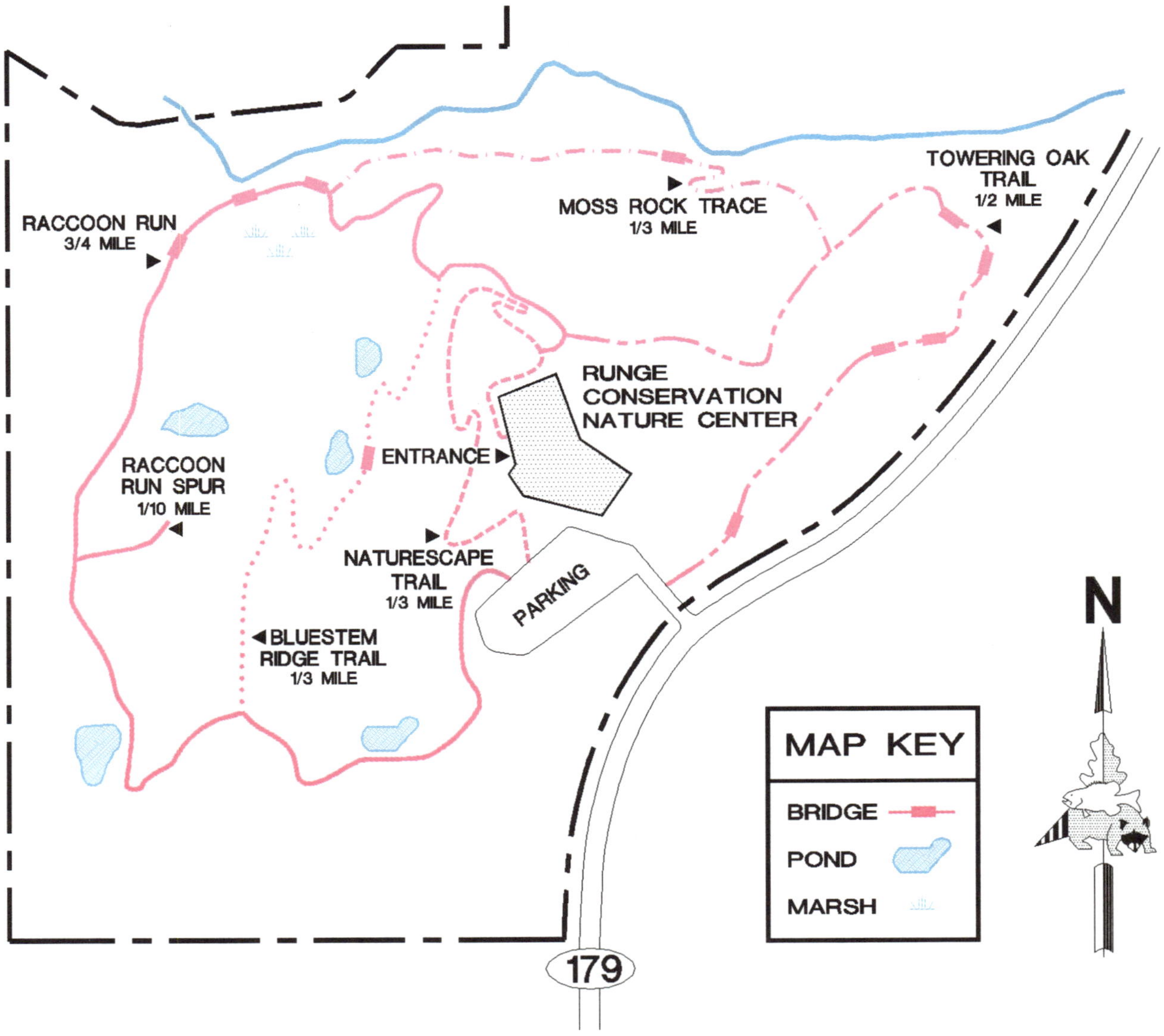

Outside

- five hiking trails, one of which is accessible to visitors who use wheelchairs
- outdoor demonstrations of landscaping and wildlife habitat
- native warm-season grass prairie
- ponds, marsh and a glade

Inside

- a 3,400-gallon aquarium, home to native fish and turtles
- hands-on exhibits
- indoor wildlife viewing area, 200-seat auditorium and library
- regularly scheduled nature programs, announced in the center's bi-monthly newsletter, *Habitat Happenings*
- sales of books, calendars and hunting and fishing permits

Building Hours

Open year-round, Monday through Saturday 8 a.m. to 5 p.m., Sundays noon to 5 p.m. Closed Thanksgiving, Christmas and New Year's Day.

Area Hours

Open daily, 8 a.m. to 8 p.m. during daylight savings; and daily, 8 a.m. to 6 p.m. during standard time. Closed Thanksgiving, Christmas and New Year's Day.

Pets, horses, hunting, fishing, collecting, littering, fires and alcoholic beverages are not permitted. Bicycling, jogging, in-line skating and motorized vehicles are not permitted on trails.

Runge Conservation Nature Center
Missouri Department of Conservation
P.O. Box 180, Mo. Hwy. 179
Jefferson City, MO 65102-0180
573/526-5544

Jim Rathert

Jim Rathert

Jim Rathert

Paul Childress

Volunteers play an important role at all four of the Conservation Department's nature centers. During one year at Runge Conservation Nature Center, adults and kids volunteered more than 6,550 hours. Volunteers give nature programs, greet visitors, help develop exhibits and help produce the center's newletter. Opportunities abound at all four nature centers, so if you are interested in becoming a volunteer, inquire at the one nearest you.

COOPER COUNTY

1 **Blackwater Bridge Access** is 3 miles north of I-70 on Highway K, near the town of Blackwater. This is a 5-acre area with a boat ramp on the Blackwater River.

2 **de Bourgmont Access** is on Highway 41, 1.5 miles north of I-70. This is a 3-acre area with a boat ramp on the Lamine River.

3 **Harriman Hill Access** is north of I-70 on Highway M. This is a 37-acre area with a boat ramp on the Lamine River.

4 **Lamine River Conservation Area** consists of eight tracts. The main tract is 1 mile east of Otterville on Highway A. This 5,692-acre area has four major bodies of water, the Lamine River (12.2 miles), Richland Creek (1.5 miles), Otter Creek (1.2 miles) and Flat Creek (.4 mile). It contains diverse natural features, such as wooded bluffs along the river, broad floodplains, riparian forest corridor and rolling hills.

5 **Prairie Home Conservation Area** is 4 miles southwest of Prairie Home on Cooper County Route W. This 1,461-acre area contains 1.5 miles of Schaaf Creek. Most of the area is forestland, though there are 100 acres of prairie. There are two trails. One is an 8-mile hiking trail, and the other is a 4-mile horseback riding trail. The area also contains a Boy Scouts campground.

6 **Roberts Bluff Access** is 4.5 miles northwest of Pilot Grove on Highway HH, then 2 miles north on Gravel Road HH-1. This 7-acre area has a boat ramp on the Lamine River.

7 **Swinging Bridge Access** is 1 mile west of Clifton City on Highway 50, then 2.5 miles north on Highway JJ, and 1.5 miles east on Swinging Bridge Drive. This is 25 acres on the Lamine River.

8 **Wooldridge Access** is 1.5 miles southwest of Wooldridge off Highway 179 on Gravel Road 179-3. The area is 30 acres with no boat ramp on the Petite-Saline Creek.

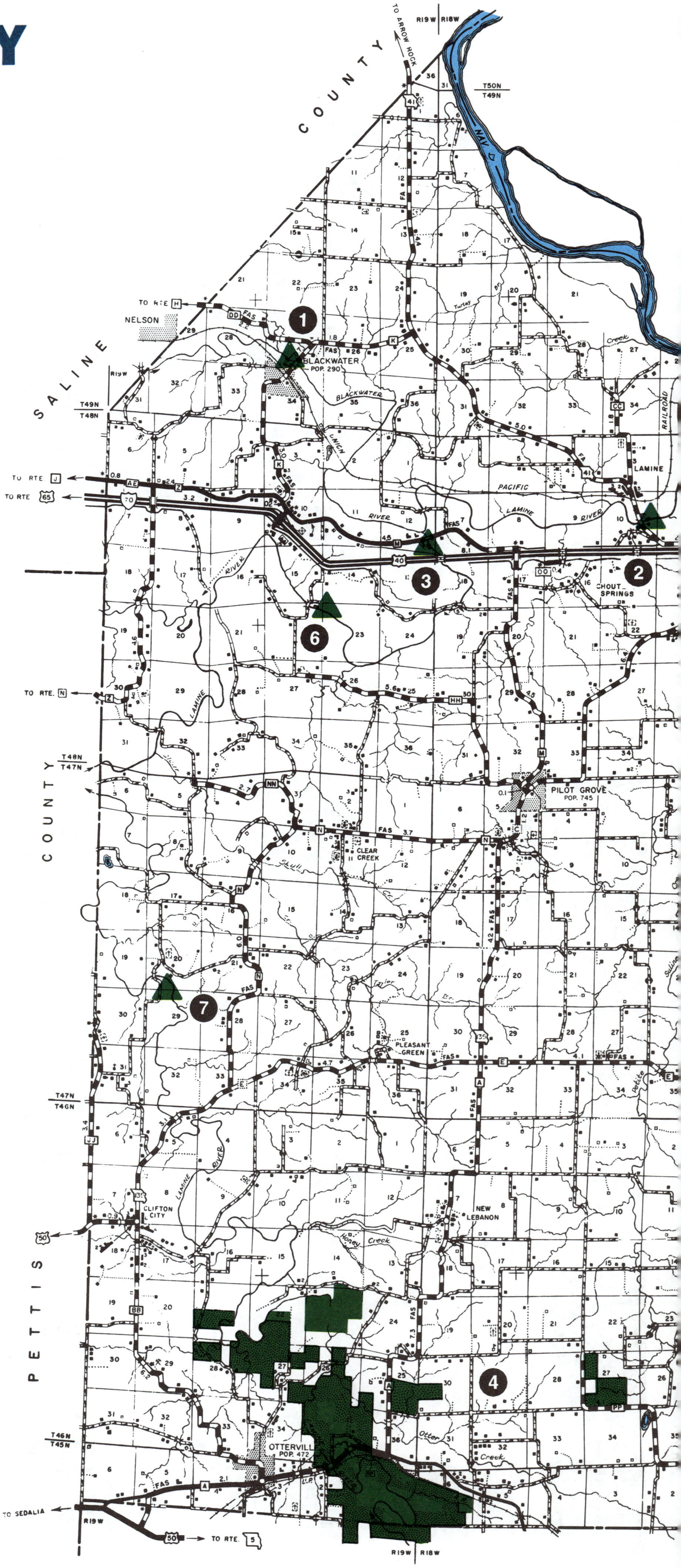

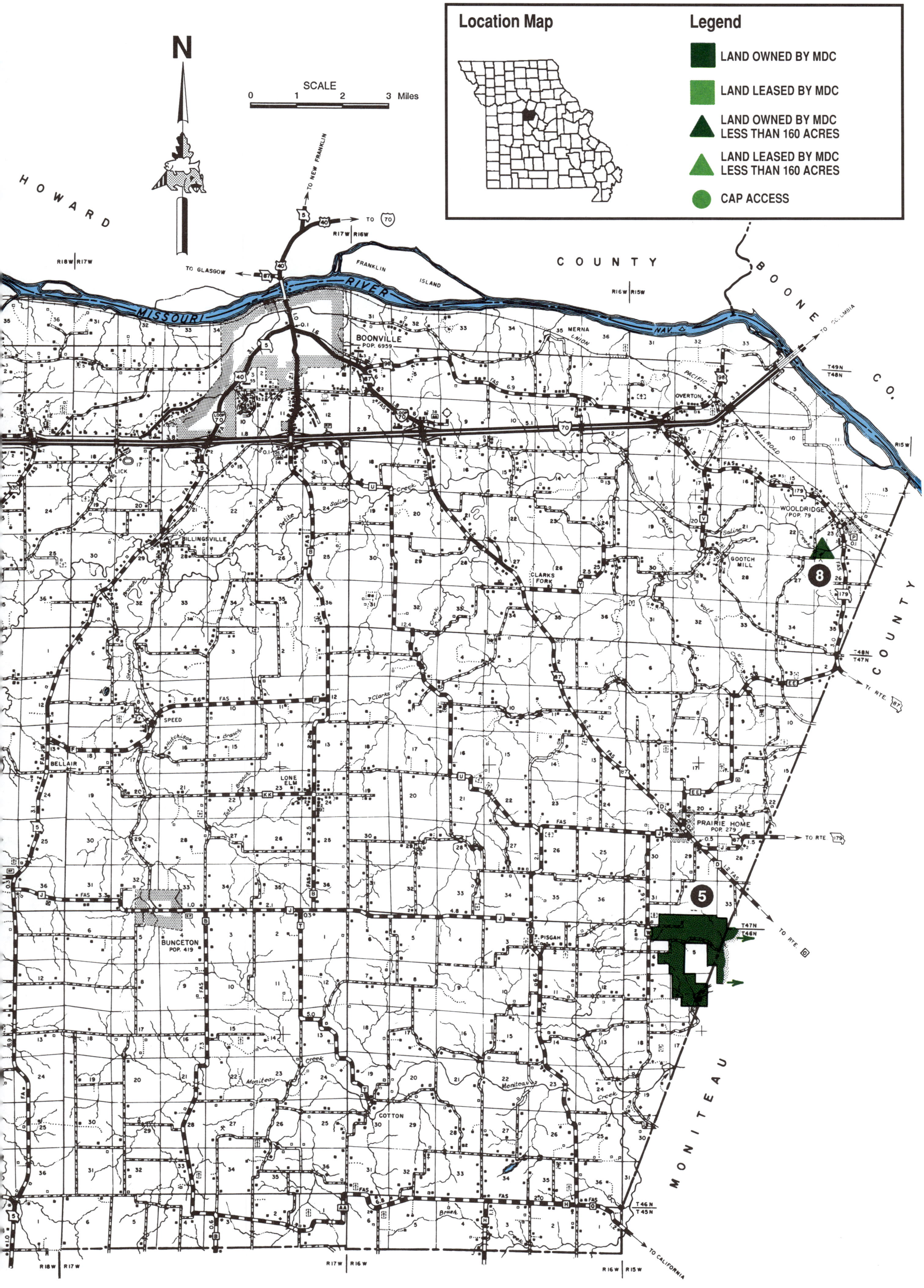
Location Map
Legend
LAND OWNED BY MDC
LAND LEASED BY MDC
LAND OWNED BY MDC LESS THAN 160 ACRES
LAND LEASED BY MDC LESS THAN 160 ACRES
CAP ACCESS
N
SCALE
0 1 2 3 Miles
HOWARD
COUNTY
BOONE
CO.
COUNTY
MONITEAU
MISSOURI RIVER
FRANKLIN ISLAND
TO NEW FRANKLIN
TO GLASGOW
TO COLUMBIA
TO CALIFORNIA
BOONVILLE
POP. 6959
MERNA UNION
OVERTON
WOOLDRIDGE
POP. 79
GOOTCH MILL
CLARKS FORK
BILLINGSVILLE
LICK
SPEED
BELLAIR
LONE ELM
PRAIRIE HOME
POP. 279
PISGAH
BUNCETON
POP. 419
COTTON
8
5

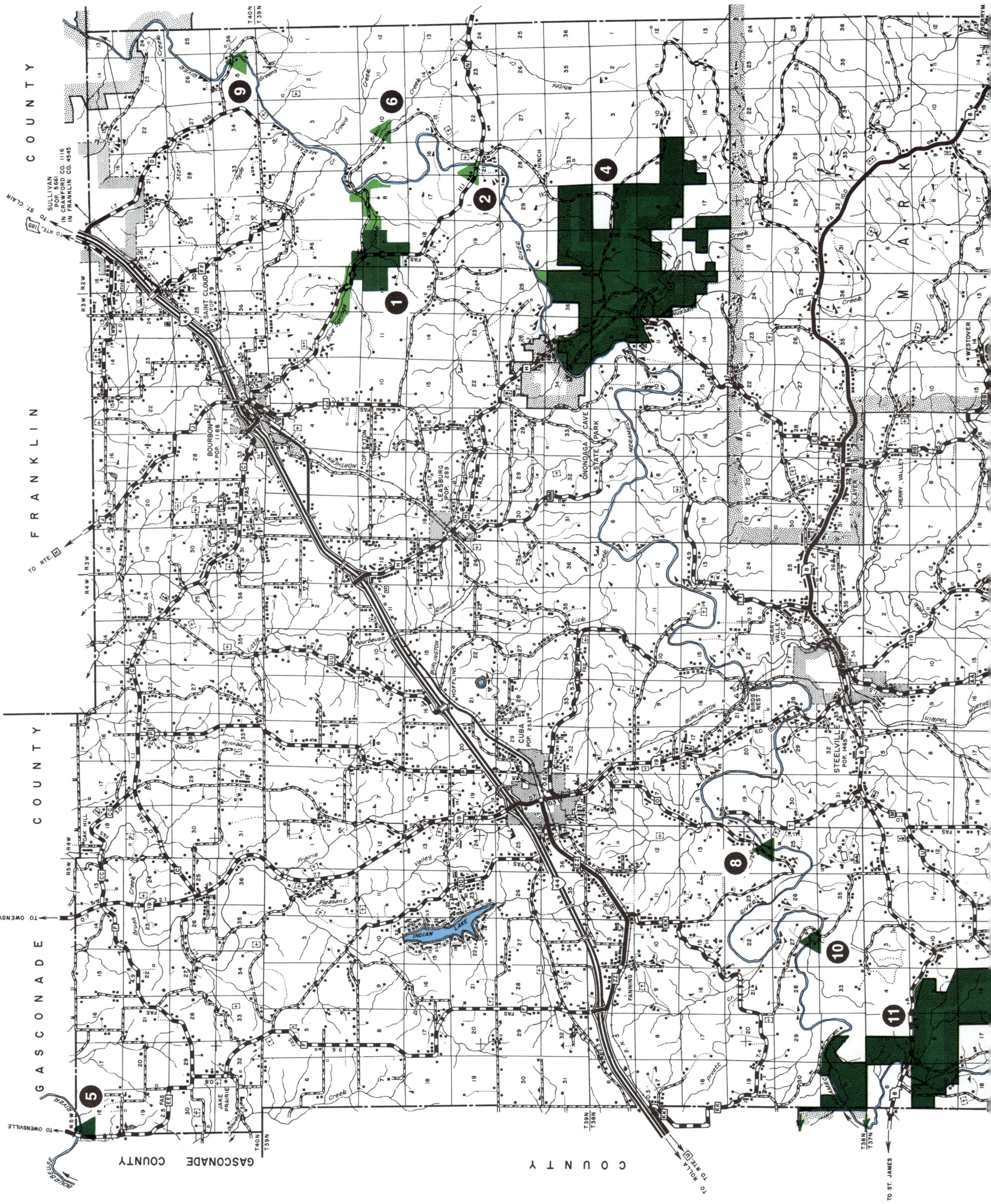

CRAWFORD COUNTY

1 **Blue Springs Creek Conservation Area** is 2.5 miles south of Bourbon on Highway N, then off Blue Springs County Road (marked by a cantilever sign). The Meramec River and Blue Springs Creek flow for about 5 miles through this forested, 857-acre area.

2 **Campbell Bridge Access** is 10 miles south of Bourbon on Highway N. This is a 10-acre area with a boat ramp and canoe launch area on the Meramec River.

3 **Crooked Creek Conservation Area** (337 acres) is at the junction of highways M and VV. It is mainly forestland.

4 **Huzzah Conservation Area** consists of two tracts. The largest can be reached from Onondaga Cave State Park, or by heading northeast of Steelville on Highway E and crossing a low water bridge at the Courtois Creek. This 6,225-acre area contains parts of the Meramec River, Huzzah Creek and Courtois Creek. Eight and a half miles of the Ozark Trail will extend through the area (approximately 4.5 has been constructed), and spelunkers often explore the area's 20 caves.

5 **Mint Spring Access** is 9 miles south of Owensville on Highway EE, just where EE crosses the Bourbeuse River. This 11-acre area on the Bourbeuse River (with a boat ramp) is near Mint Springs Natural Area.

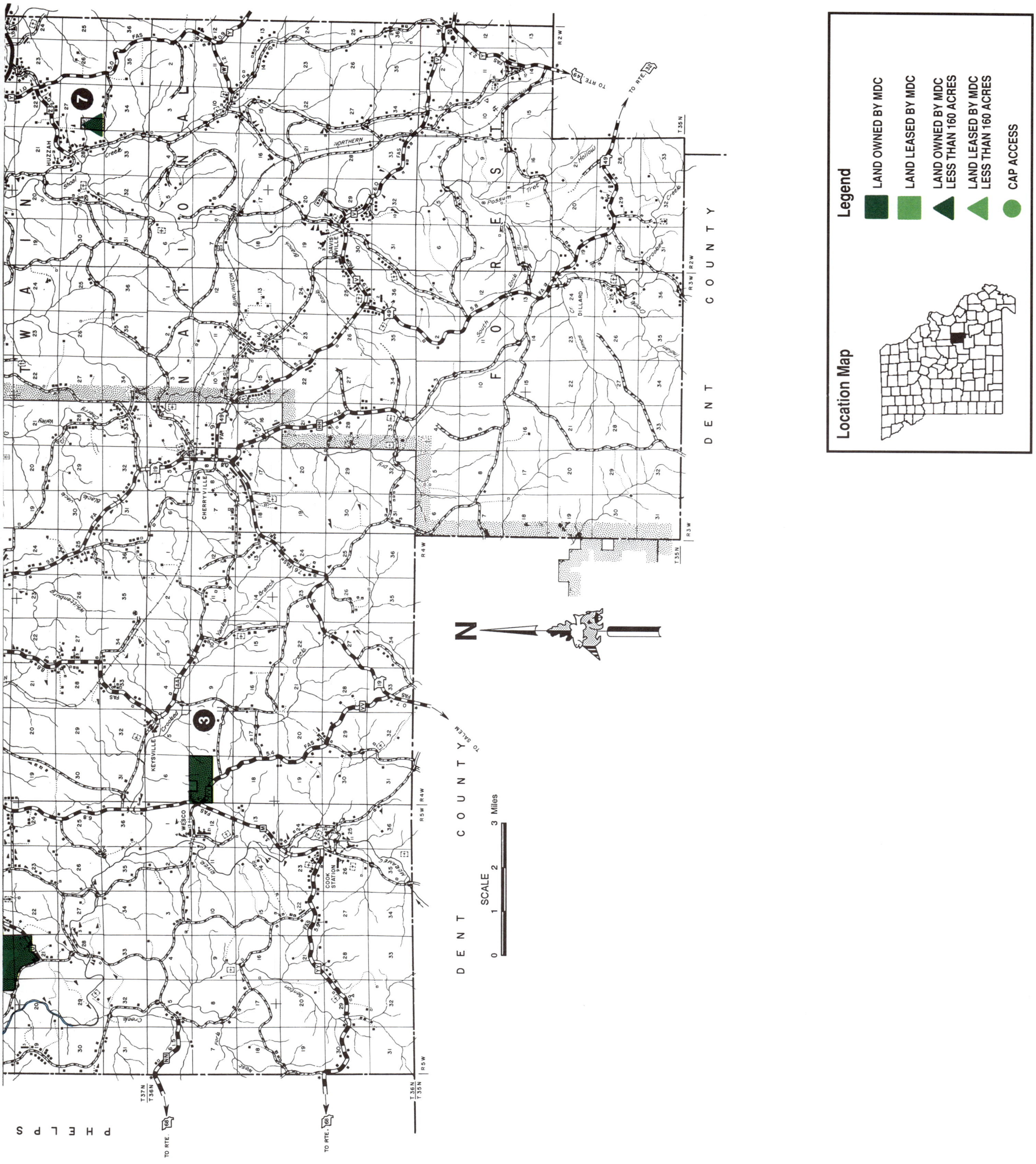

6 **Onyx Cave Conservation Area** is 6.5 miles south of Bourbon on Highway N, and 3 miles east (left) on Thickety Ford Road. There is no public access to the cave, but it is surrounded by about 35 acres of forestland. The Brazil Creek flows through the area for about .3 mile.

7 **Richter Conservation Area** is 14 miles east of Steelville on Highway 8, then 3 miles down Highway Y. This 81-acre area is all forestland.

8 **Riverview Access** is 1.5 miles south of Cuba on Highway 19, then 4 miles west on Highway O. The 5.6-acre area is .25 mile past the end of the state maintained part of Highway O. There is a boat ramp on the Meramec River.

9 **Sappington Bridge Access** is 4.5 miles south of Sullivan on Highway D, then left on Sappington Bridge Road (gravel) about 2 miles. This 10-acre area has a boat ramp and canoe launch area on the Meramec River.

10 **Scotts Ford Access** is 4 miles west of Steelville on Highway 8, and 2 miles north on Thurman Lake Road. This is a 34-acre area on the Meramec River with no boat ramp.

11 **Woodson K. Woods Memorial Conservation Area** is southeast of St. James on Highway 8. This 5,660-acre area on Dry Fork Creek and the Meramec River contains Spring's End Forest, a 30-acre natural area of bottomland hardwoods.

DADE COUNTY

1 **Fiddlers Ford Access** is 1.5 miles east of the junction of highways 39 and O (on Highway O). From Highway O, go 3.5 miles north on a county gravel road marked with an area cantilever sign. The road curves east, and after .5 mile, crosses a set of railroad tracks just before the area. This is 50 acres on Turnback Creek with no boat ramp.

2 **Horse Creek Prairie Conservation Area** (80 acres) is 2 miles west of Lockwood on Highway 160, and 1 mile north on an unnamed gravel road.

3 **Indigo Prairie Conservation Area** (40 acres) is 1 mile south of Lockwood on Highway 97, and 2.5 miles east on an unnamed gravel road.

4 **Niawathe Prairie Conservation Area** is 1 mile west of Highway 97 on Highway E, then .5 mile north on an unnamed gravel road. This is a 320-acre Designated Natural Area.

5 **Dr. O. E. and Eloise H. Sloan Conservation Area** is on Highway 160, 3 miles east of Lockwood. This 320-acre area is composed mainly of old pastures and has three ponds (1.5 acres suitable for fishing).

6 **Stockton Reservoir**–see the Cedar County map, page 44.

7 **Stony Point Prairie Conservation Area** is 4 miles west of Highway 97 on County Road E, then 1 mile north on County Road D. This 640-acre area is nearly all native prairie and has a .8-acre pond suitable for fishing.

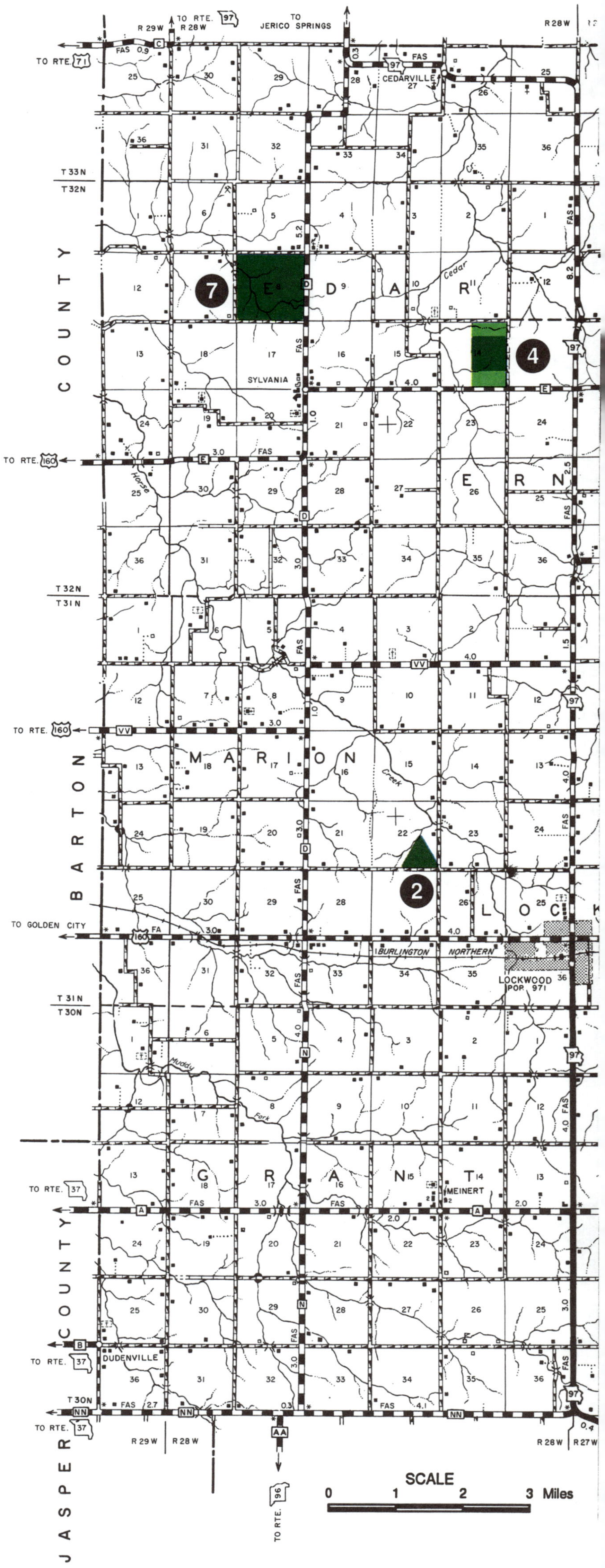

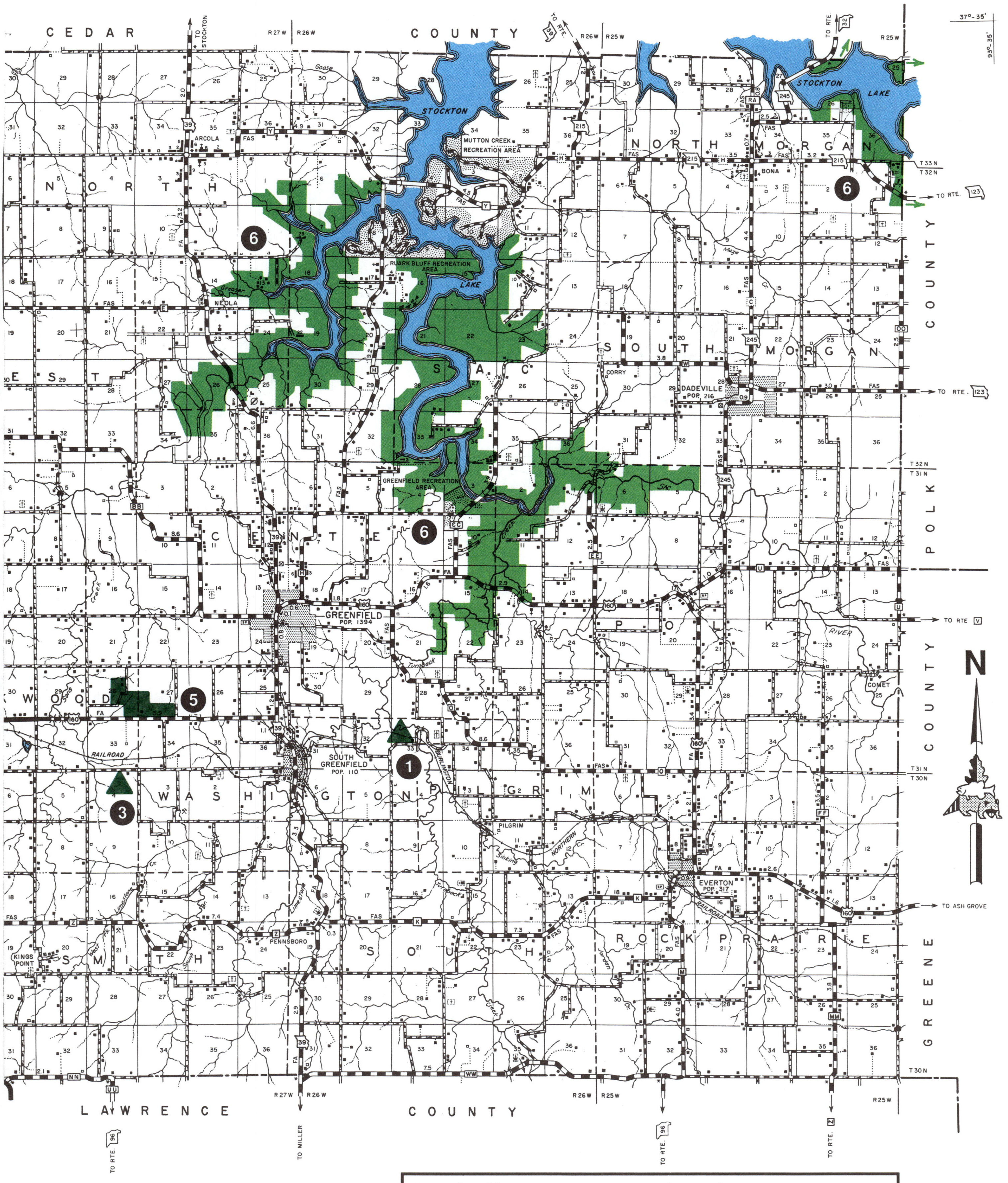

Location Map

Legend

- LAND OWNED BY MDC
- LAND LEASED BY MDC
- LAND OWNED BY MDC LESS THAN 160 ACRES
- LAND LEASED BY MDC LESS THAN 160 ACRES
- CAP ACCESS

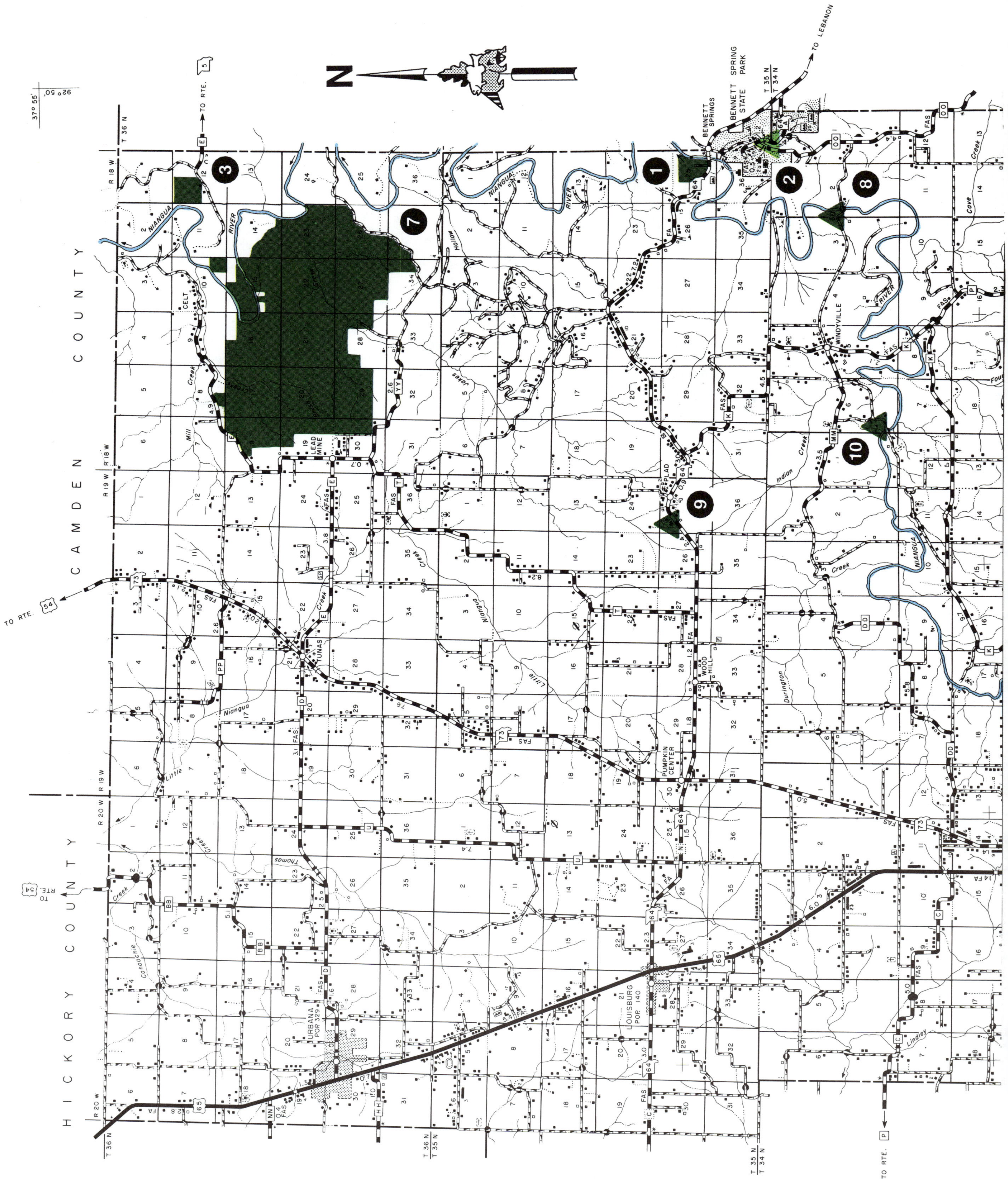

DALLAS COUNTY

1 **Bennett Spring Access** is 12 miles west of Lebanon on Highway 64, just across the Niangua River Bridge. This is a 178-acre access with two boat ramps.

2 **Bennett Spring Fish Hatchery** is inside Bennett Spring State Park and is 12 miles west of Lebanon on Highway 64. This 3,099-acre park is home to the Niangua River and Bennett Spring Branch. Natural areas of interest include the Bennett Spring Natural Tunnel, savannah and a 1-acre fen. There are five hiking trails totaling 4 miles in length and several picnic areas. The park has a Nature Interpretative Center, overnight lodging, fish hatchery tours and a store.

3 **Berry Bluff Conservation Area** is north of Lebanon off Highway E. Where Highway E is no longer paved, turn left for .25 mile on the first gravel road. This area is 159 acres of forestland located on the north (right) side of this road.

4 **Big John Access** is 2 miles east of Buffalo on Highway 32, then 1 mile north on a county gravel road marked with a cantilever sign. This 16-acre area is mainly forestland and lies along a .25-mile stretch of the Niangua River (no boat ramp).

5 **Charity Access** is 8 miles south of Buffalo on Highway 65, east on Highway 38, then 4.5 miles on Highway M. This 162-acre area is on the south side of Highway M along the Niangua River. There is a boat ramp.

6 **Goose Creek Conservation Area** is made up of three separate tracts, spanning 10

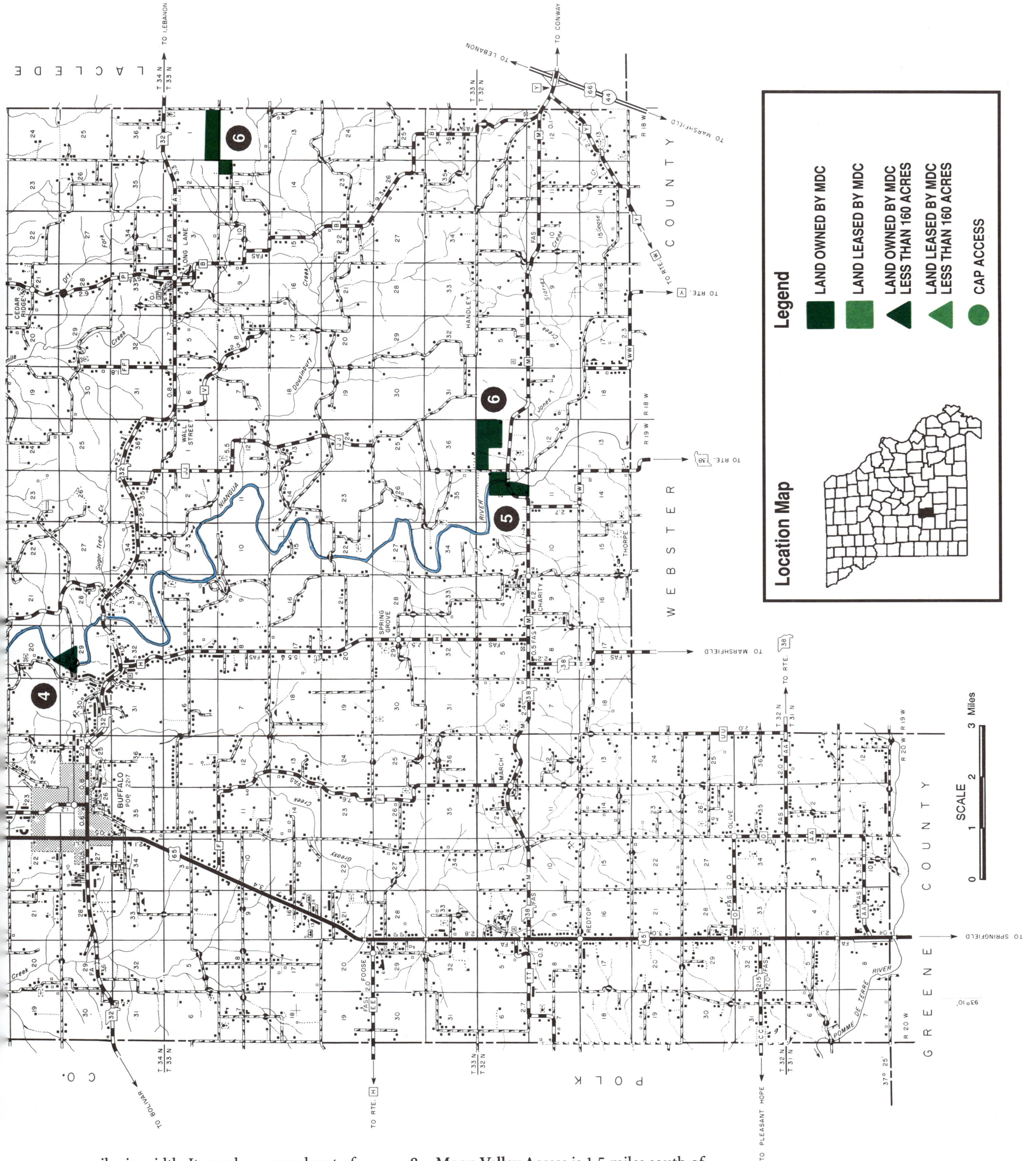

miles in width. It may be accessed east of Charity on Highway M. This is a 1,040-acre area of mainly forestland (1,000 acres).

7 **Lead Mine Conservation Area** is north of Plad on Highway 64, north on Highway T, and .5 mile east on Highway YY. At the first gravel road on the left, turn north for about 1 mile to reach the main parking lot. This 6,957-acre area has 1 mile of Niangua River frontage and a 3-mile stretch of Jakes Creek running through it. The area also has two ponds and seven springs and is composed mainly of forestland (5,788 acres).

8 **Moon Valley Access** is 1.5 miles south of Bennett Spring State Park on Highway OO, and 1.5 miles to the right on Gravel Road 00-126 just across a low-water bridge. This is a 3-acre area with a boat ramp along a short stretch of the Niangua River.

9 **Plad Towersite** is on Highway 64, 6 miles east of Highway 73. This is a 2-acre area.

10 **Williams Ford Access** is 2 miles west of Windyville on Highway MM, and 1 mile left on an unnamed gravel road. This is 40 acres on the Niangua River with no boat ramp.

DAVIESS COUNTY

1 **Gallatin Conservation Area** is 5 miles south of Gallatin on Highway 13, and .5 mile east on Route M. This 670-acre area has a firearms range and 3.5 miles of trails (field roads).

2 **Green Access** is 1.5 miles south of Pattonsburg on Highway 69. This is a 3.5-acre area with no boat ramp on the Grand River.

3 **Holmes Bend Access** is 2 miles south of Gallatin on Highway 13, then east at the cantilever sign marking the area (on a gravel road). This is 76 acres of forestland, warm season grasslands and old fields with frontage on North Grand River. There is no boat ramp.

4 **Jamesport City Lake** is north of the city of Jamesport on Highway 190, then west on State Route RA (follow the signs). The land around this 18-acre reservoir is forestland (38 acres), and there is a boat ramp. There is also a hitching post for horse drawn vehicles.

5 **Jamesport Community Lake** is northwest of Jamesport off Route RA. This is a 30-acre lake with one boat ramp. The land around the lake is mainly old fields and forestland (145 acres total).

6 **Newman Memorial Access** is 5 miles east of Gallatin on Highway 6, then 6.5 miles south on Route V. This is a 9-acre area along the Grand River. There is no boat ramp.

7 **Savage Access** on the Grand River is 3.5 miles west of Pattonsburg on Route Z. This is a 13-acre area with no boat ramp.

8 **Wabash Crossing Access** is 1 mile north-east of Gallatin on Highway 6. This 15-acre area with a boat ramp is on the Grand River.

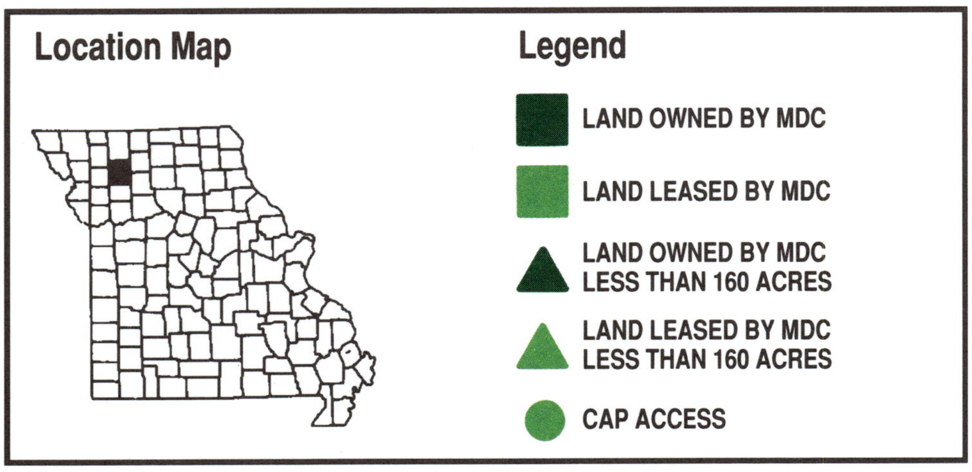

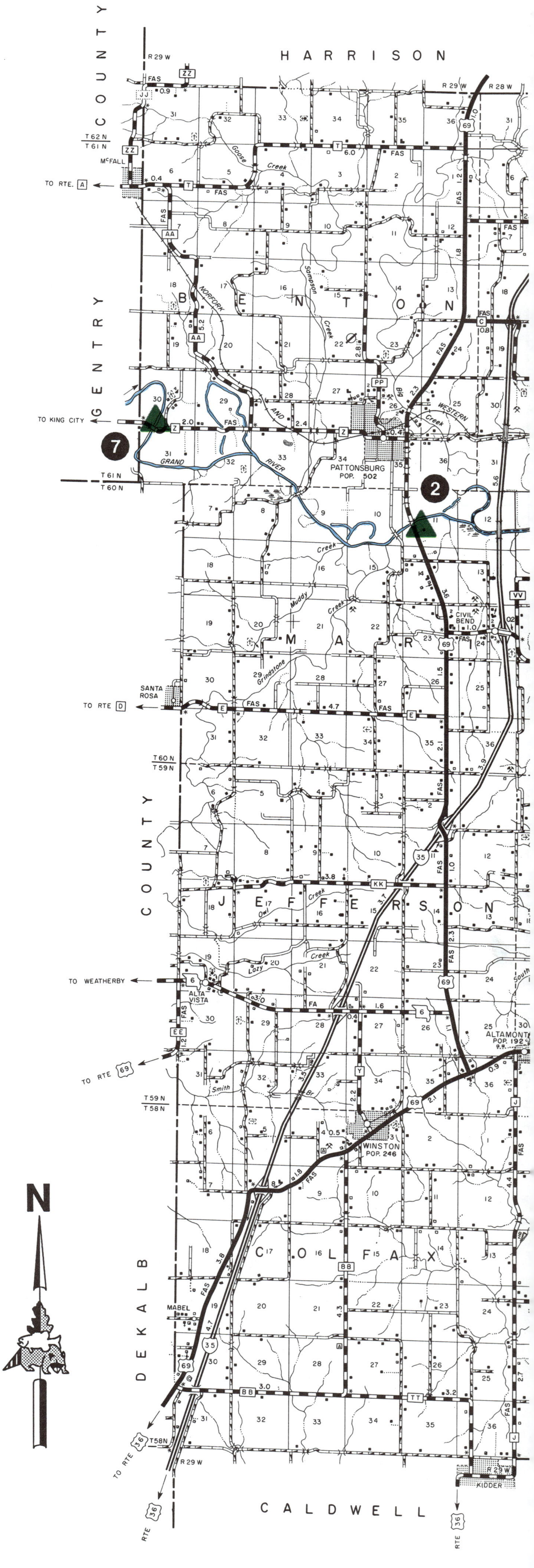

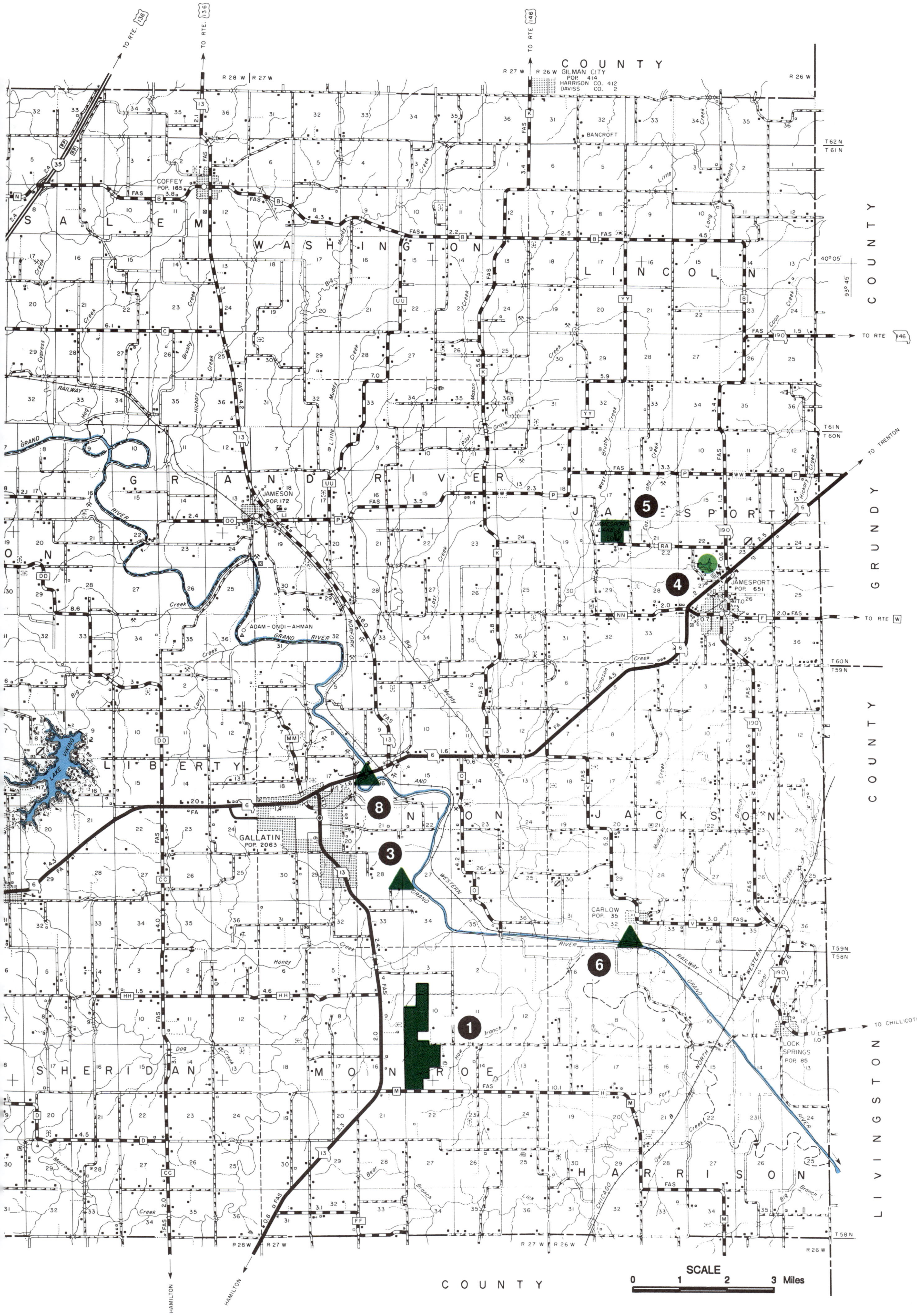
COUNTY
GILMAN CITY POP. 414 HARRISON CO. 412 DAVISS CO. 2
BANCROFT
COFFEY POP. 165
S A L E M
W A S H I N G T O N
L I N C O L N
G R A N D   R I V E R
JAMESON POP. 172
J A M E S P O R T
JAMESPORT POP. 651
ADAM-ONDI-AHMAN
L I B E R T Y
LAKE VIKING
U N I O N
GALLATIN POP. 2063
J A C K S O N
CARLOW POP. 35
LOCK SPRINGS POP. 85
S H E R I D A N
M O N R O E
H A R R I S O N
GRUNDY COUNTY
LIVINGSTON COUNTY
TO TRENTON
TO RTE W
TO CHILLICOTHE
TO RTE 146
TO RTE 136
HAMILTON
COUNTY
SCALE
0 1 2 3 Miles
40°05'
93°45'
T62N T61N
T61N T60N
T60N T59N
T59N T58N
T58N
R 28 W R 27 W
R 27 W R 26 W
R 26 W
1
3
4
5
6
8

DeKALB COUNTY

1 **Cameron City Lakes** is made up of three reservoirs. Reservoir #1 is .5 mile west of the city of Cameron on Highway 36. This 25-acre body of water is managed for fishing and is surrounded by about 23 acres of forestland and old pastureland. To reach Reservoir #2, turn west off Highway 69 on the first road that lies north of Highway 36 and follow signs. This is a 35-acre fishing lake surrounded by about 63 acres of forest. Reservoir #3 is 1 mile west of Cameron on Highway 36, 1.5 miles north on Highway C, then .75 mile east on an unnamed gravel road. This is a 96-acre area managed for fishing and is surrounded by about 100 acres of forestland. Reservoirs #1 and 2 have boat ramps.

2 To reach **Grindstone Reservoir** (also known as Cameron City Reservoir #4), turn north on the first gravel road that lies west of the junction of highways C and 36. The area totals 800 acres and the reservoir (180 acres) is managed for fishing.

3 **King Lake Conservation Area** is 5 miles east of Highway 169 on County Road 459. King Lake is 200 acres of fishable water and is surrounded by about 1,000 acres of forestland and old pastures. There is a boat ramp.

4 **Pony Express Lake Conservation Area** is 9 miles west of Cameron on Highway 36, north on Highway 33, then 1 mile on Route RA (marked with a cantilever sign). Pony Express Lake, Buffalo Bill Lake and Outpost Lake sit on this 3,150-acre area, providing anglers with 290 acres of fishable water. There is a boat rental, picnic area, boat ramps and disabled accessible fishing jetties.

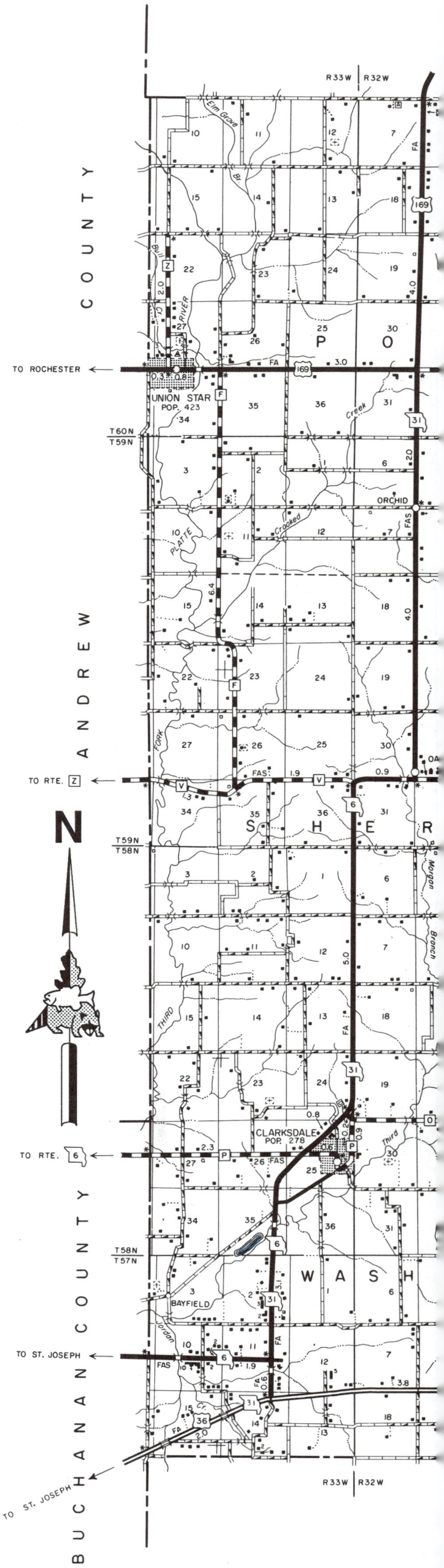

Location Map

Legend

- LAND OWNED BY MDC
- LAND LEASED BY MDC
- LAND OWNED BY MDC LESS THAN 160 ACRES
- LAND LEASED BY MDC LESS THAN 160 ACRES
- CAP ACCESS

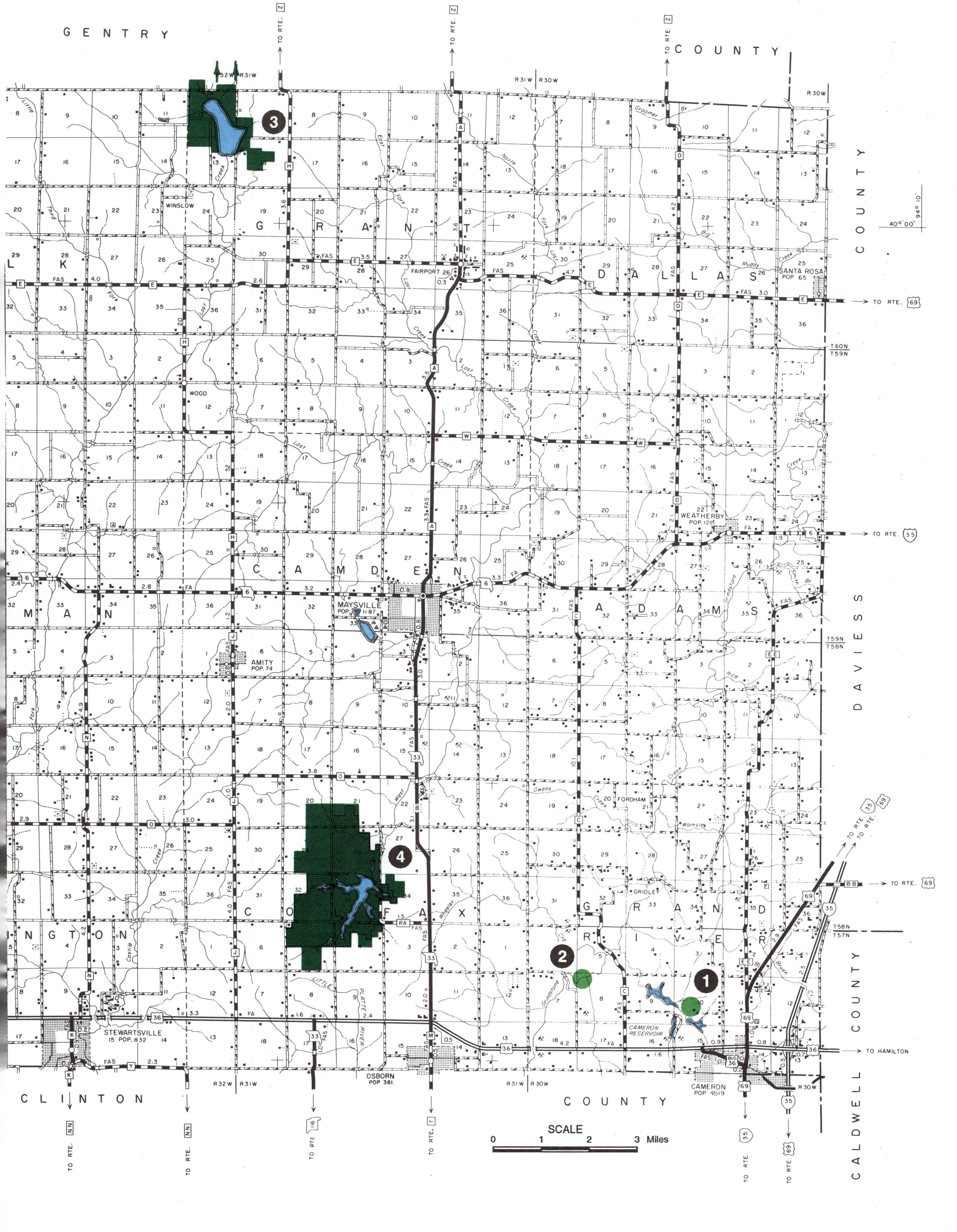
GENTRY COUNTY
DAVIESS COUNTY
CALDWELL COUNTY
CLINTON COUNTY
GRANT
DALLAS
CAMDEN
ADAMS
COLFAX
GRAND RIVER
WINSLOW
FAIRPORT
SANTA ROSA POP. 65
WOOD
WEATHERBY POP. 121
MAYSVILLE POP. 1187
AMITY POP. 74
FORDHAM
GRIDLEY
STEWARTSVILLE POP. 832
OSBORN POP 381
CAMERON POP 4519
CAMERON RESERVOIR
TO RTE. Z
TO RTE. 69
TO RTE. 35
TO HAMILTON
TO RTE. NN
TO RTE. 116
TO RTE. T
R32W
R31W
R30W
T60N
T59N
T58N
T57N
40°00'
94°10'
SCALE
0 1 2 3 Miles
1
2
3
4

DENT COUNTY

1 The main entrance to **Cedar Grove Conservation Area** is 3 miles south of Jadwin off Highway K. This is a 1,160-acre forest composed of three separate tracts.

2 **R. F. Clement Memorial Forest and Wildlife Area** is 12 miles northwest of Salem off Highway OO. This is a 520-acre forest.

3 **Hyer Woods Natural Area** is near Lake Spring on Highway 72, 15 miles south of Rolla. This is 30 acres of upland hardwood forest and is a Designated Natural Area.

4 **Indian Trail Conservation Area** is 12 miles northeast of Salem on Highway 19. This is a large area (13,383 acres) composed mainly of forestland. It contains six springs, many small ponds and wildlife watering holes. Horseback riding is permitted on forest roads, and there is a picnic area. Indian Trail Fish Hatchery is located within the area. It consists of 75 acres and is home to Blackwell Lake (35 acres) and a 1-mile stretch of Crooked Creek. There is a 1-mile hiking trail and two picnic areas.

5 **Lenox Towersite** is on Highway C, 1 mile south of the town of Lenox. The tower is surrounded by 5 acres of forestland.

6 **Montauk Fish Hatchery** is inside Montauk State Park southeast of Licking. From Highway 137, turn east on Highway VV, and go 10 miles to Highway 119. Turn right on Highway 119, and watch for signs to the parking area. The hatchery is 276 acres at the headwaters of the Current River. The river and Montauk Lake (3.5 acres) combine for 232 acres of fishable water. The area has three caves, three springs and the Montauk Upland Forest Natural Area (40 acres). There are many picnic areas and a 1.5-mile hiking trail. The area hosts exhibits and interpretive programs.

7 **Montauk Towersite** is 6 miles north of Montauk State Park off Highway 119 and is composed of 40 acres of forestland.

8 **Shawnee Mac Lakes Conservation Area** is 2 miles east of Salem on County Road 411. This 255-acre area has two lakes, providing anglers with about 48 acres of fishable water. There are boat ramps, picnic areas and a wheelchair accessible fishing dock.

9 **Short Bend Access** is on Highway 19, 9 miles northeast of Salem. This is 85 acres on the Meramec River (no boat ramp).

10 **White River Trace Conservation Area** is 8 miles west of Salem on Highway H. This 2,044-acre area has six ponds which provide anglers with 6 acres of fishable water, and a 1-mile stretch of the historic White River Trace Cherokee Trail of Tears (Hildebrand cutoff).

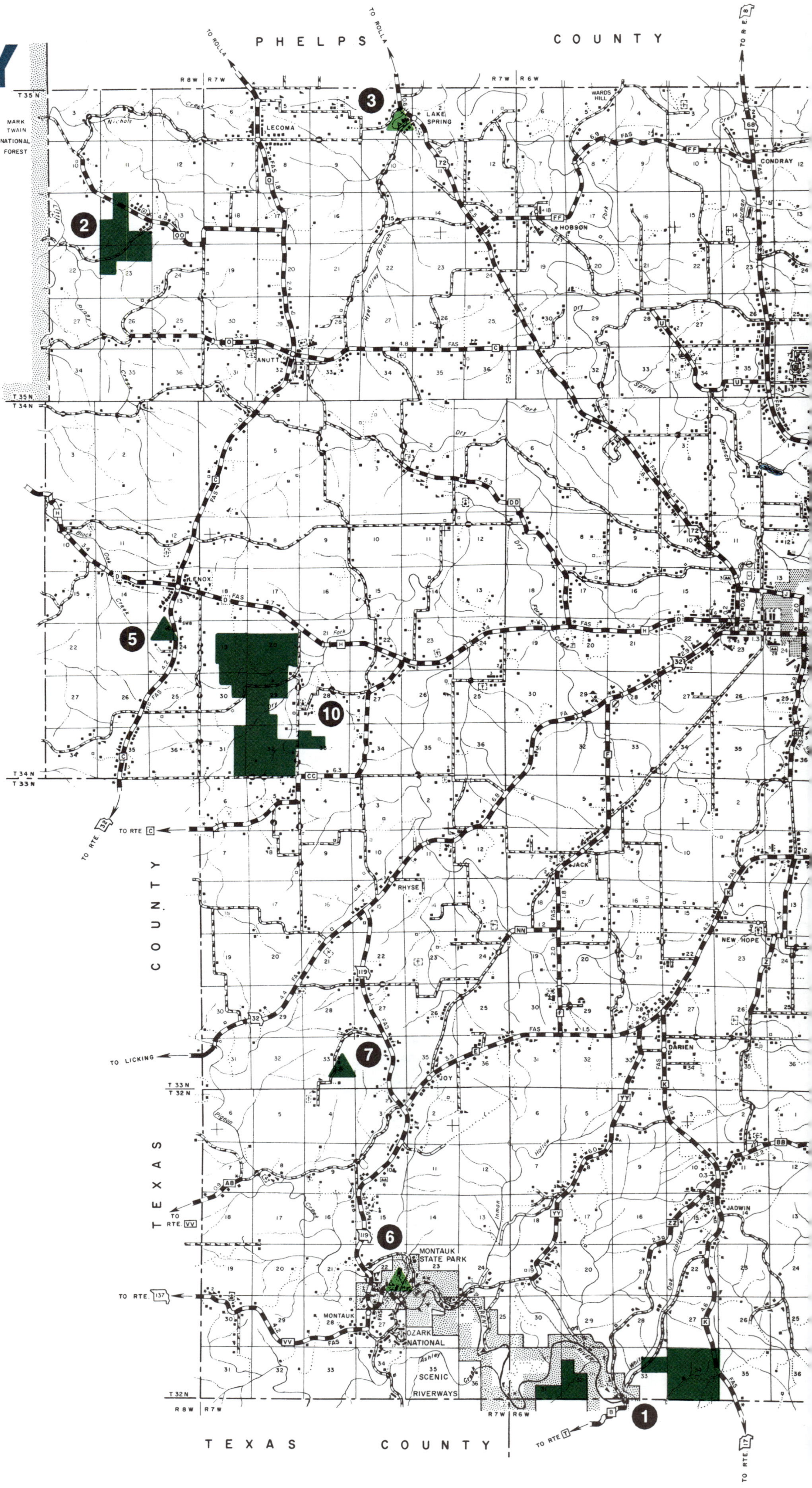

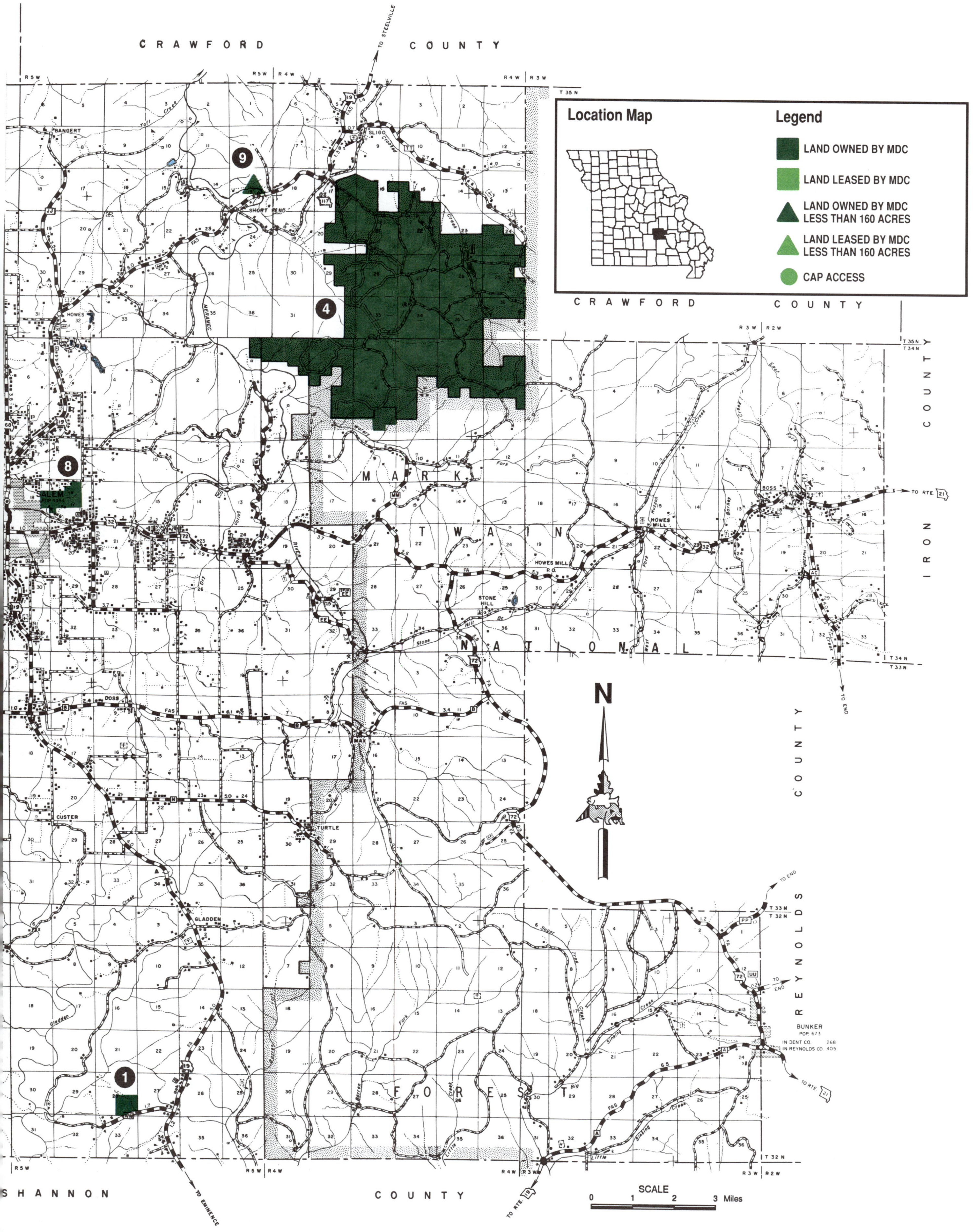
CRAWFORD COUNTY
TO STEELVILLE
Location Map
Legend
LAND OWNED BY MDC
LAND LEASED BY MDC
LAND OWNED BY MDC LESS THAN 160 ACRES
LAND LEASED BY MDC LESS THAN 160 ACRES
CAP ACCESS
CRAWFORD COUNTY
IRON COUNTY
BANGERT
SLIGO
SHORT BEND
HOWES
SALEM
MARK TWAIN NATIONAL FOREST
BOSS
HOWES MILL
STONE HILL
DOSS
MAX
CUSTER
TURTLE
GLADDEN
BUNKER
POP. 673
IN DENT CO. 268
IN REYNOLDS CO. 405
DENT COUNTY
REYNOLDS
TO RTE 21
TO END
TO EMINENCE
SHANNON COUNTY
N
SCALE
0 1 2 3 Miles

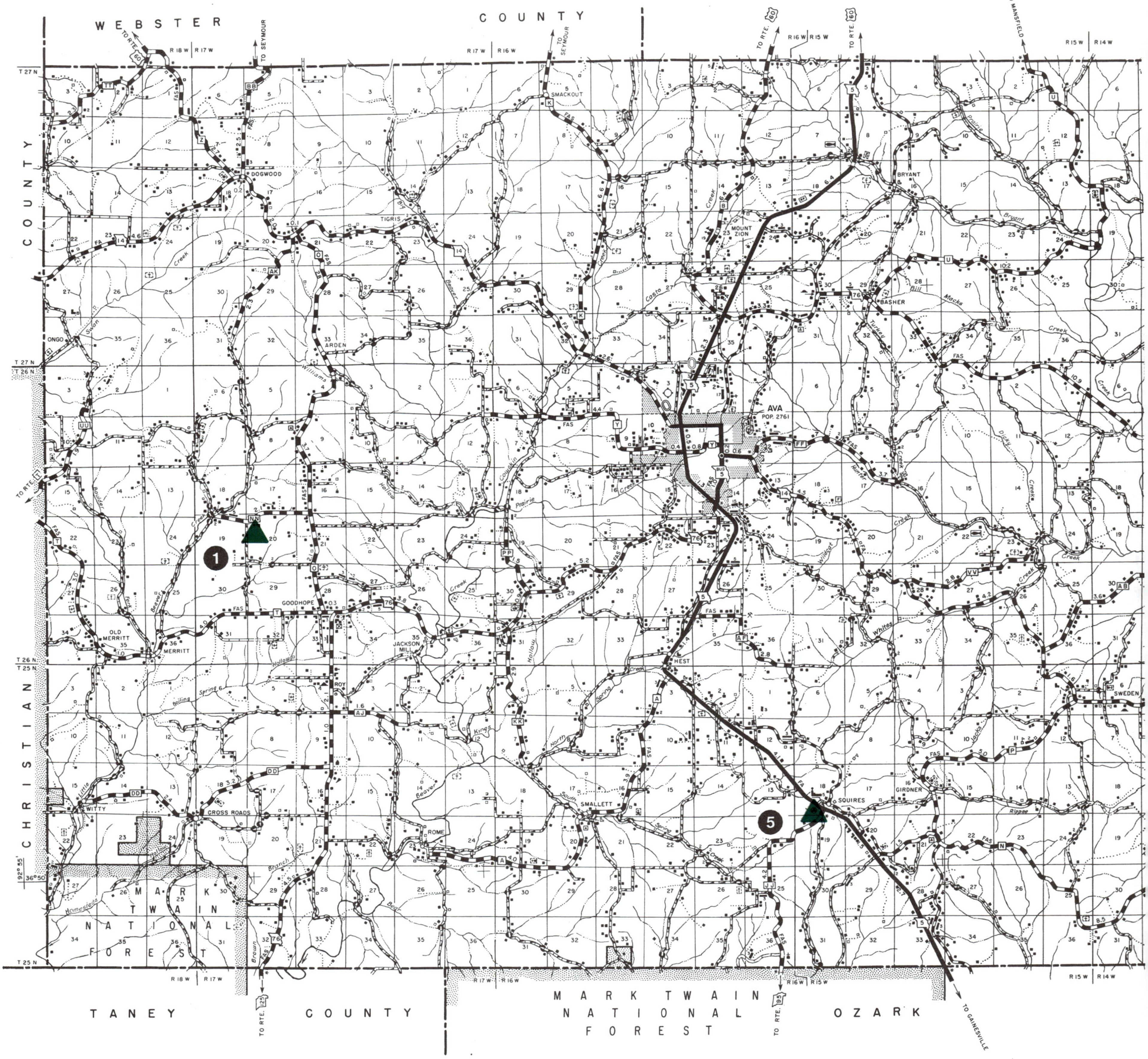

DOUGLAS COUNTY

1 **Grundy Memorial Wildlife Area** is northwest of Goodhope off Highway NN. From Highway NN, turn 1.5 miles south on County Road 537. This 40-acre area is nearly all forestland and has a .5-mile dirt nature trail.

2 **Hebron Access** is 17 miles west of the junction of highways 14 and 63. From Highway 14, turn 2 miles north on County Road 279. This is a 12-acre area on the North Fork of White River (1.5 acres of fishable water). There is a boat ramp, camping and picnic areas.

3 **Rippee Conservation Area** is 11 miles east of Ava on Highway 14, and 1.5 miles south on County Road 328. Rippee Creek and Bryant Creek flow through this 418-acre area. Bryant Creek Natural Area is about 30 acres of a 100-foot-wide riparian corridor, and the area is the site of a recorded Civil War skirmish.

4 **Shannon Ranch Conservation Area** is 1.25 miles south of Vanzant on Highway 95. This 1,325-acre area is composed of forestland (748 acres), pastureland (306 acres) and glade (268 acres).

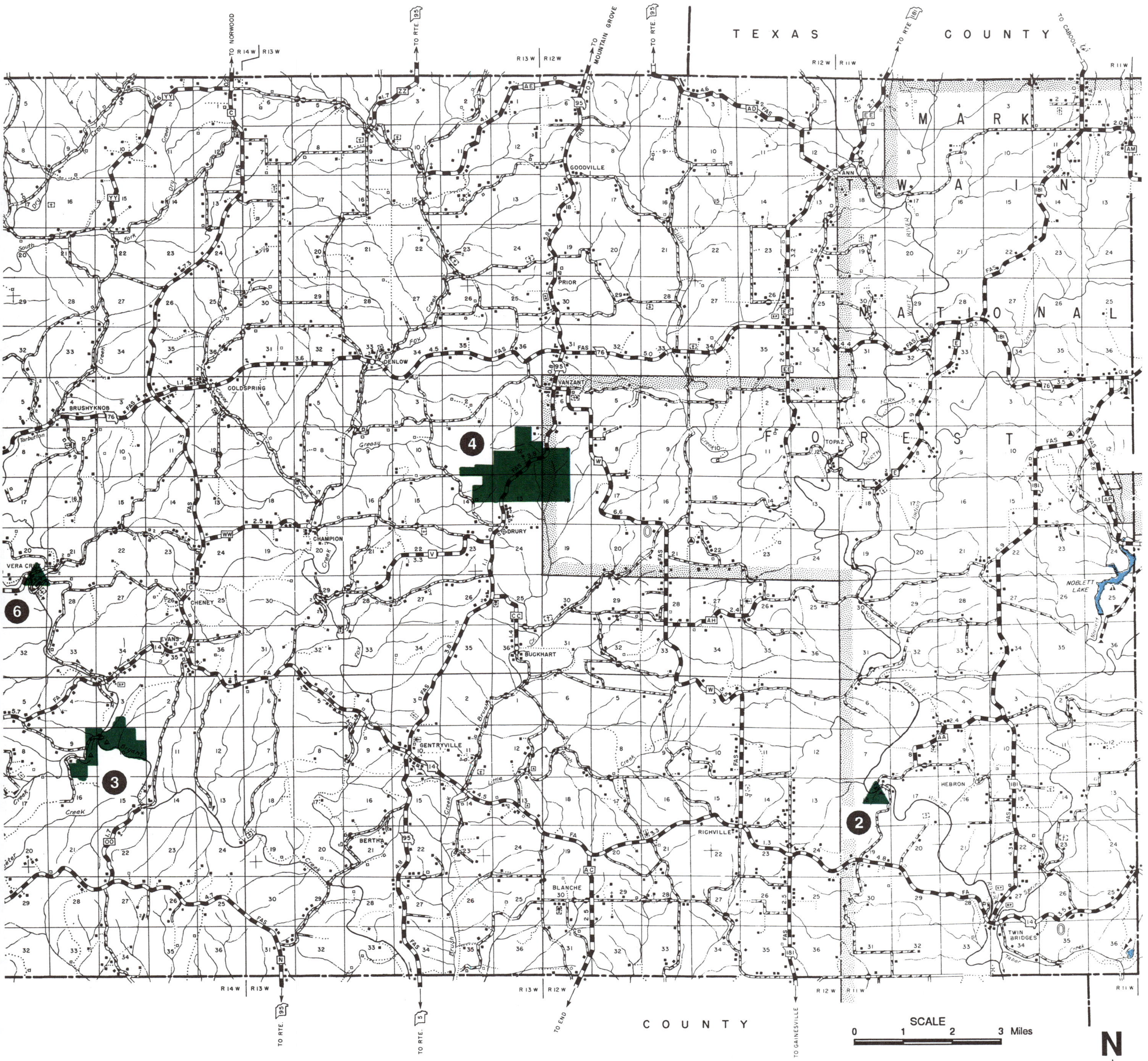

5 **Squires Towersite** is .25 mile west of the town of Squires on Highway JJ. This is a fire lookout tower on 5 acres. The access road is closed when the tower is not occupied.

6 **Vera Cruz Access** is 3 miles north of Highway 14 on Highway AB. The 80-acre area is on a gravel road at the end of the pavement. Hunter Creek and Bryant Creek flow through the area (6 acres for fishing) beneath limestone bluffs. There is a picnic area.

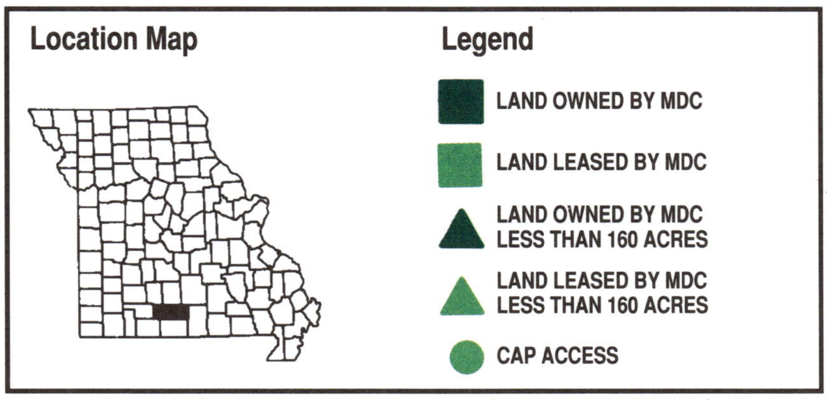

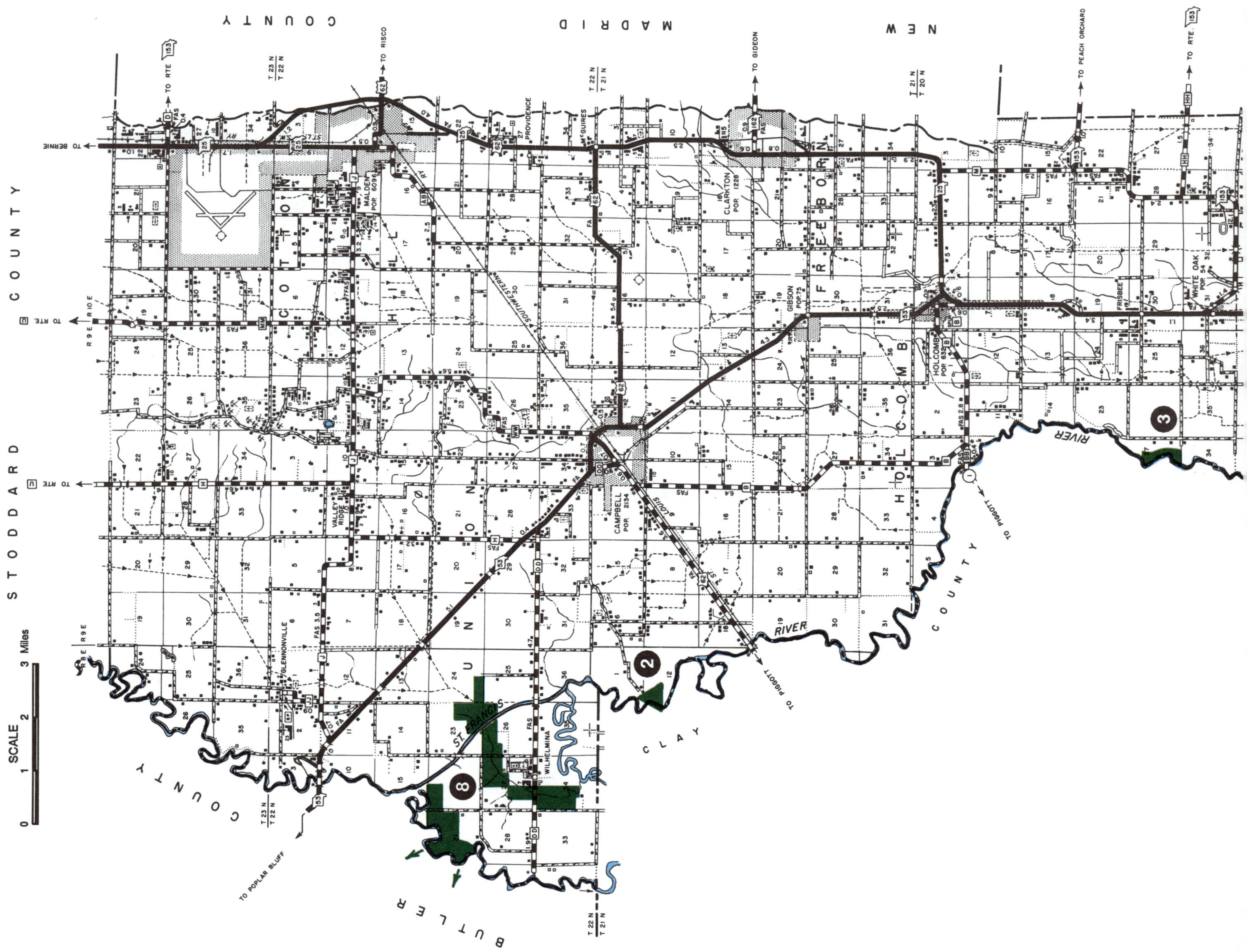

DUNKLIN COUNTY

1 **Ben Cash Memorial Conservation Area** and the Largent Annex are southwest of Kennett on Highway A. This is a 1,422-acre area on the St. Francis River. Natural areas include Cash Swamp, which contains bottomland hardwood timber and a shrub-scrub swamp.

2 **Chalk Bluff Trail Access** on the St. Francis River is south of Campbell. Take Highway 62 south to County Road 232. Turn right, and go 3 miles to County Road 228. Turn left, and the 84-acre area is in 1.25 miles. It is a Civil War battle site.

3 **Frisbee Cuttoff Access** is south of Frisbee on Highway 25, west on County Road 410, .25 mile north on County Road 411, then 1 mile west on County Road 407. This is 92 acres on the St. Francis River.

4 **Hornersville Access** (5 acres) is east of Hornersville on the old railroad line off Highway 164. There is a boat ramp.

5 To reach **Hornersville Swamp Conservation Area,** go south of Caruth on Highway 164 to the north side of the town of Hornersville. Watch for the cantilever sign pointing to the left (Pecan Street). There are five floodway ditches in this 3,381-acre area, a boat ramp and 2,851 acres of bottomland forest.

6 The main entrance to **Little River Conservation Area** and **Jerry P. Combs Lake** is in Pemiscot County. From Kennett, go 4 miles east Highway 84. Turn on Highway NN and watch for Pemiscot County Road 299. This is a 1,036-acre area.

7 **Warbler Woods Conservation Area** is southeast of Kennett off Highway TT. From TT, turn .5 mile west (right) on County Road 730, then .75 mile south on County Road 707. The area is composed of 89 acres of bottomland forest.

8 **Wilhelmina Conservation Area**–see the Butler County map, page 28.

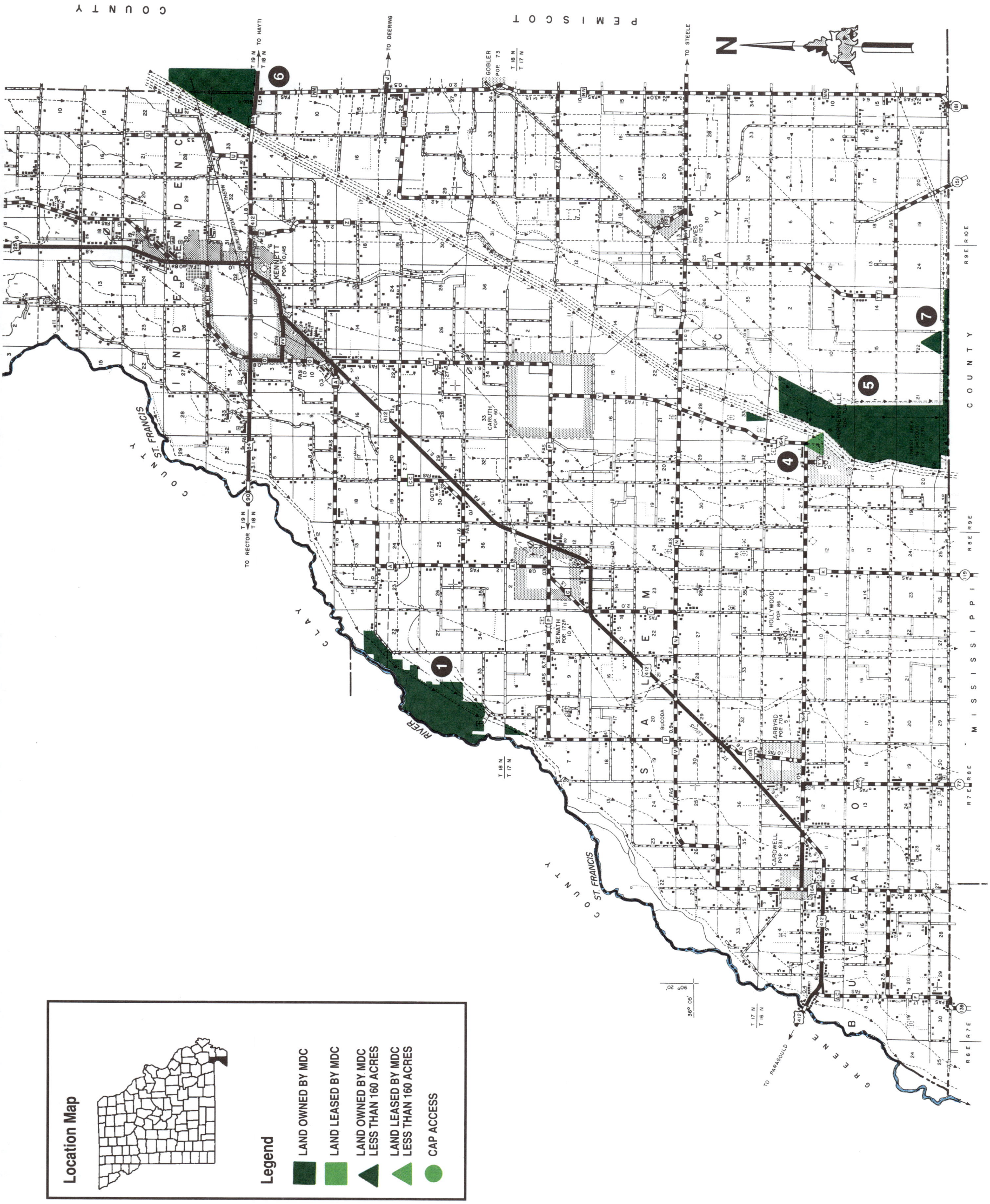
Location Map
Legend
LAND OWNED BY MDC
LAND LEASED BY MDC
LAND OWNED BY MDC LESS THAN 160 ACRES
LAND LEASED BY MDC LESS THAN 160 ACRES
CAP ACCESS
PEMISCOT
COUNTY
MISSISSIPPI
CLAY
ST. FRANCIS
RIVER
GREENE
KENNETT POP. 10,145
SENATH POP. 1,726
CARDWELL POP. 831
HOLLYWOOD POP. 86
ARBYRD POP. 704
HORNERSVILLE
CARUTH POP. 60
GOBLER POP. 73
RIVES POP. 120
OCTA
BUCODA
TO HAYTI
TO DEERING
TO STEELE
TO RECTOR
TO PARAGOULD
INDEPENDENCE
CLAY
SALEM
BUFFALO
N

FRANKLIN COUNTY

1 **Caldwell Memorial Wildlife Area** is 1.5 miles south of Washington on Highway 47, then 2 miles south on Bieker Road. Activities on this 124-acre area are limited to pre-arranged education and history trips.

2 **Catawissa Conservation Area** is north of the town of Catawissa on Highway N, then right on a county road that follows along railroad tracks. This road dead-ends at the 207-acre area. Most of this area is forestland (169 acres) and it has a boat ramp on the Meramec River.

3 **Chouteau Claim Access** is south on Highway AH off I-44. There is a cantilever sign on AH for the 15-acre area (.75 mile down a road off AH). This is an access on the Meramec and Bourbeuse rivers.

4 **Colter's Landing Access** is 7 miles west of Washington on Highway 100. This 6-acre area is on .25 mile of Boeuf Creek which accesses the Missouri River.

5 **Engelmann Woods Natural Area** is 6 miles east of Labadie on Highway T. This is a 148-acre forest and a Designated Natural Area. A .75-mile hiking trail winds through the area, and hunting is not permitted.

6 **Little Indian Creek Conservation Area** is south on Highway K (off Highway 30), across the Meramec River, and 3 miles to the left on Old Highway K to Little Indian Creek Road. Follow this road 3 miles and watch for the area sign. This is a 3,939-acre area composed mainly of forestland (3,742 acres). Little Indian Creek flows 1 mile through the area, and there are two horseback riding trails about 8 miles long.

7 **Long Ridge Conservation Area** is about 6 miles north of Sullivan on Highway AF. At the end of AF, continue on Ridge Road to reach this 1,816-acre area.

8 **Mayer's Landing Access** is 1 mile west of Union on Highway 50, then south on Highway UU (watch for the area sign). This is a 3.5-acre area with a boat ramp on the Bourbeuse River.

9 **Meramec Conservation Area** and the **Arthur G. Heyne Memorial Annex** is on Highway 185 (south at the Sullivan exit from I-44). This is a 3,879-acre area on the Meramec River (4.8 miles). It has a 12-mile horseback riding trail and an 8-mile hiking trail mainly through forestland. There is also a 1.4-mile disabled accessible trail. There are five ponds, 24 wildlife watering holes, four caves and a spring in the area.

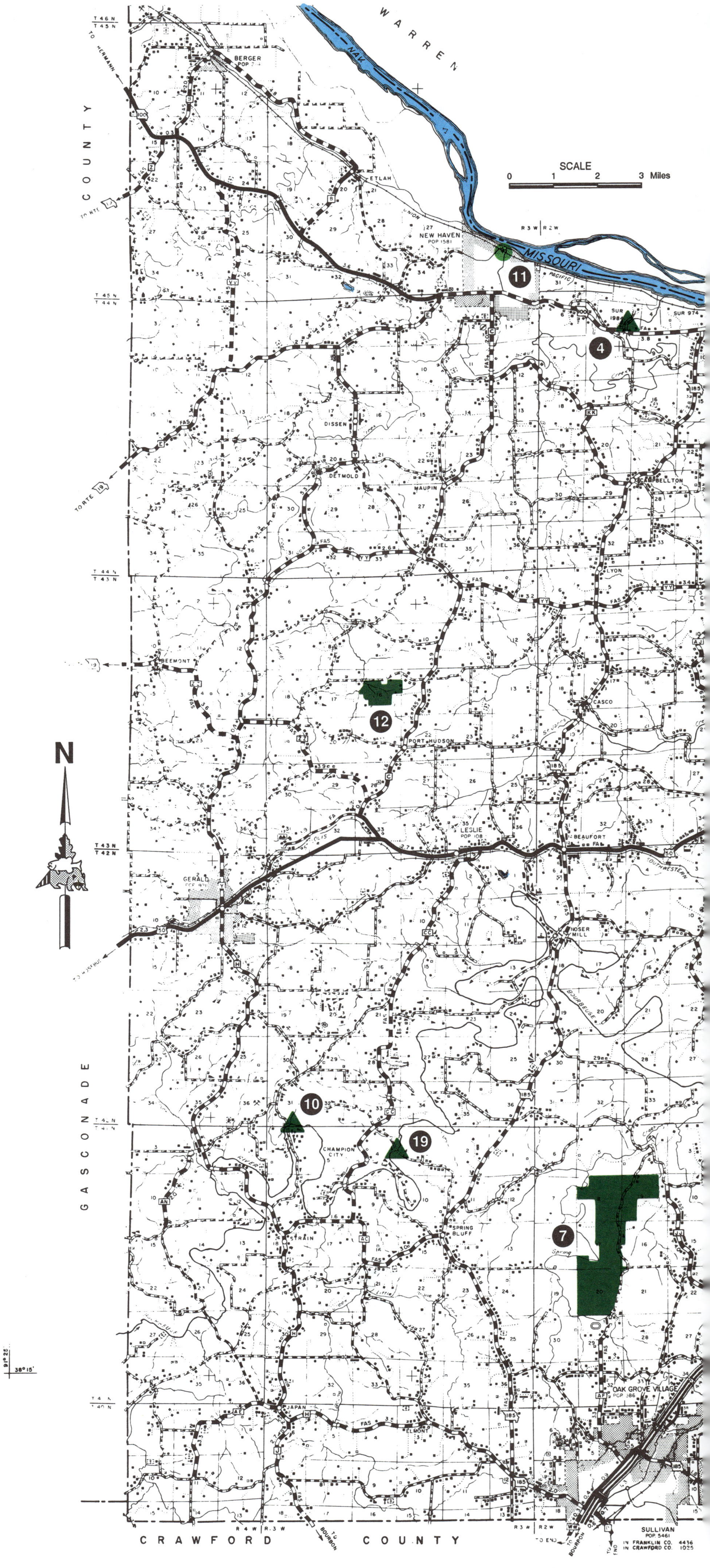

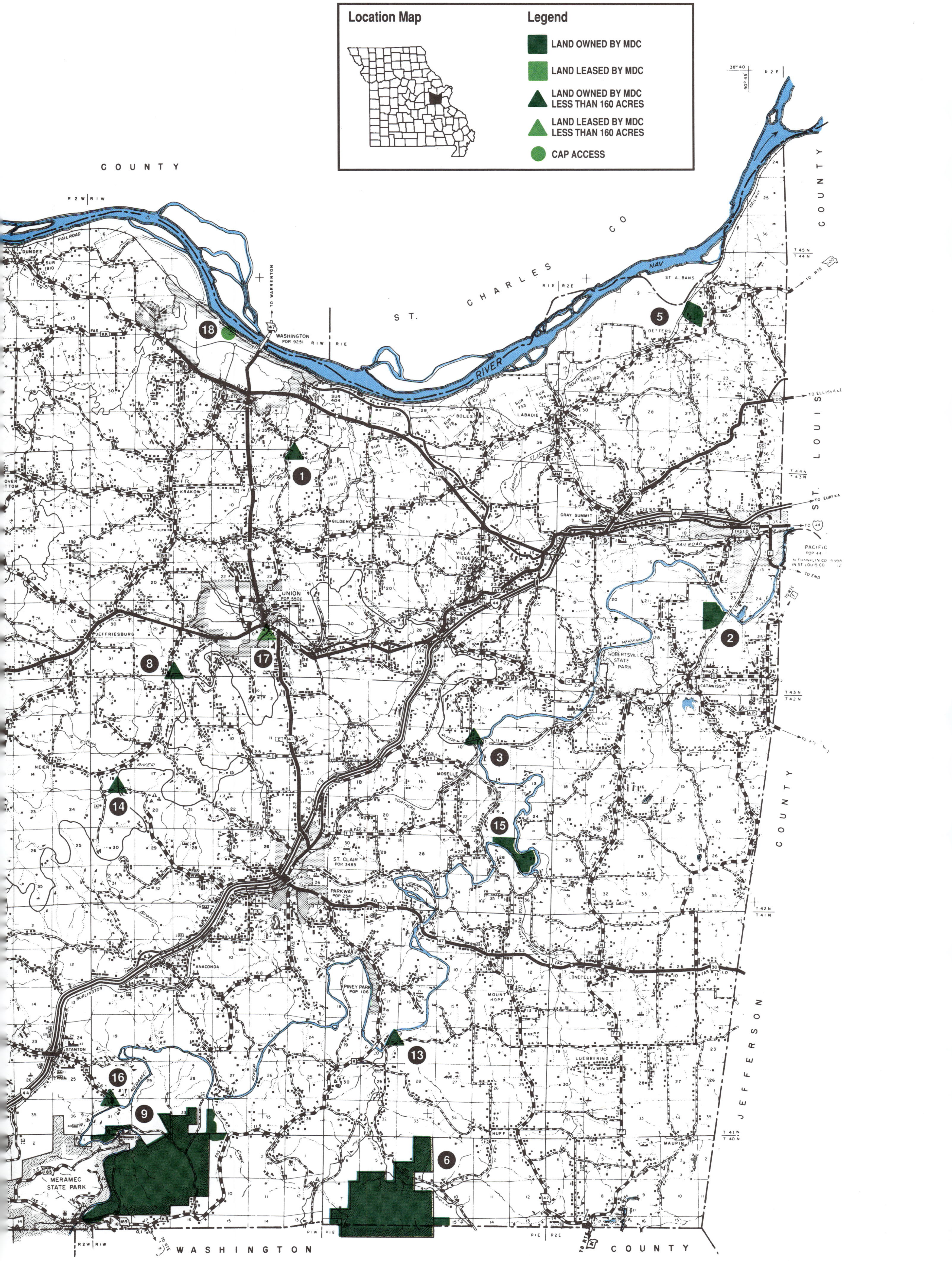

Location Map
Legend
LAND OWNED BY MDC
LAND LEASED BY MDC
LAND OWNED BY MDC LESS THAN 160 ACRES
LAND LEASED BY MDC LESS THAN 160 ACRES
CAP ACCESS
ST. CHARLES CO
ST. LOUIS COUNTY
JEFFERSON COUNTY
WASHINGTON COUNTY
RIVER
WASHINGTON POP. 9251
UNION POP. 5506
ST. CLAIR POP. 3485
PARKWAY POP. 254
PINEY PARK POP. 106
PACIFIC
GRAY SUMMIT
VILLA RIDGE
LABADIE
DUNDEE
JEFFRIESBURG
GILDEHOUSE
MOSELLE
ANACONDA
STANTON
LUEBBERING
CATAWISSA
ROBERTSVILLE STATE PARK
MERAMEC STATE PARK
MOUNT HOPE
LONEDELL
HUFF
MAUPIN
ST. ALBANS
TO WARRENTON
TO ELLISVILLE
TO EUREKA
1
2
3
5
6
8
9
13
14
15
16
17
18

FRANKLIN COUNTY continued...

10 **Mill Rock Access** on the Bourbeuse River is north of Sullivan on Highway 185, then 10 miles north on Highway H to Mill Rock Road. This 10-acre area is about 3 miles down Mill Rock Road. There is a ford but no boat ramp.

11 **New Haven Access** is in downtown New Haven on Miller Street off Highway 100. This is a 23-acre area with a boat ramp on the Missouri River.

12 **Port Hudson Lake Conservation Area** is on Bald Hill Road off Highway C. This 225-acre area is home to a 55-acre lake, disabled accessible fishing dock, a marsh and three picnic areas.

13 **Redhorse Access** is south of Sullivan on Highway 185, down Highway K to Old Highway K, and then down Project Road. The 43-acre area is then marked with a cantilever sign. This is an access on the Meramec River.

14 **Reiker Ford Access** is reached by taking Highway 50 to Highway UU, then following Snake Hill Road. After 1 mile, turn left on Reiker Ford Road and the 11-acre area is in .25 mile. One-half mile of the Bourbeuse River runs along this area.

15 To reach **River 'Round Conservation Area**, take Highway TT off I-44. From TT, turn left for .25 mile on Mill Hill, then turn north on Old Cove Road. The entrance to this 302-acre area is about .25 mile down Old Cove Road. The Meramec River flows 3.3 miles around this area. There is also a cave.

16 **Sand Ford Access** is south of Stanton on Highway W. This 30-acre area is on the Meramec River and has one boat ramp.

17 **Union Access** is marked with a sign off Highway 50 near Union. This is a 10-acre area on the Bourbeuse River with a boat ramp.

18 **Washington City Access** is northwest of Highway 47 on the Missouri River. This 10-acre area has a boat ramp and a wheelchair accessible fishing dock.

19 **Wenkel Ford Access** is just north of Spring Bluff on Highway 185, then 4 miles left on Wenkel Ford Road. This is an access on the Bourbeuse River and is made up of about 4 acres of forestland and 1 acre of open land.

Gifts That Never Go Out of Style

The Missouri Department of Conservation publishes a variety of books sure to please just about any nature lover... the botanist, hiker, canoeist, fan of reptiles, wild game cook and admirer of wildlife art.

For example, *Missouri Wildflowers*, by Edgar Denison, is a field guide to more than 450 species of Missouri wildflowers. It has 285 color photographs arranged by color and blooming time. Now in its fourth edition, the book has been extensively revised and expanded. Another favorite is by the award-winning team of Charles and Elizabeth Schwartz. *About Mammals and How They Live* is a companion to *The Wild Mammals of Missouri*, first published in 1959 by the University of Missouri Press and widely acclaimed as the definitive work on Missouri mammals.

About Mammals includes all wild, land-dwelling mammals in North America (excluding Mexico). It focuses on animal characteristics, with chapters on "Where Mammals Make Their Homes," "How Mammals Get Their Food," "How Mammals Care For Their Young," and others on habits and behavior. The 200-page book contains more than 40 pages of color illustrations and pen-and-ink drawings.

Ponds, float streams, trees, hiking trails, amphibians, eagles, wild edibles–whatever your outdoor interest, the Conservation Department has just the title for you. Our books also make wonderful presents for birthdays, holidays or anytime you just want to say, "thanks" or "thinking of you."

Send for your free copy of the Department's publications catalog, *Outdoor Library*. Write to Missouri Department of Conservation, "Outdoor Library," P.O. Box 180, Jefferson City, MO 65102. Many of our books also are available at bookstores and Conservation Department offices and nature centers throughout the state.

Publications

About Mammals and How They Live
Amphibians and Reptiles of Missouri
Butterflies and Moths
Crayfishes of Missouri
Cy Littlebee's Guide to Cooking Game and Fish
Eye of the Eagle
Fishes of Missouri
Key to Missouri Trees in Winter
Missouri Hiking Trails
Missouri Naiades
Missouri Nature Viewing Guide
Missouri Orchids
Missouri Wildflowers
Shrubs and Woody Vines of Missouri
Walk Softly Upon the Earth
(Mosses, Liverworts and Lichens Field Guide)
Water Plants for Missouri Ponds
Wildlife Drawings
Wild Edibles of Missouri

Audio Tapes and Calendar

Talking Toad and Frog Poster
Missouri Bird Calls
Voices of the Hills: A Journey to Shannon County
Natural Events Calendar

To celebrate our state's rich animal and plant life, the Missouri Department of Conservation maintains a free loan library of film and video productions. Most of these award-winning movies were produced by the Department, but we've added some titles made by outside producers which help to convey the conservation message. Any Missouri school, service club or other organization within the state can borrow them at no charge. Videos also may be purchased. To receive your free *Media Loan List*, or *Media Sales List*, write to Media Librarian, Missouri Department of Conservation, P.O. Box 180, Jefferson City, MO 65102

Jim Rathert

Found at Engelmann Woods Natural Area, celandine poppies grow on moist, wooded slopes and in valleys. Designated natural areas often are excellent places to looks for showy wildflowers and other interesting native plant species.

Engelmann Woods Natural Area

Acquired in 1983, Engelmann Woods was designated a natural area due to its high-quality mesic upland forest. The area's rugged river hills are home to mature red and white oaks, basswood, ash and sugar maple. There are many ferns, spring wildflowers, small creeks and dolomite bluffs.

Engelmann Woods Natural Area is named in honor of George Theodor Engelmann (1809–1884), a Missouri botanist who had far-reaching impact on plant taxonomy throughout the American West. Almost every botanical explorer who struck out westward in the mid 1800s paid a visit to Engelmann in his St. Louis home to seek his advice. His own field work introduced many unknown plants, and his research into grapevines led to the discovery that American vines were immune to phylloxera, a blight that later destroyed the vineyards of France. American stock replaced the devastated French vineyards.

Missouri Natural Areas are biological communities or geologic sites that are preserved and managed to perpetuate the natural character, diversity and ecological processes of Missouri's native landscapes. Natural areas are designated because they represent the best examples of each remaining terrestrial and aquatic natural community type within each natural section of the state. The Missouri Natural Areas System is coordinated by the Department of Conservation, Department of Natural Resources, Mark Twain National Forest, and Ozarks National Scenic Riverways.

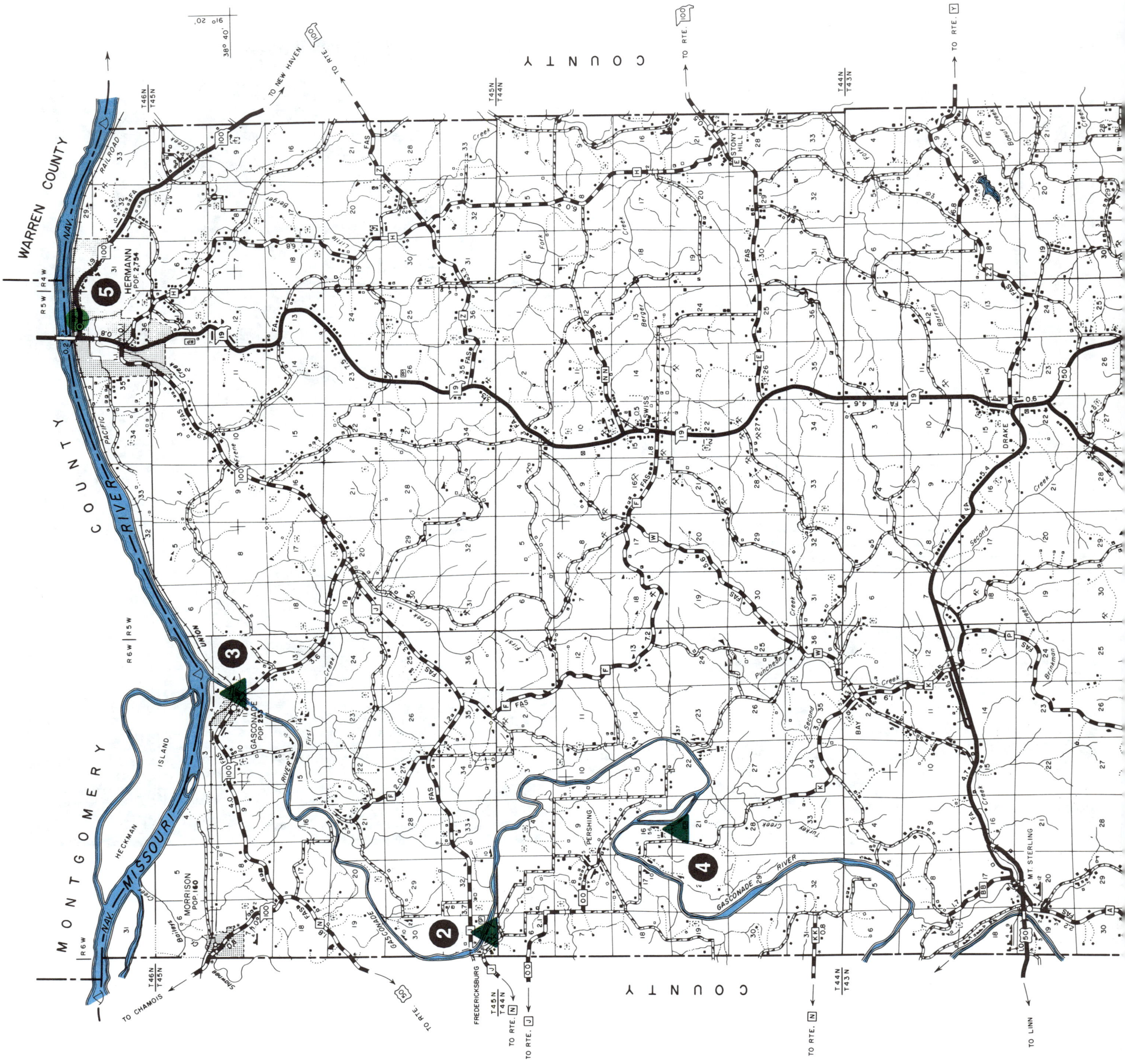

GASCONADE COUNTY

1 **Canaan Conservation Area** is 3 miles north of Bland on Highway A. This 1,435-acre area is then on Boetcher Road (gravel road off Highway A). Most of this area is forest, and it contains stretches of Sulphur Branch and Clear Fork Stream.

2 **Fredricksburg Ferry Access** is south of Morrison. From the junction of Highways 100 and N, go 5.5 miles on Highway N, 2 miles on Highway J, then 1 mile down Old Ferry Road (Gravel Road 219). This is a 5-acre lawn and parking area with a boat ramp on the Gasconade River.

3 **Gasconade Park Access** is in the city of Gasconade. From Highway 100, take Main Street east to Oak Street. This 2-acre area is about .5 mile from where the Gasconade and Missouri rivers merge. There is a boat ramp.

4 **Helds Island Access** is near Bay on Highway K. Go 4 miles on Highway K until it becomes a gravel road. The area is another 2 miles down the gravel segment of K. This is a 10-acre area of mowed lawn, brush and parking along the Gasconade River.

5 **Hermann Riverfront Park** is in downtown Hermann along the Missouri River. This 14-acre area has a boat ramp and wheelchair accessible fishing dock.

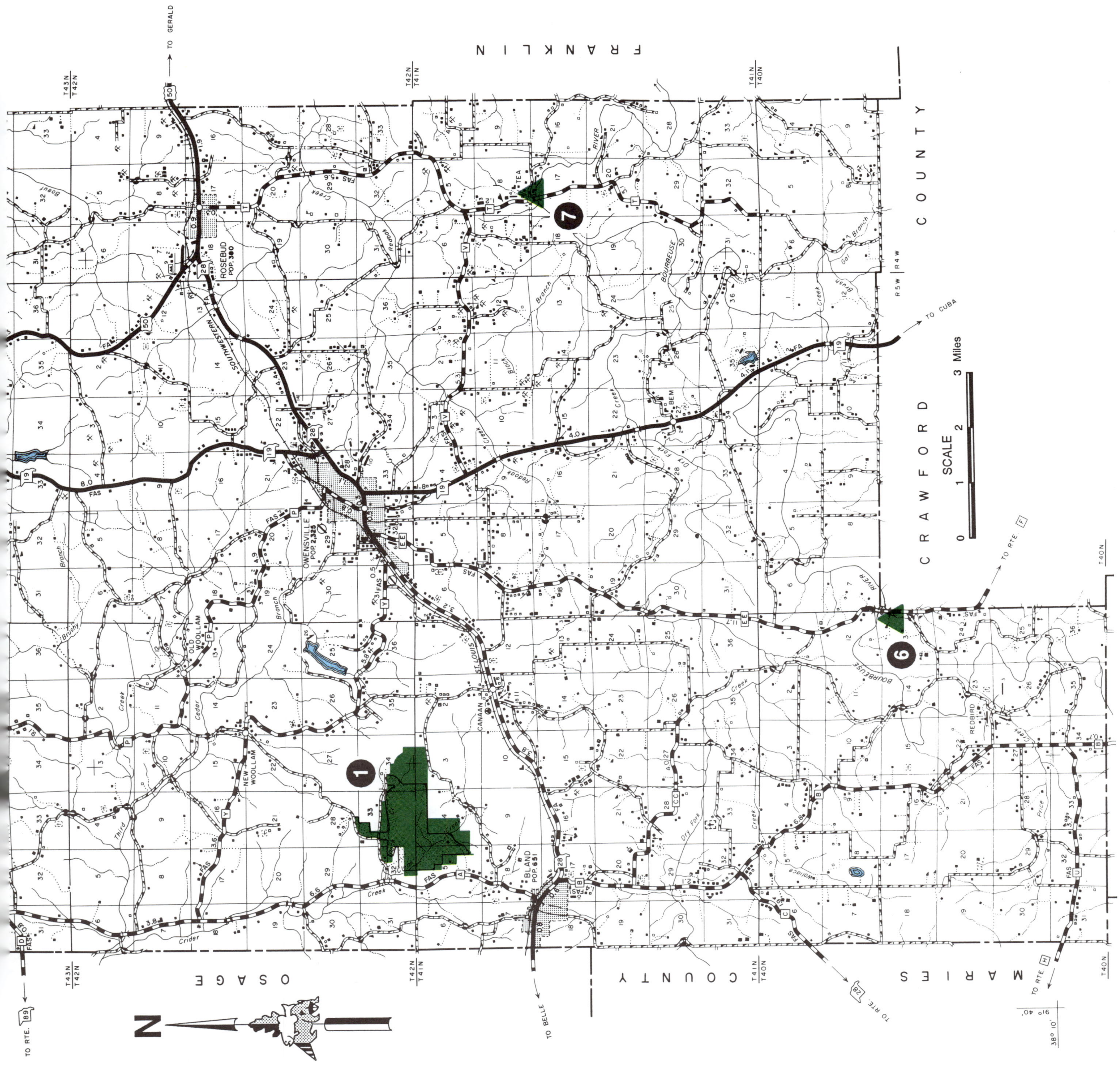

6 **Mint Spring Conservation Area and Access** is 9 miles south of Owensville on Highway EE. This is a 42-acre area on Livingwell Creek adjacent to Mint Spring. Mint Spring Seep Natural Area (10 x 60 feet) contains some rare vegetation.

7 **Tea Access** is southeast of Owensville. Take Highway 19 south 2 miles to Highway V. Turn 5 miles east on V to Highway T, then head 1.5 miles south on Highway T to Tea Road (gravel). This 4-acre area is 100 yards east on Tea Road along the Bourbeuse River. There is no boat ramp.

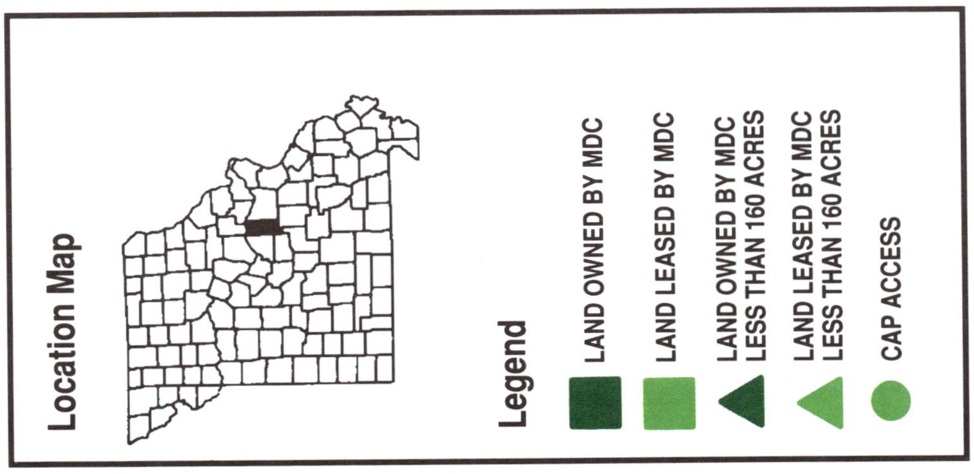

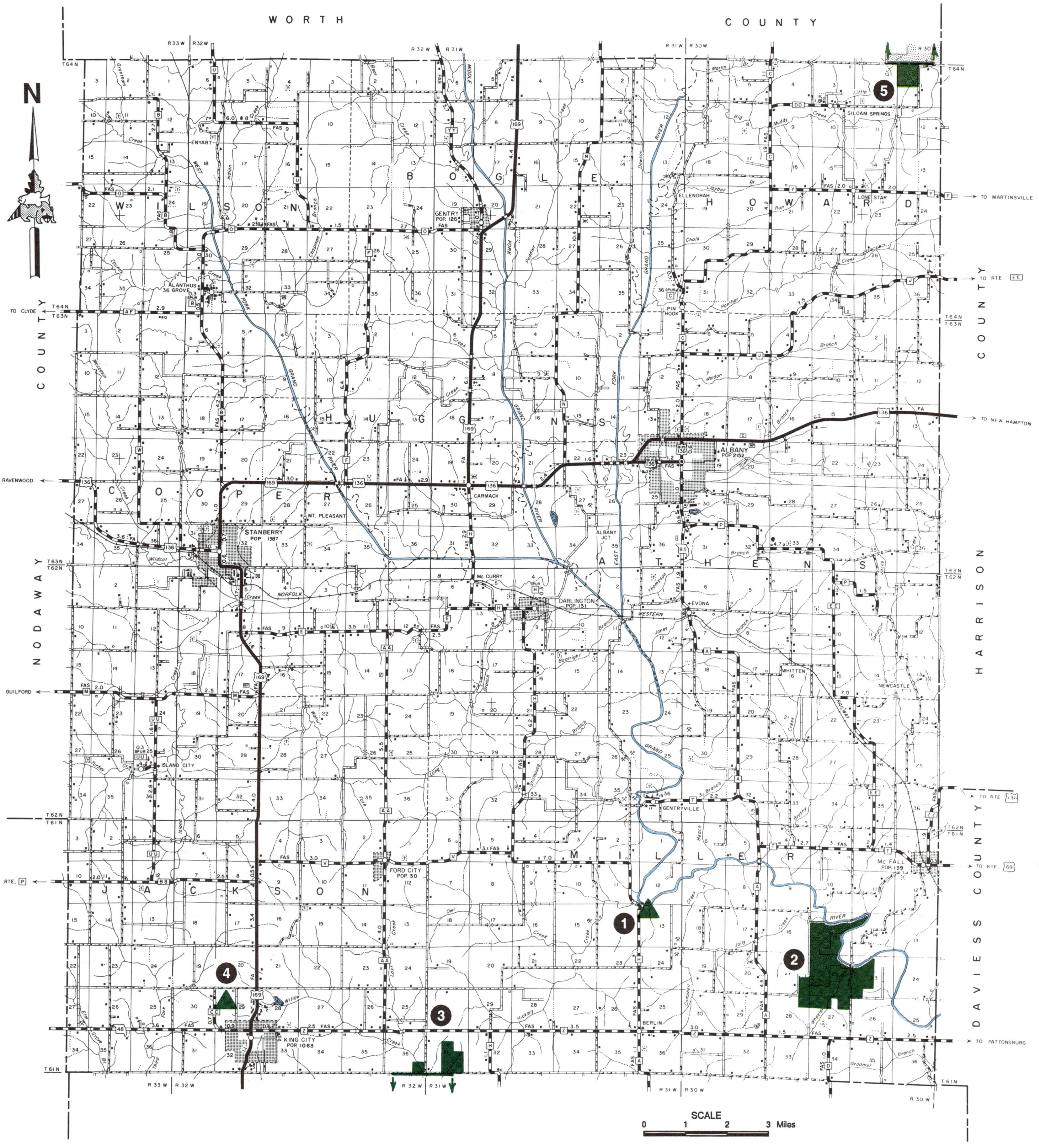

GENTRY COUNTY

1 **Andy Denton Access** (121 acres) is 3 miles north of Berlin on Highway H.

2 **Elam Bend Conservation Area** is southwest of McFall off Highway A. This 1,483-acre area is mainly forestland.

3 **King Lake Conservation Area**–see the DeKalb County map, page 68.

4 **Limpp Community Lake** (29 acres) is on Highway CC, north of King City off Highway 48.

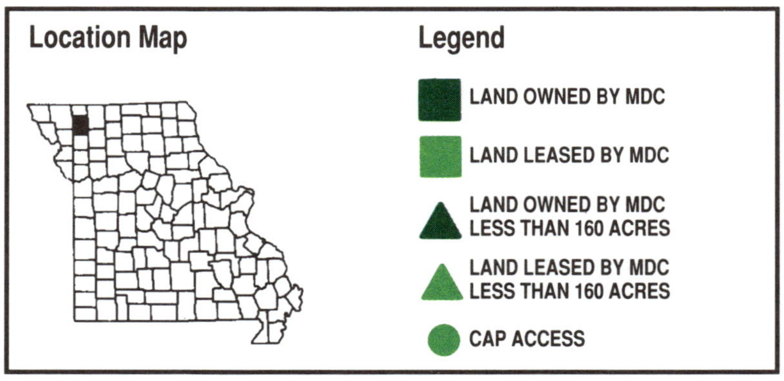

5 Most of **Emmett and Leah Seat Memorial Conservation Area** (3,164 acres) lies in Worth County, though a small portion is in Gentry. The main entrance is in Worth County, 3 miles east of Denver on Highway M.

While in Greene County, visit

Springfield Conservation Nature Center

There is always something happening at the Springfield Conservation Nature Center, which is located .5 mile from the intersection of highways 65 and 60. There are trails to walk, exhibits to explore and films and programs to enjoy. Come on your own, or call in advance to schedule a naturalist-conducted program or hike for your school class, scout troop or other tour group.

Outside

- eighty acres of forests, fields, creeks and frontage on Lake Springfield
- Ozark wildlife, including deer, foxes, mink, muskrat, raccoons, squirrels, birds, turtles and frogs
- six trails, some as short as .2 mile, others as long as 2 miles
- a boardwalk over the marshy shallows of the lake

Inside

- multi-sensory exhibit room
- regularly scheduled public programs, guided walks and nature films, announced in the center's monthly newsletter, *The Harbinger*
- classrooms and a 150-seat auditorium
- sales of books, calendars, and fishing and hunting permits
- covered sack lunch area
- indoor wildlife viewing area

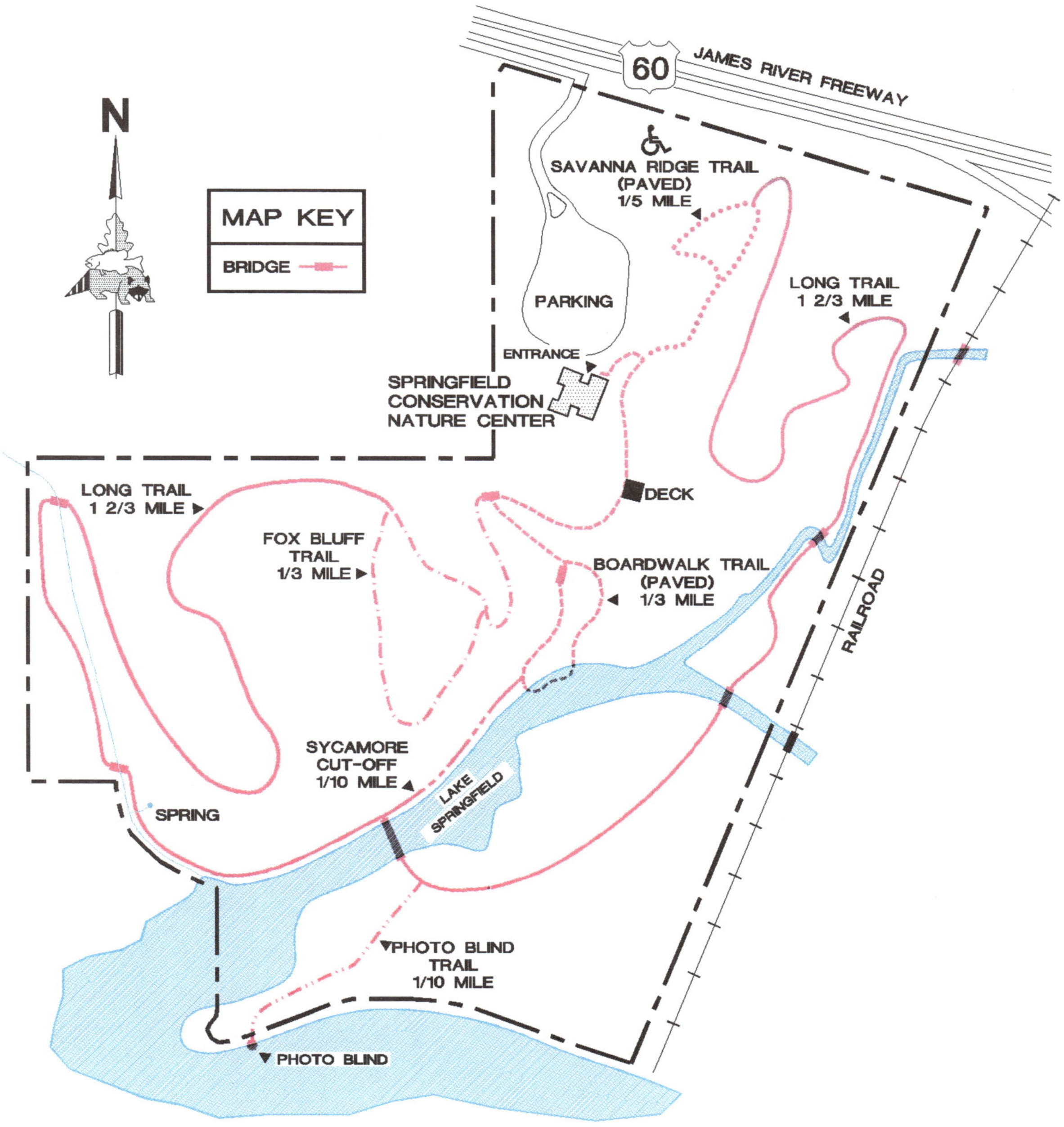

Building Hours

Open year-round, seven days a week from 8 a.m. to 5 p.m. Closed Thanksgiving, Christmas and New Year's Day

Area Hours

Open daily, 8 a.m. to 8 p.m. during daylight savings; and daily, 8 a.m. to 6 p.m. standard time. Closed Thanksgiving, Christmas and New Year's Day.

Pets, horses, hunting, fishing, collecting, littering, fires and alcoholic beverages are not permitted. Bicycling and motorized vehicles are restricted to paved roads.

Springfield Conservation Nature Center
Missouri Department of Conservation
4600 S. Chrisman
Springfield, MO 65804
417/888-4237

Paul Childress

Springfield Conservation Nature Center is one of four nature centers operated by the Missouri Department of Conservation. Nature centers offer visitors of all ages the chance to explore and learn more about the natural world. There is no charge for admission, and visitors will find plenty to see both indoors and outdoors.

GREENE COUNTY

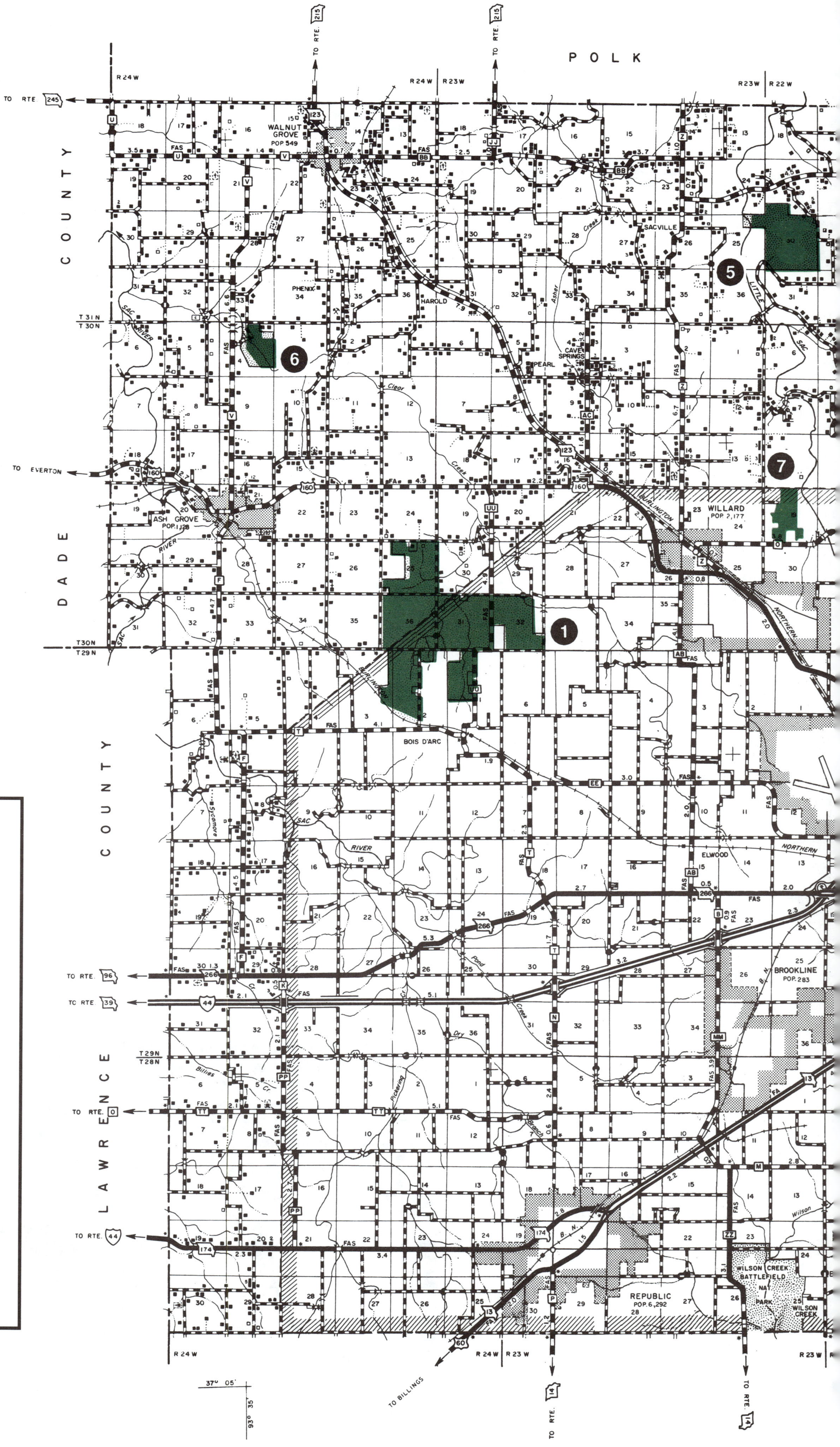

Location Map

Legend

LAND OWNED BY MDC

LAND LEASED BY MDC

LAND OWNED BY MDC LESS THAN 160 ACRES

LAND LEASED BY MDC LESS THAN 160 ACRES

CAP ACCESS

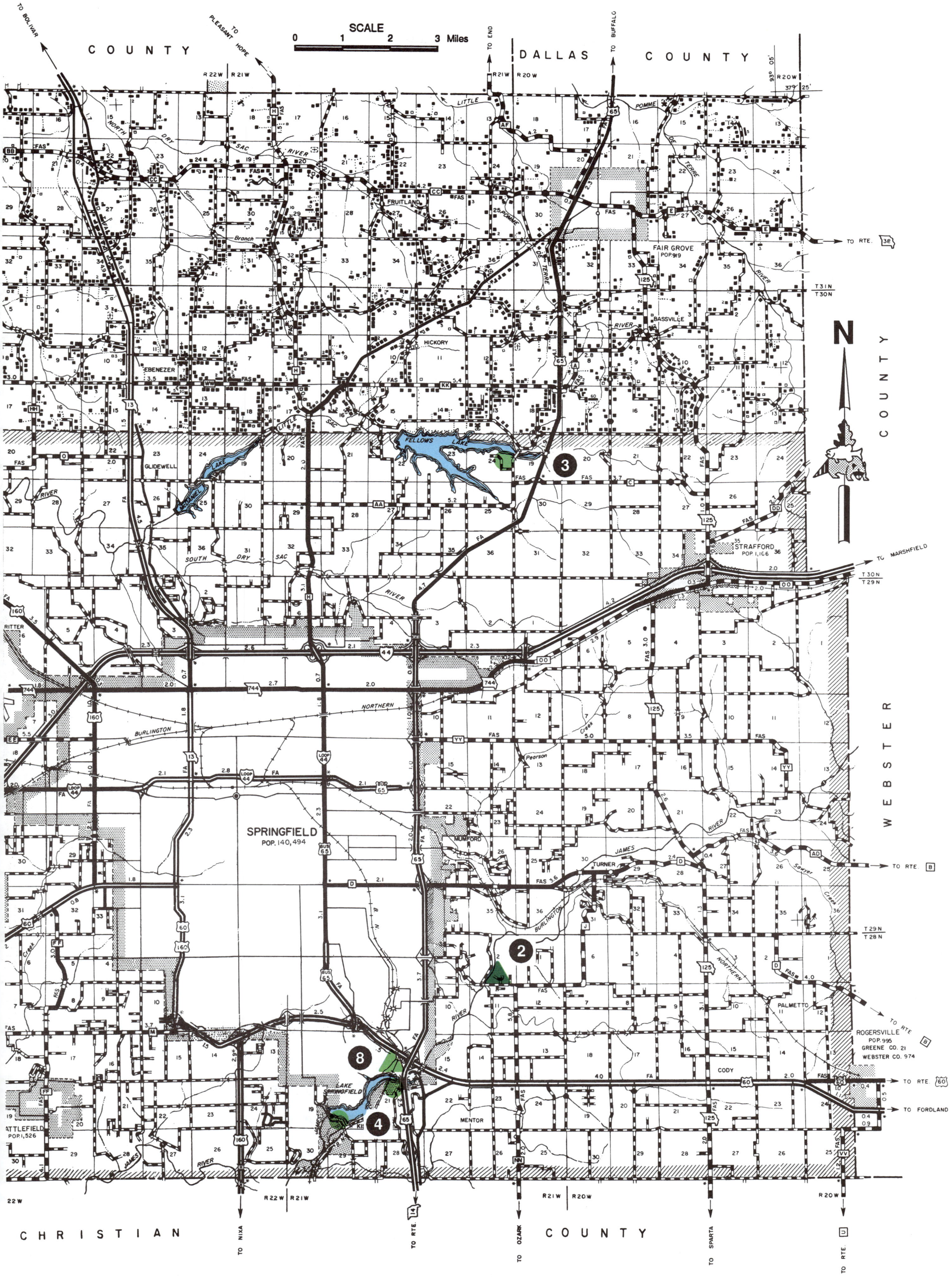
SCALE
0 1 2 3 Miles
DALLAS COUNTY
WEBSTER COUNTY
CHRISTIAN COUNTY
N
SPRINGFIELD
POP. 140,494
FAIR GROVE
POP. 919
STRAFFORD
POP. 1,166
ROGERSVILLE
POP. 995
GREENE CO. 21
WEBSTER CO. 974
FELLOWS LAKE
LAKE SPRINGFIELD
FRUITLAND
HICKORY
BASSVILLE
EBENEZER
GLIDEWELL
RITTER
MUMFORD
TURNER
PALMETTO
CODY
MENTOR
TO BOLIVAR
TO PLEASANT HOPE
TO BUFFALO
TO MARSHFIELD
TO FORDLAND
TO NIXA
TO OZARK
TO SPARTA
T31N
T30N
T29N
T28N
R22W R21W
R21W R20W
2
3
4
8

GREEN COUNTY continued...

1 **Bois D'Arc Conservation Area** is off Highway UU, southwest of Willard. This 2,892-acre area has eight ponds with 4.5 acres of fishable water and is the site of an educational Urban Fishing Program. A natural feature of interest includes Watkins Cave (Speakeasy Cave), which is accessible with a Special Use Permit only. There is an archery range and a staffed firearms range (fee charged).

2 To reach **Joe Crighton Access**, go east on Battlefield Road (off Highway 65 south of Springfield) to FR-164. Turn left on this road and travel to FR-193. This 8-acre area is a narrow tract between FR-193 and the James River.

3 **Fellows Lake** is 4 miles north of Springfield on Highway 65, .5 mile west on Highway AA, then north on County Road 197. The lake is 820 acres and suitable for fishing. There is a boat ramp and picnic area. Some visitors use the lake for sailing. The land surrounding the lake is regulated by the city of Springfield.

4 **Lake Springfield** is south of Springfield on Highway 160, 1 mile east on County Road 178, then south on County Road 169 (across the James River to the facilities along the south shore). This is a 318-acre area of fishable water with boat ramps, jetties and picnic areas. **Clay Henshaw Memorial Access** is at the dam of Lake Springfield on County Road 169. This is an 8-acre area with several wheelchair accessible fishing docks. **Southwood Access** is just west of the Highway 65 crossing of Lake Springfield. This 2-acre area also has wheelchair accessible fishing docks. **Tailwaters Access** (2 acres) is on County Road 169, across the James River below the Lake Springfield dam.

5 **Little Sac Woods Conservation Area** is northwest of Springfield on Highway BB (2.6 miles west of Highway 13) and .6 mile south on FR-115. This is a 772-acre area on the Little Sac River, though access to the water is difficult. A 3-mile hiking trail winds through the area.

6 **Phenix Access** is 2.6 miles north of Ash Grove on Highway V, then .3 mile northeast on FR-33. This is a 155-acre area on Clear Creek (5 acres for fishing). Much of the land is old pastures or cropland, and the rest is forestland (35 acres).

7 **Rocky Barrens Conservation Area** is east of Willard on Highway O, then north on County Road 105. This is a 191-acre area.

8 There are signs for **Springfield Conservation Nature Center** on both US-60 and US-65, near their intersection. This is 80 acres with frontage on Lake Springfield. There are six trails and a nature center building with exhibits, auditorium and classroom space. See page 83 for additional information.

Bois D'Arc Conservation Area and the Andy Dalton Shooting Range and Training Center Complex

When the first French trappers explored the Ozarks, they noted that the Indians used a very tough yellow wood for making bows. The French called the tree that provided this wood bois d'arc or "wood of the bow." The English name is Osage orange. Settlers found other uses for the bois d'arc tree and planted it for fences and windbreaks throughout the prairie region of Missouri.

In 1985, the Department purchased land in Greene County, northwest of the town of Bois D'Arc. Because of the name of the nearby down and the numerous Osage orange fence rows on the land, the area was appropriately named Bois D'Arc Conservation Area.

Originally acquired and managed to improve habitat for bobwhite quail, the land has benefited many other wildlife species as well. Visitors may drive through the area on county roads and spot hawks, rabbits, coyotes, doves and sometimes deer.

The area also has one of the best MDC shooting facilities in the state. It has a pistol/rifle range, a skeet and trap range and a 50-seat classroom. Baffles and earth berms help contain the bullets and noise. The range is always supervised by an MDC staff member.

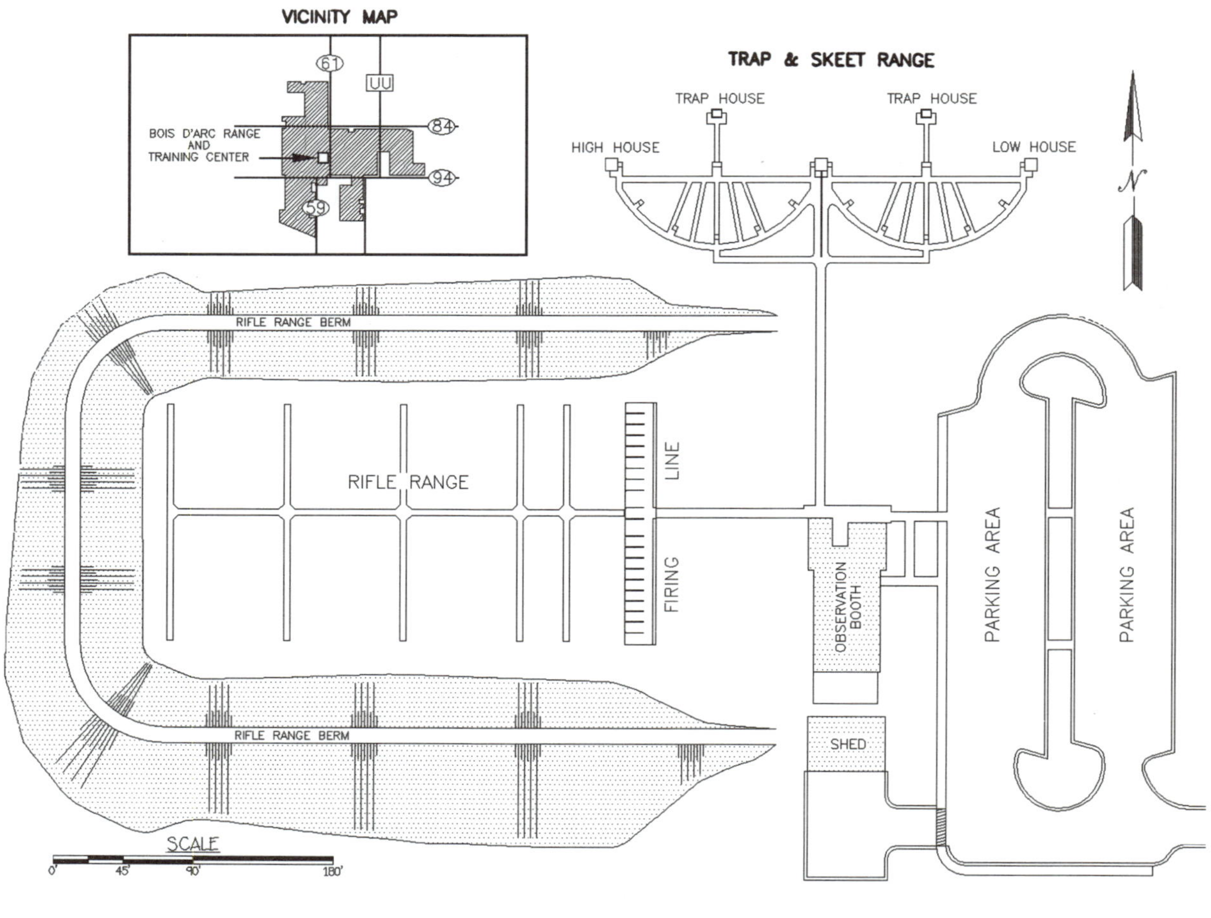

Facilities

The Andy Dalton Shooting Range is a 20-point, covered firing range with target sockets at 15, 25, 50, 75 and 100 yards. Paper targets and target holders provided. Bench rest and seats available. Ammunition is not available on site.

Hours

Open noon to 4:30 p.m. Monday, Friday, Saturday and Sunday; 9 a.m. to 4:30 p.m. Thursdays. Between April 15 and September 5, range remains open to 7:30 p.m. on Thursdays. Range hours may be subject to change.

Group Use

Open 8 a.m. to 4:30 p.m. Tuesday and Wednesday, and 8 a.m. to noon Friday through Sunday. Closed Monday. Group reservations must be made two weeks in advance.

Fees

$2 per hour per shooting booth for the pistol/rifle range. $3 per 25 clay birds for the trap and skeet range. $20 fee for competive shooting for adult groups. There is no charge for competitive or educational shooting groups of youths age 18 and under.

For more information about the Andy Dalton Shooting Range, call 417/742–4361 or 417/895–6880.

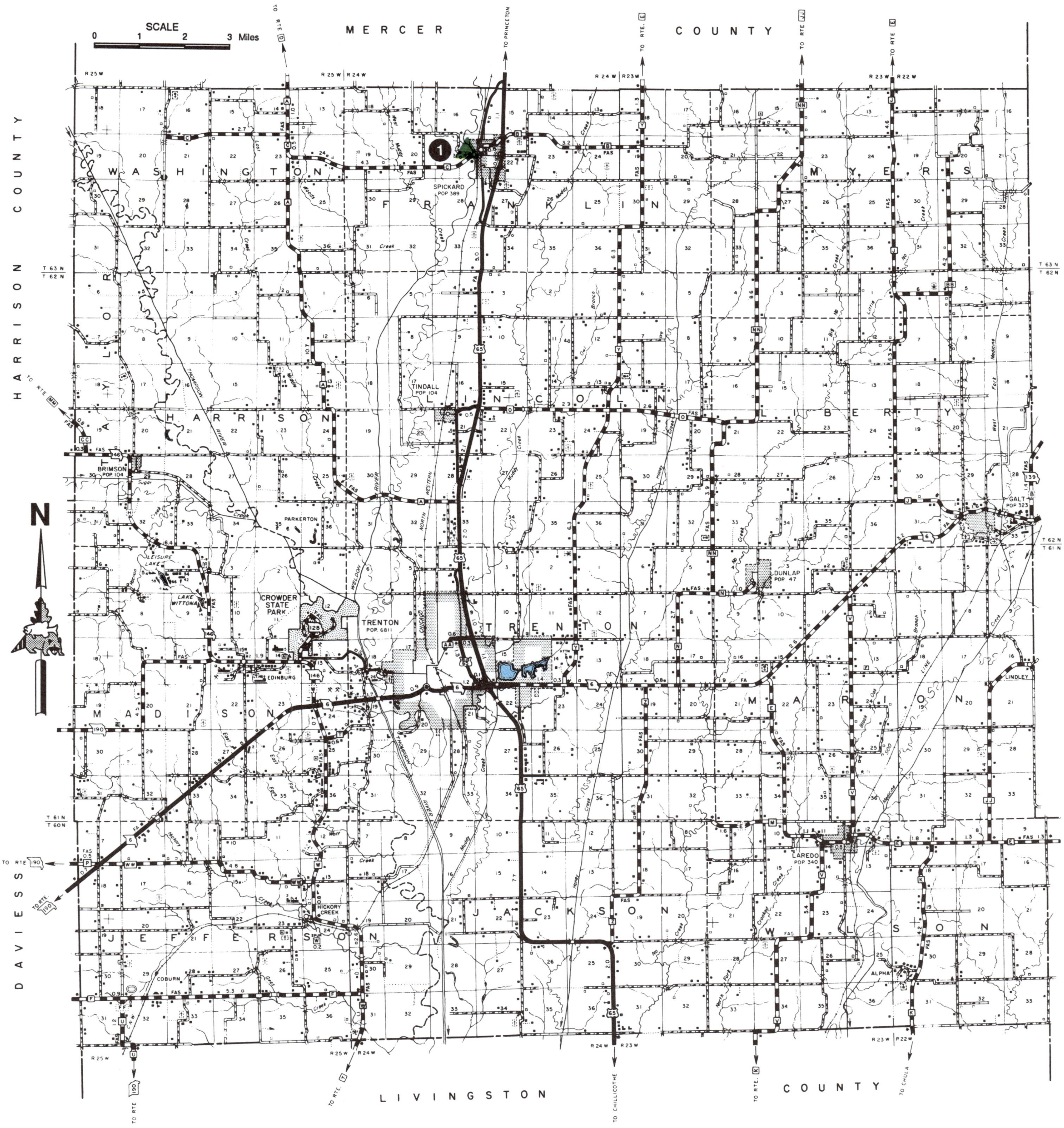

GRUNDY COUNTY

1 **Crusader's Access** is west of the city limits of Spickard, immediately north of Route C. This is a 3-acre access on the Weldon Fork of the North Grand River.

Location Map

Legend

- LAND OWNED BY MDC
- LAND LEASED BY MDC
- LAND OWNED BY MDC LESS THAN 160 ACRES
- LAND LEASED BY MDC LESS THAN 160 ACRES
- CAP ACCESS

HARRISON COUNTY

1 **Bethany City Lakes** consists of two reservoirs. North Bethany City Reservoir is about 3 miles north of the town of Bethany on Highway 69. The reservoir is 72 acres and sits on 71 additional acres of park land. There is one boat ramp. Old Bethany City Reservoir is 1.5 miles north of Bethany on Highway 69. This is a 97-acre area. Most of it is a golf course (managed by a different agency) that lies adjacent to the 16-acre reservoir. There is a wheelchair accessible fishing pad and restrooms.

2 **Grand Trace Conservation Area** is 6 miles northwest of Bethany on Highway F. This 1,497-acre area has 33 ponds (6.7 acres of fishable water) and 4.5 miles of interior field trails suitable for hiking.

3 **Wayne Helton Memorial Wildlife Area** is on Route CC, 3 miles south of Highway 136. This 2,560-acre area contains Helton Prairie Natural Area (36 acres) and 10 miles of field roads open for hiking. There are many ponds providing anglers with 10 acres of fishable water.

4 **Lotts Creek Conservation Area** is .5 mile north on Route HH off Highway 46. This area is composed of 20 acres of forestland.

5 **Pawnee Prairie Conservation Area** is .5 mile south of Hatfield on an unnamed gravel road. This 436-acre area has a 2-mile hiking trail and primitive camping. Three ponds and a stream provide anglers with about 2 acres of fishable water.

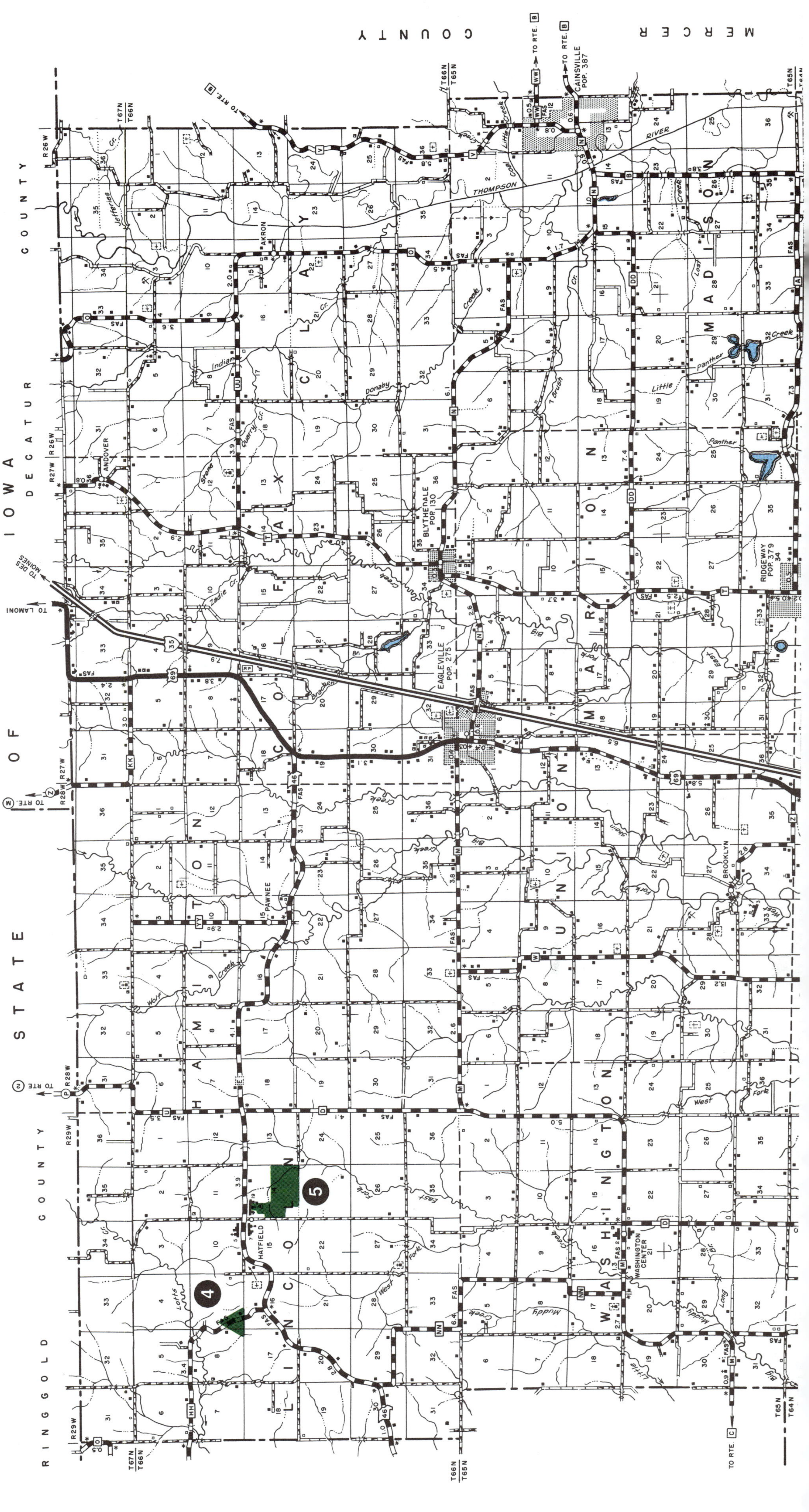

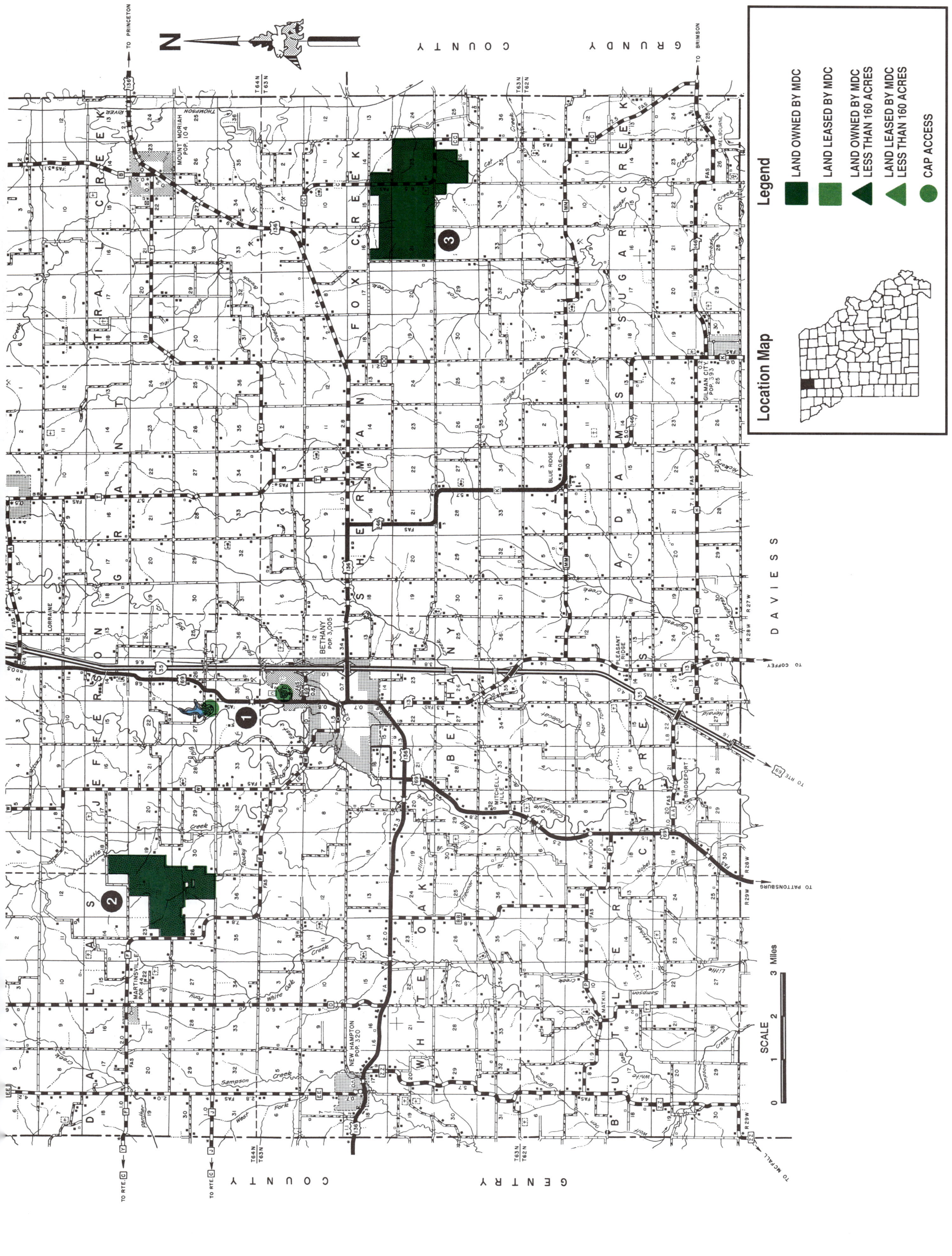

Legend
LAND OWNED BY MDC
LAND LEASED BY MDC
LAND OWNED BY MDC LESS THAN 160 ACRES
LAND LEASED BY MDC LESS THAN 160 ACRES
CAP ACCESS
Location Map
SCALE
0 1 2 3 Miles
GRUNDY COUNTY
DAVIESS
GENTRY COUNTY
TO PRINCETON
TO BRIMSON
TO COFFEY
TO PATTONSBURG
TO MCFALL
BETHANY POP. 3,005
GILMAN CITY POP. 393
MOUNT MORIAH POP. 104
MARTINSVILLE POP. 422
NEW HAMPTON POP. 320
LORRAINE
MELBOURNE
BLUE RIDGE
PLEASANT RIDGE
BRIDGEPORT
MITCHELLVILLE
WILDWOOD
MATKIN

HENRY COUNTY

1 **Chapel View Prairie Conservation Area** is 2.5 miles west of Deepwater on Highway 52, 2 miles south on Highway F, then .5 mile west on County Road SW 1000. This 384-acre area is nearly all native prairie, and the western portion is known for sprouting a diversity of prairie flowers throughout the year. There are five ponds providing 2 acres of water suitable for fishing.

2 **Farrington Park Lake** is 1 mile south on County Line Road off Highway 52 near the town of Windsor. Watch for signs leading to the area. The lake is 15 acres and is surrounded by about 46 acres of land. There is a 1-mile hiking trail and picnic areas.

3 **Connor O. Fewel Wildlife Area** consists of two tracts. The main tract is 4 miles north of Lewis on Route AC. This 320-acre area has three lakes (providing 10 acres of fishable water) and two springs.

4 **Ladue Bottoms Conservation Area** is made up of two tracts. The largest is 1.5 miles east of Ladue on Highway T, then .5 mile on County Road SW 501. This 361-acre area is mainly cropland and contains a .5-mile stretch of Deepwater Creek.

5 **Montrose Conservation Area** is 12 miles west of Clinton on Highway 18, then 4 miles south on Highway RA. The largest feature on this 2,750-acre area is Montrose Lake. The lake is about 1,600 acres and has three boat ramps. A 3-mile hiking trail winds through part of the area.

6 **Haysler A. Poague Conservation Area** is on Highway 7, 4 miles west of Clinton. There are 14 lakes on this 879-acre area, providing anglers with 80 acres of fishable water.

7 **Harry S Truman Reservoir Wildlife Management Lands, Golden Valley Shooting Range**–see the Benton County map, page 20.

8 **Urich Access** is .25 mile south of the town of Urich on Highway K. This is 4.6 acres on the Grand River.

9 To reach **Urich Conservation Area**, go .75 mile east off Highway 7 on County Road NW 700, then 1.25 miles north on County Road NW 1101. Big Creek flows through this 480-acre area for .25 mile.

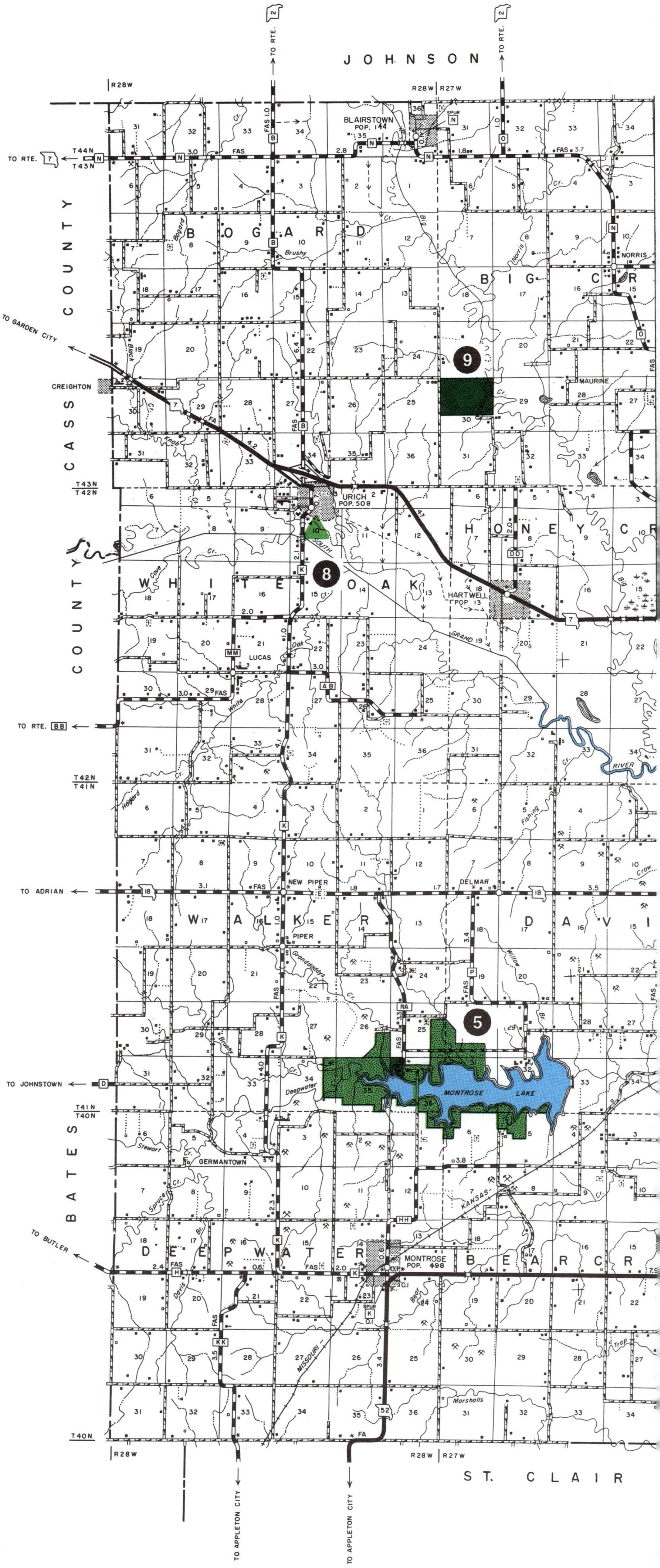

Location Map

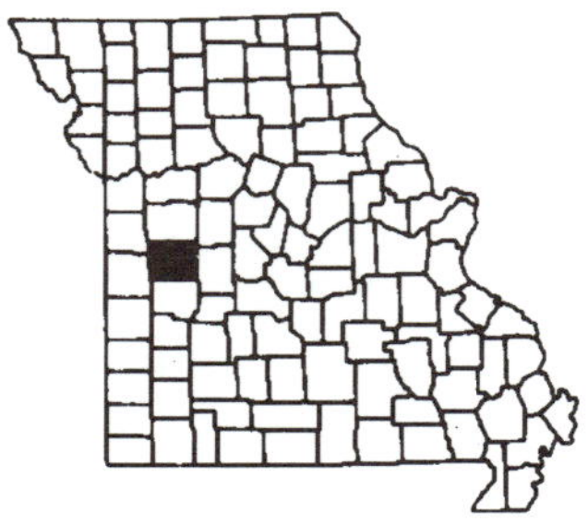

Legend

- LAND OWNED BY MDC
- LAND LEASED BY MDC
- LAND OWNED BY MDC LESS THAN 160 ACRES
- LAND LEASED BY MDC LESS THAN 160 ACRES
- CAP ACCESS

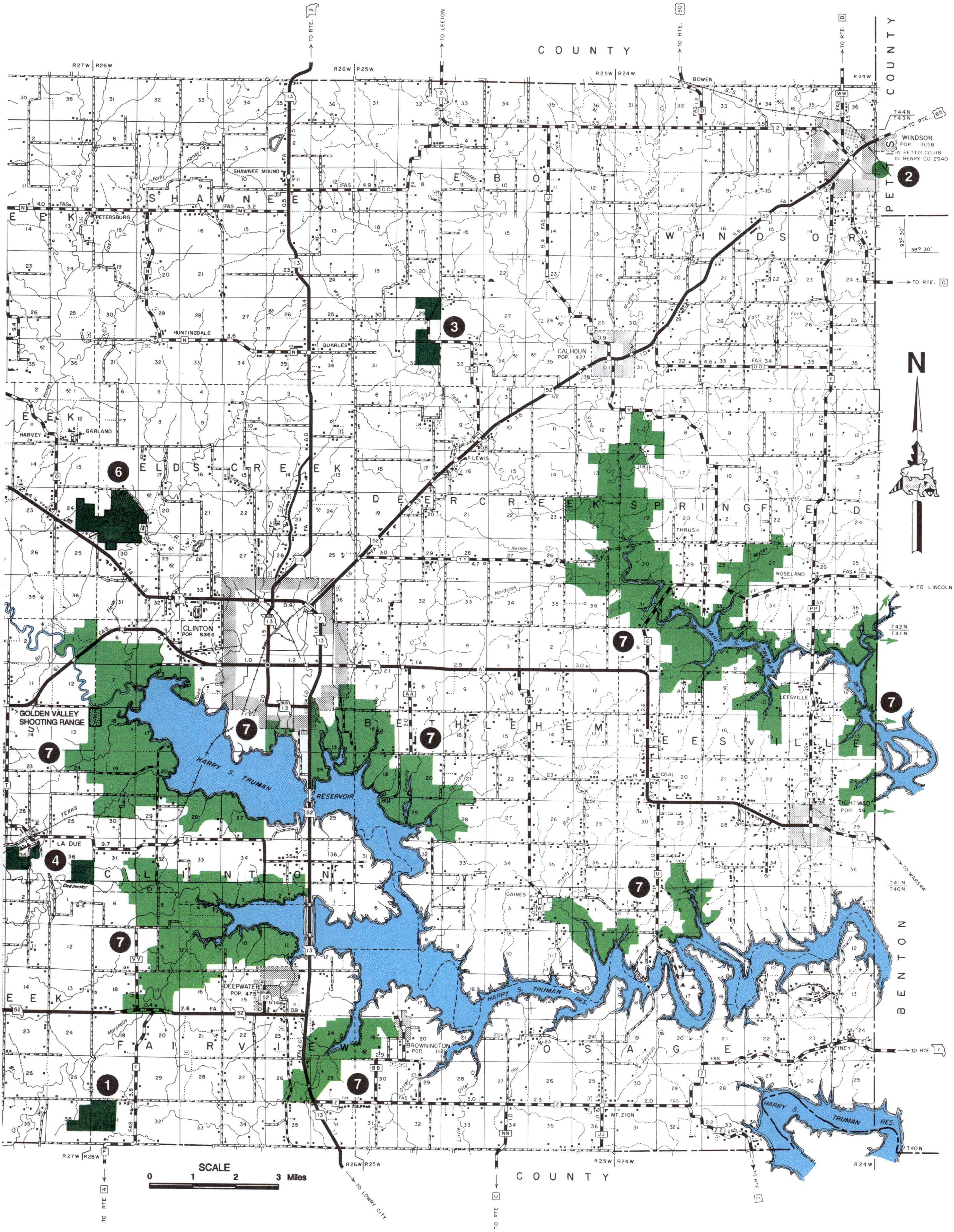
COUNTY
PETTIS COUNTY
BENTON COUNTY
N
SCALE
0 1 2 3 Miles
TO RTE. 2
TO LEETON
TO RTE. 50
TO RTE. D
TO RTE. 65
TO RTE. C
TO LINCOLN
TO WARSAW
TO RTE. 7
TO RTE. A
TO LOWRY CITY
TO RTE. C
R27W R26W
R26W R25W
R25W R24W
R24W
T44N
T43N
T42N
T41N
T40N
93° 30'
38° 30'
WINDSOR
POP. 3058
IN PETTIS CO. 118
IN HENRY CO. 2940
BOWEN
SHAWNEE MOUND
SHAWNEE
TEBO
WINDSOR
PETERSBURG
HUNTINGDALE
QUARLES
CALHOUN
POP. 427
HARVEY
GARLAND
LEWIS
FIELDS CREEK
DEER CREEK
SPRINGFIELD
THRUSH
ROSELAND
CLINTON
POP. 8366
GOLDEN VALLEY
SHOOTING RANGE
HARRY S. TRUMAN
RESERVOIR
BETHLEHEM
LEESVILLE
COAL
TIGHTWAD
POP. 56
TEXAS
LA DUE
CLINTON
GAINES
DEEPWATER
POP. 475
FAIRVIEW
BROWNINGTON
POP. 112
OSAGE
FINEY
MT. ZION
HARRY S. TRUMAN RES.
1
2
3
4
6
7

HICKORY COUNTY

1 **Cross Timbers Access** is 2 miles north of the junction of highways 54 and B, 5.5 miles east on Highway Y (until the pavement ends), then .75 south on the gravel road. This 44-acre area has a boat ramp on Pomme de Terre River.

2 **Hermitage Access** is .25 mile south of the town of Hermitage on Highway 254. This is a 2-acre area with 200 yards of frontage on Pomme de Terre River.

3 To reach **Mule Shoe Conservation Area,** go 2.5 miles east of Preston on Highway 54, turn 3.25 miles north on Highway F, then east on County Road F. The area lies just beyond the low water bridge over the Little Niangua River. A 2.7-mile stretch of the Little Niangua flows through the area, and there are four hiking trails (.75 to 3 miles in length) over old road and dirt surfaces.

4 **John F. Murphy Memorial State Forest and Wildlife Area** is on the north side of Highway 54, .75 mile east of Wheatland. This 325-acre area is nearly all forestland (295 acres) and has four caves and two springs. Five trails wind through the area.

5 There are several tracts of Conservation Department land around **Pomme de Terre Reservoir** (4,019 acres). The largest accessible tract is 3 miles east of Elkton on Route H, then 1 mile north on an unnamed gravel road. The Corps of Engineers manages adjacent land that offers a variety of facilities to boaters, campers and anglers.

6 **Harry S Truman Reservoir Wildlife Management Lands**–see the Benton County map, page 20.

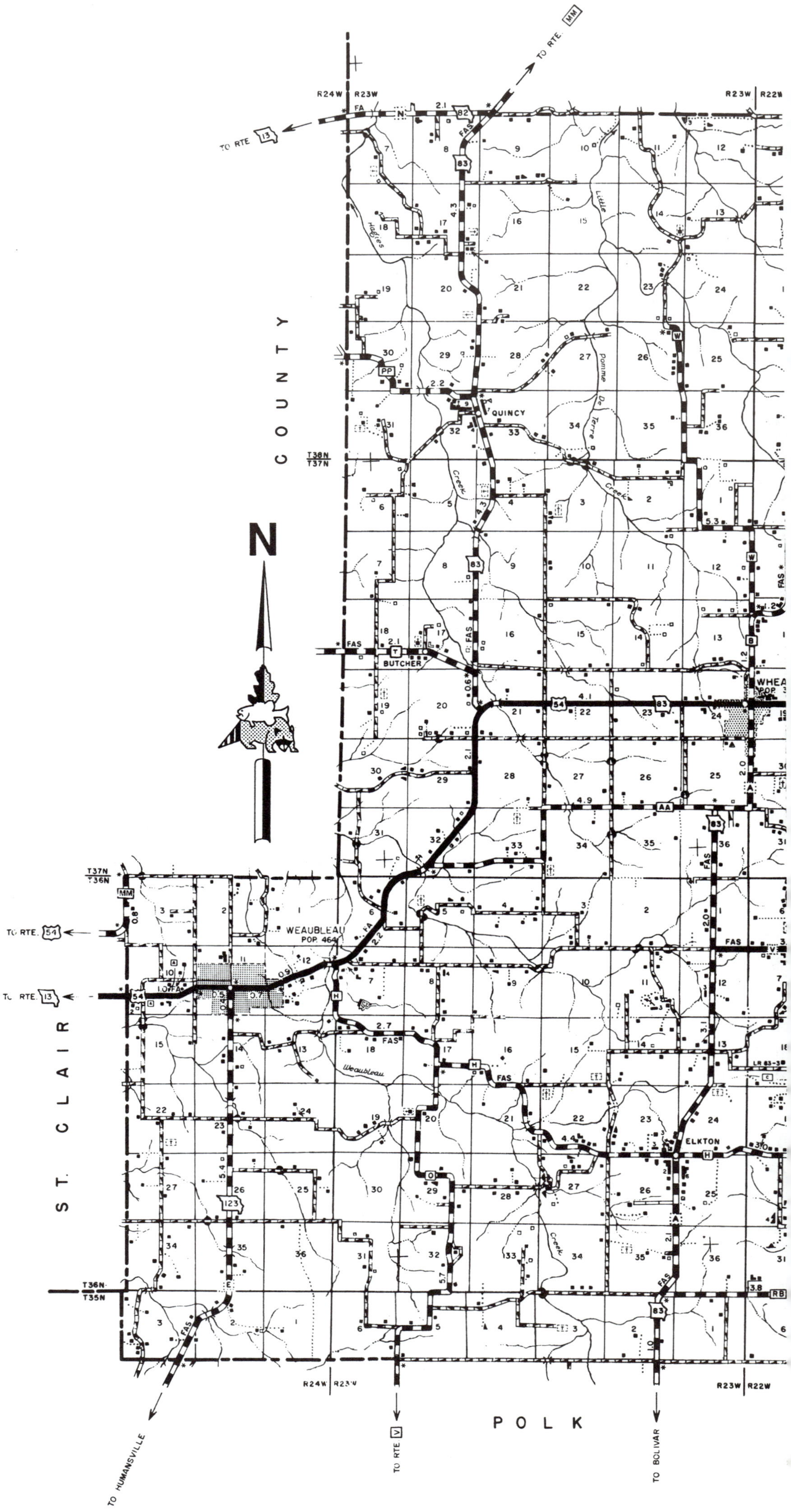

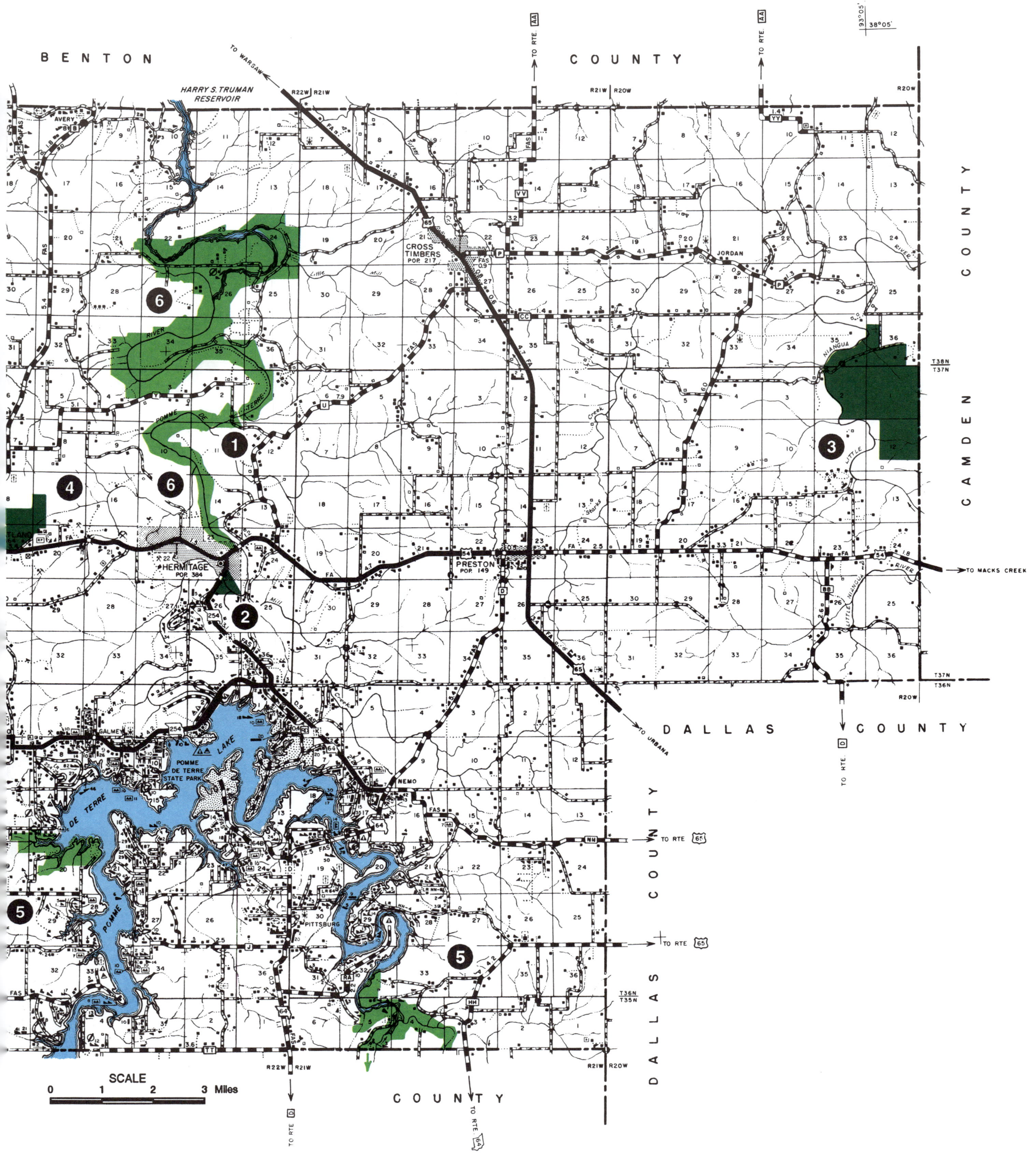

Location Map

Legend

- LAND OWNED BY MDC
- LAND LEASED BY MDC
- LAND OWNED BY MDC LESS THAN 160 ACRES
- LAND LEASED BY MDC LESS THAN 160 ACRES
- CAP ACCESS

HOLT COUNTY

1 **Bob Brown Conservation Area** is 2.5 miles west of Forest City on Old Highway 111. This 3,302-acre area contains portions of the Missouri River, Squaw Creek, Cannon Creek and Kimsey Creek. Together with two ponds, there are 13 acres of fishable water. There are many field roads open for hiking (totaling about 12 miles).

2 **Jamerson C. McCormack Conservation Area** is on Highway 159 (west off I-29, exit 79) and is adjacent to Squaw Creek National Wildlife Refuge. This 225-acre area contains 67 acres of loess hill prairie natural area and a 1.2-mile hiking trail.

3 **Maitland Access** is near the town of Maitland where Highway A crosses the Nodaway River. This is a 1-acre access and has no boat ramp.

4 **Monkey Mountain Conservation Area** is 3 miles west of I-29 on Highway 59, 5 miles south on Highway U, then 1.5 miles east on a gravel road. This 787-acre area is on the Nodaway River.

5 **Nodaway Valley Conservation Area**–see the Andrew County map, page 8.

6 **Payne Landing Access** is 2.25 miles west of Oregon on Highway 111, 2.5 miles south on Highway T, then 2.5 miles right on a gravel road (watch for signs). This is a 3-acre area on the Missouri River with a boat ramp.

7 To reach **Riverbreaks Conservation Area** (two tracts), take U.S. 59 to Highway O. Go 3.25 miles south on Highway O and watch the west side of the road for the cantilever sign to the main area. Interesting features on this 2,307-acre area include a mesic upland forest along Whales Creek and a cemetery with 12 headstones dating from the 1860s. The area has a firearms range accessible to wheelchairs.

8 **H. F. Thurnau Conservation Area** is on Highway 111, west of Craig. A stretch of the Missouri River flows through this 366-acre forested area, and there are two hiking trails.

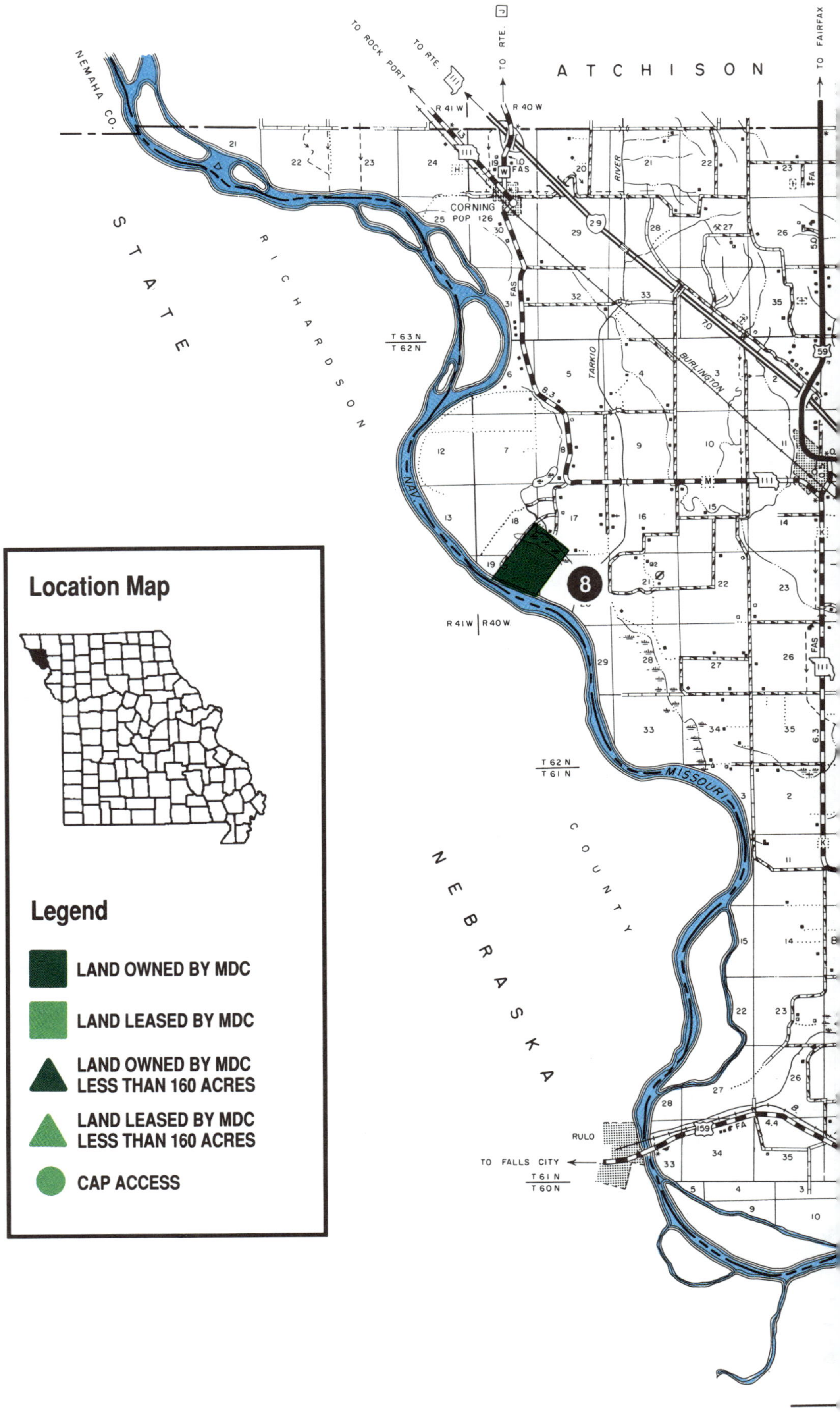

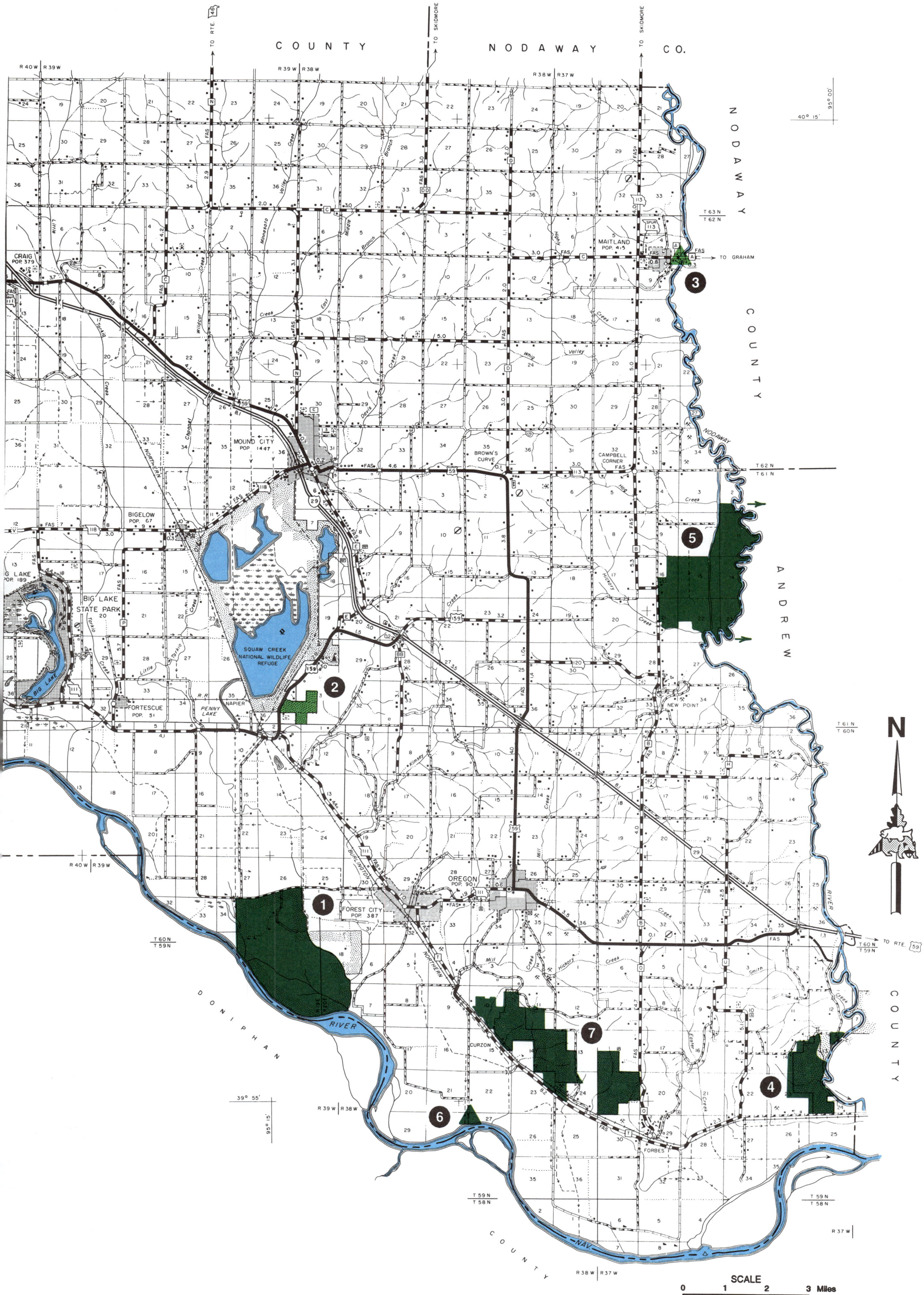

NODAWAY COUNTY
NODAWAY CO.
ANDREW COUNTY
DONIPHAN COUNTY
CRAIG POP. 379
MAITLAND POP. 415
MOUND CITY POP. 1447
BIGELOW POP. 67
BIG LAKE STATE PARK
SQUAW CREEK NATIONAL WILDLIFE REFUGE
FORTESCUE POP. 51
NAPIER
NEW POINT
BROWN'S CURVE
CAMPBELL CORNER
OREGON POP. 901
FOREST CITY POP. 387
CURZON
FORBES
NODAWAY RIVER
RIVER
TO GRAHAM
TO SKIDMORE
TO RTE. 46
TO RTE. 59
SCALE
0 1 2 3 Miles
N

HOWARD COUNTY

1 **Rudolf Bennitt Conservation Area** is 19 miles north of Columbia on Highway 63, 5 miles west on Highway F, and 2.75 miles north on Highway T. This area is 3,514 acres. It has a 12-mile horseback riding trail, a 7-mile hiking trail maintained by the Moniteau Boy Scouts and a shooting range. Natural features of interest include a dry mesic upland area. There are also heart-leaved plantains.

2 **Davisdale Conservation Area** is 15 miles west of Columbia on Highway 40. This 2,691-acre area contains Missouri River bottomland and loess hills, a portion of Salt Creek and many ponds. Wade Eagle Trail is 6 miles long over a natural surface.

3 **Fayette City Lakes**–D.C. Rogers Lake (185 acres) and Fayette City Lake #2 (80 acres) –are west of Fayette on Highway E, 1 mile and 3.5 miles west, respectively. Only boat motors 10 horsepower or less are permitted on the lakes, and there is a fishing jetty accessible to wheelchairs. There is a 7-mile primitive hiking trail, archery and firearms range and a picnic area.

4 **Franklin Island Conservation Area** is 2 miles east of New Franklin on Highway 40. This is a 1,616-acre area on the Missouri River and Bonne Femme Creek.

5 **Hungry Mother Conservation Area** is south of Higbee on Route A, then 2.5 miles west on County Road 127. This 274-acre area is mainly forestland.

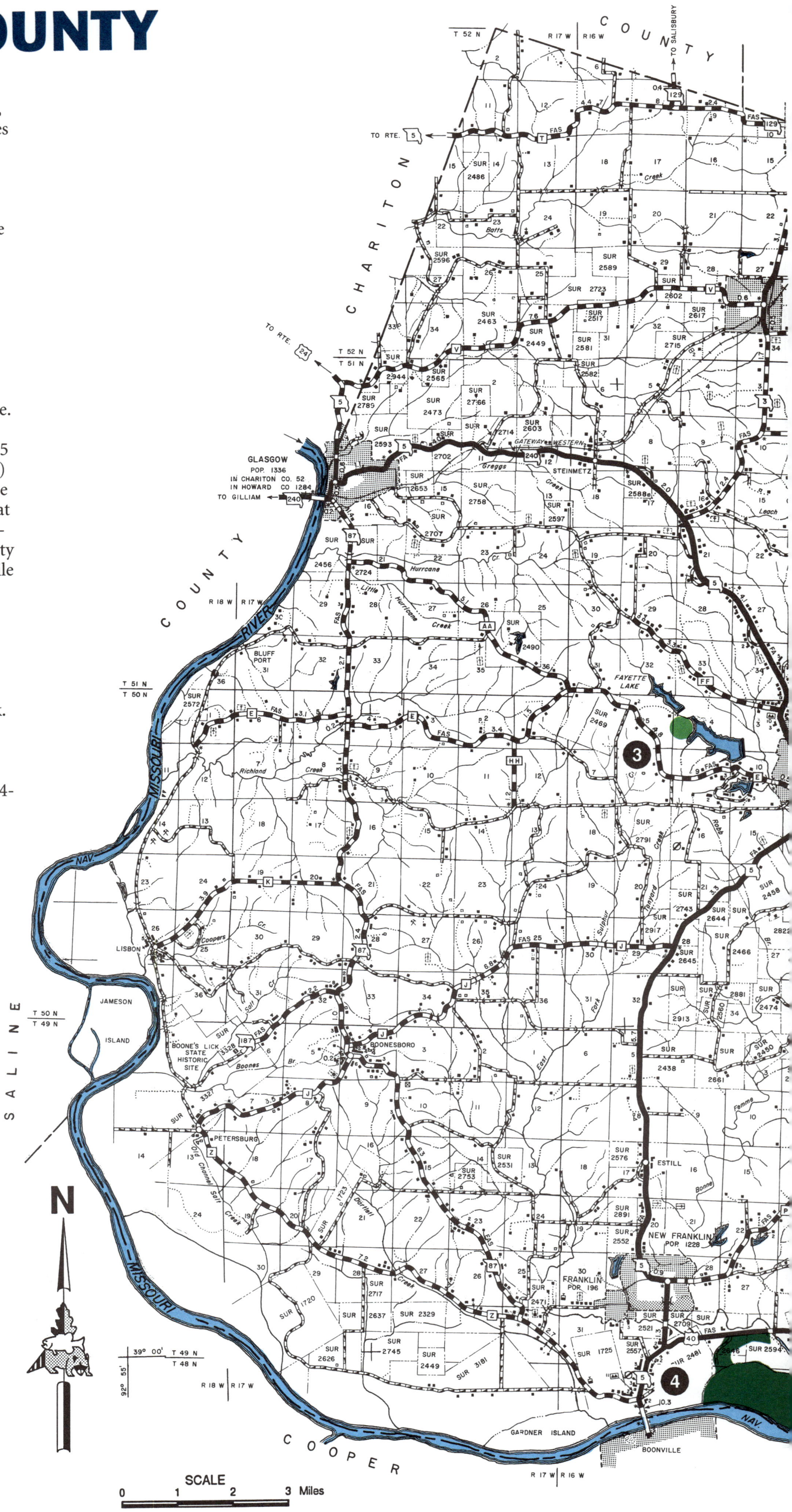

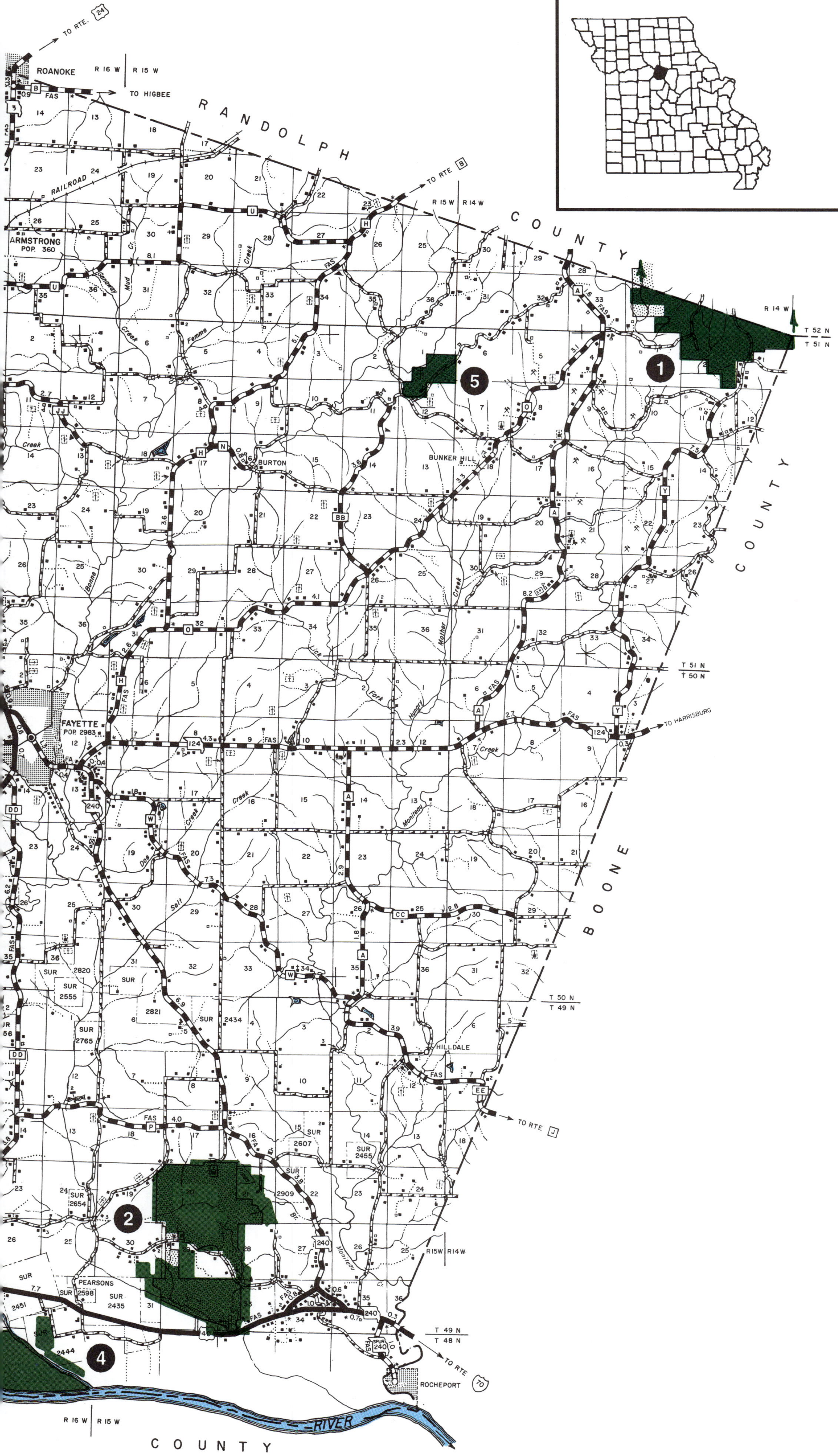

Location Map

Legend

- LAND OWNED BY MDC
- LAND LEASED BY MDC
- LAND OWNED BY MDC LESS THAN 160 ACRES
- LAND LEASED BY MDC LESS THAN 160 ACRES
- CAP ACCESS

HOWELL COUNTY

1 **Dean Davis Conservation Area** is 10 miles north of West Plains on Highway 63 and .5 mile east on County Road 2420. This is a 177-acre area with a small fishing pond and a 1-mile hiking trail.

2 To reach **Mountain View Towersite**, go 1.5 miles east of Mountain View on Highway 60, turn .5 mile north on County Road 2990, then go .25 mile on County Road 3400. This area is 30 acres with a fire lookout tower.

3 **Sims Valley Community Lake** is 7 miles east of Willow Springs on Highway 60, then 2.5 miles left on Highway RA. The land surrounding the 41-acre lake is mainly forestland and old pastures (106 acres total).

4 **Tingler Lake Conservation Area** is south of West Plains. Take Highway 17 south to County Road 910 (about 6 miles), then turn south .3 mile on County Road 811 to the area. There is a sinkhole pond, wet mesic prairie, and part of the South Fork of Spring River on this 240-acre area. There are also two 2-mile hiking trails over mowed grass.

5 To reach **Archie and Gracie VanDerhoef Memorial State Forest**, go 3 miles west of West Plains on Highway 60. Turn 8 miles south on Highway JJ, .5 mile east on County Road 9620, and .5 mile south on County Road 8110. The 140-acre area is on this road. It contains .5-mile stretch of the South Fork of Spring River and many small springs and seeps. There is also a .5-mile hiking trail with an observation deck.

6 **West Plains Towersite** is 5 miles west of the city of West Plains on Highway 160 and .25 mile east on County Road 6920. The fire lookout tower sits on 1.9 acres. The area is closed to vehicles when the tower is not occupied.

7 **White Ranch Conservation Area** consists of three tracts. The main tract is 7.5 miles south of West Plains on Highway 17, then 5 miles south (right) on County Road 8390. Parts of the west and south forks of Spring River and Hunt Branch lie in this 6,614-acre area. The area contains a shooting range and picnic area.

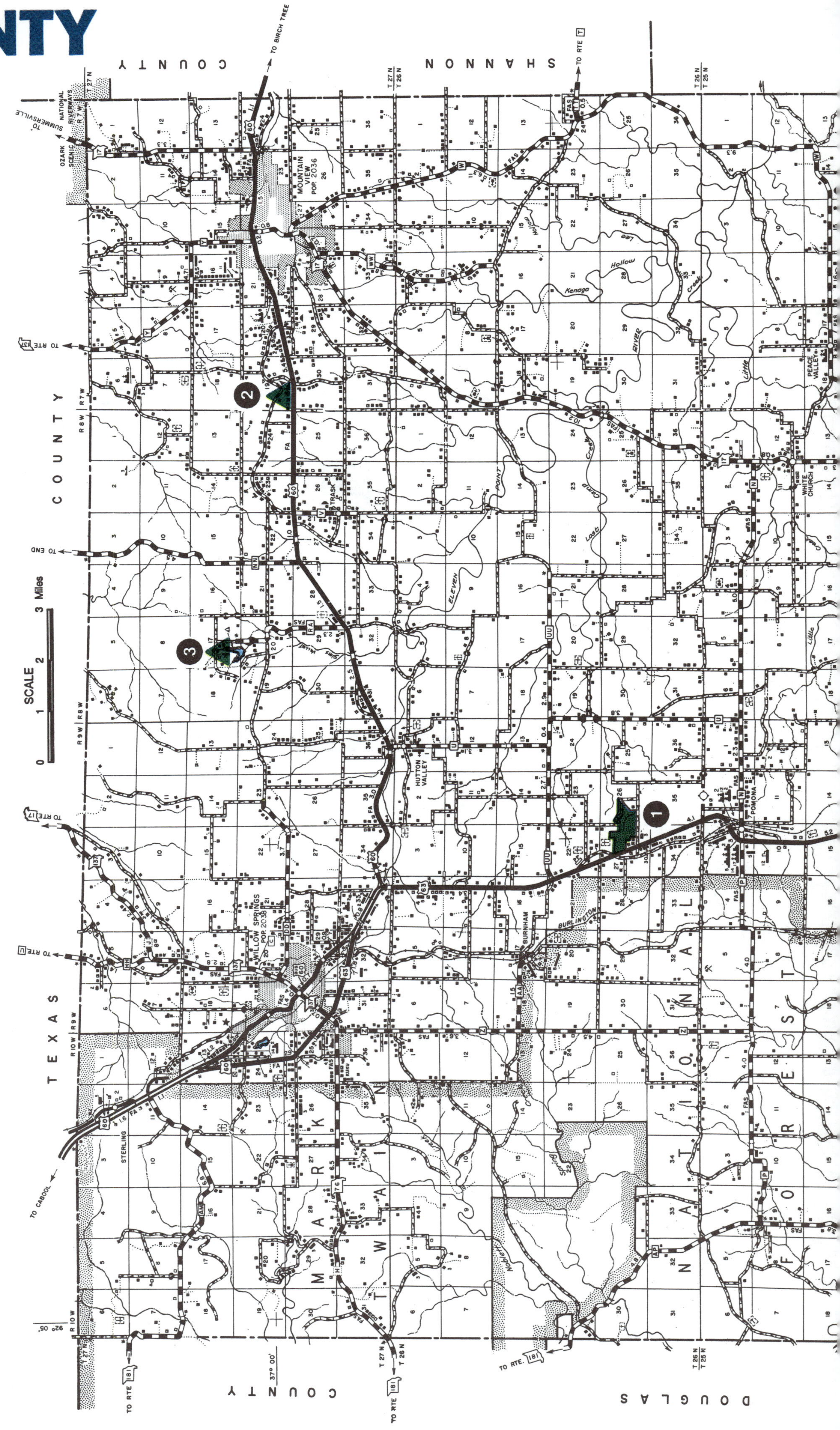

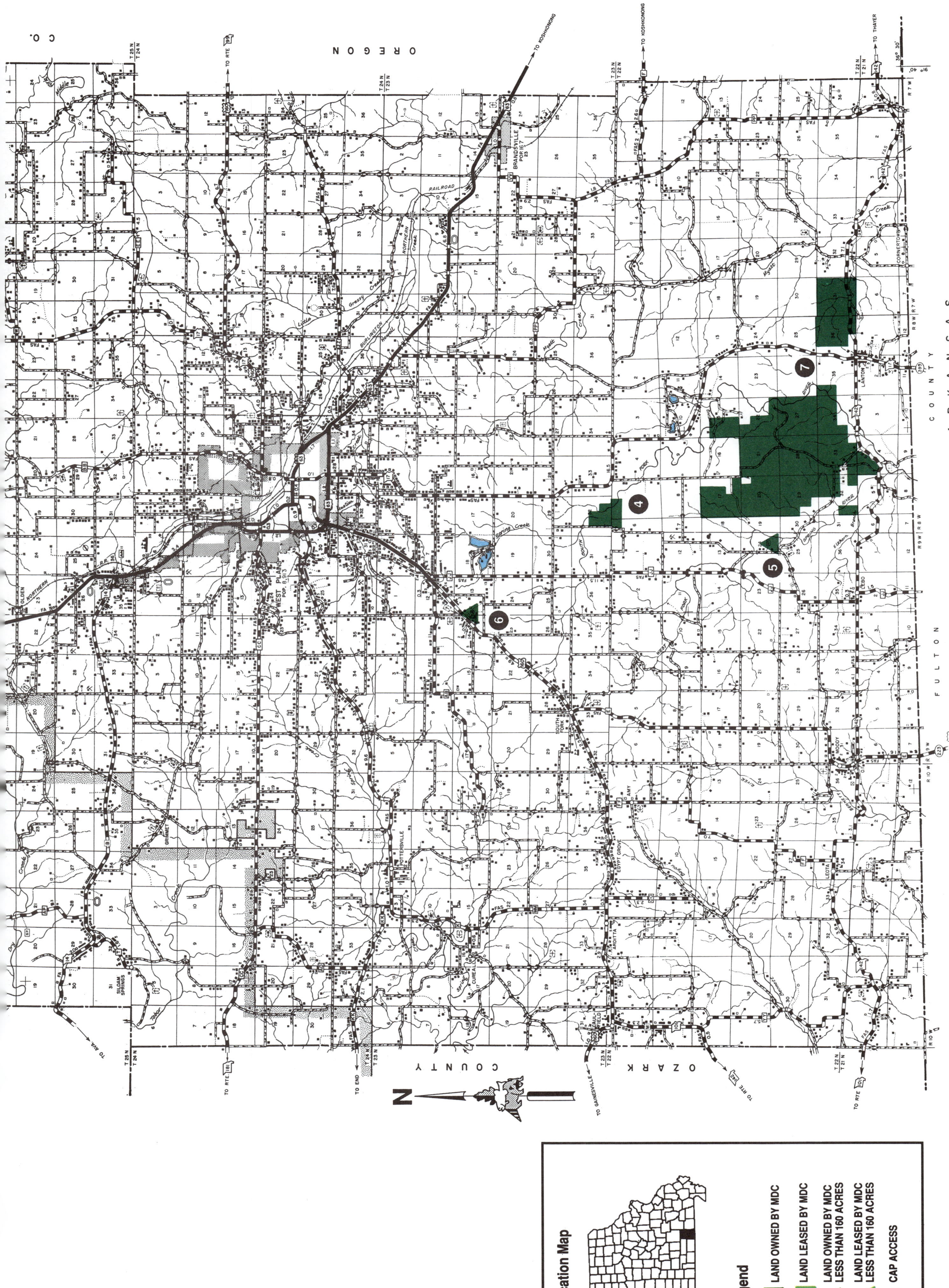

OREGON
CO.
COUNTY
OZARK
STATE OF ARKANSAS
FULTON COUNTY
WEST PLAINS
BRANDSVILLE
TO KOSHKONONG
TO THAYER
TO AVA
TO GAINESVILLE
Location Map
Legend
LAND OWNED BY MDC
LAND LEASED BY MDC
LAND OWNED BY MDC LESS THAN 160 ACRES
LAND LEASED BY MDC LESS THAN 160 ACRES
CAP ACCESS

IRON COUNTY

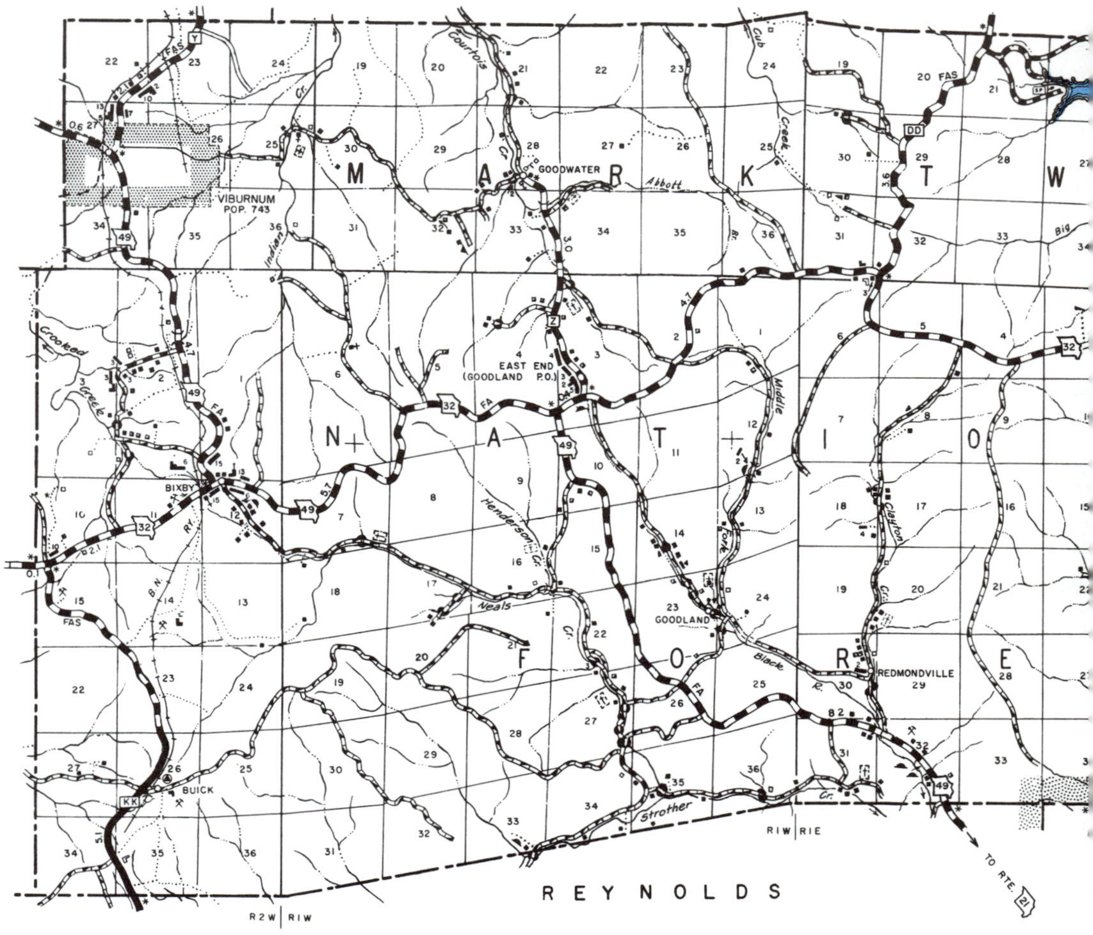

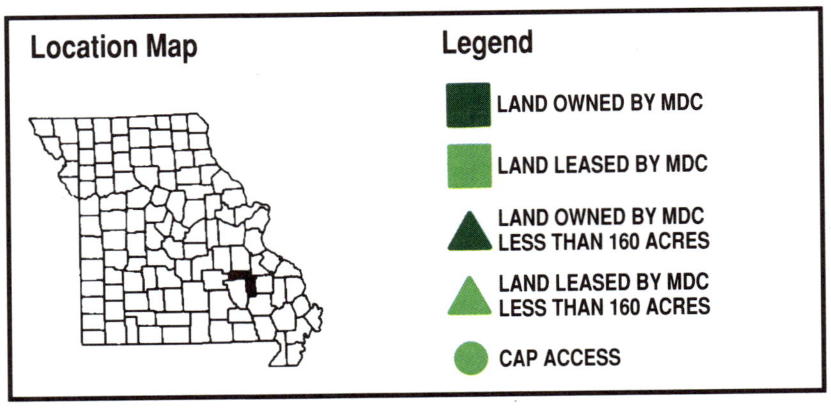

1 The entrance to **Bismarck Lake Conservation Area** is in St. Francois County, south of the town of Bismarck on Highway N, then right on Grider Road. This 1,188-acre area is off Grider Road on Lakeview Road. Bismarck Lake has a boat ramp and wheelchair accessible fishing jetty. The area is also at the head of the St. Francis River.

2 **Buford Mountain Conservation Area** is 8 miles north of Ironton on Highway 21, and 1 mile north on Highway U. This is a 3,743-acre forest with a 6-mile hiking trail.

3 To reach **Funk Memorial State Forest and Wildlife Area** (180 acres), take Highway 49 to Vulcan, then take Highway BB 4 miles to an unnamed gravel road. Watch for signs. This area borders part of Riverside Conservation Area.

4 **Graves Mountain Conservation Area** consists of several tracts in Iron and Wayne Counties. The main tract is in Wayne County, about 3 miles north of Sam A. Baker State Park on Highway 143. This is a 3,678-acre area composed of forestland (3,521 acres) and glade (157 acres).

5 **Ketcherside Mountain Conservation Area** is 4 miles south of Ironton on Highway 21, and 3 miles down Highway CC. This 3,356-acre forest contains Royal Gorge Natural Area (80 acres), that has rhyolite glades and pine-oaks. A picnic area is available to visitors, and a 3.5-mile stretch of the Ozark Trail runs through the area.

6 **Riverside Conservation Area** consists of five tracts in Iron, Reynolds and Wayne Counties. The main tract is in Wayne County. Go 6 miles north of Piedmont on Highway 49, turn 1.8 miles left on Highway CC, then go .5 mile north on Gravel Road 462. The forest boundary is at the top of the hill. The McKenzie Creek headwaters are on this 2,528-acre area as well as an igneous glade complex. There are also five forested sinkholes.

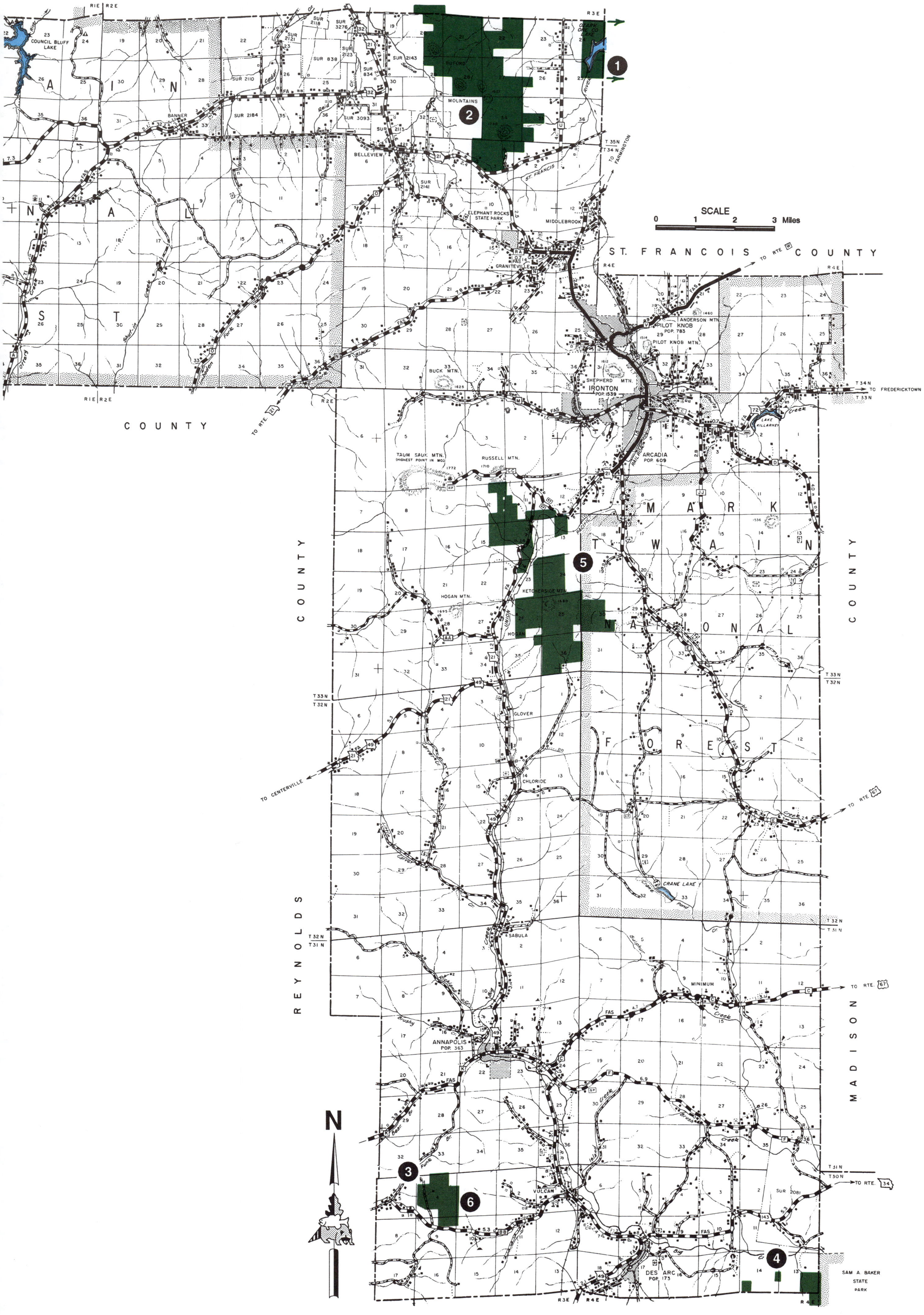
COUNCIL BLUFF LAKE
BANNER
BUFORD
MOUNTAINS
BELLEVIEW
ELEPHANT ROCKS STATE PARK
MIDDLEBROOK
GRANITEVILLE
TO FARMINGTON
SCALE
0 1 2 3 Miles
ST. FRANCOIS COUNTY
TO RTE. W
ANDERSON MTN.
PILOT KNOB POP. 783
PILOT KNOB MTN.
SHEPHERD MTN.
IRONTON POP. 1539
BUCK MTN.
TO FREDERICKTOWN
LAKE KILLARNEY
TO RTE. 21
COUNTY
TAUM SAUK MTN. (HIGHEST POINT IN MO.)
RUSSELL MTN.
ARCADIA POP. 609
MARK TWAIN NATIONAL FOREST
HOGAN MTN.
KETCHERSIDE MTN.
HOGAN
GLOVER
TO CENTERVILLE
CHLORIDE
TO RTE. 67
CRANE LAKE
SABULA
REYNOLDS COUNTY
MINIMUM
MADISON COUNTY
ANNAPOLIS POP. 363
VULCAN
TO RTE. 34
DES ARC POP. 173
SAM A. BAKER STATE PARK
N

JACKSON COUNTY

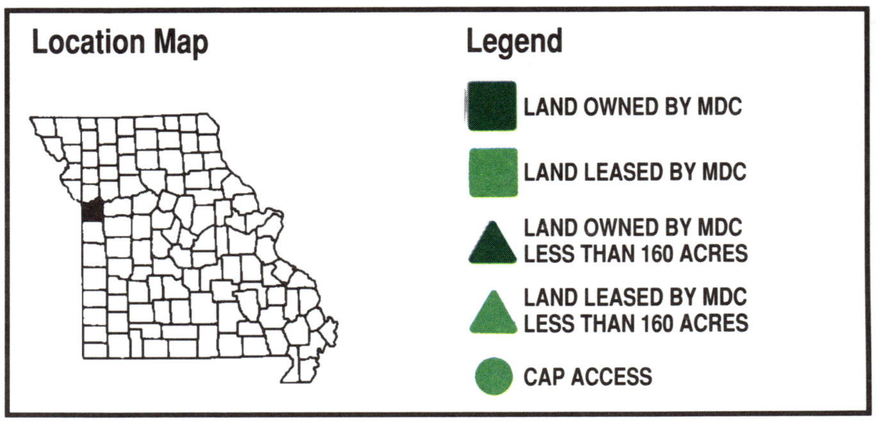

1 To reach **Alex George Lake**, go south on Highway 71 (off I-435), turn west on Red Bridge Road, then go north on Blue River Road. This 8-acre lake is about 1 mile down Blue River Road on the west side. The lake is within the Blue River Parkway (managed by Jackson Co. Parks & Recreation) and is stocked with trout in Nov., Feb. and Mar.

2 **Bergan Lake** (4 acres) is 5 miles north of Highway 50 on Buckner-Tarsney Road, east on Arnold Road, then .1 mile north on Brown Road.

3 To reach **Blue River Parkway**, take 1-435 west to Holmes Road, turn 3.5 miles south to Blue Ridge Boulevard, then go .25 mile east. There is a picnic area and boat ramp at this 5-acre area.

4 To reach the 720-acre **Blue Springs Lake** (Upper Lake Access), go south of I-70 on M-291. Take the Highway 40 exit 3.5 miles east to Woods Chapel Road. A county boating permit is required.

5 **Bowlin Road Lake** (4 acres) is south of I-70 on M-291, .3 mile east on Bowlin Road, and left on the first campground road. Camping facilities are administered by Jackson Co. Parks & Recreation.

6 **Jim Bridger Urban Conservation Area** is 3.5 miles south of the junction of highways 40 and 7 on Highway 7, 1.2 miles west on Cowherd Road, then .5 mile north on Jasper Bell Road. Use the entrance to the Kemper Outdoor Education Center. This 320-acre area (100 acres forestland, 220 acres pasture) is used in conjunction with educational activities at the Kemper Outdoor Education Center.

7 **Burr Oak Woods Conservation Area** is 1.1 miles north of I-70 on Highway 7, then west on Park Road. Burr Oak Creek flows through this 1,071-acre area for 2.4 miles, and there are 15 ponds. Natural features include Bethany Falls limestone exposures and Burr Oak Woods Natural Area–an old growth upland forest. There is a nature center with exhibits and interpretative programs. See page 104 for additional information.

8 **Lake Jacomo** is on Woods Chapel Road, 1.5 miles east of M-291. This 970-acre area contains hiking trails and picnic areas as well as the Kemper Outdoor Education Center (administered by Jackson County).

9 **Lone Jack Lake Conservation Area** is east of Kansas City on Highway 50, .25 mile north on Buckner-Tarsney Road, 1 mile east on Lone Jack-Lee's Summit Road, and .75 mile north on Brown Road. Watch for the area sign on the right.

10 **Prairie Lee Lake** is east of M-291 on Colbern Road, then 1 mile south on Blackwell Road to Scruggs Road. The 150-acre lake is .8 mile west on Scruggs Road. There is a boat ramp and a small fishing dock.

11 **James A. Reed Memorial Wildlife Area** is 1 mile east of Lee's Summit on Ranson Road and 1.5 miles south of Highway 50. The 2,602-acre area is known for its 12 lakes and 23 ponds (252 acres for fishing). Natural features include a great blue heron rookery and an old-growth forest stand. Activities in the area include a self-guided auto tour, horseback riding and a 3-mile hiking trail.

12 **Saeger Woods Conservation Area** is adjacent to Jerry Smith Farm Park. Go north from 139th Street on Prospect Road. Kansas City Parks and Recreation manages the park.

13 **Scherer Lake** is reached by traveling south of I-470 on Raytown Road, and then east on Scherer Road. The 15-acre lake is beyond two bridges over neighboring Longview Lake on the north side of the road.

14 **Tarsney Lake** (17 acres) and **Wood Lake** (8 acres) are 4 miles south of I-70 on Highway BB, then east on Tarsney Road. There is a boat ramp and fishing jetty.

15 **Wyatt Lake** is 2 miles north on Highway BB (from I-70 at Grain Valley) to Pink Hill Road. Turn 1 mile west on Pink Hill to Dillingham Road. Go 1 mile north on Dillingham to Argo Road, and take Argo a short distance west to the 3-acre lake. The area is available for organized group camping, and arrangements may be made with Jackson Co. Parks & Recreation.

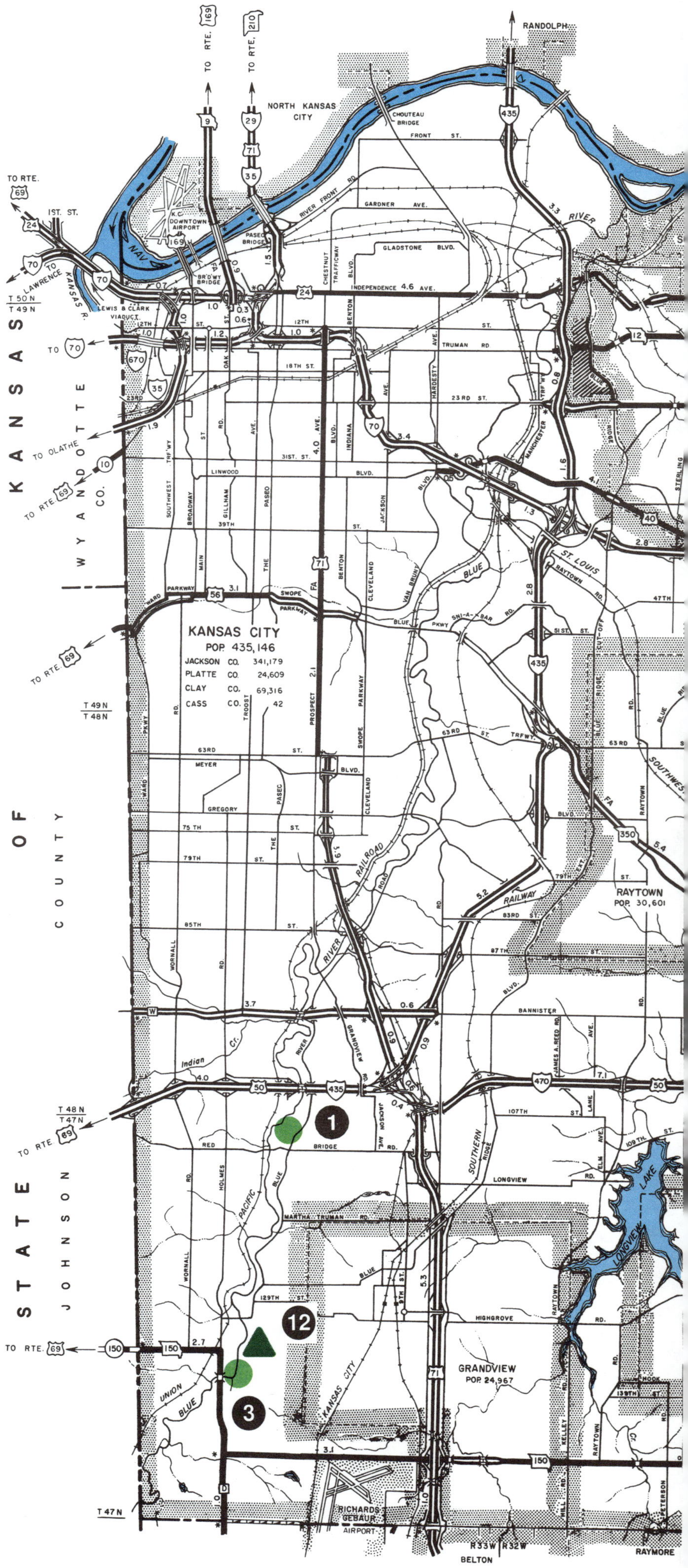

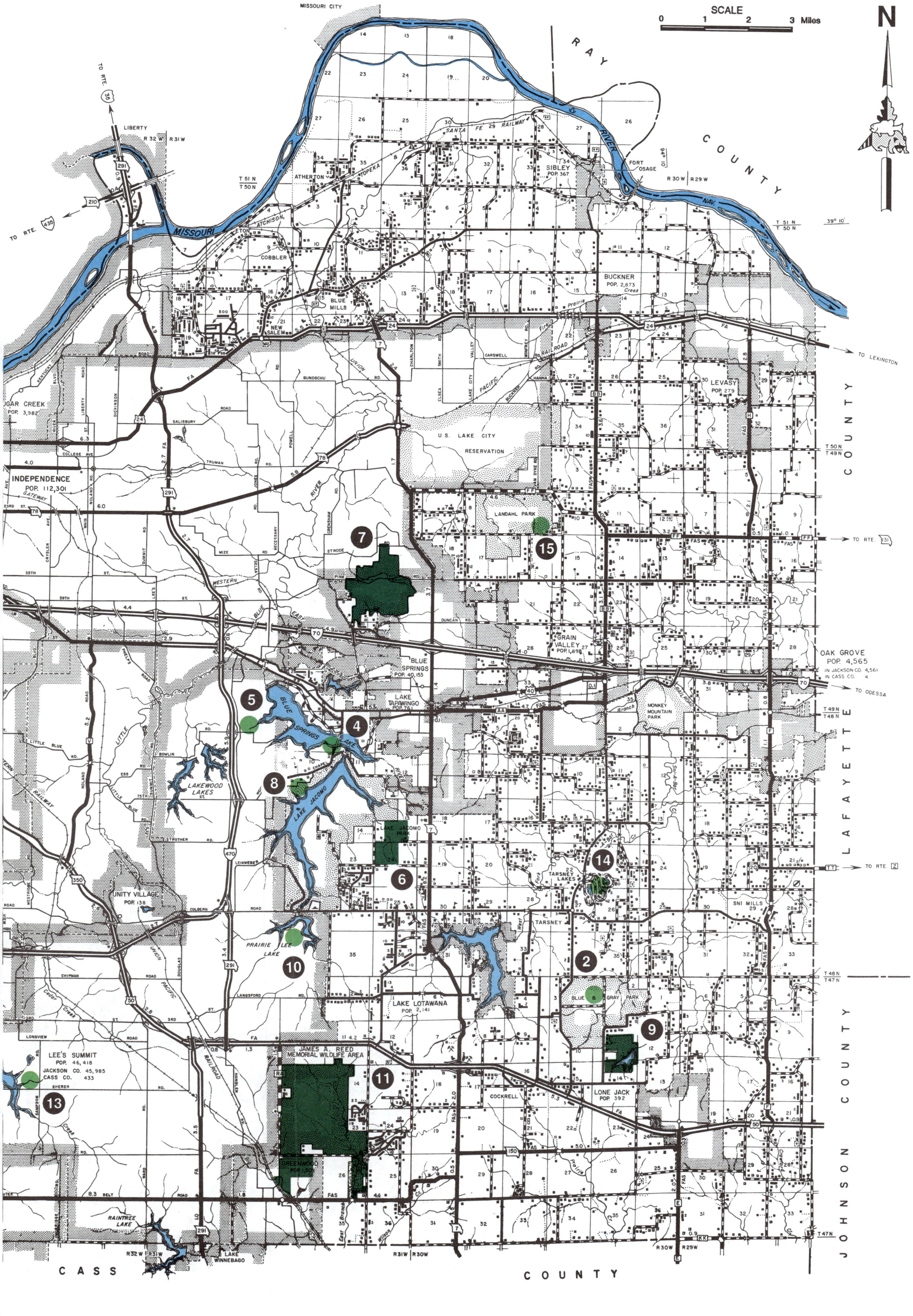
SCALE
0 1 2 3 Miles
N
MISSOURI CITY
RAY COUNTY
MISSOURI RIVER
LIBERTY
ATHERTON
SIBLEY
POP. 367
FORT OSAGE
COBBLER
BLUE MILLS
NEW SALEM
BUCKNER
POP. 2,873
LEVASY
POP. 279
TO LEXINGTON
SUGAR CREEK
POP. 3,982
INDEPENDENCE
POP. 112,301
U.S. LAKE CITY RESERVATION
LANDAHL PARK
TO RTE. 131
GRAIN VALLEY
POP. 1,898
BLUE SPRINGS
POP. 40,153
LAKE TAPAWINGO
POP. 761
OAK GROVE
POP. 4,565
IN JACKSON CO. 4,561
IN CASS CO. 4
TO ODESSA
MONKEY MOUNTAIN PARK
BLUE SPRINGS LAKE
LAKEWOOD LAKES
LAKE JACOMO
LAKE JACOMO PARK
TARSNEY LAKES
TARSNEY
TO RTE. Z
SNI MILLS
UNITY VILLAGE
POP. 138
PRAIRIE LEE LAKE
LAKE LOTAWANA
POP. 2,141
BLUE & GRAY PARK
LEE'S SUMMIT
POP. 46,418
JACKSON CO. 45,985
CASS CO. 433
JAMES A. REED MEMORIAL WILDLIFE AREA
LONE JACK
POP. 392
COCKRELL
GREENWOOD
POP. 1,315
RAINTREE LAKE
LAKE WINNEBAGO
LAFAYETTE COUNTY
JOHNSON COUNTY
CASS COUNTY
1
2
4
5
6
7
8
9
10
11
13
14
15

While in Jackson County, visit

Burr Oak Woods Conservation Nature Center

Burr Oak Woods Conservation Nature Center is 1.1 miles north of I-70 (exit 20) on Highway 7, then west 1 mile on Park Road—only minutes from downtown Kansas City. It is a showplace for Missouri's fish, forests and wildlife. Drop by on your own or call in advance to schedule a naturalist-conducted program or hike for your scout or tour group.

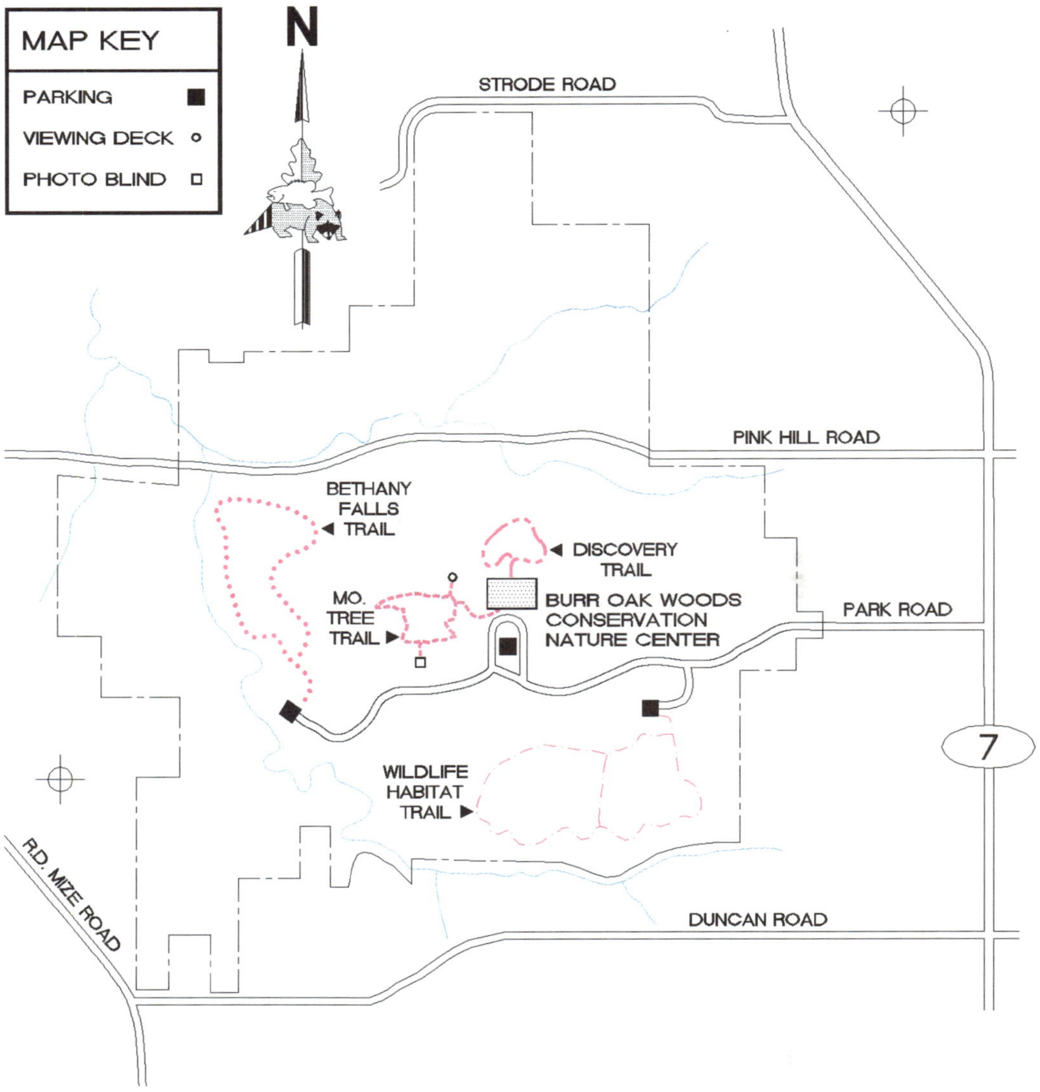

Outside

- more than a thousand acres of oak and hickory forests, prairies, ponds and glades
- four trails let you wind through tall grass prairie or a crevice of limestone rock outcrops; one is accessible to people who use wheelchairs
- two picnic areas

Inside

- hands-on nature exhibits
- regularly scheduled nature programs, announced in the center's monthly newsletter, *The Nutshell*
- indoor wildlife viewing area
- a 3,000-gallon aquarium, home to native fish and turtles
- sales of books, calendars and fishing and hunting permits

Building Hours

Open year-round, Monday through Saturday, 8 a.m. to 5 p.m.; Sunday, noon to 5 p.m. Closed Thanksgiving, Christmas and New Year's Day

Area Hours

Open daily, 8 a.m. to 8 p.m. during daylight savings; and daily, 8 a.m. to 6 p.m. standard time. Closed Thanksgiving, Christmas and New Year's Day.

Pets, horses, hunting, fishing, collecting, littering, fires and alcoholic beverages are not permitted. Mushroom collecting is permitted for personal use. Bicycles and motorized vehicles are restricted to paved roads.

Burr Oak Woods
Conservation Nature Center
Missouri Department of Conservation
1401 NW Park Road
Blue Springs, MO 64015
816/228-3766

Paul Childress

Paul Childress

Jim Rathert

Paul Childress

Just a short drive from downtown Kansas City, Burr Oak Woods Conservation Nature Center is a convenient destination for people eager to get away from the city bustle and into Missouri's great outdoors. Free admission makes it a great family outing, and people of all ages will enjoy exploring the center's many trails, exhibits and scheduled programs.

JASPER COUNTY

1 **Carl Junction Access** is on Highway JJ, 1.5 miles south of the junction of highways Z and JJ (near the village of Carl Junction). This is a 3-acre area with a boat ramp on Center Creek.

2 **Kellogg City Lake** is off Highway 96 near the Carthage eastern city limits. This 25-acre lake is equipped with a boat ramp, picnic area and a disabled accessible fishing jetty.

3 **La Russell Access** is 1 mile north of the town of La Russell on Highway U. This is a small area (1 acre) on Spring River with no boat ramp.

4 **Stones Corner Access** is 2.5 miles north of the junction of highways 43 and 171 on Highway 43. The access is a 9-acre area located on the first road (to the left off Highway 43) past the river bridge. It has a boat ramp on Center Creek.

5 **Wah-Sha-She Prairie** is on Highway M, 1 mile east of Highway 171. This is a 160-acre Designated Natural Area (hunting is not permitted).

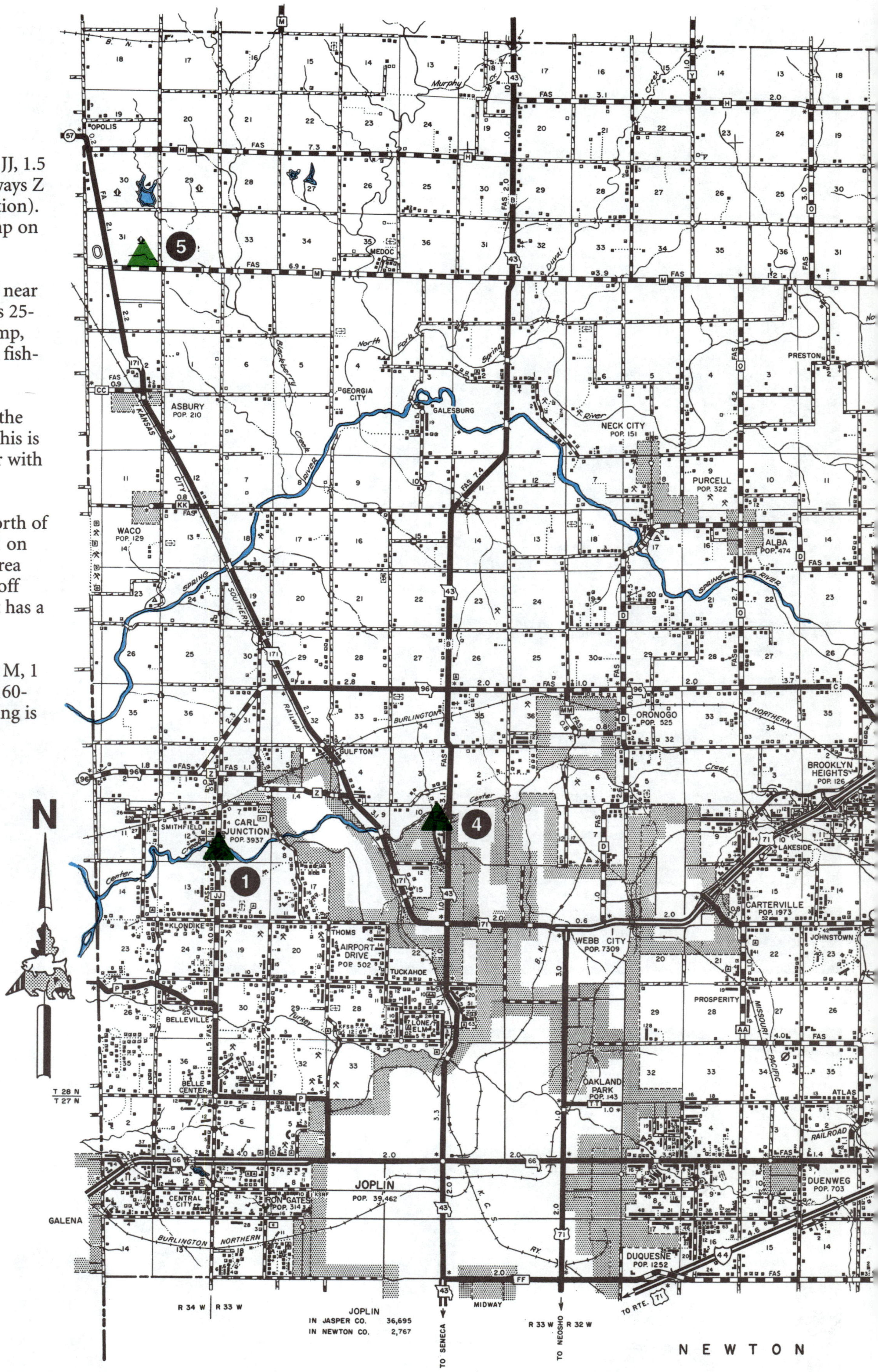

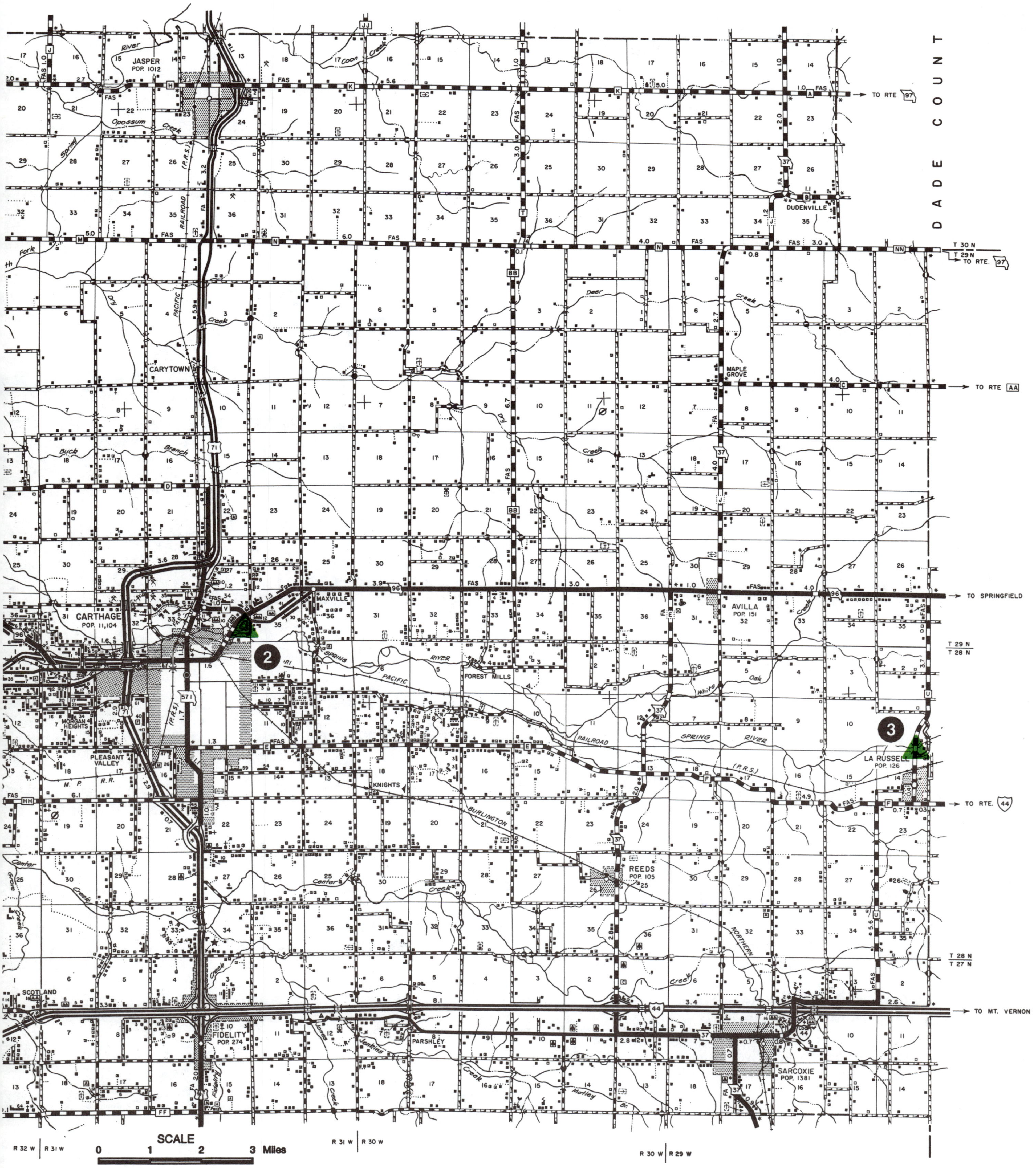
JASPER
POP. 1012
CARYTOWN
CARTHAGE
POP. 11,104
MAXVILLE
FOREST MILLS
AVILLA
POP. 151
MAPLE GROVE
DUDENVILLE
LA RUSSELL
POP. 126
MORGAN HEIGHTS
PLEASANT VALLEY
KNIGHTS
REEDS
POP. 105
SCOTLAND
FIDELITY
POP. 274
PARSHLEY
SARCOXIE
POP. 1381
DADE COUNT
TO RTE. 97
TO RTE AA
TO SPRINGFIELD
TO RTE. 44
TO MT. VERNON
T 30 N
T 29 N
T 28 N
T 27 N
R 32 W
R 31 W
R 30 W
R 29 W
SCALE
0 1 2 3 Miles
COUNTY

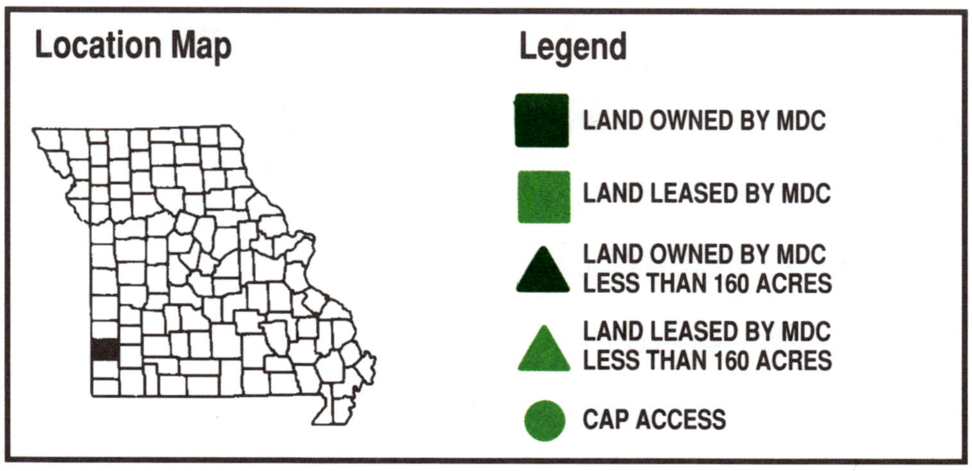
Location Map
Legend
LAND OWNED BY MDC
LAND LEASED BY MDC
LAND OWNED BY MDC LESS THAN 160 ACRES
LAND LEASED BY MDC LESS THAN 160 ACRES
CAP ACCESS

JEFFERSON COUNTY

1 To reach **Jeremiah Blackwell Access,** go south of De Soto on Highway 21 to Highway CC. Turn east to the Big River bridge, and go .5 mile east on Blackwell Road. This 3-acre access on Big River is on this road and has .2 mile of river frontage. It has no boat ramp.

2 **Brown's Ford Access** is southwest of Cedar Hill. Take Highway 30 west to Highway Y. Turn south on Highway Y, then west on Brown's Ford Road. This 1-acre access is on Big River near the bridge. It has a boat ramp.

3 **Cedar Hill Access** is .5 mile east of Highway 30 on Highway BB to Cedar Hill Road. Take Cedar Hill Road to the access, which is just south of the bridge over Big River. The area is 7 acres with no boat ramp. Jefferson County charges a fee to use the area.

4 **Flamm City Access** is on the Meramec River on Highway 231. This is a 10-acre area with a boat ramp.

5 **House Spring Access** is about .25 mile west of the town of House Springs on Highway W near the Big River bridge. The area is 8 acres and has about .6 mile of frontage on Big River and two boat ramps. Jefferson County charges a fee to use the area.

6 **Mammoth Access** is west of De Soto on Highway H and 1.5 miles south on Mammoth Road. This 5-acre access on Big River has a boat ramp.

7 **Merrill Horse Access** is 5 miles west of De Soto on Highway H. This is a 74-acre area with .5 mile frontage and boat ramp on Big River. This is a smallmouth bass management area.

8 **Pacific Palisades Conservation Area** is 1.5 miles south of Pacific on Highway F, then .1 mile east on English Road. The boat ramp is reached by taking Highway 44 to Allenton-Six Flags exit. Go 1.8 miles west on Old Rt. 66. Turn left on Dozier Crossing and make a right on Franklin Rd. This is a 695-acre area on the Meramec River (2.9 miles) composed mainly of forestland (465 acres) and cropland (120 Acres). Palisades Gravel Mines Lakes provide about 60 acres for fishing, and there is an archery range.

9 **Teszars Woods Conservation Area** is east of Arnold on Telegraph (off Highway 61-67), then east on Rockwood Drive. The 90-acre area is at the end of Rockwood, adjacent to the Meramec River. There is a boat ramp on the river in the neighboring Flamm City Park.

10 **Valley View Glades Natural Area** is on Highway B between Cedar Hill and Hillsboro. The entire 225-acre area is a dolomite glade and a Designated Natural Area.

11 **Victoria Glades Conservation Area** is 1.5 miles south of Hillsboro on Victoria Hillsboro Road. This 239-acre area is glade (100 acres) and forestland (139 acres).

12 **Young Conservation Area** is south on Highway W off I-44 and 3 miles west on Highway FF. This 970-acre area is mostly forestland and contains a 1.3-mile stretch of LaBarque Creek. There are two ponds (1.2 acres for fishing), demonstration sites, and a hiking trail 2.5 miles over a soil and chat surface.

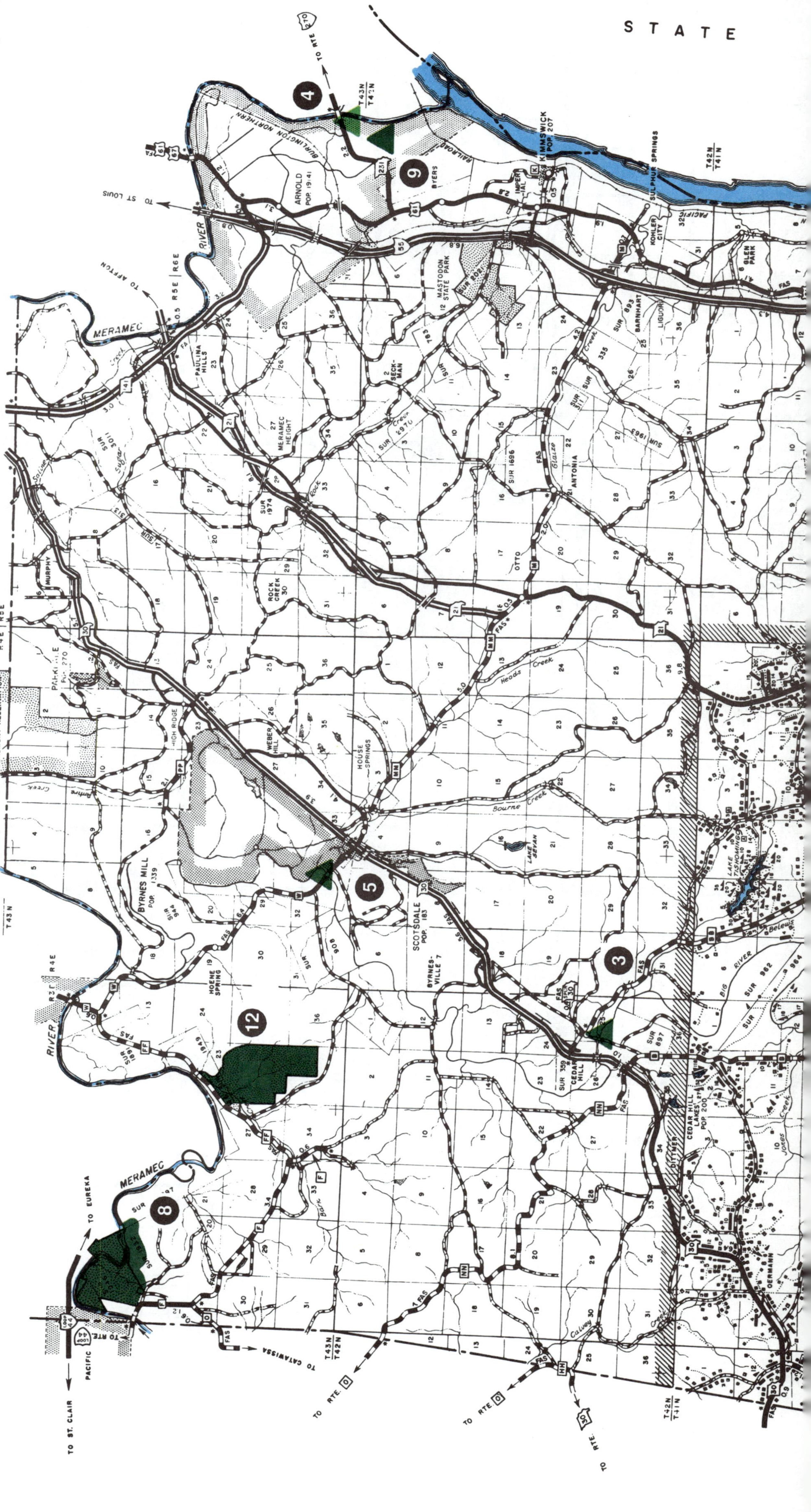

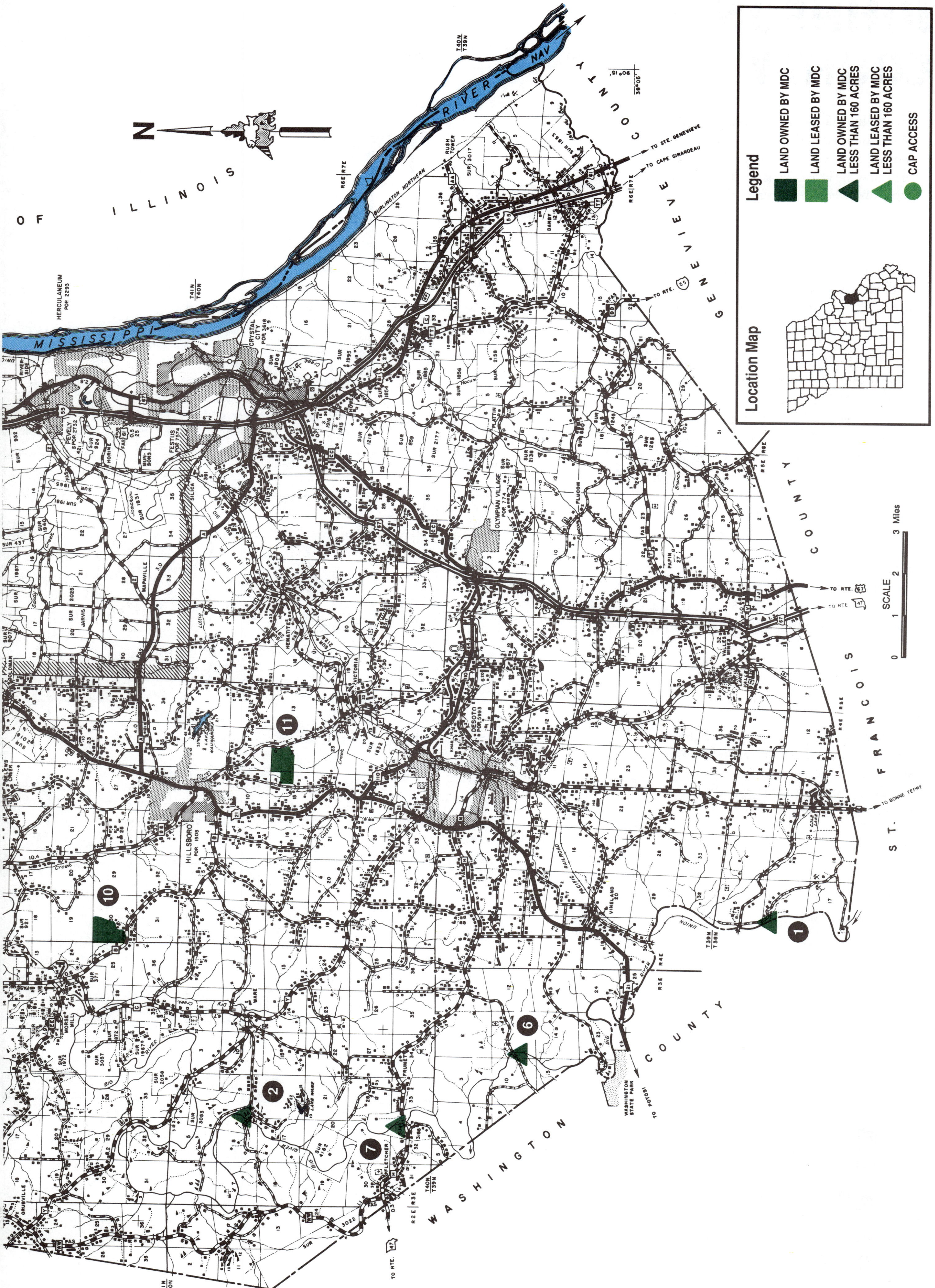
Legend
LAND OWNED BY MDC
LAND LEASED BY MDC
LAND OWNED BY MDC LESS THAN 160 ACRES
LAND LEASED BY MDC LESS THAN 160 ACRES
CAP ACCESS
Location Map
SCALE
0 1 2 3 Miles
N
OF ILLINOIS
MISSISSIPPI RIVER
HERCULANEUM POP. 2293
CRYSTAL CITY POP. 3618
FESTUS POP. 7574
PEVELY POP. 2732
HILLSBORO POP. 1508
DESOTO POP. 5993
OLYMPIAN VILLAGE POP. 774
STE. GENEVIEVE COUNTY
ST. FRANCOIS COUNTY
WASHINGTON COUNTY
WASHINGTON STATE PARK
TO STE. GENEVIEVE
TO CAPE GIRARDEAU
TO BONNE TERRE
TO POTOSI
BURLINGTON NORTHERN

JOHNSON COUNTY

1 **Hazel Hill Lake** is 5 miles northwest of Warrensburg on Highway 13, and 3 miles west on Highway OO. This is a 502-acre area, and the lake is 72 acres. It has a boat ramp and wheelchair accessible fishing jetty.

2 **Holden City Lake** is 3.25 miles west of the town of Holden on Highway 58, and north on a gravel entrance road marked with an area cantilever sign. This is a 600-acre area with 380 acres of fishable water. There is a boat ramp and picnic area.

3 **J. N. "Turkey" Kearn Memorial Wildlife Area** is southeast of Knob Noster on Highway 23 (1 mile south of the junction of highways 23 and Y), then east on County Road 459. This 1,674-acre area is composed mainly of old fields (860 acres) and forestland (423 acres). Ten ponds and a section of Muddy Creek provide anglers with about 4 acres of fishable water. Additional tracts lie in Pettis County.

4 **Lions Lake** is off Highway 13, .75 mile west on Hale Lake Road, and .33 mile north on Rotary Road. This 8-acre lake sits on about 30 acres of land and has a picnic area and a disabled accessible fishing jetty.

5 **Ralph and Martha Perry Memorial Conservation Area** consists of two tracts. The largest tract is in Johnson County, 9 miles north of Knob Noster on Highway 23. This is a 4,053-acre area along a 5-mile stretch of the Blackwater River. It has 19 ponds (16 acres for fishing) and a firearms range.

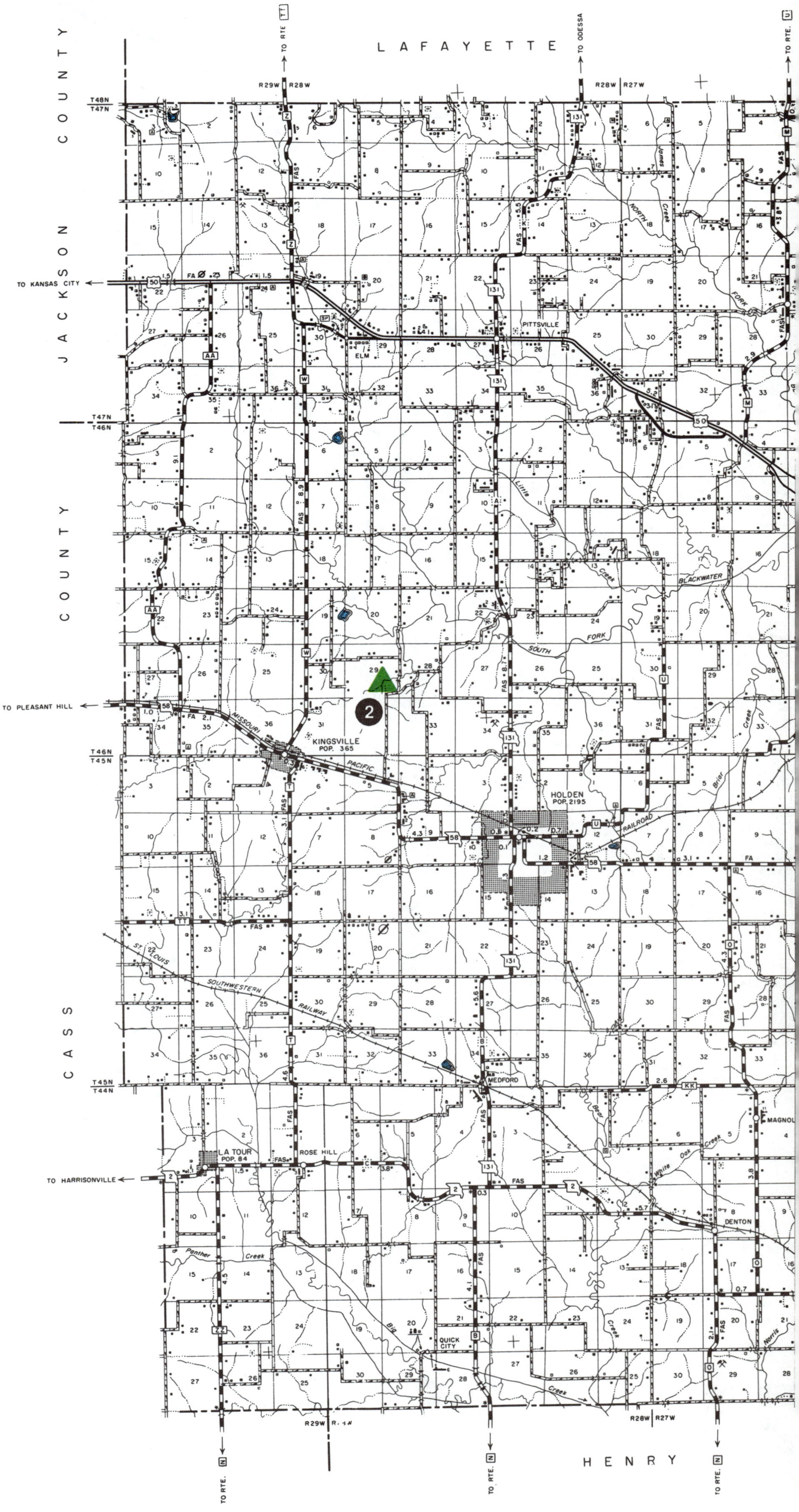

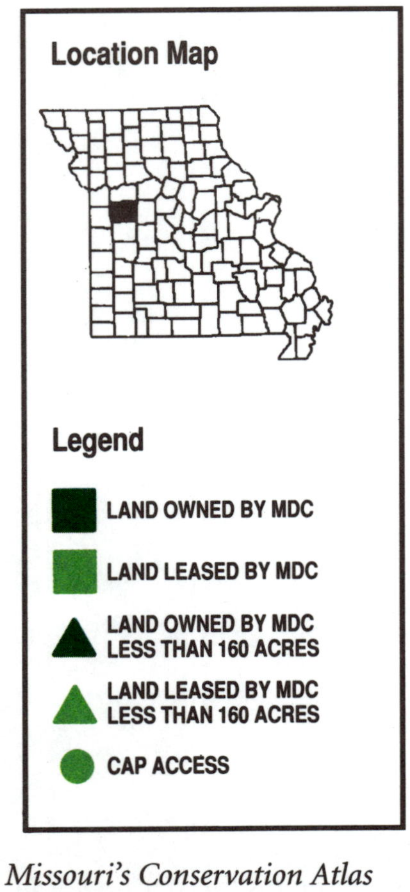

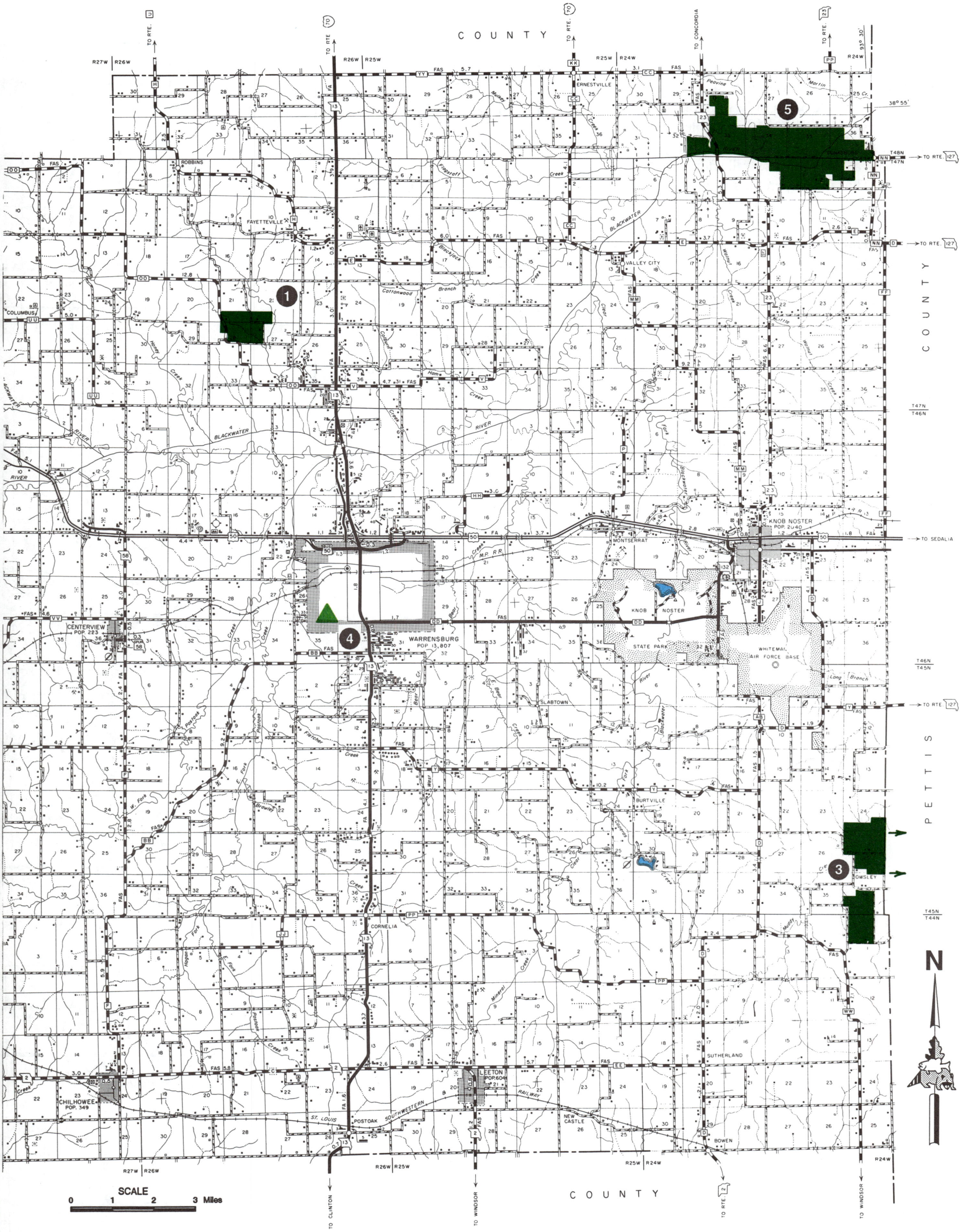
COUNTY
COUNTY
PETTIS
COUNTY
TO RTE. 13
TO RTE. 20
TO CONCORDIA
TO RTE. 23
TO RTE. 127
TO SEDALIA
TO CLINTON
TO WINDSOR
TO RTE. 2
R27W R26W
R26W R25W
R25W R24W
R24W
T48N T47N
T47N T46N
T46N T45N
T45N T44N
93° 30'
38° 55'
ROBBINS
FAYETTEVILLE
COLUMBUS
ERNESTVILLE
VALLEY CITY
BLACKWATER RIVER
CENTERVIEW POP. 223
WARRENSBURG POP. 13,807
KNOB NOSTER POP. 2,040
MONTSERRAT
KNOB NOSTER STATE PARK
WHITEMAN AIR FORCE BASE
SLABTOWN
BURTVILLE
OWSLEY
CORNELIA
SUTHERLAND
CHILHOWEE POP. 349
LEETON POP. 604
POSTOAK
NEW CASTLE
BOWEN
ST. LOUIS SOUTHWESTERN RAILWAY
M.P. RR
SCALE
0 1 2 3 Miles
N
1
3
4
5

KNOX COUNTY

1 **Henry Sever Lake Conservation Area** (1,115 acres) is 1 mile north of Newark on Route KK. Sever Lake is 162 acres suitable for fishing. There is one boat ramp, five fishing docks, a 7.5-mile horse trail and picnic areas. The lake is open for swimming, and there is a concession stand.

2 **White Oak Bend Access** is 3 miles west of Newark on Highway 156, then 1 mile north on an unnamed gravel road. The access consists of 160 acres on South Fabius River (no boat ramp).

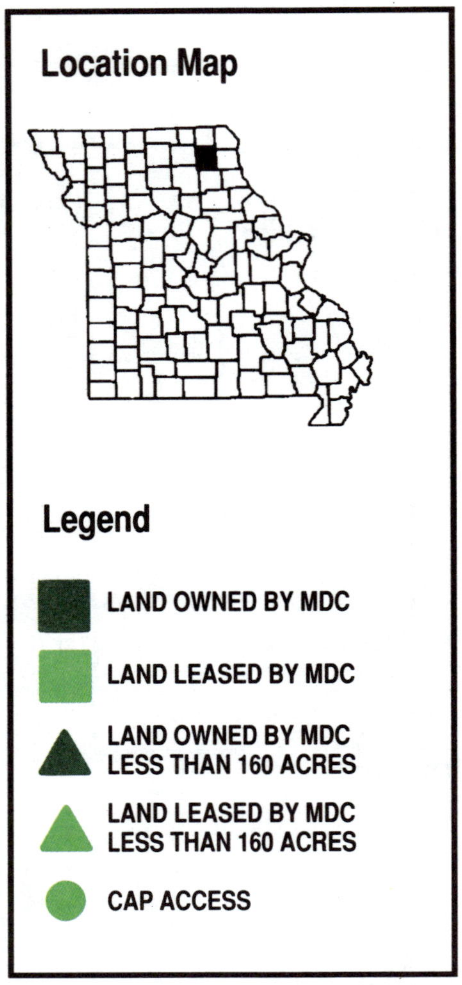

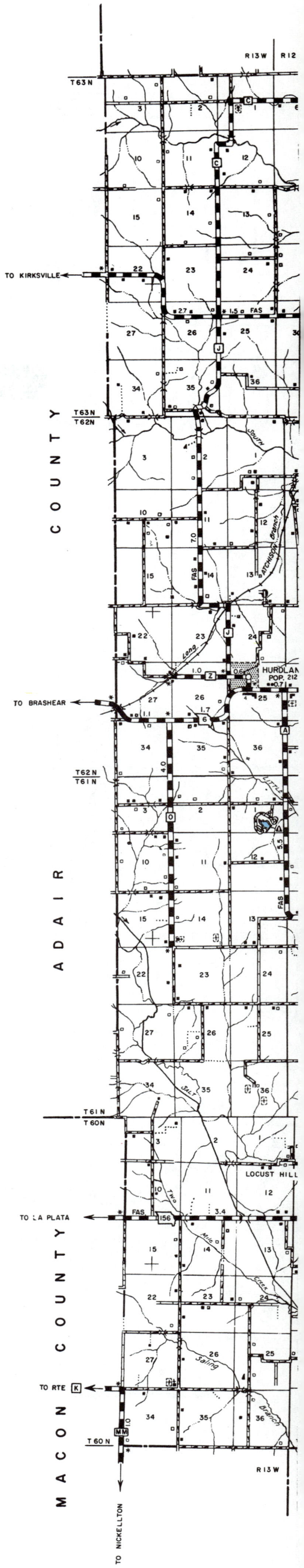

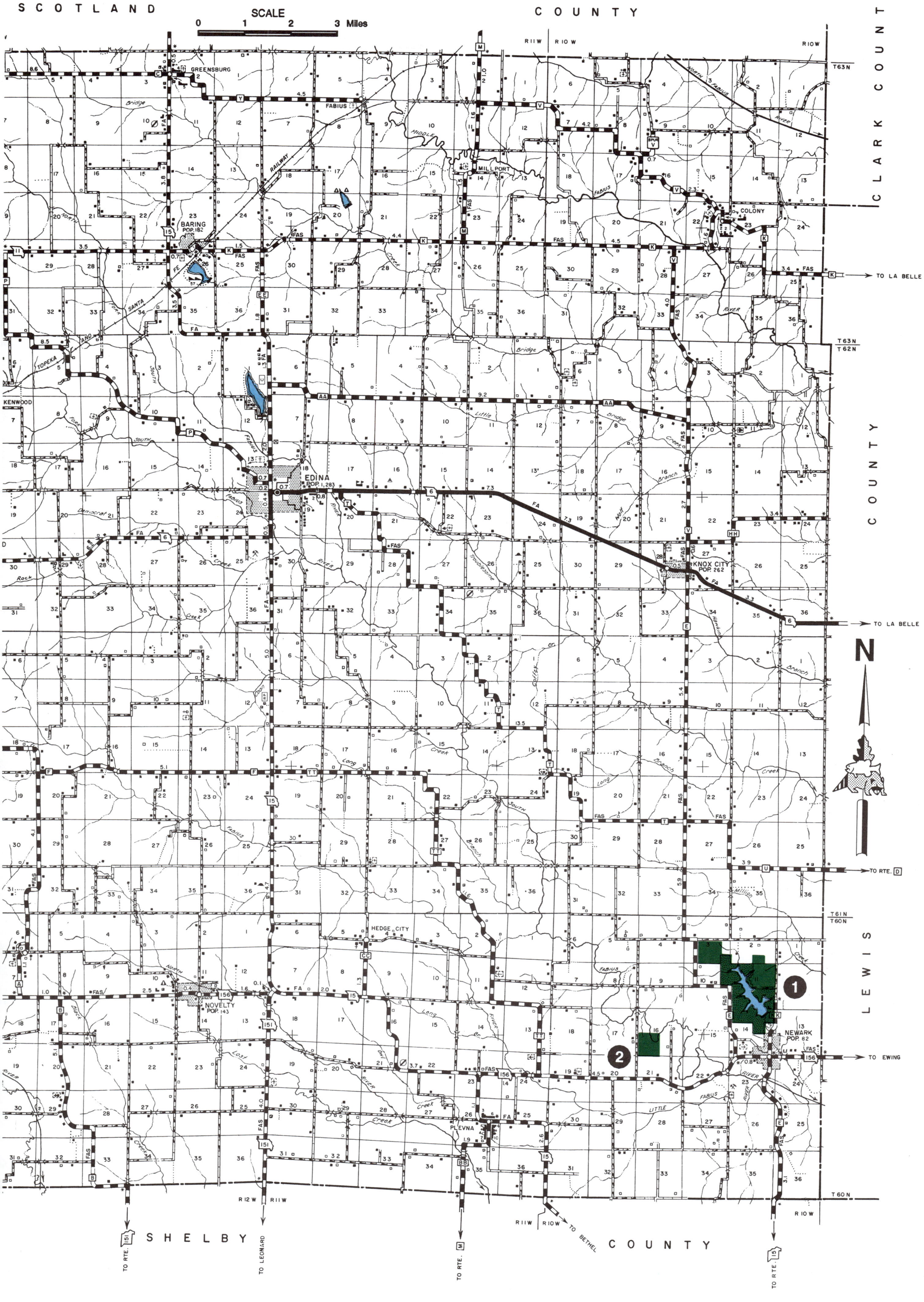

SCOTLAND COUNTY
SCALE
0 1 2 3 Miles
CLARK COUNTY
LEWIS COUNTY
SHELBY COUNTY
N
GREENSBURG
FABIUS
MILLPORT
COLONY
BARING POP. 182
KENWOOD
EDINA POP. 1,283
KNOX CITY POP. 262
HEDGE CITY
NOVELTY POP. 143
NEWARK POP. 82
PLEVNA
TO LA BELLE
TO LA BELLE
TO RTE. D
TO EWING
TO RTE. 151
TO LEONARD
TO RTE. M
TO BETHEL
TO RTE. 15
T63N
T62N
T61N
T60N
R12W
R11W
R10W

LACLEDE COUNTY

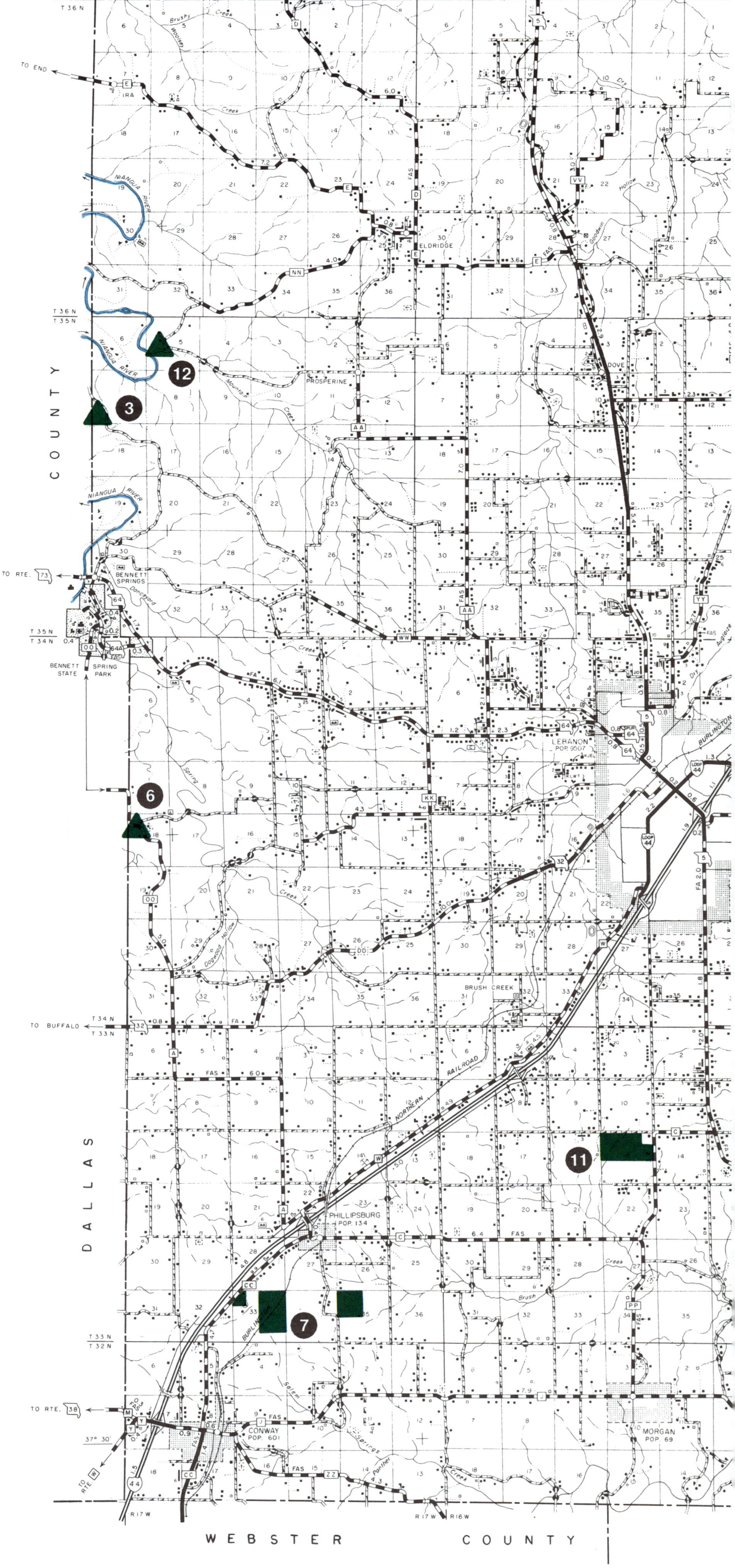

1 **Anna M. Adams Access** is 18 miles east of Lebanon on Highway 32, 3 miles north on Highway K, then 3 miles east on Dawn Road. This is a 17-acre area on the Gasconade River.

2 **Bear Creek Conservation Area** (three tracts) is 5.5 miles east of Lebanon on I-44. Take exit 135 and go east on South Outer Road .1 mile, then turn 1.5 miles south on Rustic Drive to reach the main parking area. This is a 729-acre forest.

3 **Coffin Cave Conservation Area** is 12 miles west of Lebanon on Highway 64, and 4.5 miles north on Marigold Drive. The cave on this 60-acre forested area is closed to the public except to those who obtain a special permit.

4 **Davis Ford Access** is 6 miles east of Lebanon on Highway 32, 1.5 miles north on Highway N, and 1.5 miles down Highway AC. This is a 24-acre area on the Osage Fork River with a boat ramp.

5 **Drynob Access** is 15 miles east of Lebanon on Highway 32. This is a 17-acre area on a .25-mile stretch of the Osage Fork River with no boat ramp.

6 **Flatwood Church Conservation Area** is 3.5 miles south of Bennett Spring State Park on Highway OO, and left on Flatwood Church Road. This is a 71-acre area of mainly forestland (60 acres) and some pastureland (11 acres).

7 **Goose Creek Conservation Area**–see the Dallas County map, page 64.

8 **Hazelgreen Access** is 13 miles east of Lebanon on I-44, 1.2 miles east on Highway N (exit 140), then 2 miles east on South Outer Road. This is a 1-acre area with a boat ramp on the Gasconade River.

9 **Hull Ford Access** is 6 miles east of Lebanon on Highway 32, then 4 miles down Highway N. This is a 10-acre area on the Osage Fork River with no boat ramp.

10 **Lebanon Towersite** is east of Lebanon on Highway 32, left on Tower Road, then right on Tower Lane. This is a 5-acre area.

11 **Osage Fork Conservation Area** is 4 miles south of Lebanon on Highway 5, and 1.25 miles west on Highway C. This is a 280-acre area with an archery range.

12 To reach **Prosperine Access**, go 2 miles west of Lebanon on Highway 64 to Highway AA. Take AA until the blacktop ends. Turn 4.5 miles left on Kinfolk Road, and watch for the cantilever sign on the left. This is an 8-acre area on the Niangua River with no boat ramp.

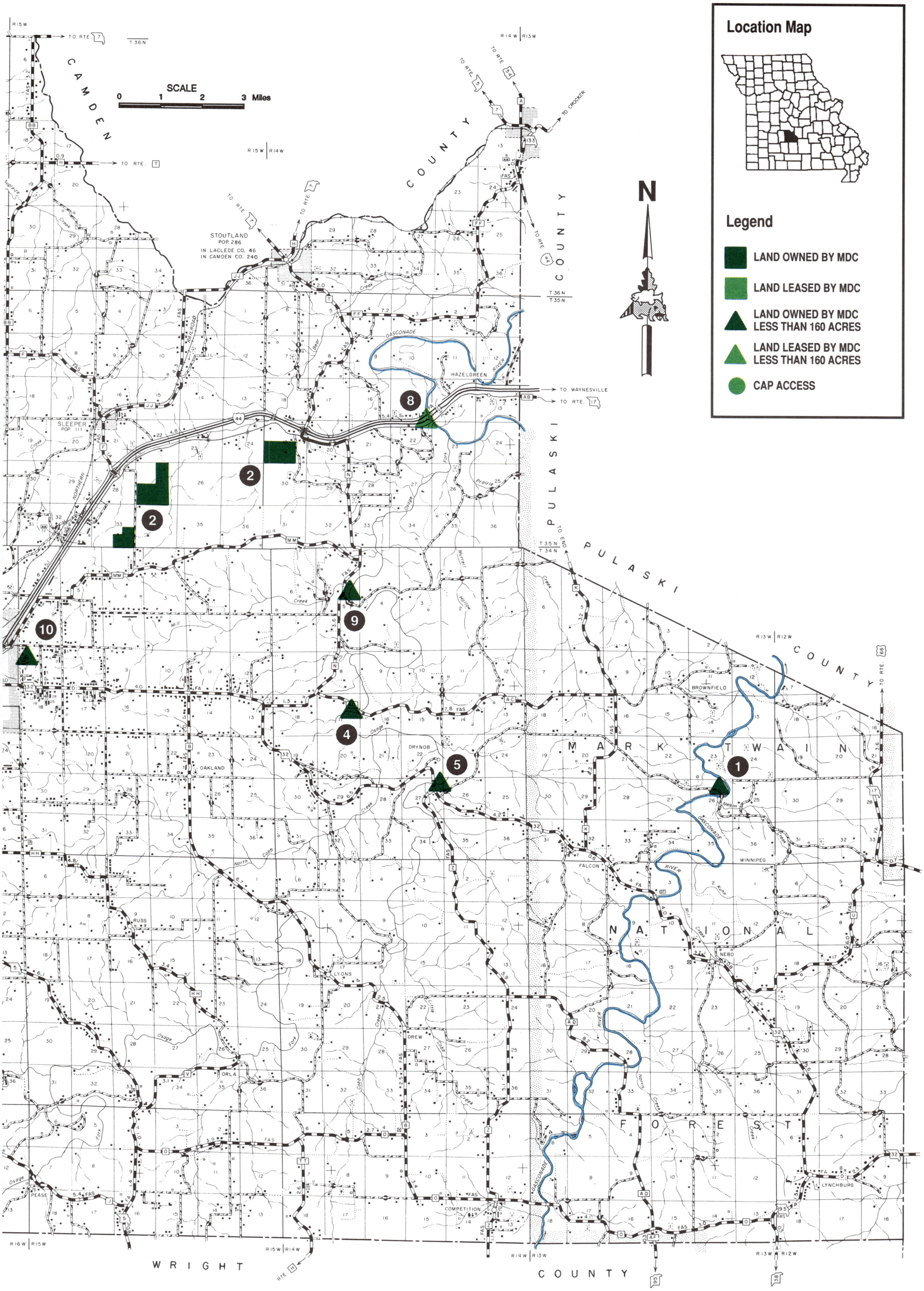

Location Map
Legend
LAND OWNED BY MDC
LAND LEASED BY MDC
LAND OWNED BY MDC LESS THAN 160 ACRES
LAND LEASED BY MDC LESS THAN 160 ACRES
CAP ACCESS
N
SCALE
0 1 2 3 Miles
CAMDEN
COUNTY
PULASKI
COUNTY
PULASKI COUNTY
WRIGHT
COUNTY
MARK TWAIN
NATIONAL
FOREST
STOUTLAND
POP. 286
IN LACLEDE CO. 46
IN CAMDEN CO. 240
SLEEPER
POP. 111
HAZELGREEN
TO WAYNESVILLE
TO CROCKER
BROWNFIELD
DRYNOB
OAKLAND
FALCON
WINNIPEG
RUSS
NEBO
LYONS
DREW
ORLA
PEASE
COMPETITION
LYNCHBURG
GASCONADE
RIVER

LAFAYETTE COUNTY

1 **Baltimore Bend Conservation Area** consists of two tracts. The largest tract is 3 miles west of Waverly on Highway 24. This 1,192-acre area has seven ponds, providing 1.5 acres for fishing. It is mainly forestland (867 acres) and cropland.

2 **Cecil G. Grove Memorial Conservation Area** is 1.25 miles east of Aullville on County Road 219, then .5 mile north (left) on County Road 216. This is 40 acres of forestland and old pastures.

3 **Higginsville City Lake** is 3 miles east of the town of Higginsville off Highway AA. This 190-acre facility has two lakes. The upper lake is 40 acres, and the main lake is 150 acres. There are boat ramps and a wheelchair accessible dock.

4 **Maple Leaf Lake Conservation Area** is west of the junction of Highway 13 and I-70, then 3 miles west on the South I-70 Outer Access Road. This 826-acre area is home to Maple Leaf Lake (140 acres).

5 **Odessa City Lakes** (105 acres) are south of Odessa on Highway 131, then 3 miles west on Highway OO. There is a boat ramp and disabled accessible fishing jetty.

6 **Edwin A. Pape Lake** is 3 miles south of Concordia off Highway 23. This is a 245-acre lake equipped with a boat ramp, fishing jetty (wheelchair accessible) and picnic areas.

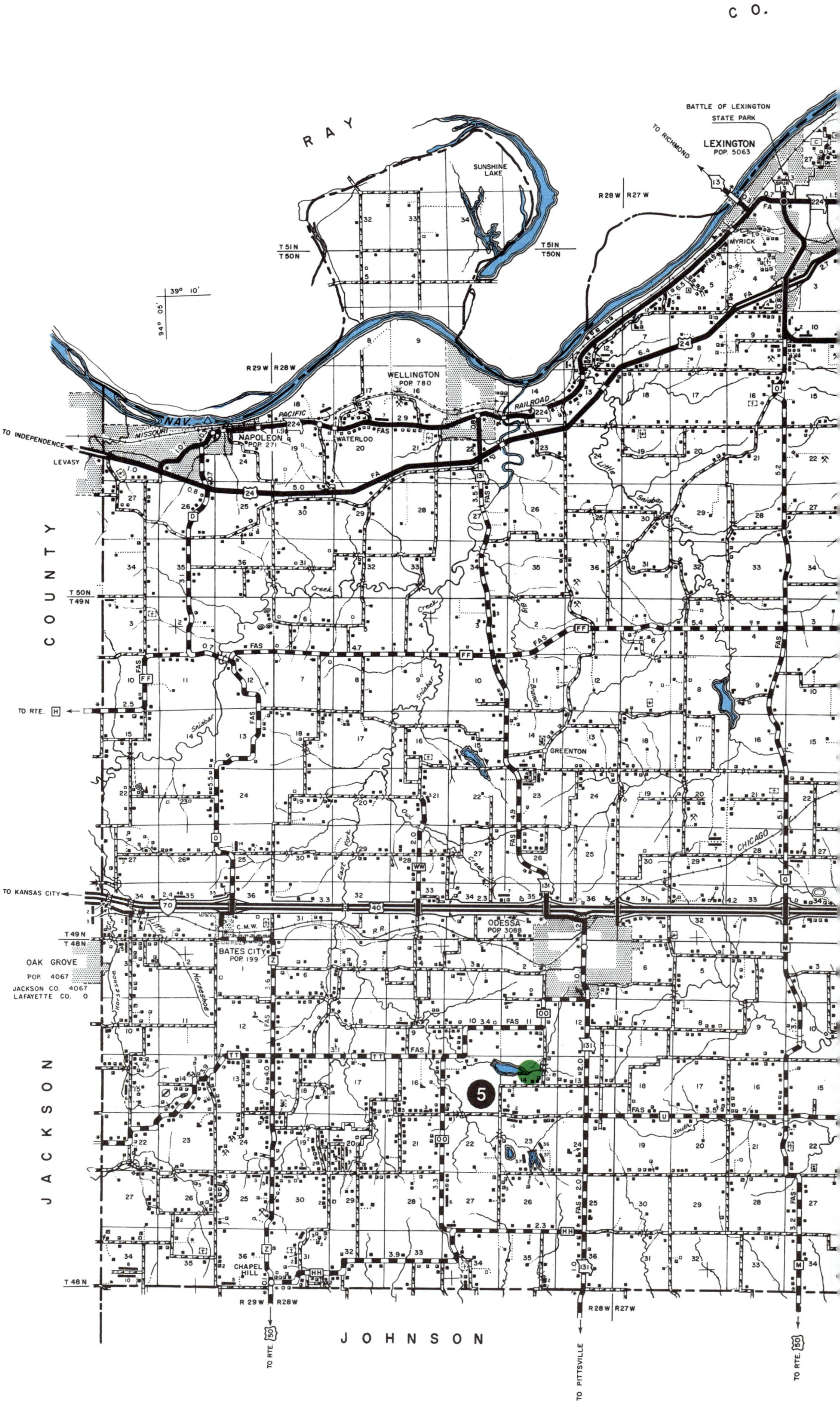

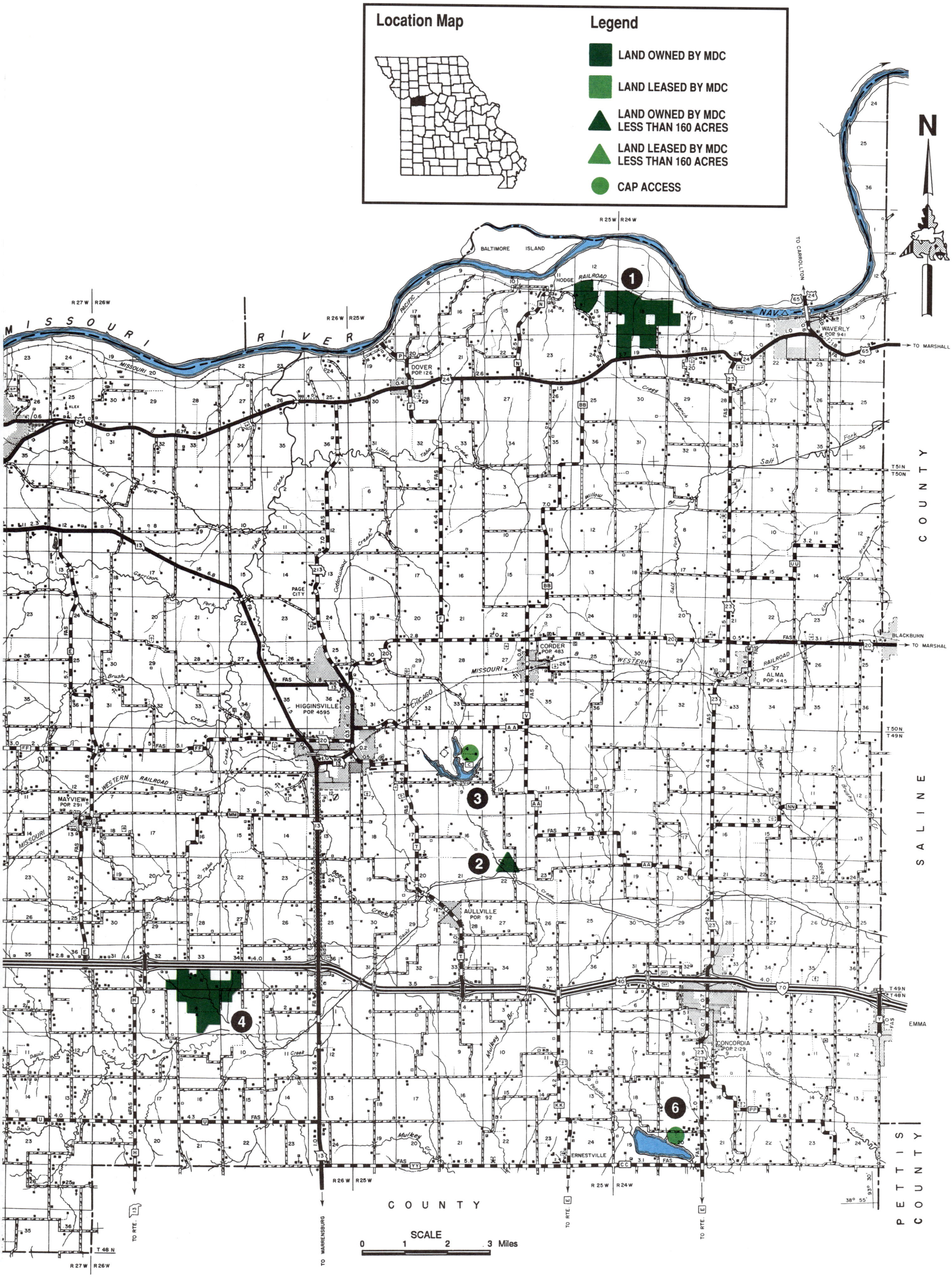
Location Map
Legend
LAND OWNED BY MDC
LAND LEASED BY MDC
LAND OWNED BY MDC LESS THAN 160 ACRES
LAND LEASED BY MDC LESS THAN 160 ACRES
CAP ACCESS
N
MISSOURI RIVER
BALTIMORE ISLAND
HODGE
DOVER POP 126
WAVERLY POP 941
TO CARROLLTON
TO MARSHALL
PAGE CITY
CORDER POP 483
ALMA POP 445
BLACKBURN
HIGGINSVILLE POP 4595
MAYVIEW POP 291
AULLVILLE POP 92
CONCORDIA POP 2129
EMMA
ERNESTVILLE
SALINE COUNTY
PETTIS COUNTY
COUNTY
TO RTE. 13
TO WARRENSBURG
TO RTE. E
SCALE
0 1 2 3 Miles
R 27 W R 26 W
R 26 W R 25 W
R 25 W R 24 W
T 51 N T 50 N
T 50 N T 49 N
T 49 N T 48 N
T 48 N
38° 55'
93° 30'
1
2
3
4
6

LAWRENCE COUNTY

1 **Chesapeake Fish Hatchery** (118 acres) is 8 miles east of Mt. Vernon on Highway 174. The hatchery rears a variety of warm water fish species, including channel catfish, largemouth bass, walleye, mukellunge and flathead catfish. Tours are available, and there is a picnic area.

2 To reach **Kickapoo Prairie Conservation Area**, take the Stott City exit off I-44 and go 8 miles north on Highway 97. The turn-off for this 160-acre area is 2 miles north of the junction of highways 97 and 96 (1 mile west on a county gravel road). It is a 160-acre native prairie.

3 **Mount Vernon Prairie** is 1 mile west of Highway M on Farm Road 2100, then .5 mile north on Farm Road 1150. This is 40 acres of dry-mesic sandstone and shale prairie.

4 **Paris Springs Access** on Turnback Creek is 3 miles west of Halltown on Highway 96. This 208-acre area provides anglers with 3 acres of fishable water (three ponds and the creek). Turnback Cave, which is closed to the public, is home to three endangered species: Ozark cavefish, cave crayfish and the gray bat.

5 **Providence Prairie Conservation Area** is 1.5 miles west of the junction of highways 97 and NN, then 1 mile south on County Road 1040. This is 197 acres of mostly prairie.

6 **Robert E. Talbot Conservation Area** is near Mt. Vernon on Highway 96, 1 mile east of the junction of highways 96 and 97. This 4,321-acre area is home to a 100-nest great blue heron rookery along Spring River. The river flows 2 miles through the area, and there are 51 ponds (45 total acres for fishing). There is a 1.5-mile hiking trail, and the Talbot Demonstration Farm shows visitors various management practices that are beneficial to wildlife.

7 **Williams Creek Park** is about .25 mile north of the Mount Vernon square on the east side of Main Street along Williams Creek. It is 4 acres and has a disabled accessible fishing jetty.

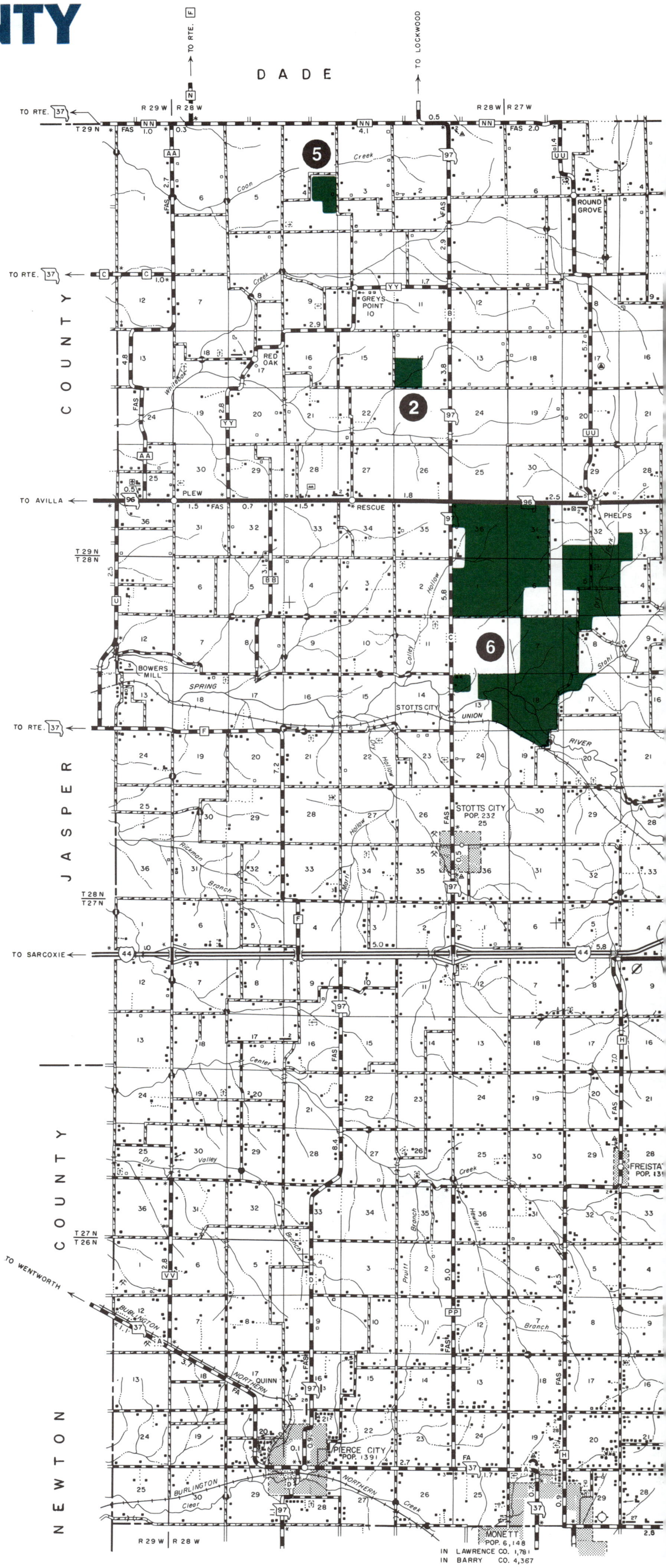

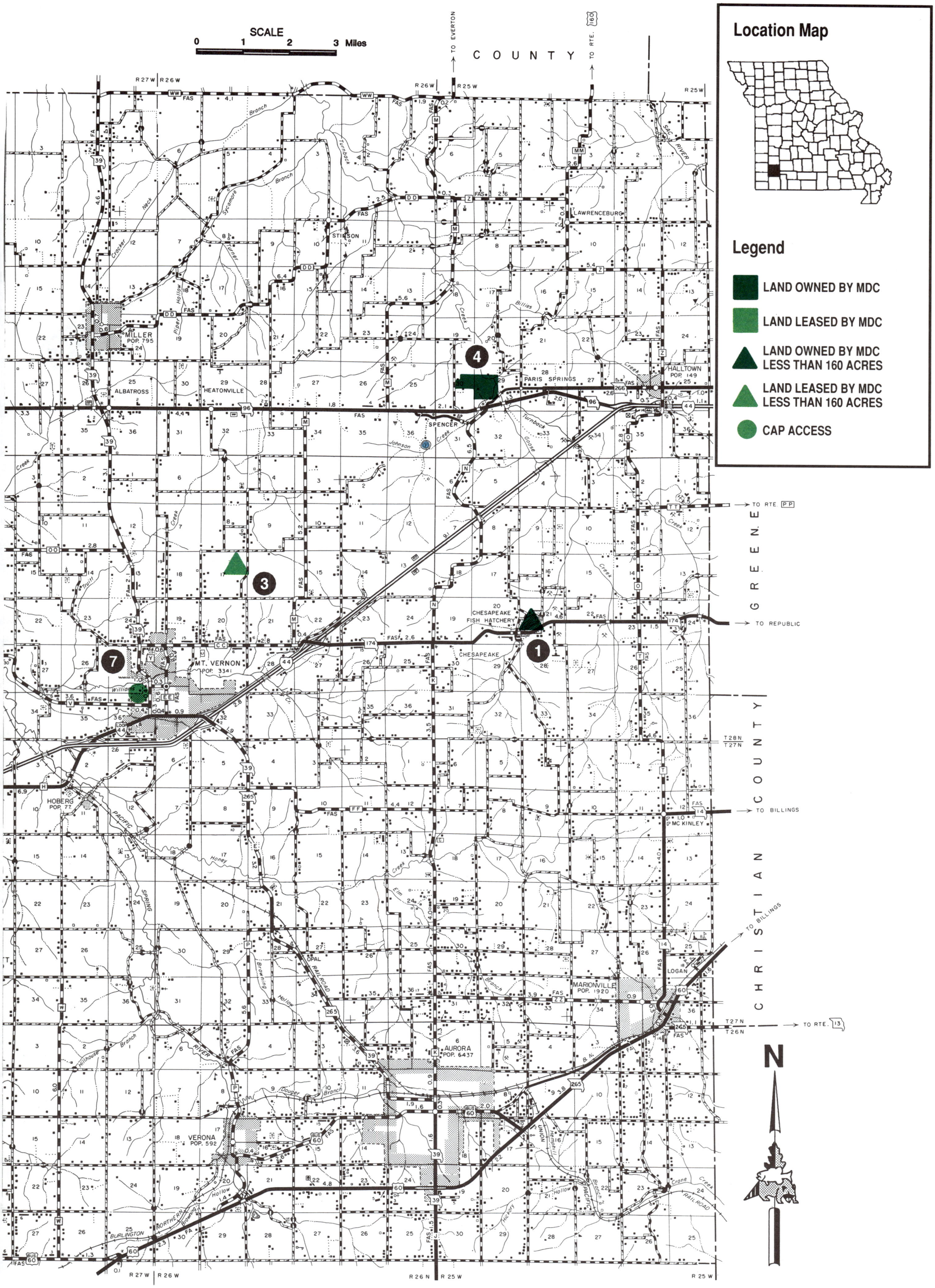
SCALE
0 1 2 3 Miles
COUNTY
TO EVERTON
TO RTE. 160
GREENE
CHRISTIAN COUNTY
TO RTE. PP
TO REPUBLIC
TO BILLINGS
TO RTE. 13
Location Map
Legend
LAND OWNED BY MDC
LAND LEASED BY MDC
LAND OWNED BY MDC LESS THAN 160 ACRES
LAND LEASED BY MDC LESS THAN 160 ACRES
CAP ACCESS
MILLER POP. 795
STINSON
LAWRENCEBURG
ALBATROSS
HEATONVILLE
PARIS SPRINGS
HALLTOWN POP. 149
SPENCER
CHESAPEAKE FISH HATCHERY
CHESAPEAKE
MT. VERNON POP. 3341
HOBERG POP. 77
MC KINLEY
LOGAN
MARIONVILLE POP. 1920
OPAL
AURORA POP. 6437
VERONA POP. 592
BURLINGTON
R 27 W
R 26 W
R 25 W
T 28 N
T 27 N
T 26 N
N

LEWIS COUNTY

1 **Canton Ferry Access** is marked with a sign at the north edge of the city of Canton, east of Main Street. This is a 2-acre area on the Mississippi River with a boat ramp. There is a city-managed campground .25 mile away.

2 **Deer Ridge Conservation Area** consists of three tracts. The main tract is 5.5 miles north of Lewistown on Highway H, and 2 miles west on Highway Y. North Fabius River flows through this 4,755-acre area, and there are 25 ponds and a lake with a boat ramp. There is a 9-mile horseback riding trail, firearms range, trap range and picnic pavillions.

3 **Ewing Lake** is west off Highway C, 1 mile north of Ewing. This 47-acre lake is equipped with a boat ramp and picnic area.

4 **Fenway Landing Access** is 4 miles north of Canton on Highway 61, then 1 mile east on an unnamed gravel road. This is a 10-acre area with a boat ramp on the Mississippi River.

5 **Halls Mill Access** is 1 mile north of Ewing on Highway C. This is a 6-acre area with no boat ramp on the Middle Fabius River.

6 **LaBelle City Lake** is 1.5 miles south of LaBelle on Highway D, then about 1 mile down the first county road heading east (it is near the old City Lake). This is a 112-acre lake with a boat ramp and picnic area. The city of LaBelle regulates activities on the surrounding land.

7 **Sunnyside Access** is 2 miles north of LaGrange on Highway B, then 2.5 miles west on County Road 552. This is a 114-acre area.

8 **Tolona Access** is 1 mile north of Ewing on Highway C, 2 miles north on Highway BB, and 1 mile west on Gravel Road 508. This is 176 acres on the Middle Fabuis River.

9 **Upper Mississippi Conservation Area** consists of 87 scattered tracts (14,906 acres total) adjacent to the Mississippi River and some river islands. Within these areas, there are three boat ramps and 112 waterfowl blinds. Westport Island Natural Area is 400 acres of bottomland wet-mesic forest. Three levees are suitable for hiking (ranging 1 to 10 miles) and there are two service access roads near Westport Island. For additional information, contact the Hannibal Conservation Service Center, Box 428, Hannibal, MO 63401, phone: 573/248-2530.

10 **Wyaconda Crossing Conservation Area** is 3 miles northwest of Canton on Highway 81, and 1.25 miles west on County Road 436. The Wyaconda River flows through this 148-acre area, most of which is forestland.

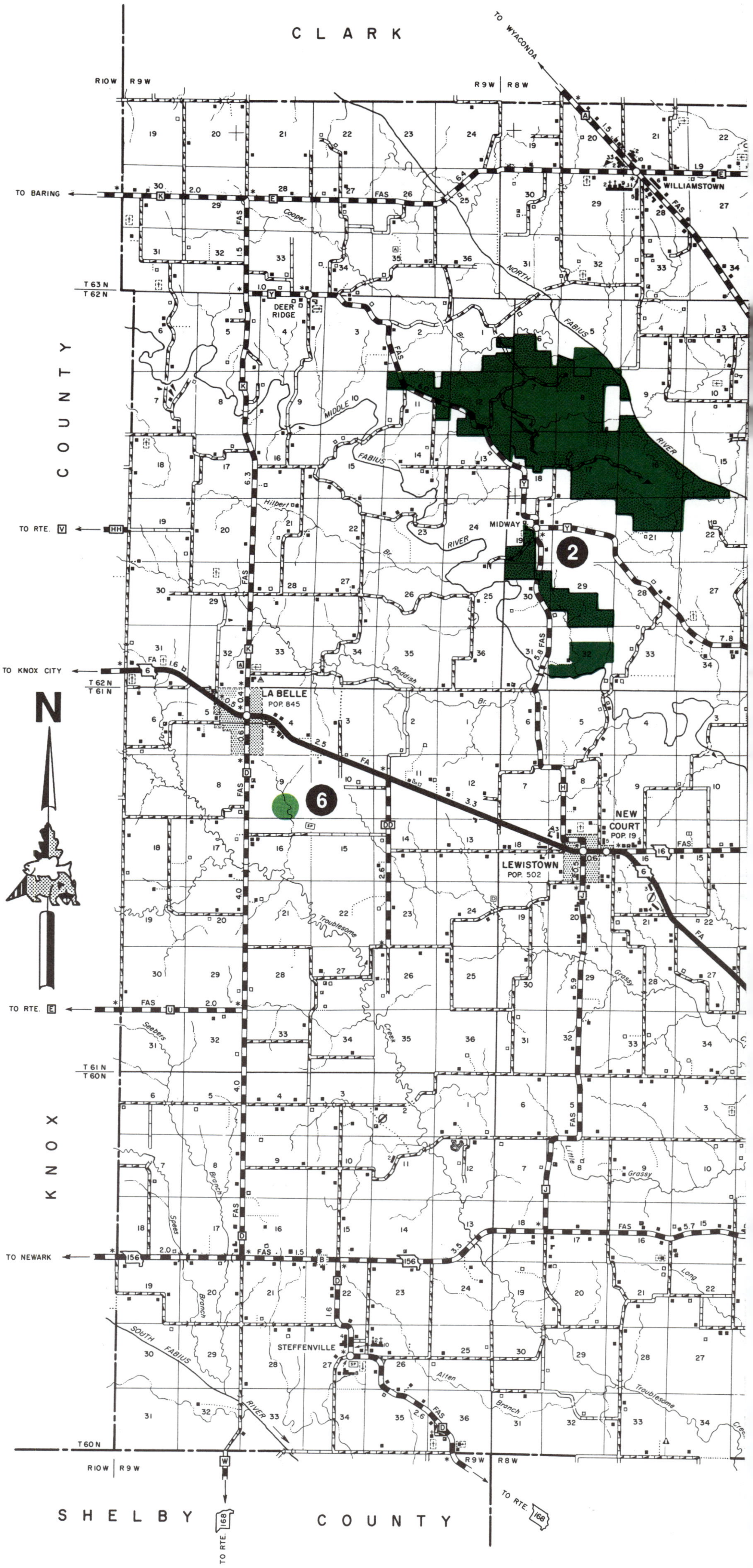

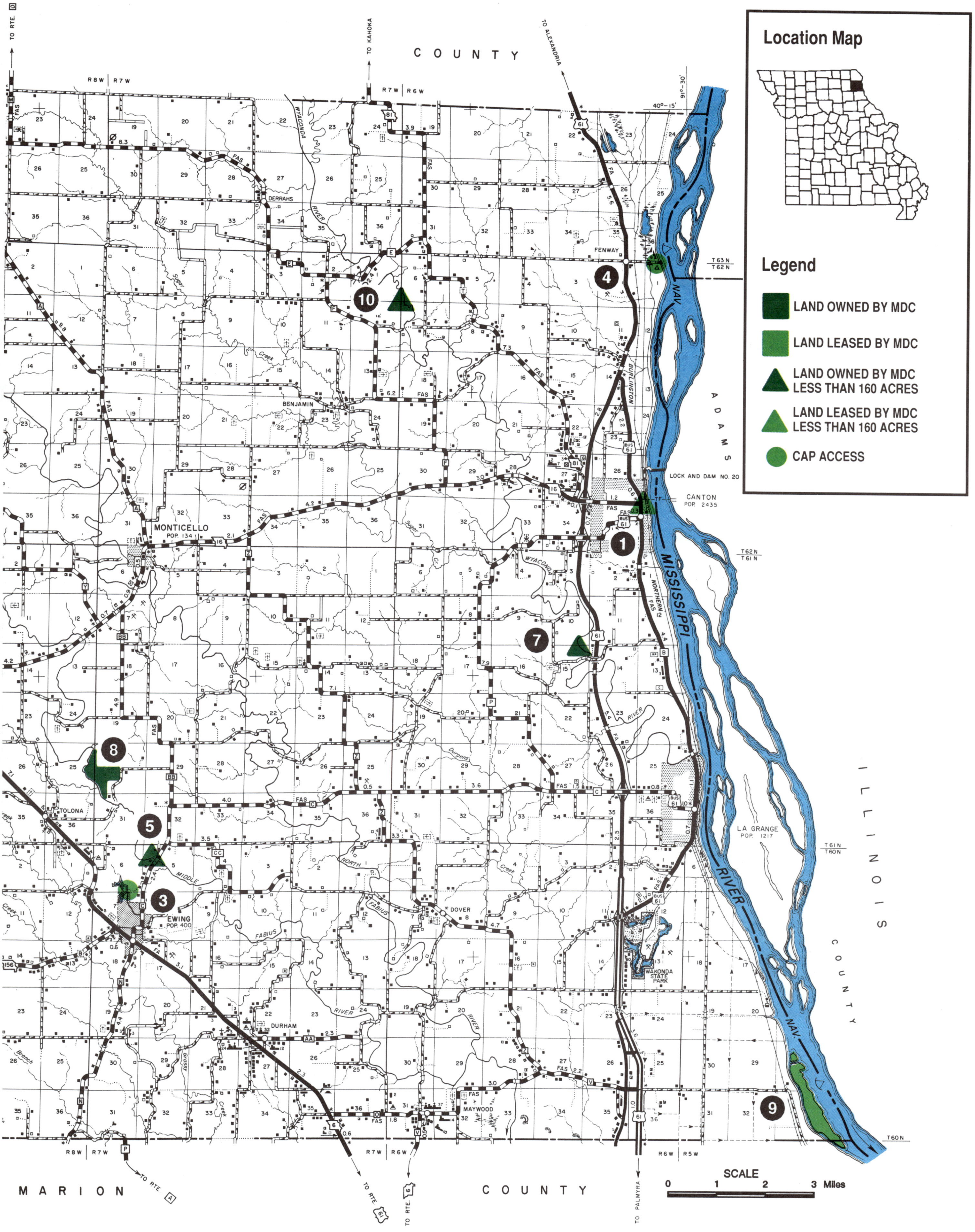
Location Map
Legend
LAND OWNED BY MDC
LAND LEASED BY MDC
LAND OWNED BY MDC LESS THAN 160 ACRES
LAND LEASED BY MDC LESS THAN 160 ACRES
CAP ACCESS
COUNTY
ADAMS
ILLINOIS COUNTY
MARION COUNTY
MISSISSIPPI RIVER
NAV
LOCK AND DAM NO. 20
CANTON POP. 2435
LA GRANGE POP. 1217
MONTICELLO POP. 134
EWING POP. 400
FENWAY
DERRAHS
BENJAMIN
TOLONA
DOVER
DURHAM
MAYWOOD
WAKONDA STATE PARK
BURLINGTON
NORTHERN
WYACONDA RIVER
FABIUS
NORTH
MIDDLE
Durgens
Sugar
TO RTE. 2
TO KAHOKA
TO ALEXANDRIA
TO RTE. A
TO RTE. 61
TO RTE. 6
TO PALMYRA
40°–15'
91°–30'
T63N
T62N
T61N
T60N
R8W
R7W
R6W
R5W
SCALE
0 1 2 3 Miles

LINCOLN COUNTY

1 **Crouch Access** is north of Troy on Route H. This is a 2-acre area with no boat ramp on the Cuivre River.

2 **Cuivre Island Conservation Area** is 2 miles south of Old Monroe on Highway 79, and 5 miles east on Dalbow Road. The Mississippi River and Cuivre Slough are along this 1,341-acre area. There is a boat ramp and three trails totaling 6 miles in length.

3 **Kessler Memorial Wildlife Area** is 5 miles north of Troy on Highway 61, and 3 miles east on Highway KK. This area is 157 acres composed mainly of forestland and old pastures.

4 **B. K. Leach Memorial Conservation Area** consists of two tracts. The main tract is 3 miles south of Elsberry on Highway 79, and 3 miles east on Highway M. This 1,413-acre area borders the Mississippi River. Bates Slough, Houghmaster Slough and Burns Pond provide 30 acres of fishable water. It is noted for its freshwater marsh. The area hosts a waterfowl hunting program and has service roads and levees suitable for hiking. Access is restricted from October 15 through the close of waterfowl season.

5 **William R. Logan Conservation Area** consists of two tracts. The largest tract is 10 miles north of Troy on Highway 61, .25 mile west on Highway E, then 3 miles north on Highway RA. This 1,798-acre area is mostly forestland (1,180 acres) or old pastures.

6 **Norton Woods Access** (part of the Upper Mississippi Wildlife Area) is 4 miles south of Elsberrry on Highway 79, east on Highway M, then 1 mile left on County Road 927.

7 **Prairie Slough Conservation Area** is 5 miles northeast of Elsberry on Highway P. This 584-acre area has 50 acres for fishing and 406 acres of wet-mesic bottomland forest. It is a Designated Natural Area.

8 **Riggs Ferry Access** is 3 miles north of Troy on the frontage road adjacent to Highway 61. This is 2 acres with no boat ramp on the Cuivre River.

9 To reach **Sandy Island Conservation Area,** take Highway N east off Highway 79 to Lock and Dam #25. Visitors must then walk 300 yards to reach the 308-acre area.

10 **Upper Mississippi Conservation Area**–see the Lewis County map, page 120.

11 **Vonaventure Memorial Forest and Wildlife Area** is north of Silex on Highway UU. This is a 203-acre forest with one small pond (.25 acre for fishing).

12 The entrance to **Wm. G. and E. P. White Memorial Wildlife Area** is on Highway Z at Whiteside. This 810-acre area contains Lake Whiteside and nine ponds, providing anglers with 30 acres of fishable water. There is a boat ramp and disabled accessible fishing jetty.

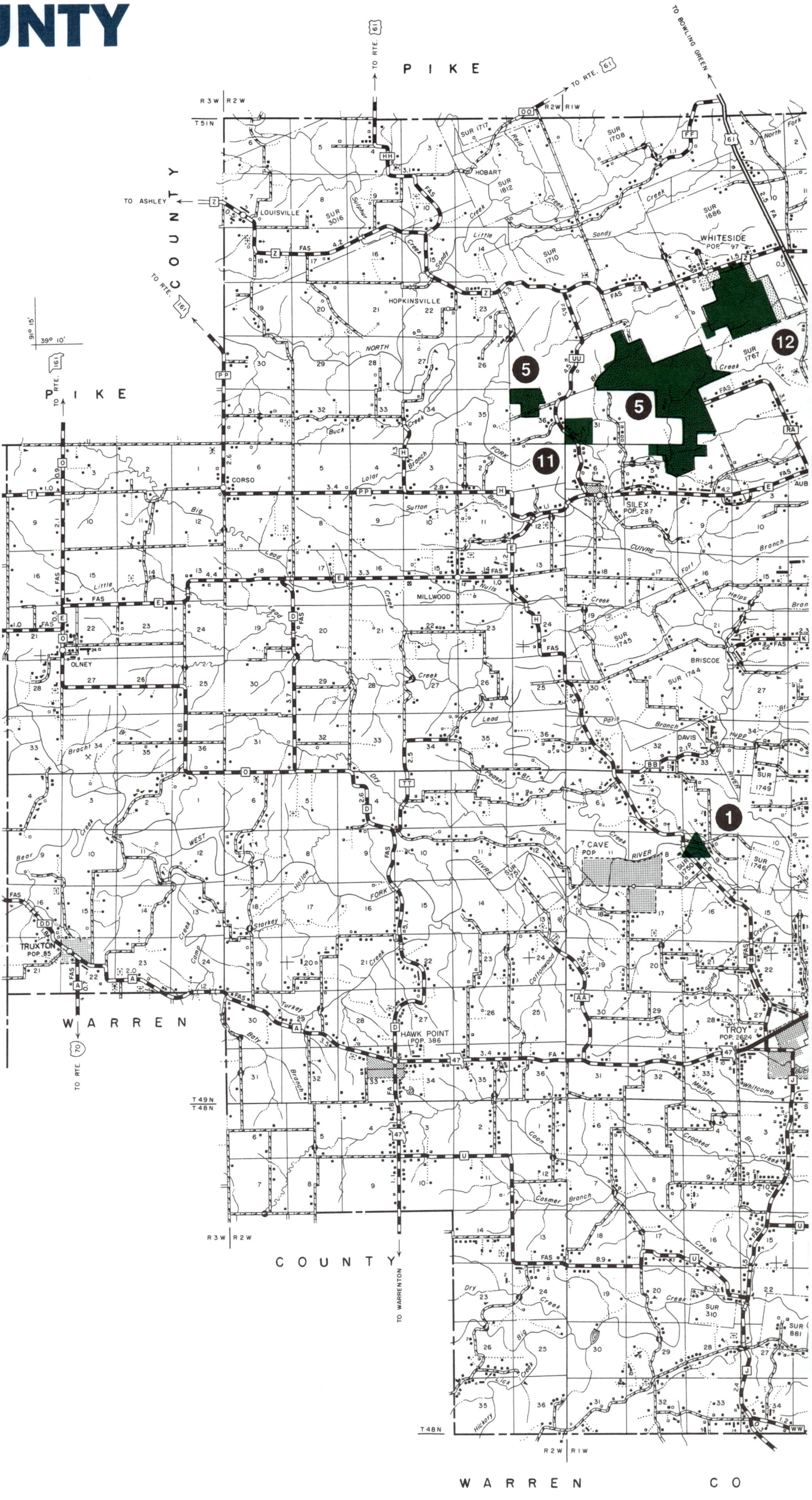

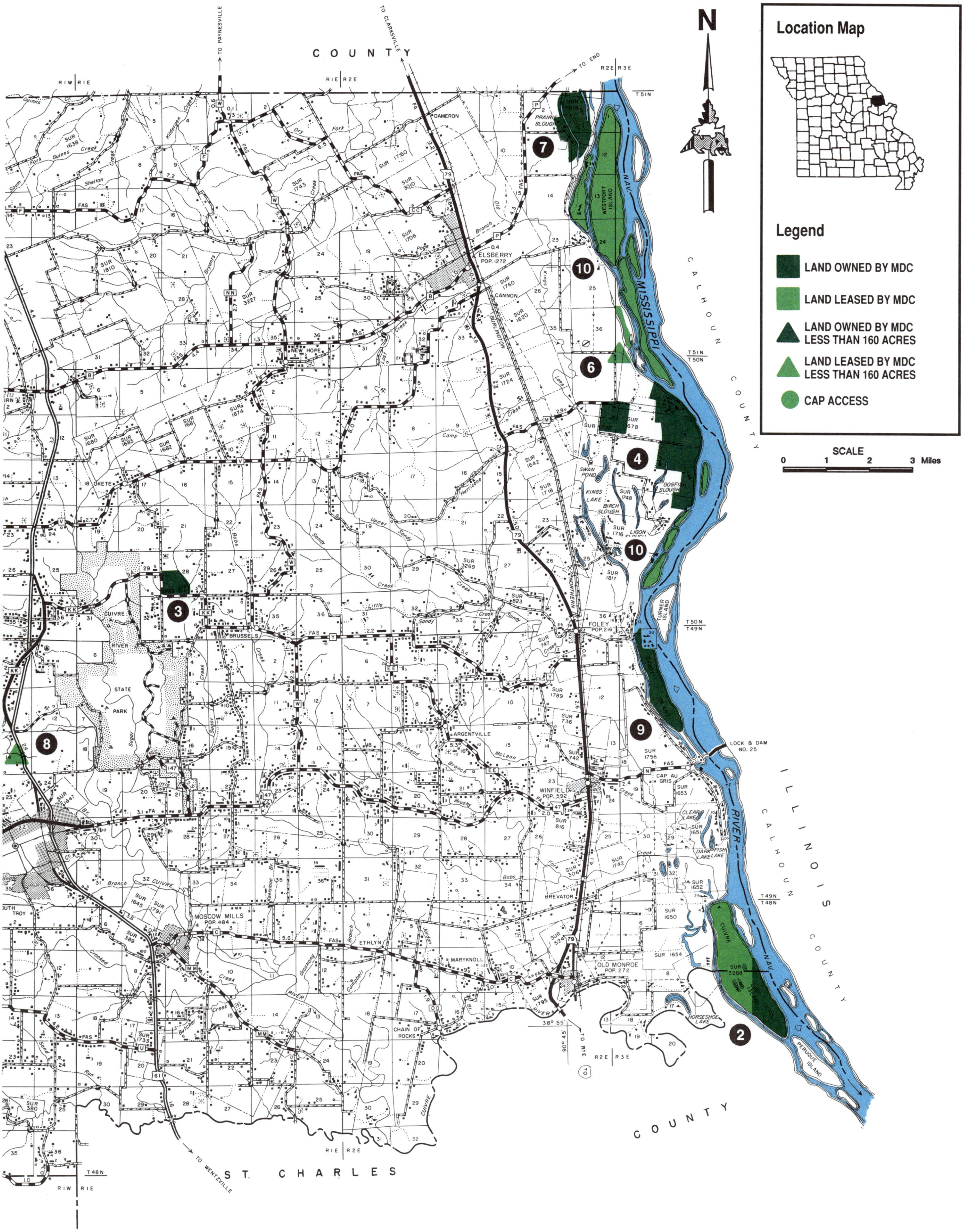
Location Map
Legend
LAND OWNED BY MDC
LAND LEASED BY MDC
LAND OWNED BY MDC LESS THAN 160 ACRES
LAND LEASED BY MDC LESS THAN 160 ACRES
CAP ACCESS
SCALE
0 1 2 3 Miles
N
MISSISSIPPI
RIVER
CALHOUN COUNTY
ILLINOIS
ST. CHARLES COUNTY
ELSBERRY POP. 1272
FOLEY POP. 216
WINFIELD POP. 592
MOSCOW MILLS POP. 484
OLD MONROE POP. 272
CUIVRE RIVER STATE PARK
LOCK & DAM NO. 25

LINN COUNTY

1 **Brookfield City Lake** is marked by a cantilever sign on Highway 11, 2 miles east of the city of Brookfield. The lake is 118 acres (with a boat ramp) and facilities on the surrounding land, such as picnic and camping areas, are regulated by the city of Brookfield.

2 **Fountain Grove Conservation Area** is on Route W, 5 miles south of the Meadville junction off Highway 36. This 7,154-acre area has three major bodies of water; Che-Ru Lake (160 acres), Bittern Marsh (60 acres) and Jo Shelby Lake (30 acres). There is a heron rookery, hiking trails, auto tour and picnic areas.

3 **Mussel Fork Conservation Area** is about 10 miles east of Brookfield on Highway 36. Mussel Fork Creek flows through this 2,491-acre area, and there are four ponds (12 acres total for fishing). The surrounding land is mainly pasture and forestland.

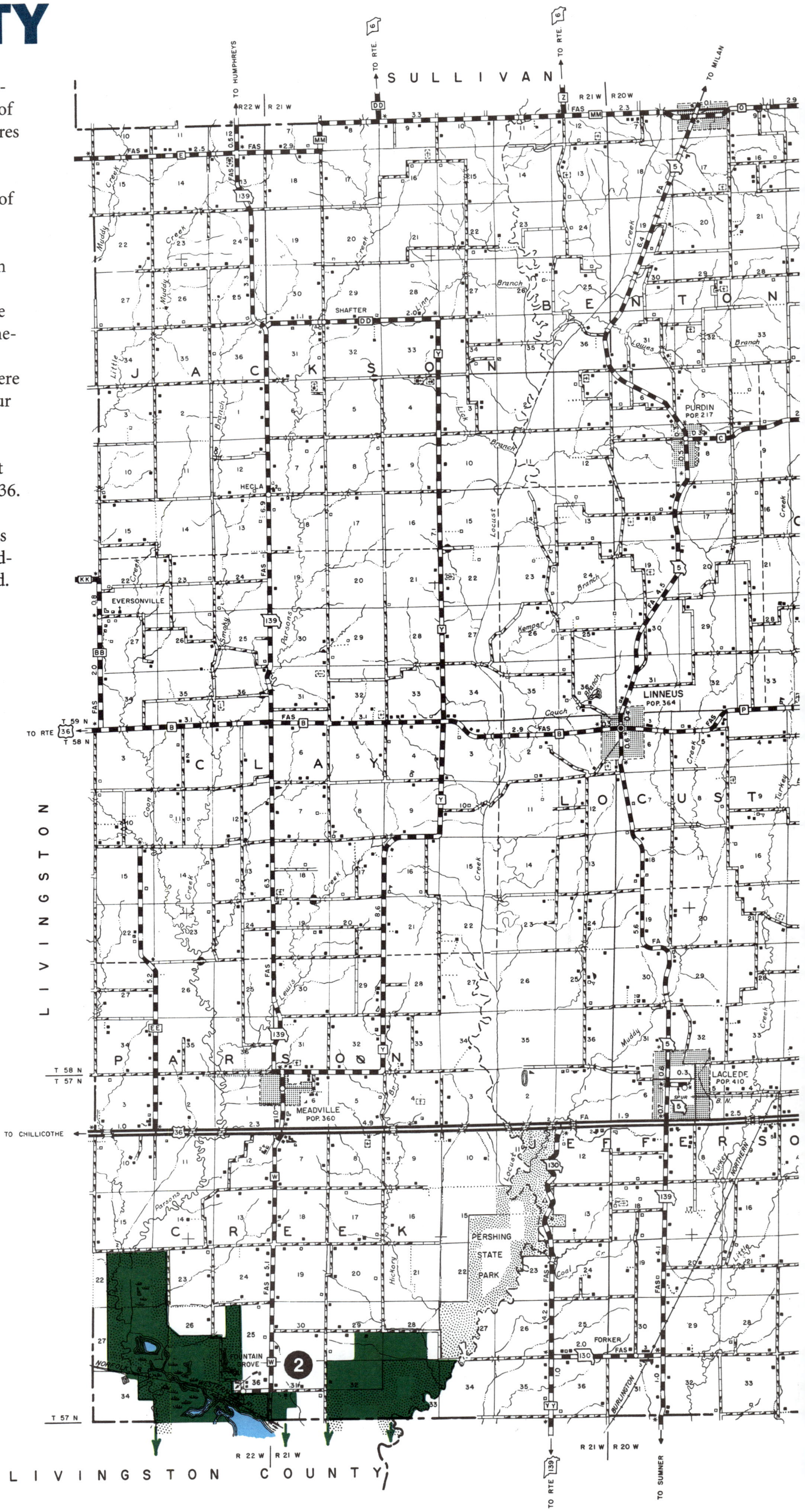

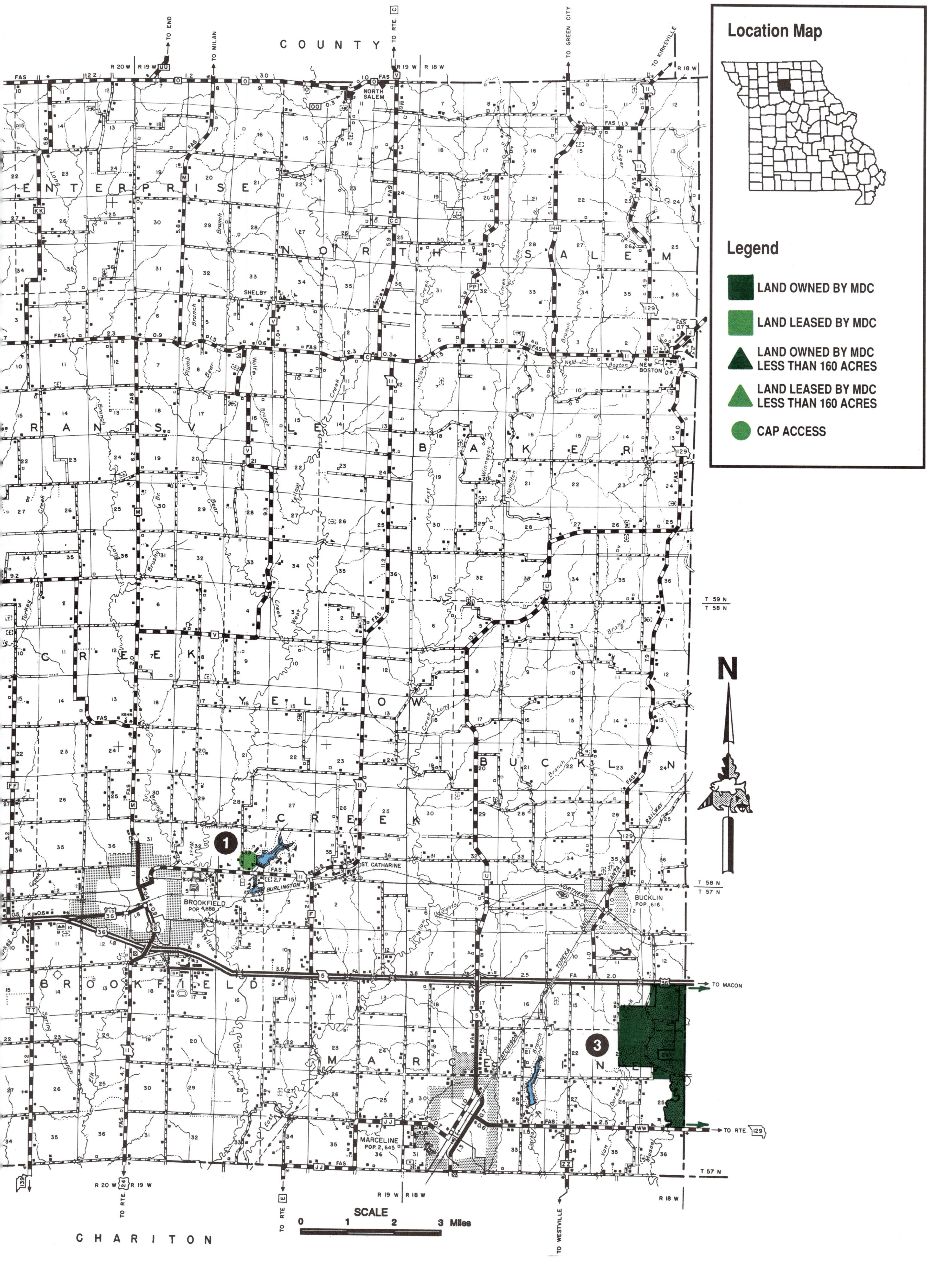

Location Map
Legend
LAND OWNED BY MDC
LAND LEASED BY MDC
LAND OWNED BY MDC LESS THAN 160 ACRES
LAND LEASED BY MDC LESS THAN 160 ACRES
CAP ACCESS
COUNTY
CHARITON
TO END
TO MILAN
TO RTE. C
TO GREEN CITY
TO KIRKSVILLE
TO MACON
TO RTE. 129
TO RTE. 24
TO RTE. E
TO WESTVILLE
R 20 W
R 19 W
R 18 W
T 59 N
T 58 N
T 57 N
ENTERPRISE
NORTH SALEM
BAKER
YELLOW CREEK
BUCKLIN
BROOKFIELD
MARCELINE
NORTH SALEM
SHELBY
NEW BOSTON
ST. CATHARINE
BROOKFIELD POP. 4,888
BUCKLIN POP. 616
MARCELINE POP. 2,645
BURLINGTON
NORTHERN
ATCHISON TOPEKA AND SANTA FE RAILWAY
N
SCALE
0 1 2 3 Miles

LIVINGSTON COUNTY

1 **Chillicothe Access** is off Highway 65, 1.25 miles south of the junction of highways 65 and 36 near the Grand River Bridge. The road to the access is north of the bridge on the east side of the highway. This is a 7-acre area with a boat ramp on the Grand River.

2 **Fountain Grove Conservation Area**–see the Linn County map, page 124.

3 **Poosey Conservation Area** is marked by a cantilever sign on Highway A, 1 mile east of the junction of highways A and U. This 5,591-acre area has 17 ponds and Indian Creek Community Lake, providing a total of 266 acres of fishable water. There is a 16.3-mile horseback riding trail and a retriever trial course available for use with a permit. A natural feature of interest includes a moist sandstone cliff.

4 To reach **Punkin Center Access**, go 2 miles north of the junction of highways W and Y (on Highway W). Take the second gravel road 1.25 miles east and watch for the area sign. This is a 74-acre area along the Thompson River.

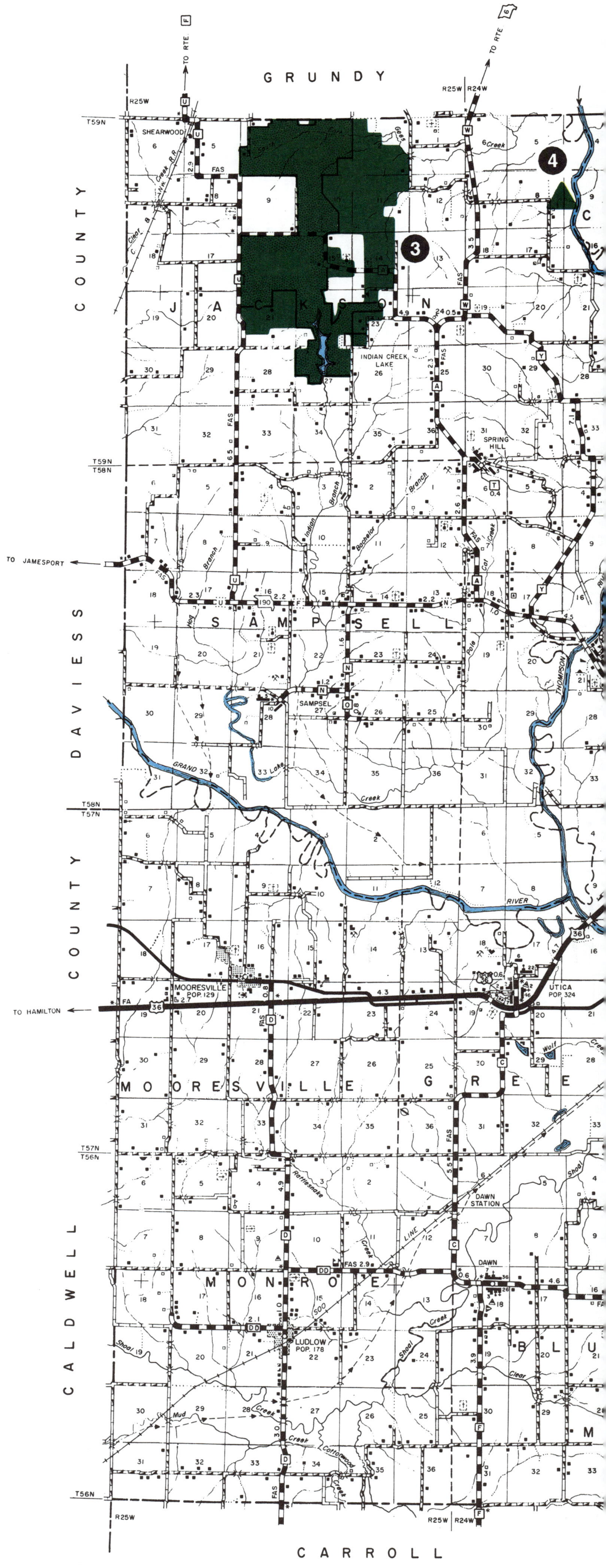

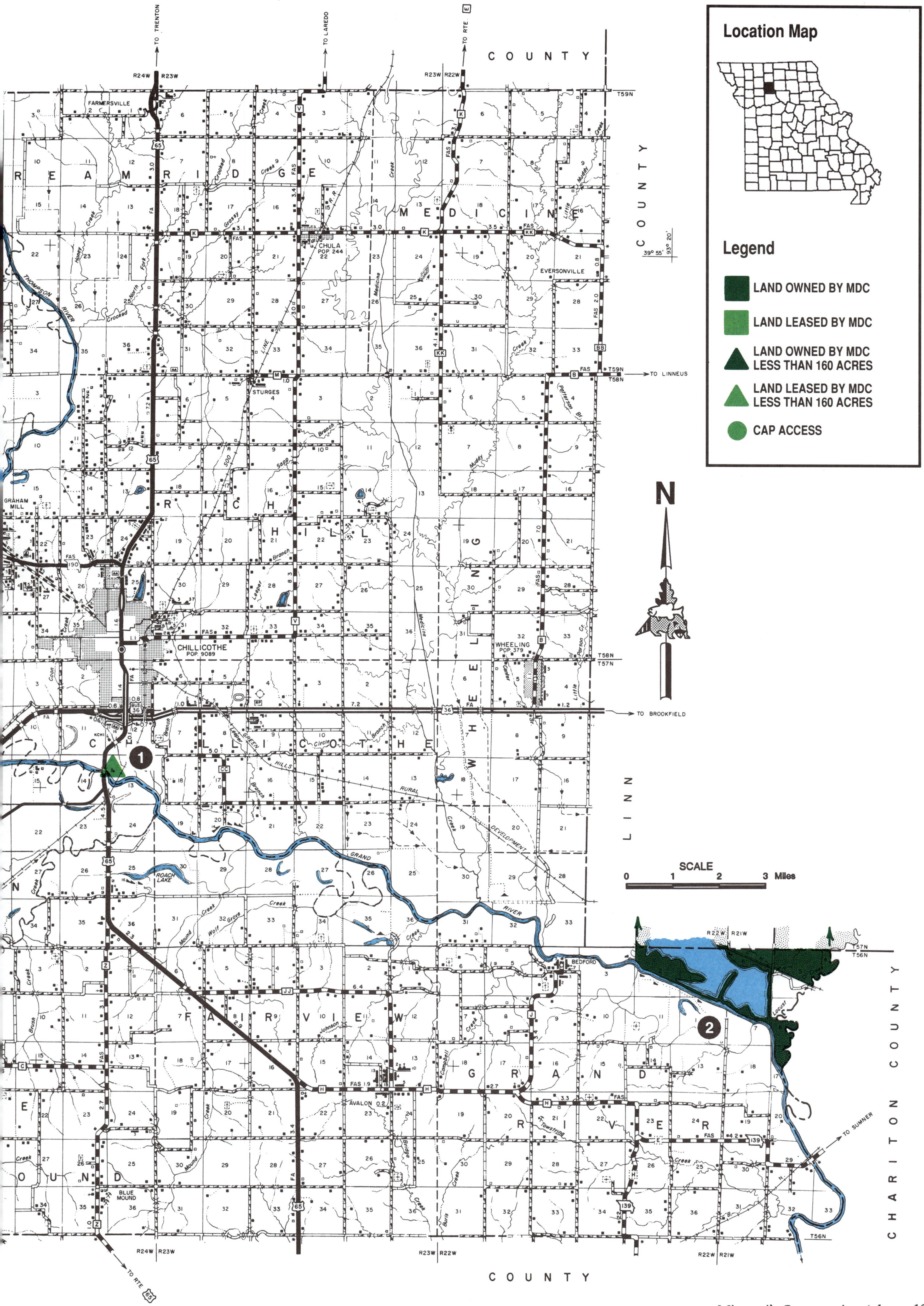
Location Map
Legend
LAND OWNED BY MDC
LAND LEASED BY MDC
LAND OWNED BY MDC LESS THAN 160 ACRES
LAND LEASED BY MDC LESS THAN 160 ACRES
CAP ACCESS
SCALE
0 1 2 3 Miles
N
TO TRENTON
TO LAREDO
TO RTE E
TO LINNEUS
TO BROOKFIELD
TO SUMNER
TO RTE 65
COUNTY
LINN COUNTY
CHARITON COUNTY
FARMERSVILLE
CHULA POP 244
EVERSONVILLE
STURGES
GRAHAM MILL
CHILLICOTHE POP 9089
WHEELING POP 379
BEDFORD
AVALON
BLUE MOUND
MEDICINE
RICH HILL
CHILLICOTHE
WHEELING
FAIRVIEW
GRAND RIVER
THOMPSON RIVER
GRAND RIVER
ROACH LAKE
GREEN HILLS RURAL DEVELOPMENT
R24W R23W
R23W R22W
R22W R21W
T59N
T58N
T57N
T56N
39° 55'
93° 20'

McDONALD COUNTY

1 **Buffalo Hills Natural Area** is 2.5 miles west of the junction of highways Y and B, on Highway B. About two-thirds of this 486-acre area is a Designated Natural Area of white oak, black oak and post oak. There is a 2-mile hiking trail, four ponds and a spring.

2 **Cowskin Access** is 8 miles north of Southwest City on Highway 43. This is a 41-acre area made up mainly of forestland on the Elk River (with a boat ramp).

3 To reach **Deep Ford Access**, take Highway W to Pineville. At the elementary school in town, turn east on the gravel road (unnamed) that runs parallel to Big Sugar Creek. The 42-acre area is 3.5 miles east on this road. The area has a boat ramp and stream stabilization demonstration site.

4 **Flag Spring Conservation Area**–see the Barry County map, page 14.

5 **Huckleberry Ridge Conservation Area** is southeast of Pineville on Highway K (4 miles east of Highway 71). This is a 2,106-acre forest with five ponds, a cave and a spring. Eighteen miles of interior trails and old roads frequently are used by horseback riders.

6 **Lanagan Towersite** is on Highway EE, 1 mile west of the junction of highways EE and 59, near the town of Lanagan. The area is 57 acres of forestland.

7 **Mount Shira Access** is on Highway 59, 2 miles north of Noel. This is a 26-acre area on the Elk River without a boat ramp.

8 **Powell Towersite** is on Highway 90, .5 mile west of the junction of highways 90 and E. This is an 80-acre forest.

9 **Town Hole Access** is on Main Street in the town of Anderson. The area consists of lawn and parking (2 acres total) on Indian Creek. There is no boat ramp.

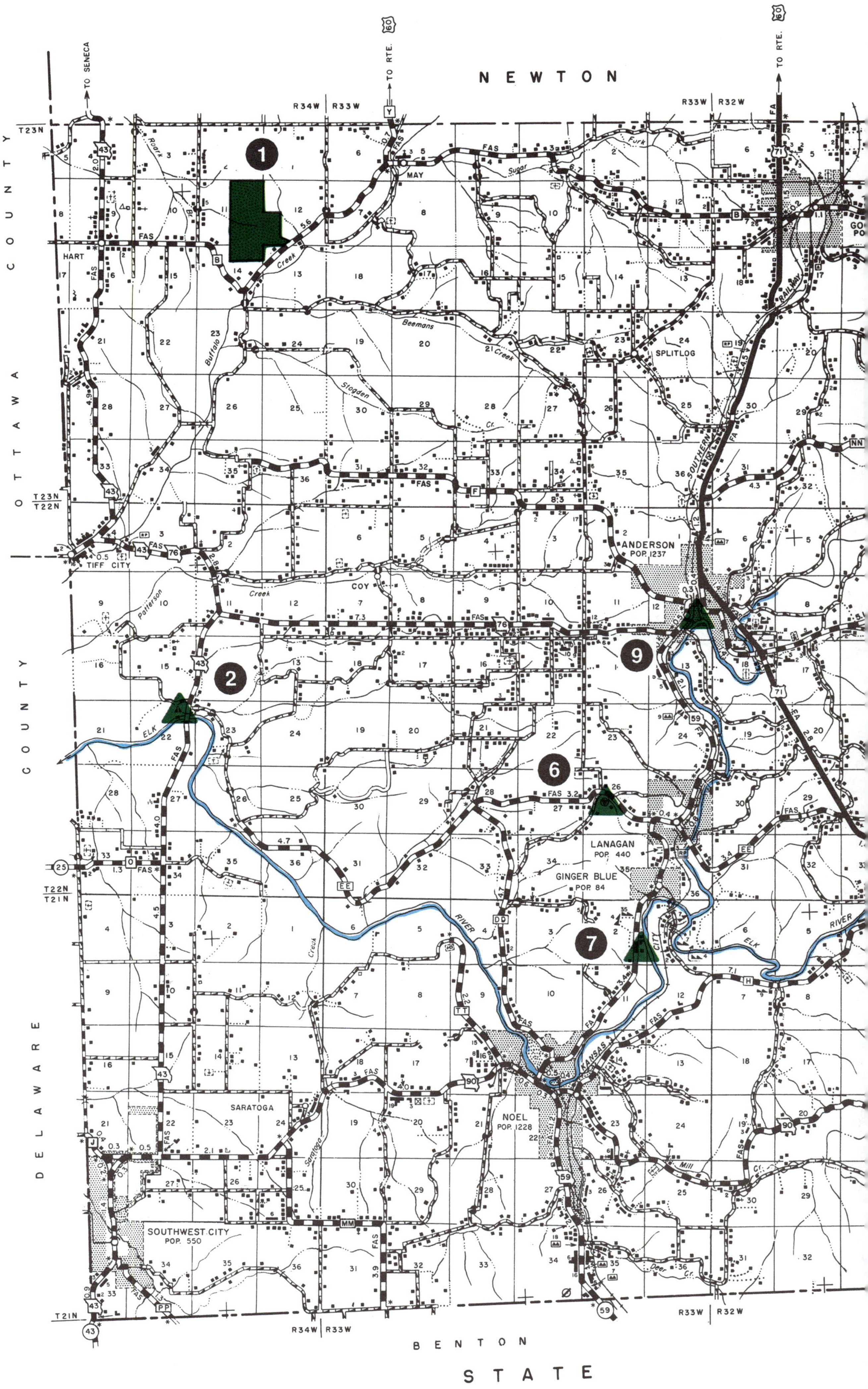

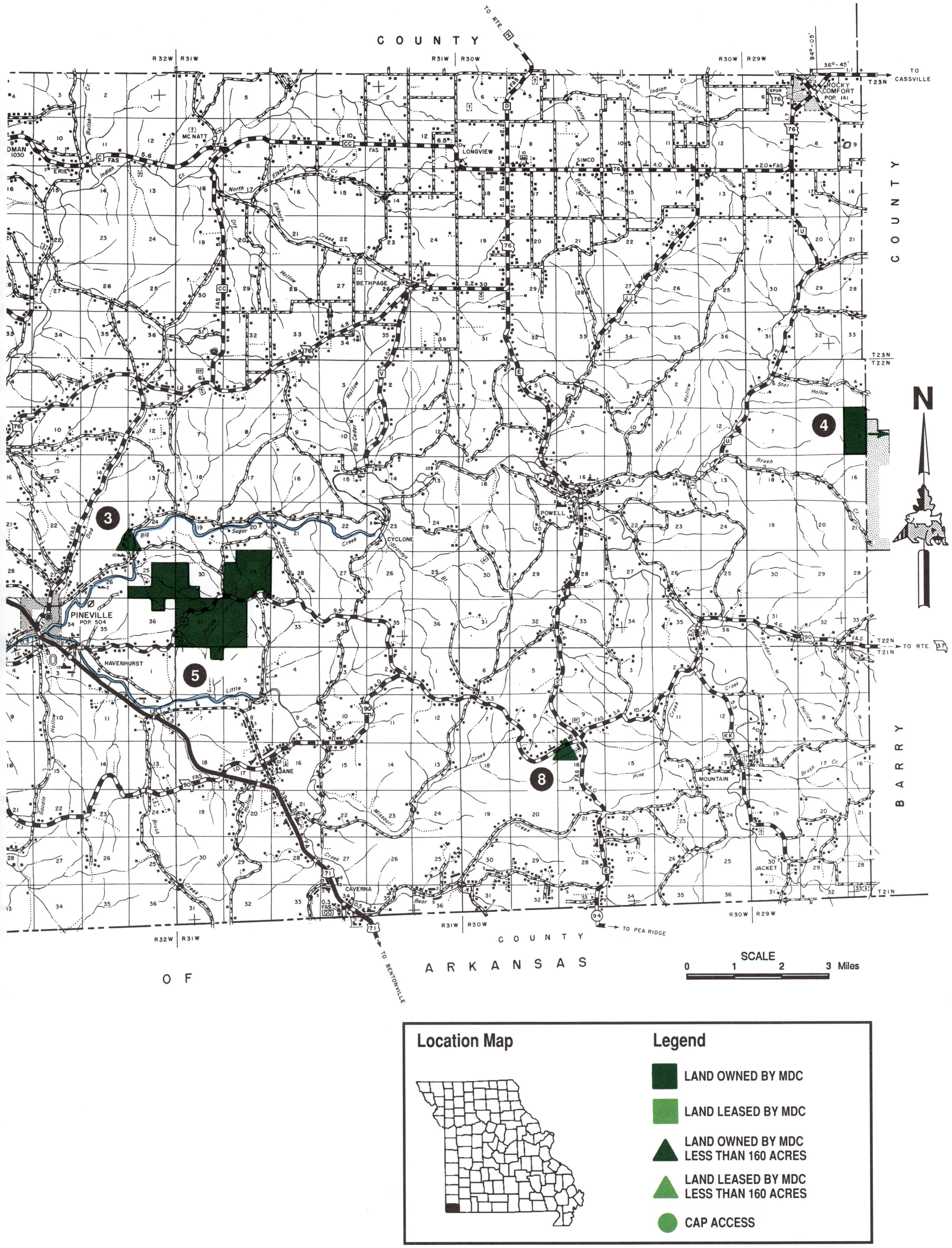

COUNTY
COUNTY
BARRY
ARKANSAS
OF
COUNTY
TO RTE. H
TO CASSVILLE
TO RTE. 37
TO PEA RIDGE
TO BENTONVILLE
R32W R31W
R31W R30W
R30W R29W
T23N
T22N
T21N
94°-05'
36°-45'
ROCKY COMFORT
POP. 141
MC NATT
ERIE
LONGVIEW
SIMCO
BETHPAGE
PINEVILLE
POP. 504
HAVENHURST
CYCLONE
POWELL
JANE
CAVERNA
MOUNTAIN
JACKET
N
SCALE
0
1
2
3 Miles
Location Map
Legend
LAND OWNED BY MDC
LAND LEASED BY MDC
LAND OWNED BY MDC LESS THAN 160 ACRES
LAND LEASED BY MDC LESS THAN 160 ACRES
CAP ACCESS

MACON COUNTY

1 **Atlanta City Lake** can be reached by turning west on Highway M at the flashing yellow light on Highway 63. At the Atlanta town square, turn south and follow this road about 1 mile. It will become a gravel road near the city limits and lead to the parking area. This is an 18-acre lake with a boat ramp. Only electric boat motors are permitted. Activities on the surrounding land are regulated by the city.

2 **Atlanta Conservation Area** is 8 miles north of Macon on Highway 63, 2.5 miles west on Axtell Road, then 2.5 miles north on Jupiter Road. There is a heron rookery, nine ponds and a wet-mesic bottomland forest on this 2,193-acre area.

3 **Bee Hollow Conservation Area** is 9 miles south of Macon on Highway 63. Turn 2.5 miles west on Jackpot Street and watch for the area on the south side of the road. This is a 271-acre area on the East Fork of the Little Chariton River. There are also oxbows and strip pits.

4 **Blees Lake** (10 acres) is 3 miles south of the junction of highways 63 and 36. Turn west at Lakeview Apartments.

5 To reach **Dodd Access**, take Highway 36 west of Callao, and turn north on Highway UU. Go north 3 miles, then take Echo Avenue 1 mile to the 40-acre area. This is an access with no boat ramp on the Chariton River.

6 **Griffith Memorial Conservation Area** is 2.5 miles north of New Cambria on Highway 149, 1 mile west on Derby, and 1 mile north on Crystal. This 128-acre area is on the west side of the road. It is mainly forestland (68 acres) and pastureland (60 acres).

7 **Hidden Hollow Conservation Area** is 7 miles west of La Plata on Highway 156, 1.5 miles south on Highway 3, then 3 miles west on Fox Road. This 1,360-acre area is nearly all forestland.

8 **LaPlata City Lake** (81 acres) is 2.5 miles east of the city of LaPlata on Highway 156, then 200 yards south on County Road 217 (marked with a cantilever sign). The city regulates activities on the surrounding land.

9 **Long Branch Lake** is 5 miles north of Macon on Highway 63, then 2 miles west on Axtell Road. The area is 2,454 acres, and the lake has a boat ramp and 940 acres of fishable water.

10 **Macon City Lake** is near County Road 606 off Highway 63, 1 mile north of Macon. Take 606 about 1 mile to the second road heading south. Follow this road to the 200-acre lake. The city of Macon maintains picnic areas and manages other activities on the surrounding land.

11 **Montgomery Woods Conservation Area**–see the Adair County map, page 6.

12 **Mussel Fork Conservation Area**–see the Linn County map, page 124.

13 **Redman Conservation Area** is .5 mile south of the town of Redman on Highway KK. The 120-acre area is on the west side of the road.

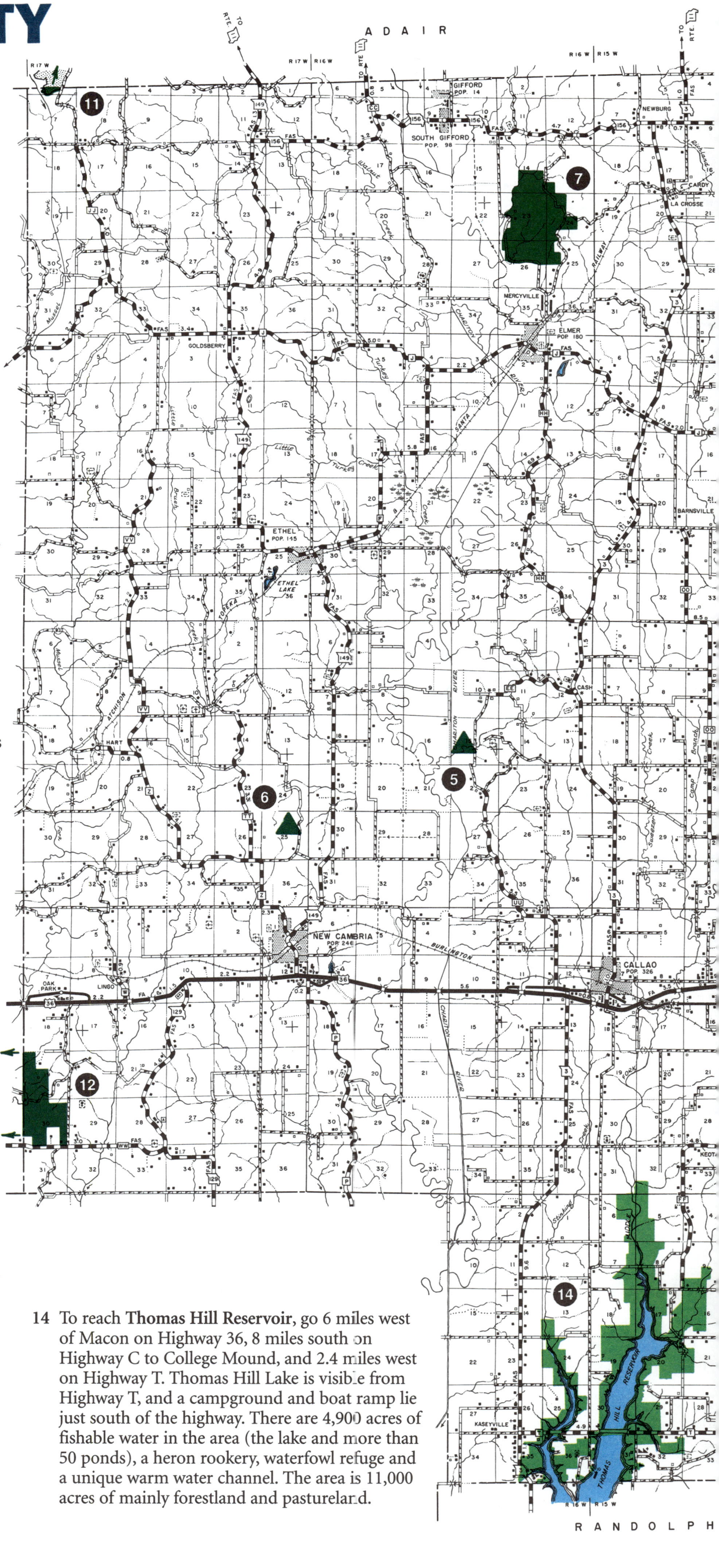

14 To reach **Thomas Hill Reservoir**, go 6 miles west of Macon on Highway 36, 8 miles south on Highway C to College Mound, and 2.4 miles west on Highway T. Thomas Hill Lake is visible from Highway T, and a campground and boat ramp lie just south of the highway. There are 4,900 acres of fishable water in the area (the lake and more than 50 ponds), a heron rookery, waterfowl refuge and a unique warm water channel. The area is 11,000 acres of mainly forestland and pastureland.

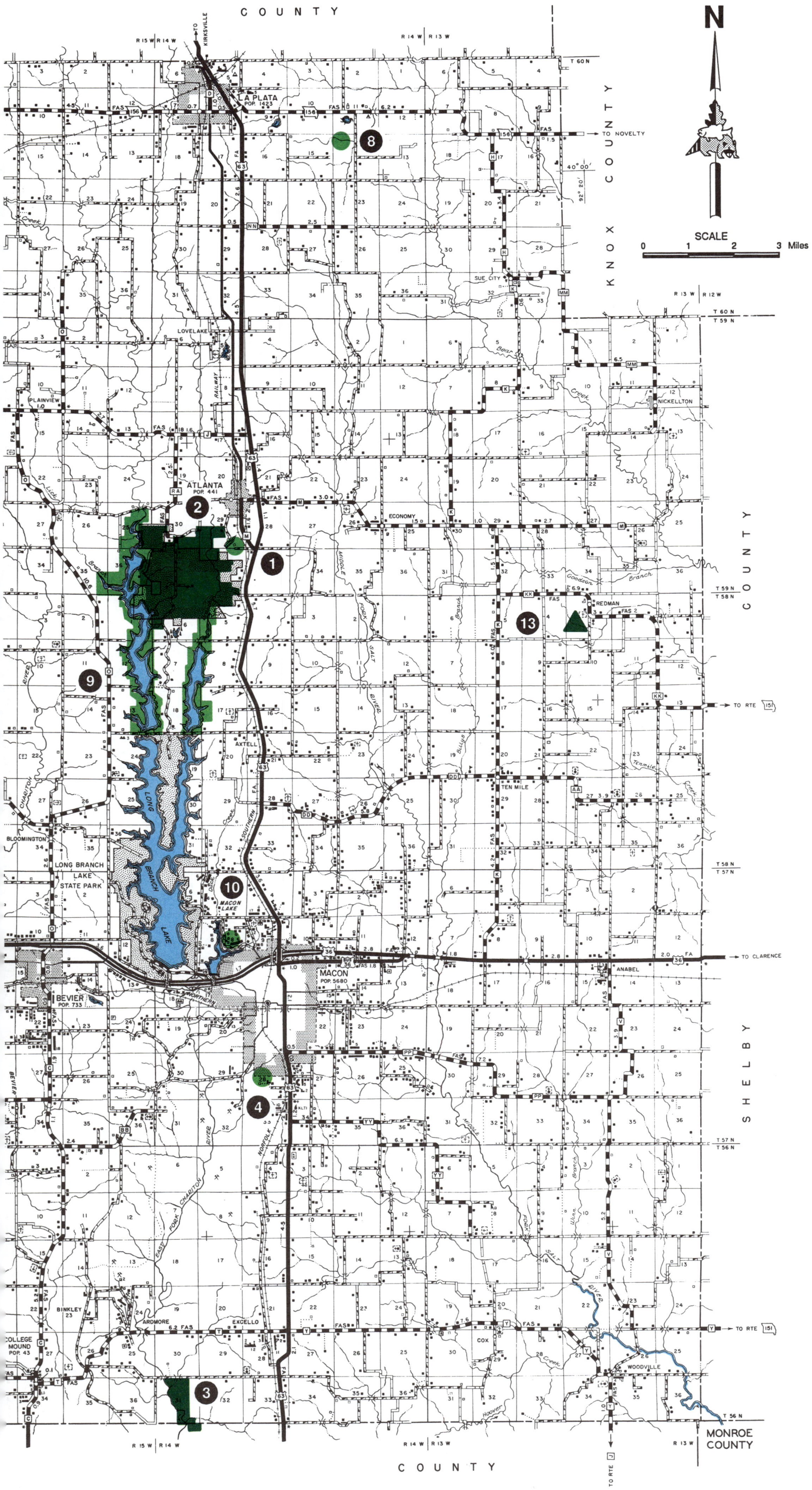

Location Map

Legend

- LAND OWNED BY MDC
- LAND LEASED BY MDC
- LAND OWNED BY MDC LESS THAN 160 ACRES
- LAND LEASED BY MDC LESS THAN 160 ACRES
- CAP ACCESS

MADISON COUNTY

1 **Amidon Memorial Conservation Area**–see the Bollinger County map, page 22.

2 **Fredericktown City Lake** (140 acres) is north of Fredericktown on Highway OO, then 1 mile west on County Road 209. The lake sits on 290 acres of land composed mainly of pasture and forestland. Anglers can reach Little St. Francis River above the lake by boat, and there is a 100-yard long shut-ins located below the dam. The area is equipped with a boat ramp, picnic areas, playground and a .25-mile hiking trail.

3 **Marquand Access** is 3.4 miles south of the city of Marquand on Highway DD. This is a 63-acre area on .5 mile of the Castor River.

4 **Millstream Gardens Conservation Area** is 8 miles west of Fredericktown on Highway 72. A natural area of interest on this 684-acre area includes 20 acres of shut-ins on the St. Francis River. There is a 1-mile asphalt hiking trail (wheelchair accessible), picnic areas, demonstration sites, archery range and a hand launch boat ramp. The river is suitable for kayaking.

5 **Roselle Access** is 12 miles west of Fredericktown on Highway 72. This is a 22-acre area on the St. Francis River with one boat ramp.

6 **Thompson Ford Access** is marked by signs on Highway 72, west of Fredericktown. This is an 84-acre area with a hand launch and public contact office.

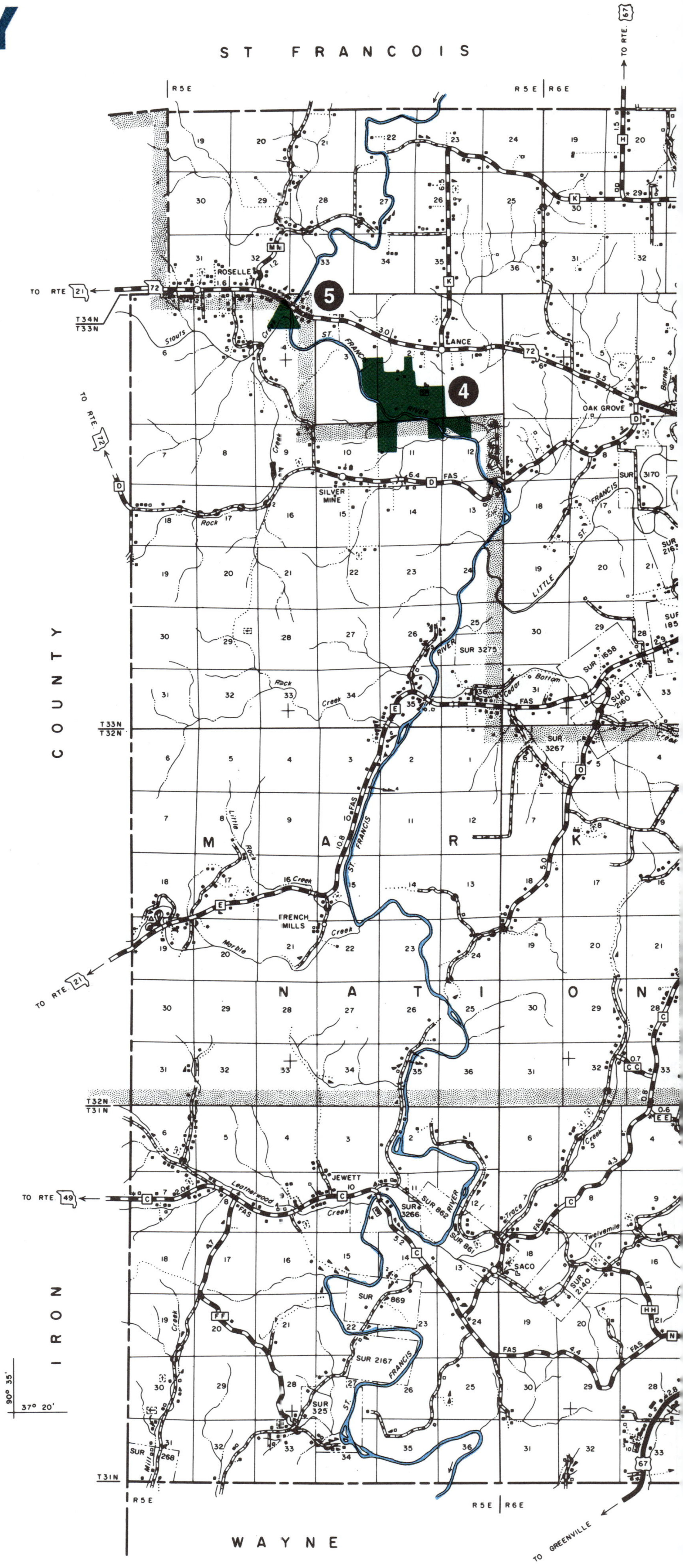

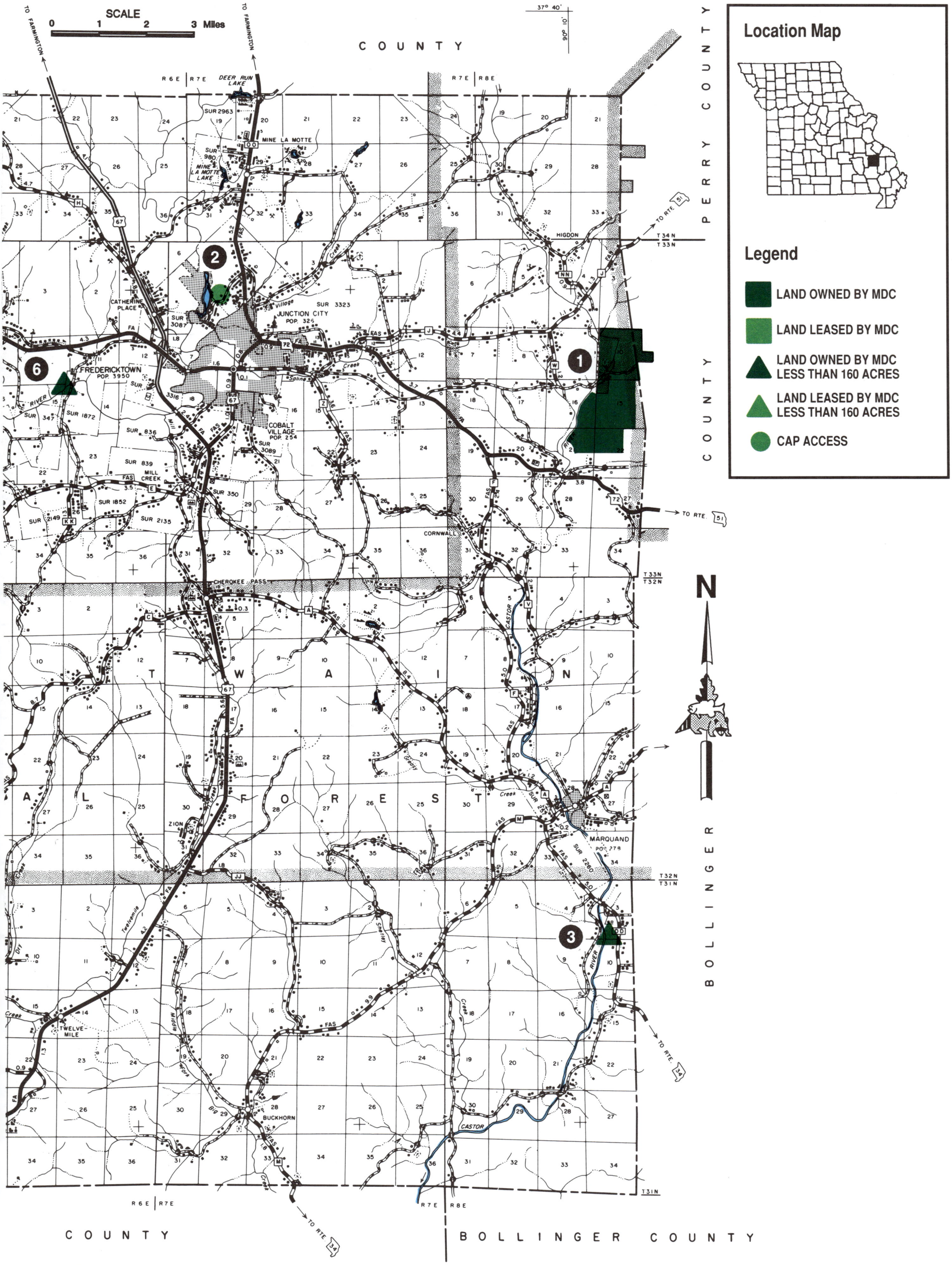

Location Map
Legend
LAND OWNED BY MDC
LAND LEASED BY MDC
LAND OWNED BY MDC LESS THAN 160 ACRES
LAND LEASED BY MDC LESS THAN 160 ACRES
CAP ACCESS
SCALE
0 1 2 3 Miles
TO FARMINGTON
TO FARMINGTON
COUNTY
PERRY COUNTY
COUNTY
BOLLINGER
BOLLINGER COUNTY
COUNTY
N
DEER RUN LAKE
MINE LA MOTTE
MINE LA MOTTE LAKE
HIGDON
CATHERINE PLACE
JUNCTION CITY POP. 325
FREDERICKTOWN POP. 3950
COBALT VILLAGE POP. 254
MILL CREEK
CORNWALL
CHEROKEE PASS
TWELVE-MILE
ZION
BUCKHORN
MARQUAND POP. 278
CASTOR
RIVER
TO RTE. 51
TO RTE. 51
TO RTE. 34
TO RTE. 34
T34N
T33N
T33N
T32N
T32N
T31N
T31N
R6E R7E
R7E R8E
37° 40'
90° 10'
MADISON
MADISON
FOREST
1
2
3
6

MARIES COUNTY

1 **Bell Chute Access** is 10 miles south of Vienna on Highway Y, then down County Road 513. Watch for the area sign. This is a 6-acre access with a boat ramp on the Gasconade River.

2 **Clifty Creek Conservation Area** and **Clifty Creek Natural Area** are both 6 miles northeast of Dixon at the end of Highway W. The natural area is 230 acres and closed to camping. It contains high bluffs and a natural bridge over a stream. The conservation area is a 256-acre forest and is open to campers.

3 **Freeburg Towersite** is 6 miles north of Vienna on Highway 63. The area is 12 acres.

4 **Paydown Access** is northeast of Vienna off Highway 63 (watch for the cantilever sign). This is a 5-acre area with a boat ramp and picnic area on the Gasconade River.

5 **Rinquelin Trail Community Lake** is 12 miles southwest of Vienna off Highway 133. Watch for signs to the 285-acre area. The area has about 29 acres of fishable water and is equipped with a boat ramp and two wheelchair accessible fishing docks.

6 **Spring Creek Gap Conservation Area** is 4 miles north of Vichy on Old Highway 63. This 1,819-acre area contains Spring Creek Gap Natural Area (40 acres) and a picnic area.

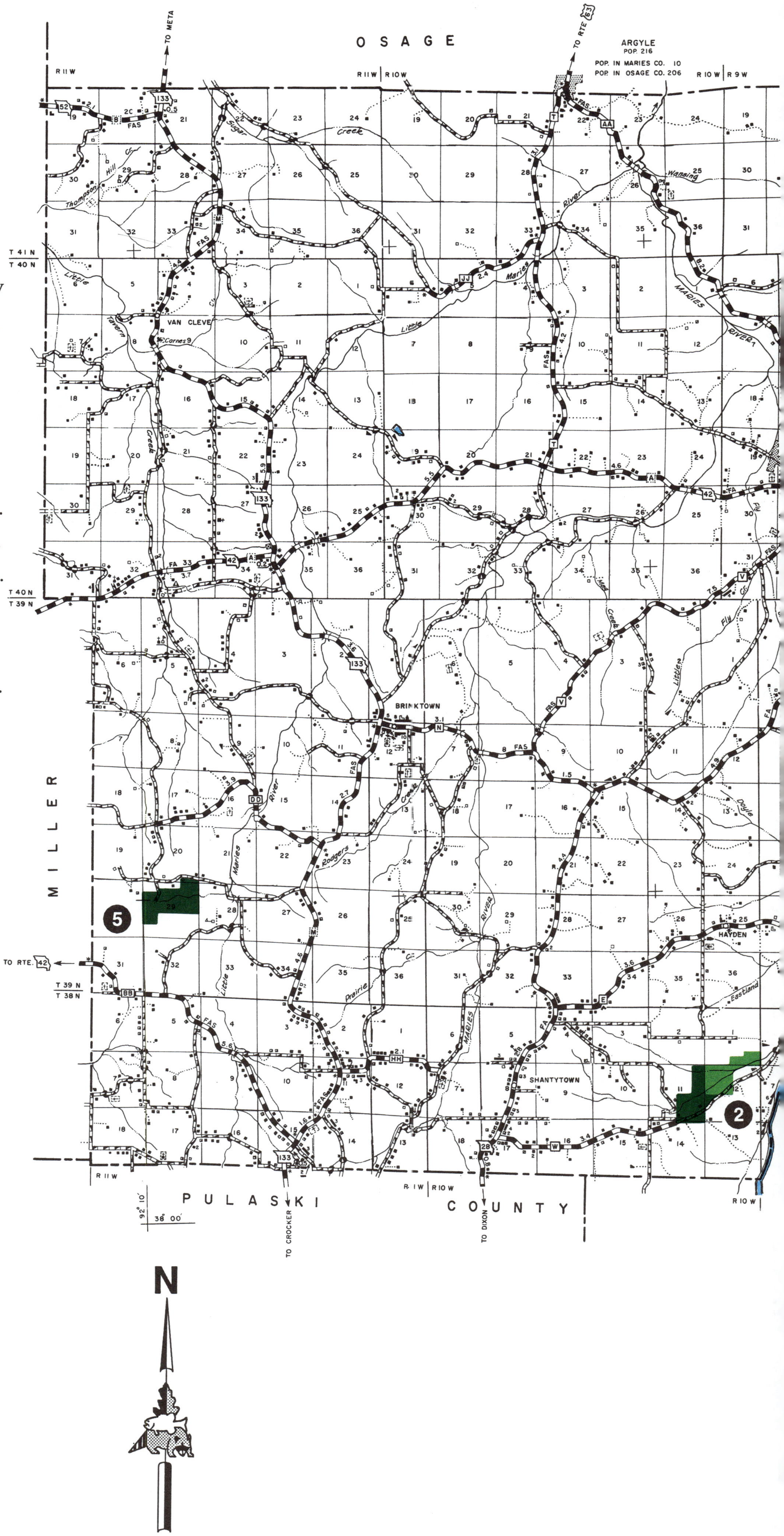

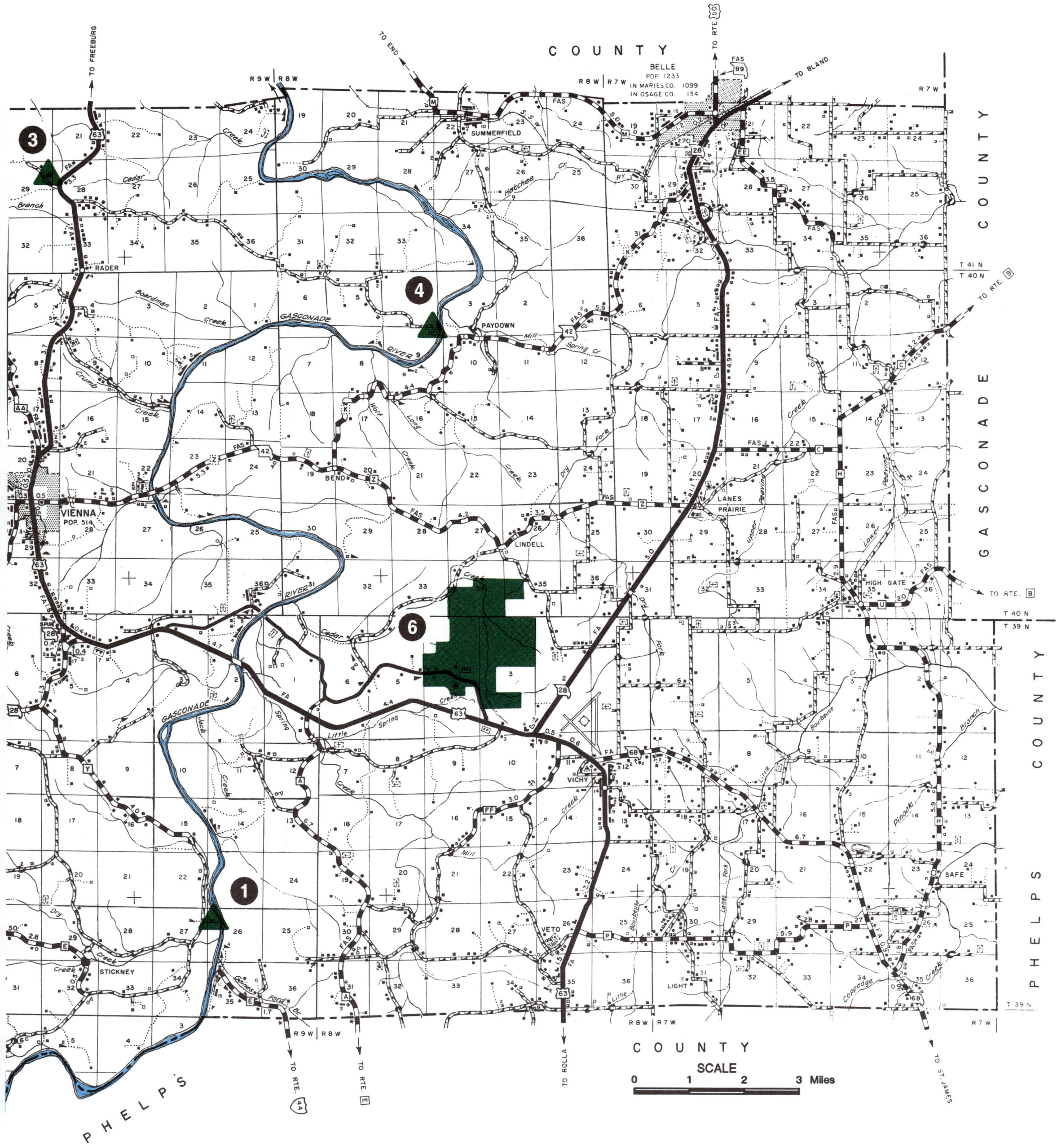
COUNTY
BELLE
POP. 1233
IN MARIES CO. 1099
IN OSAGE CO. 134
TO FREEBURG
TO END
TO RTE 50
TO BLAND
SUMMERFIELD
RADER
PAYDOWN
BEND
VIENNA
POP. 514
LANES PRAIRIE
LINDELL
HIGH GATE
VICHY
SAFE
STICKNEY
VETO
LIGHT
GASCONADE RIVER
GASCONADE COUNTY
PHELPS COUNTY
PHELPS
COUNTY
TO RTE. B
TO RTE. 44
TO RTE. E
TO ROLLA
TO ST. JAMES
T 41 N
T 40 N
T 39 N
R 9 W
R 8 W
R 7 W
SCALE
0 1 2 3 Miles
1
3
4
6

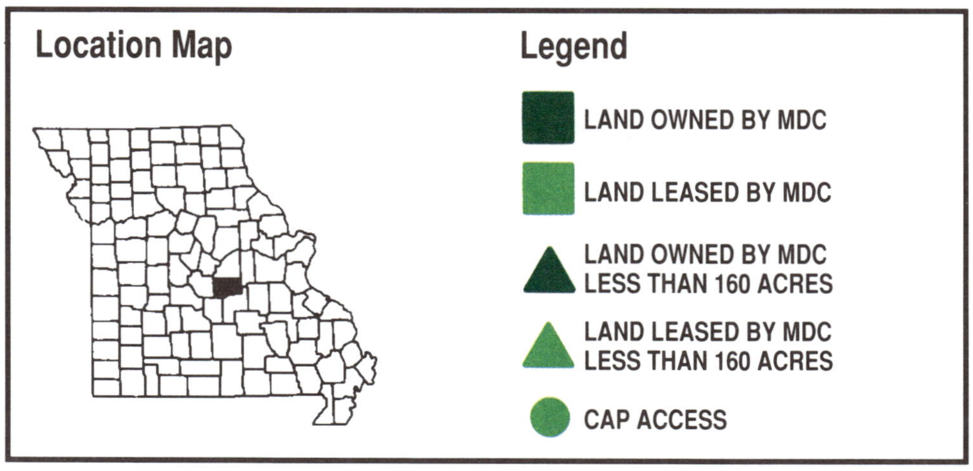
Location Map
Legend
LAND OWNED BY MDC
LAND LEASED BY MDC
LAND OWNED BY MDC LESS THAN 160 ACRES
LAND LEASED BY MDC LESS THAN 160 ACRES
CAP ACCESS

MARION COUNTY

1 To reach **Black Hawk Access**, go 3 miles north of Philadelphia on Highway D, 2 miles east to Emerson on Highway M, 1 mile south on County Road 149, and 1.5 miles east on County Road 134. About 1.4 miles of the South Fabius River flows along this 141-acre area. There is seasonal canoeing on the river.

2 **Callaghan Mound Access** is 2.5 miles west of Philadelphia on Highway 168, and .25 mile south on County Road 197 (gravel). This is a 127-acre area on the North River. There is no boat ramp.

3 **Dunn Ford Access** is 2 miles north of Philadelphia on Highway D, .5 mile west on Highway J, then 1 mile north on County Road 123 (gravel). One and a half miles of the South Fabius River flows along this 132-acre area. There is no boat ramp.

4 **Elmslie Memorial Conservation Area** is 1 mile east of Warren on Highway C, and 1.5 miles south on County Road 229 (gravel). The South Fork of North River flows .5 mile through this 238-acre area. Natural areas include 100-acre Elmslie Forest made up of northern hardwoods, bottomland hardwoods, upland white oak, cove hardwoods and bluffs.

5 **Fabius Chute Access** is 2.5 miles north of Palmyra on Highway 61, 2 miles east on County Road 332, then 2 miles northeast on County Road 342. This 38-acre area is an access on the Mississippi River and North River. There is a boat ramp.

6 **McPike Access** is 1.5 miles northeast of Emerson on County Road 45. This is a 78-acre area.

7 **J. Thad Ray Memorial Wildlife Area** is near the northwest edge of the Hannibal city limits on highways W and 61. There is one archery range on this 147-acre forested area and a spring.

8 **Soulard Access** is 6 miles north of Palmyra on Highway 61, then 1.5 miles east on County Road 344 (gravel). There are several bottomland sloughs on this 270-acre area, and 2.6 miles of the Fabius River flows through it (about 10 acres for fishing). There is a boat ramp and five fishing jetties.

9 **Julian Steyermark Woods Conservation Area** is just north of Hannibal on Highway 168, then 1 mile east on County Road 404. This is a 76-acre area made up of forestland (64 acres) and old fields (12 acres).

10 **Sunrise Access** is 3 miles west of Highway 61 on Highway 6, then 1 mile west on County Road 150 (gravel). This is a 40-acre area on the South Fabius River.

11 **Upper Mississippi Conservation Area**–see the Lewis County map, page 120.

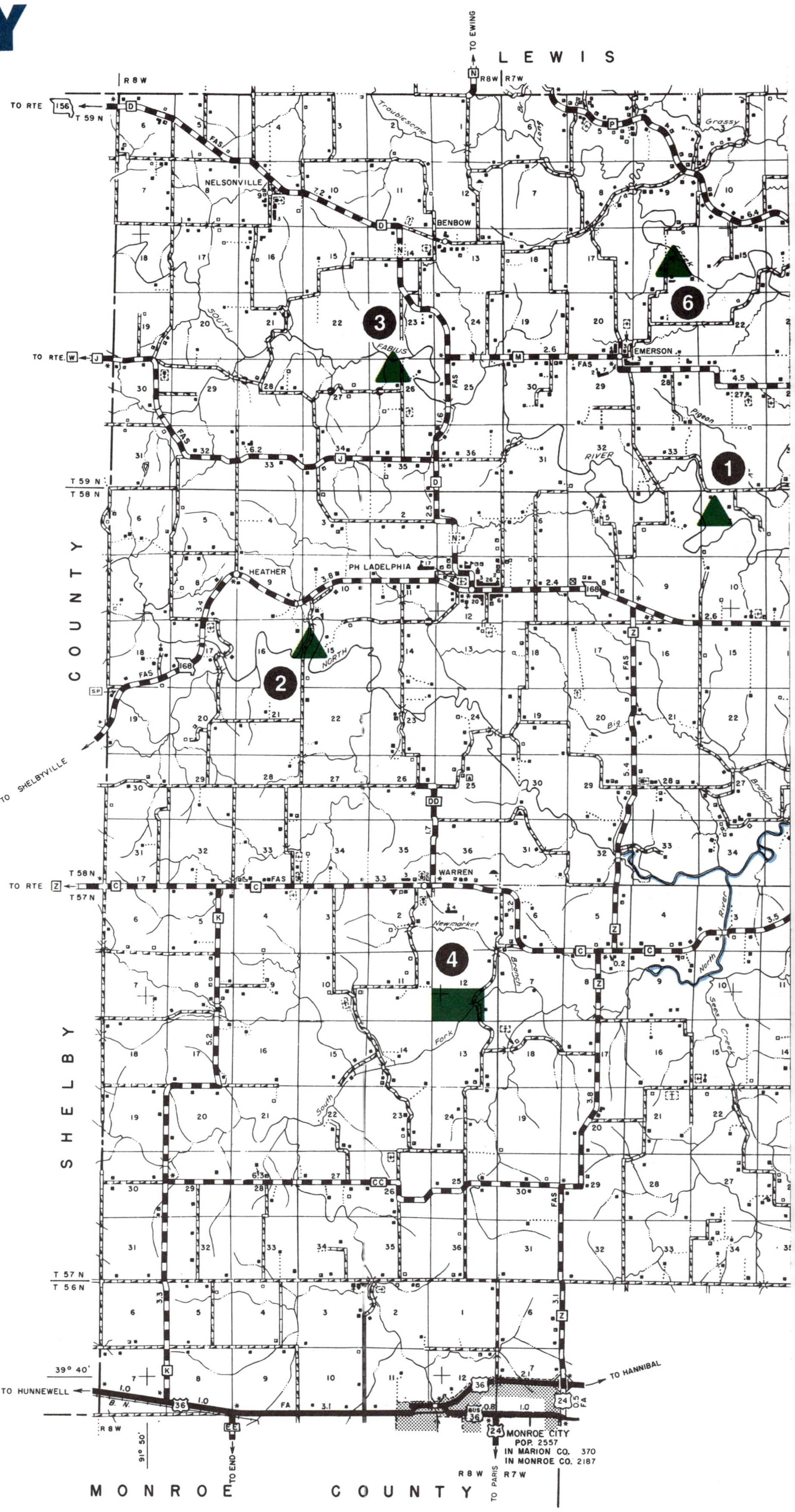

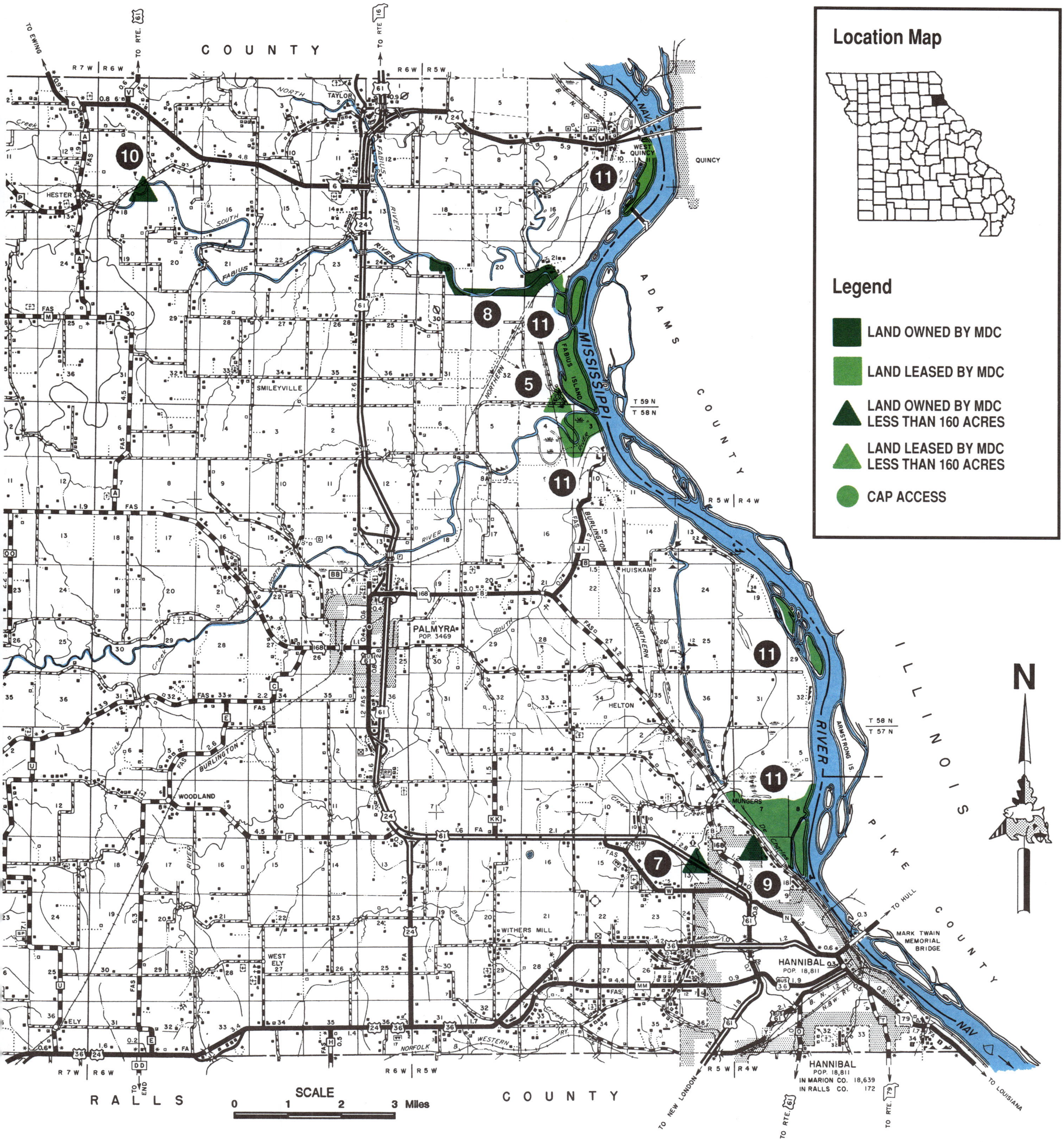
Location Map
Legend
LAND OWNED BY MDC
LAND LEASED BY MDC
LAND OWNED BY MDC LESS THAN 160 ACRES
LAND LEASED BY MDC LESS THAN 160 ACRES
CAP ACCESS
N
SCALE
0 1 2 3 Miles
COUNTY
ADAMS COUNTY
ILLINOIS
PIKE COUNTY
RALLS COUNTY
MISSISSIPPI
RIVER
FABIUS ISLAND
ARMSTRONG IS.
QUINCY
WEST QUINCY
TAYLOR
HESTER
SMILEYVILLE
PALMYRA
POP. 3469
HUISKAMP
HELTON
WOODLAND
WITHERS MILL
WEST ELY
ELY
MUNGERS
HANNIBAL
POP. 18,811
IN MARION CO. 18,639
IN RALLS CO. 172
MARK TWAIN MEMORIAL BRIDGE
TO EWING
TO RTE. 61
TO RTE. 16
TO NEW LONDON
TO RTE. 79
TO LOUISIANA
TO HULL
T 59 N
T 58 N
T 57 N
R 7 W
R 6 W
R 5 W
R 4 W
NAV
10
11
8
5
7
9

MERCER COUNTY

1 **Chloe Lowry Marsh Conservation Area** is near Princeton. Go 2 miles north on Road FF, then 2 miles on a gravel road that leads to Lowry Memorial Airport. This 115-acre Designated Natural Area is past the airport where the road becomes a dead end. There are several species of rare plants found here, including tufted loosestrife and marsh skullcap, which are are not found anywhere else in the state. A .5-mile hiking trail (an old railroad bed) runs through the area.

2 **Lake Paho Conservation Area** (1,782 acres) is 4 miles west of Princeton on Highway 136. There are 50 ponds in the area and, together with Lake Paho, provide anglers with 283 acres of fishable water. There are two boat ramps, one wheelchair accessible fishing jetty, 50 camping pads and about 5 miles of field roads available to hikers. Interpretive programs and demonstration sites are open, and tours of the area's catfish rearing facility are available upon request.

Location Map

Legend

- LAND OWNED BY MDC
- LAND LEASED BY MDC
- LAND OWNED BY MDC LESS THAN 160 ACRES
- LAND LEASED BY MDC LESS THAN 160 ACRES
- CAP ACCESS

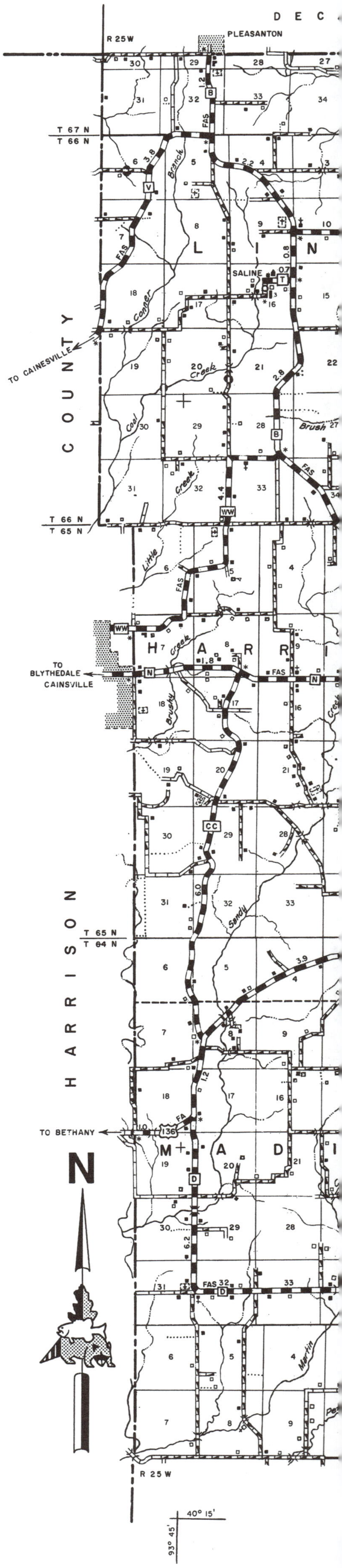

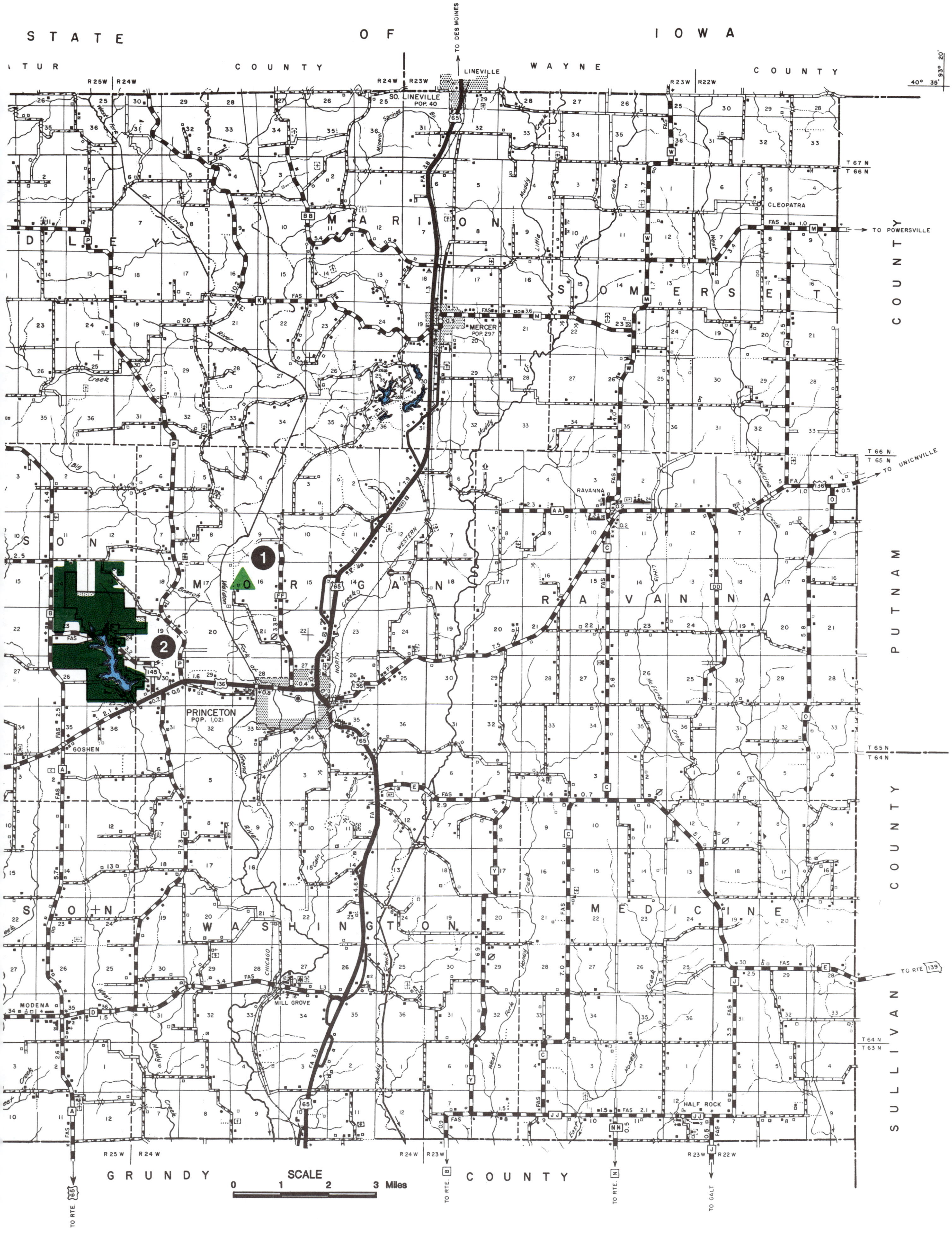
STATE OF IOWA
WAYNE COUNTY
TO DES MOINES
LINEVILLE
SO. LINEVILLE POP. 40
MARION
SOMERSET
CLEOPATRA
TO POWERSVILLE
MERCER POP. 297
RAVANNA
TO UNIONVILLE
MORGAN
PRINCETON POP. 1,021
GOSHEN
WASHINGTON
MEDICINE
MILL GROVE
MODENA
HALF ROCK
PUTNAM COUNTY
SULLIVAN COUNTY
TO RTE. 139
GRUNDY COUNTY
SCALE
0 1 2 3 Miles
TO RTE. 65
TO RTE. B
TO RTE. N
TO GALT
T 67 N
T 66 N
T 65 N
T 64 N
T 63 N
R 25 W
R 24 W
R 23 W
R 22 W

MILLER COUNTY

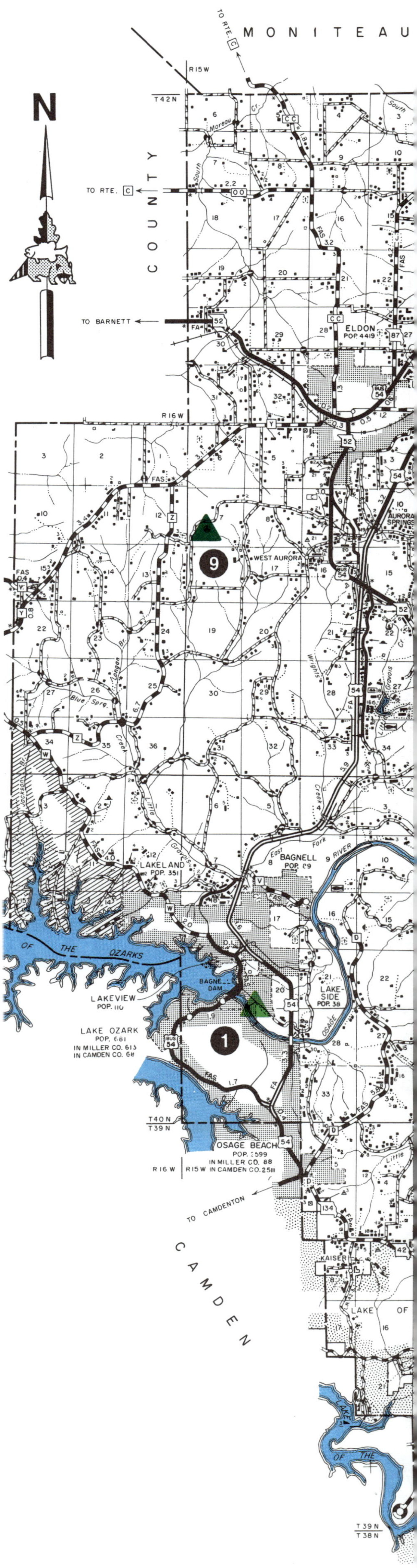

1 **Bagnell Dam Access** is immediately past the Osage River Bridge on Highway 54. Turn across from Lake Ozark Construction and Quarries, and the area is about .3 mile down the highway access road. This is a 10-acre area on the Osage River and is within .5 mile of Bagnell Dam and Lake of the Ozarks. There are two boat ramps.

2 **Bat Cave Conservation Area** is near St. Elizabeth. Take State Road E north from town to the end of the blacktop. Go left to the Osage-Tavern Access when the blacktop becomes gravel. At the access, cross the bridge and go left .1 mile. This 30-acre Designated Natural Area (forest) is indicated with signs. It borders the Osage River and is wintering and breeding grounds for the gray bat.

3 **Boeckman Bridge Access** is just west of St. Elizabeth on Highway 52. Watch for the area cantilever sign on the left. Take this gravel road to the second gravel road to the right (access sign). Turn right and follow signs for about 3 miles to the area. This is a 57-acre area with about a mile of frontage on Big Tavern Creek. There is no boat ramp.

4 To reach **Brays Access**, go east of Iberia on Highway 42 to the second gravel road on the left (about 2 miles). Take this road .75 mile, and watch for area boundary signs on the right. Tavern Creek flows .5 mile along this 30-acre area.

5 To reach **Eugene Towersite**, exit Highway 54 at the Eugene-Highway 17 exit. Drive to the end of the ramp, and take an immediate right on Old Highway 54. This 4-acre area is about a mile down this road.

6 **Kings Bluff Access** is near Eugene. Take Highway 17 south to Highway H. Take H to Marys Home, and continue straight ahead on a gravel road for .75 mile. Watch for signs to the area, and turn right on Gravel Road H8A. This road is also marked with signs and will lead to the area in about 3.5 miles. This is 6 acres on the Osage River and has a boat ramp.

7 **Madden Ford Access** is south of Iberia on Highway 17, and right on County Road 17-42. Stay on this road, cross the second low water bridge, and watch for boundary signs on the left. This is a 164-acre area on Tavern Creek. There is no boat ramp.

8 **Osage-Tavern Access** is on Route E outside of St. Elizabeth. When the Route E blacktop ends, go left on the gravel road about 1 mile to the 28-acre area. The area has a .25 mile of frontage on the Osage River and a boat ramp.

9 **Rocky Mount Towersite** is off Route Z, west of Eldon. From Route Z, turn left on the gravel road marked with signs for Dooley Cemetary. When this road ends, turn to the left, go .5 mile, and the tower is on the right. This is a 40-acre area of forestland.

10 **Saline Valley Conservation Area** is near Eldon, 2 miles west of the junction of highways M and 87, on Highway M. This 4,782-acre area contains sections of the Osage River, Big Saline Creek, Jack Buster Creek, Little Saline Creek and Jim Henry Creek, providing anglers with about 12 acres of fishable water.

11 **Tuscumbia Access** is marked by a cantilever sign on Highway 52, just south of the Tuscumbia city limits and north of the highway 52 and 17 junction. This is 1 acre on .10 mile of Osage River frontage. There is a boat ramp.

12 **Ulman Towersite** is off Highway 17 south of Highway 52, then right on Route C. This is a 15-acre area.

13 **Wilson Camp Access** is south of Tuscumbia on Highway 17, 3 miles east on Route A, then a .5 mile down a gravel road that is marked with a cantilever sign. This is 150 acres on Tavern Creek with no boat ramp.

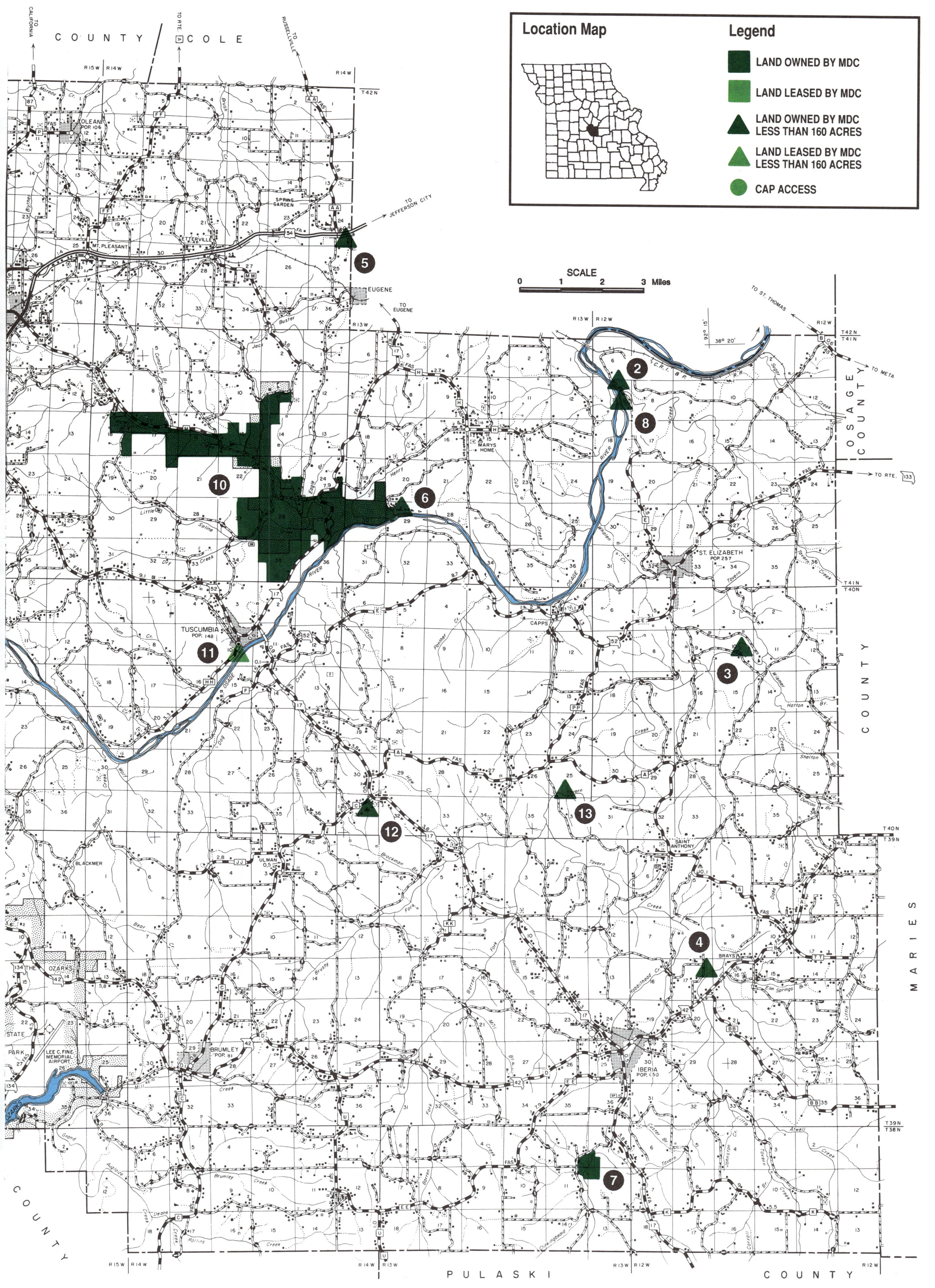
Location Map
Legend
LAND OWNED BY MDC
LAND LEASED BY MDC
LAND OWNED BY MDC LESS THAN 160 ACRES
LAND LEASED BY MDC LESS THAN 160 ACRES
CAP ACCESS
SCALE
0 1 2 3 Miles
COLE COUNTY
OSAGE COUNTY
MARIES COUNTY
PULASKI COUNTY
TO CALIFORNIA
TO RUSSELLVILLE
TO JEFFERSON CITY
TO EUGENE
TO ST. THOMAS
TO META
TO RTE. 133
OLEAN POP. 106
SPRING GARDEN
MT. PLEASANT
ETTERVILLE
EUGENE
MARYS HOME
ST. ELIZABETH POP. 257
TUSCUMBIA POP. 148
CAPPS
SAINT ANTHONY
ULMAN
BLACKMER
BRAYS
IBERIA POP. 730
BRUMLEY POP. 81
LEE C. FINE MEMORIAL AIRPORT
STATE PARK
THE OZARKS
OSAGE RIVER
T42N
T41N
T40N
T39N
T38N
R15W
R14W
R13W
R12W
1
2
3
4
5
6
7
8
10
11
12
13

MISSISSIPPI COUNTY

1 **Big Oak Lake** (22 acres) is south of East Prairie off Highway 102. This 1,004-acre area is designated a National Natural Landmark. There is 80 acres of virgin wet-mesic bottomland hardwoods and two trails. One trail is accessible to wheelchairs. The area is home to several plants and animals on the Endangered Species List. A nature center with a variety of exhibits and programs is also open to visitors.

2 **Bird's Blue Hole** is at the junction of County Road 301 and a levee, off Highway 62. This is a 77-acre area with a pond (27 acres of fishable water and a boat ramp), and is within a mile of the Mississippi River.

3 **Robert G. Delaney Lake Conservation Area** is north of Charleston on Highway N, then .5 mile west on County Road 222. This is a 159-acre area with a boat ramp, disabled accessible fishing jetty and picnic area.

4 **Dorena Access** is near the levee at the junctions of county roads 507 and 510. This is a 4-acre area with a boat ramp on the Mississippi River.

5 **Joseph Hunter Moore Access** is east of Wyatt at the junction of Highway HH and County Road 303. This is a 10-acre area with a boat ramp.

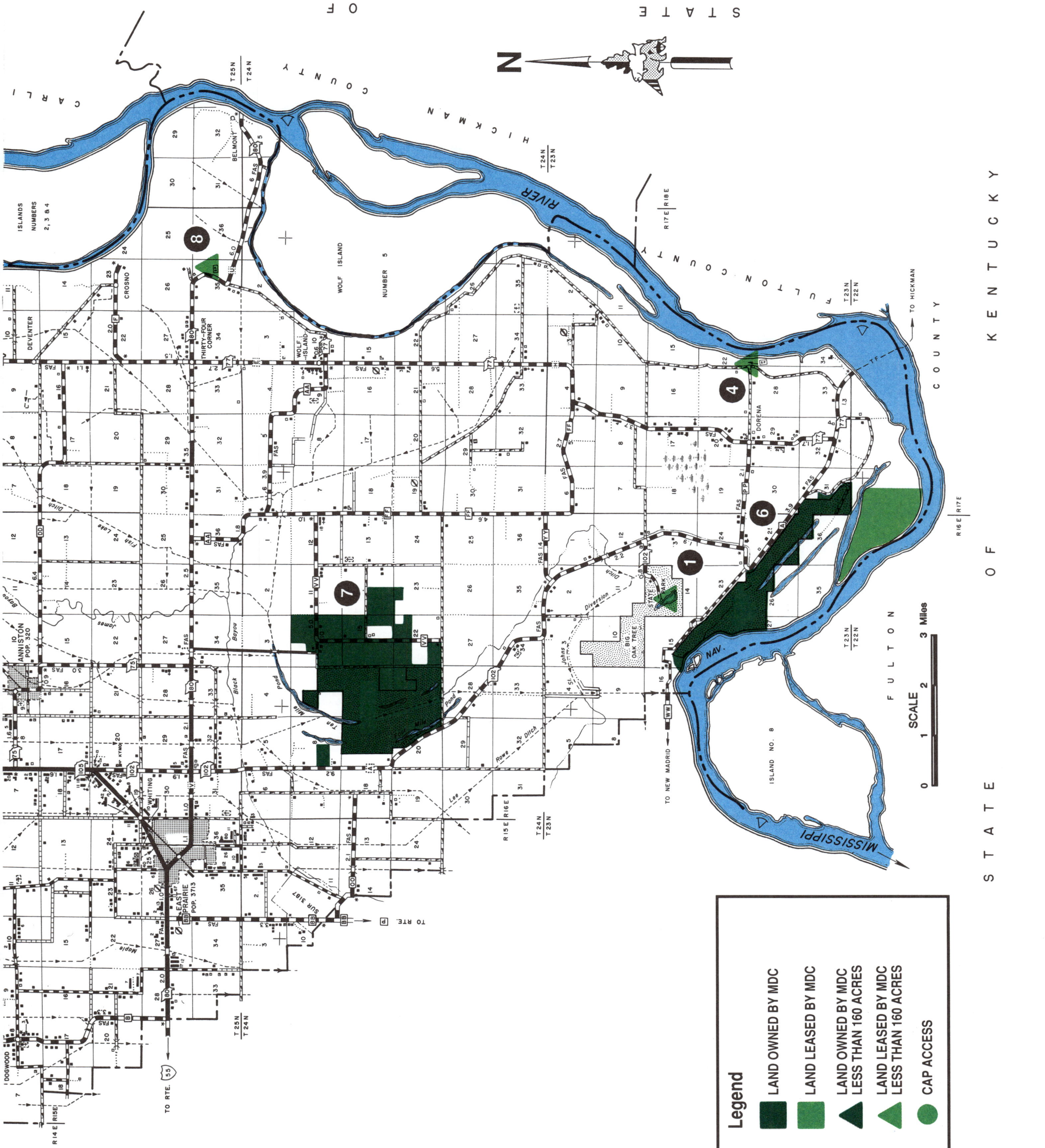

6 **Seven Island Conservation Area** is east of East Prairie on Highway 80, south on Highway 102, then east on Highway A. This 1,376-acre area is home to swamp rabbits and has had nesting Mississippi kites. It contains a 1.75-mile stretch of the Mississippi River and two lakes (20 total acres for fishing). There are two boat ramps.

7 **Ten Mile Pond Conservation Area** is east of East Prairie on Highway 80, then 5 miles south on Highway 102. This 3,793-acre area contains a 2.5-mile stretch of the Ten Mile Ditch (with boat ramps). The area is composed of cropland (2,882 acres), wetlands (811 acres) and forestland (100 acres).

8 **Thirty-Four Corner Blue Hole** is 10 miles east of East Prairie on Highway 80. This 30-acre area has a 9-acre lake, and is 6 miles west of the Mississippi River. There is a boat ramp on the lake.

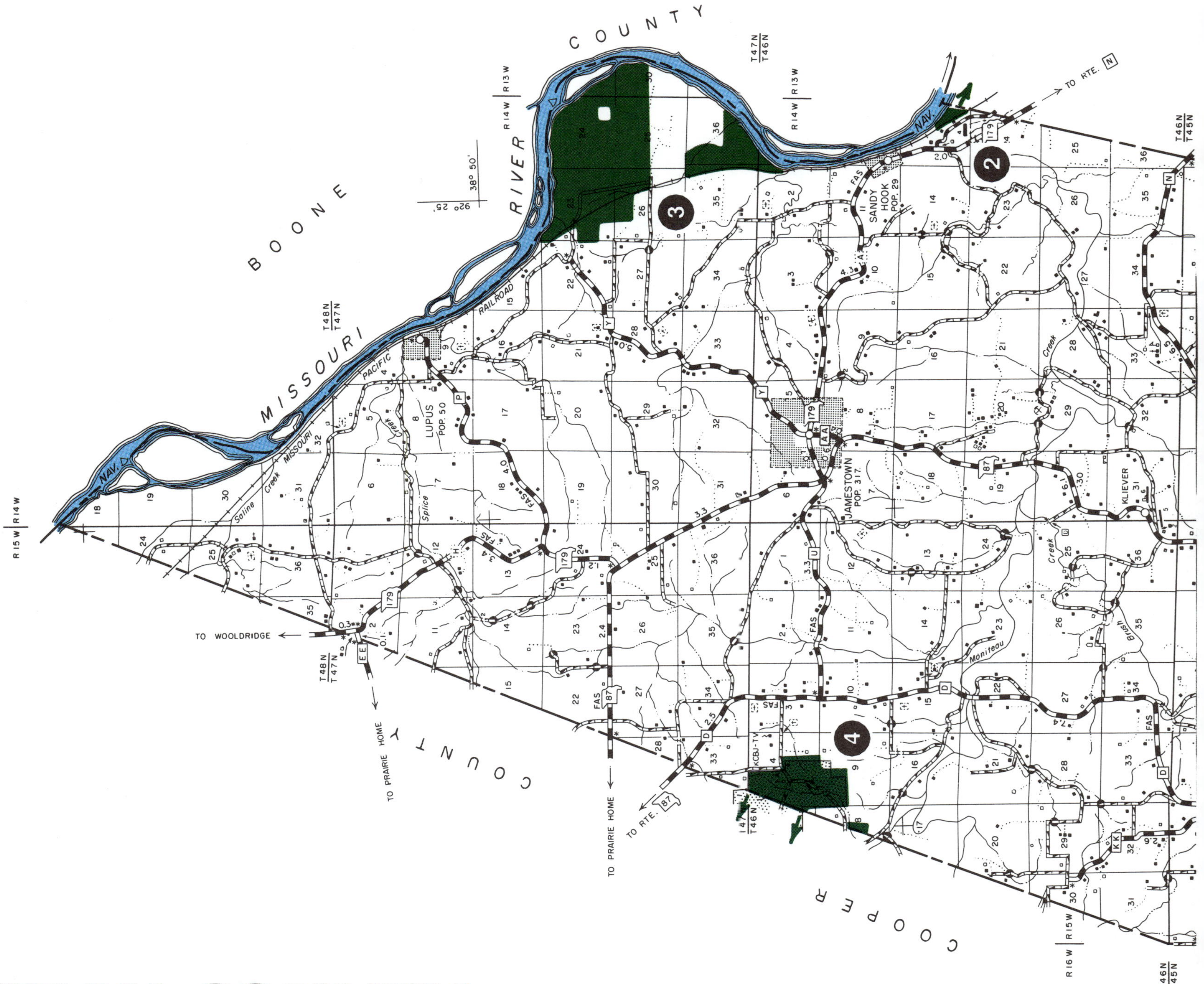

MONITEAU COUNTY

1 **Manito Lake Conservation Area** is 5 miles south of Tipton on Highway 5, then .5 mile west on an unnamed gravel road (marked with an area cantilever sign on Highway 5). This 851-acre area is home to Manito Lake (77 acres of fishable water), which is equipped with a boat ramp and wheelchair accessible jetties. There is also a picnic area and a .25-mile hiking trail.

2 **Marion Bottoms Conservation Area**–see the Cole County map, page 54.

3 **Plowboy Bend Conservation Area** is 5 miles north of Jamestown on Route Y. When the road turns to gravel, continue another .5 mile to reach this 2,073-acre area. The area runs along 3 miles of the Missouri River and has bottomlands, bluffs and blue holes.

4 **Prairie Home Conservation Area**–see the Cooper County map, page 58.

5 **Proctor Park Lake** is .5 mile south of the junction of highways 50 and 87, immediately east of Highway 87 in California. This 7-acre lake is open for fishing, picnicking and has a short trail.

6 **Roger V. and Viola Wachal Smith Conservation Area**–see the Cole County map, page 54.

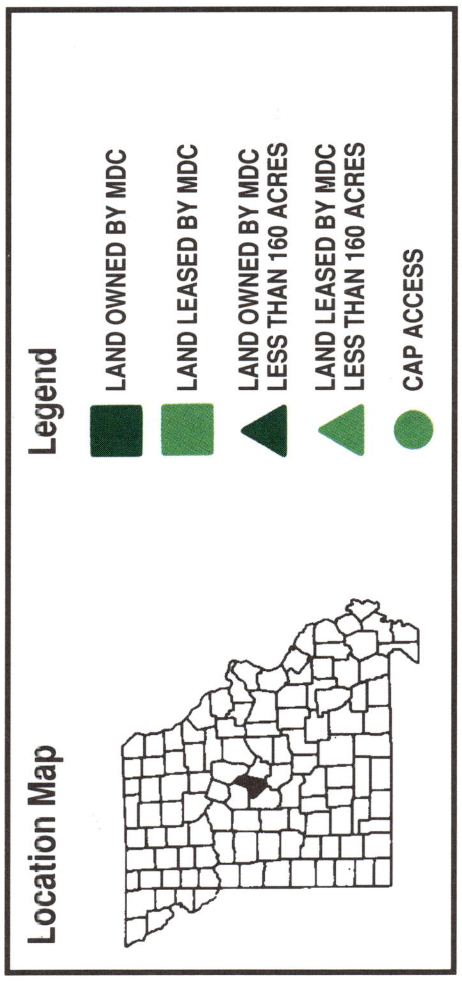

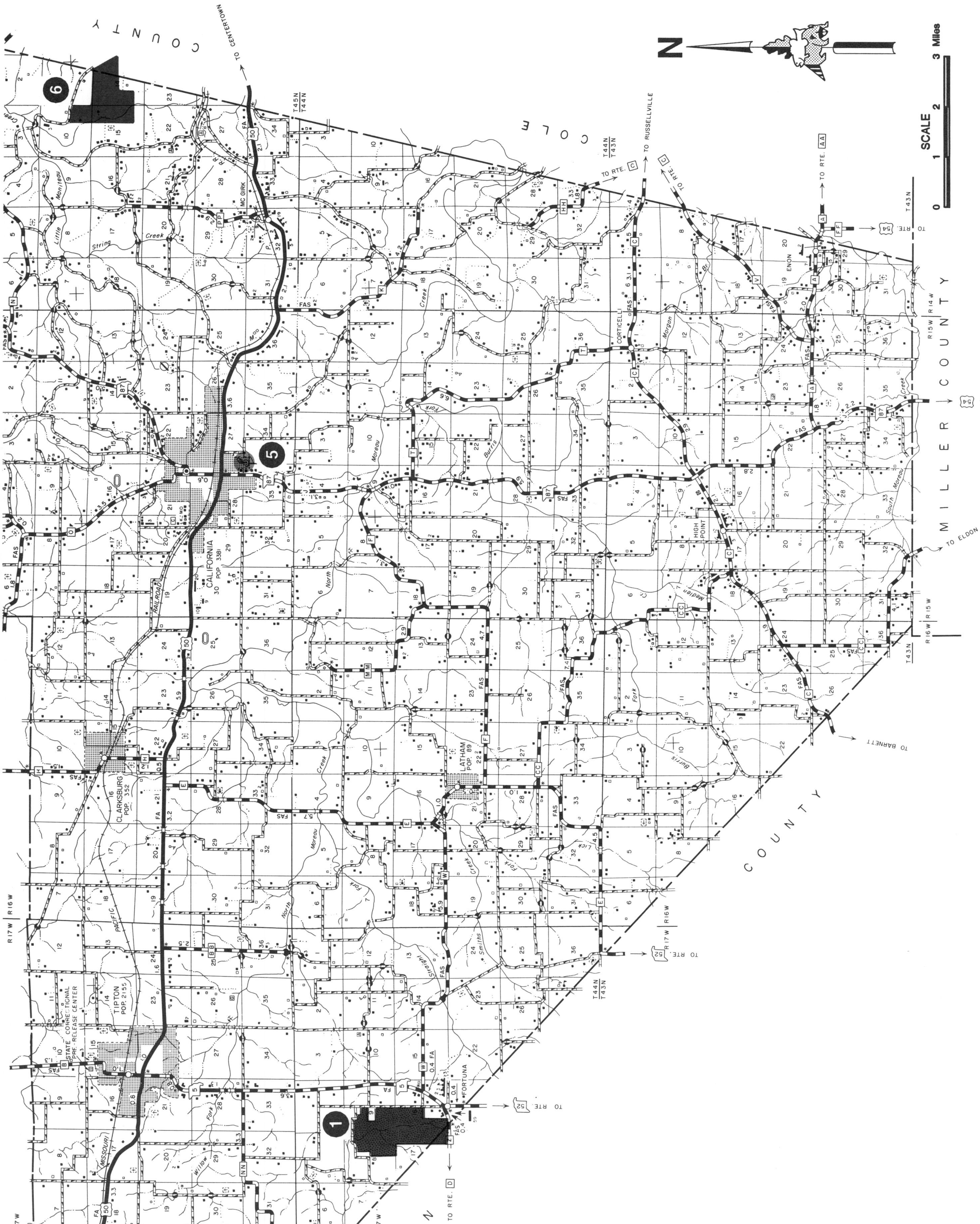

COLE COUNTY
MILLER COUNTY
MORGAN COUNTY
CALIFORNIA POP. 3381
TIPTON POP. 2155
CLARKSBURG POP. 352
LATHAM POP. 89
FORTUNA
HIGH POINT
CORTICELLI
ENON
MC GIRK
STATE CORRECTIONAL PRE-RELEASE CENTER
TO CENTERTOWN
TO RUSSELLVILLE
TO ELDON
TO BARNETT
TO SYRACUSE
SCALE
0 1 2 3 Miles
N

MONROE COUNTY

1 **Cedar Bluff Access** is 1 mile west of Paris on Highway 24 and 3 miles south on County Road 209. This is a 40-acre access to the Elk Fork of Salt River. (The river does not carry enough water to float year-round.)

2 **Paris Access** is on Business 24, near the north edge of the town of Paris. It is a 9-acre area on the Middle Fork of Salt River. There is a boat ramp.

3 **Santa Fe Access** is .5 mile east of the town of Santa Fe just off Highway D. This is a 7-acre area on the South Fork of Salt River.

4 **Robert M. White II Conservation Area**– see the Audrain County map, page 12.

5 **Ruby Clark Willingham Memorial Wildlife Area** is near Holliday. From Highway 24, go .5 mile north on County Road 253, .5 mile east on County Road 251, then .5 mile north on County Road 252. This is 70 acres of mainly forestland.

6 **Woodlawn Access** is on Highway 151, 4.5 miles north of Madison. This is a 65-acre area with no boat ramp on the Middle Fork of Salt River.

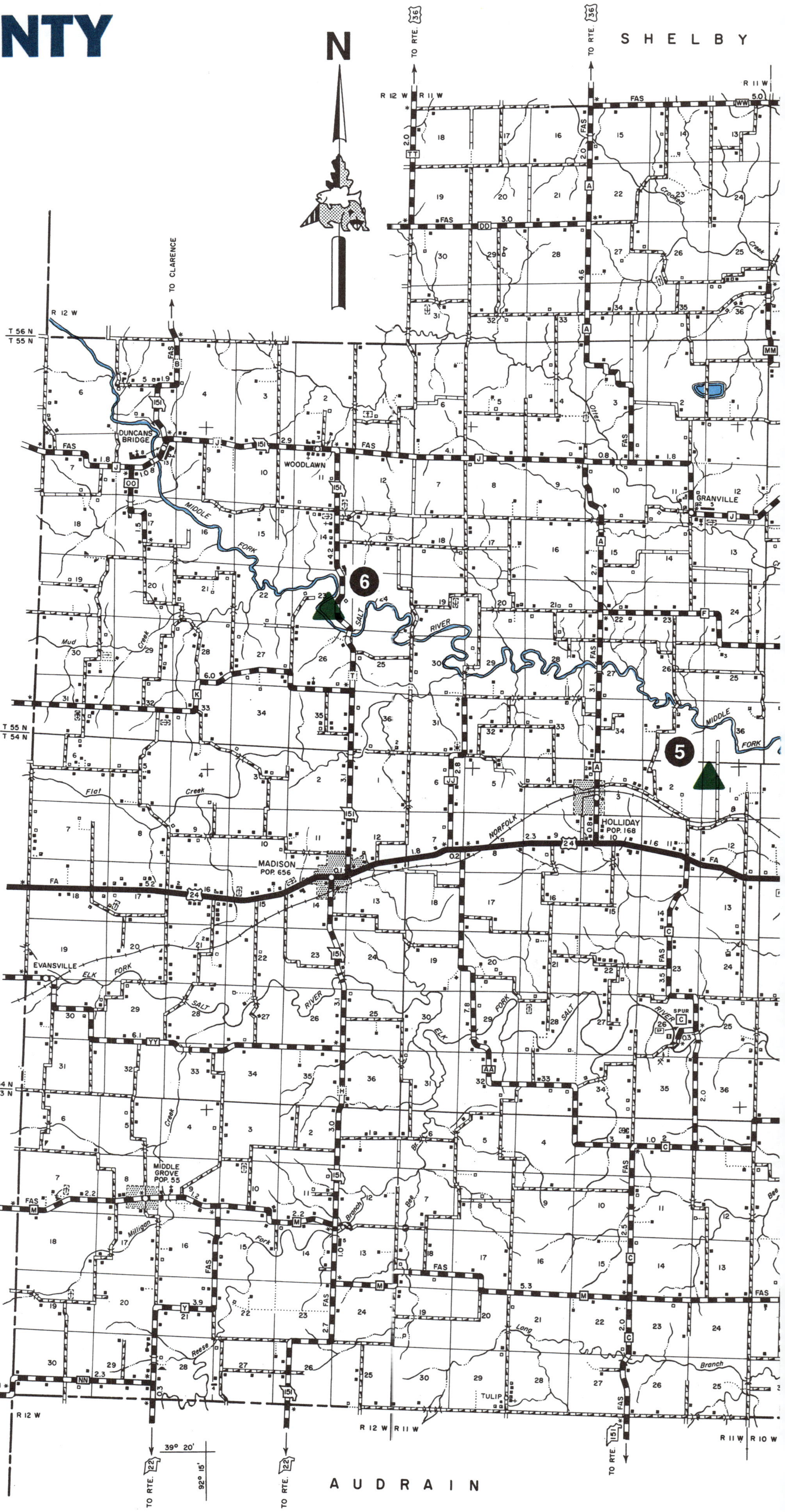

Location Map

Legend

LAND OWNED BY MDC

LAND LEASED BY MDC

LAND OWNED BY MDC LESS THAN 160 ACRES

LAND LEASED BY MDC LESS THAN 160 ACRES

CAP ACCESS

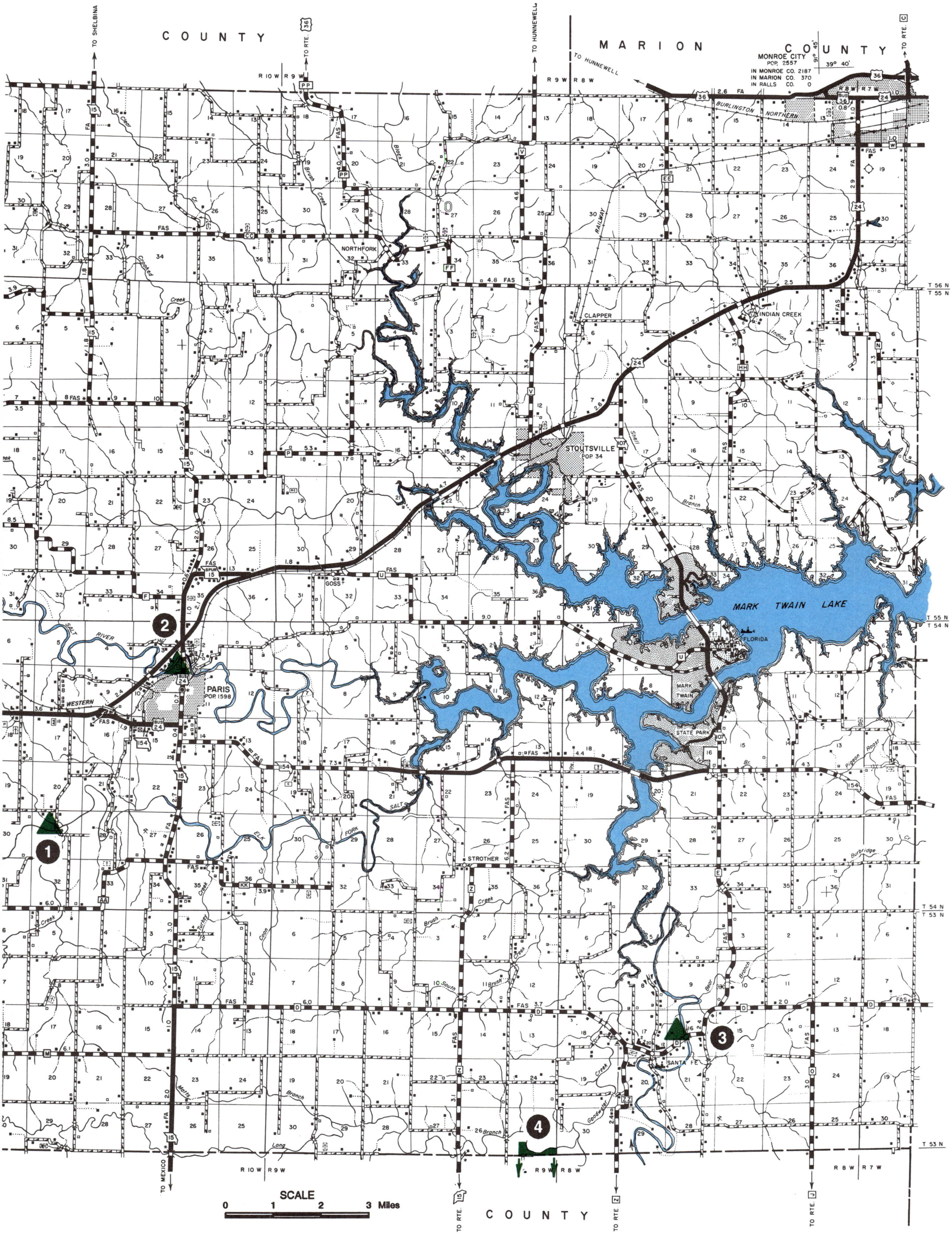
COUNTY
MARION COUNTY
MONROE CITY
POP. 2557
IN MONROE CO. 2187
IN MARION CO. 370
IN RALLS CO. 0
TO SHELBINA
TO HUNNEWELL
BURLINGTON NORTHERN
RAILWAY
NORTHFORK
CLAPPER
INDIAN CREEK
STOUTSVILLE
POP 34
GOSS
MARK TWAIN LAKE
FLORIDA
MARK TWAIN STATE PARK
PARIS
POP. 1598
SALT RIVER
ELK FORK
STROTHER
SANTA FE
T 56 N
T 55 N
T 54 N
T 53 N
R 10 W
R 9 W
R 8 W
R 7 W
TO MEXICO
SCALE
0 1 2 3 Miles
COUNTY

MONTGOMERY COUNTY

1 **Danville Conservation Area** is south of I-70 at the Highway 161 exit, then 2 miles east on the outer road, and 2 miles south on County Road RB (gravel). This 2,655-acre area contains 12 ponds and a 1-mile stretch of Clear Creek (8 acres total for fishing). Danville Glades Natural Area is 113 acres, and a hiking trail winds 3 miles through the area.

2 **Marshall I. Diggs Conservation Area**–see the Audrain County map, page 12.

3 **Grand Bluffs Conservation Area** is .25 mile north on County Road 291, off Highway 94. This 223-acre area provides access to the top and part ownership of a 300-foot dolomite bluff overlooking the Missouri River.

4 **Loutre Lick Access** is 3 miles south of I-70 on Highway N (Mineola Exit), and .75 mile on County Road 278. This is a 163-acre area on the Loutre River with a boat ramp.

5 **Wellsville Lake Conservation Area** is north on Highway 19 (toward Wellsville), 1.5 miles south on County Road 21, then .25 mile west on County Road 22. This is a 14-acre lake mainly surrounded by forestland and old fields (120 acres total). There is a boat ramp and firearms range.

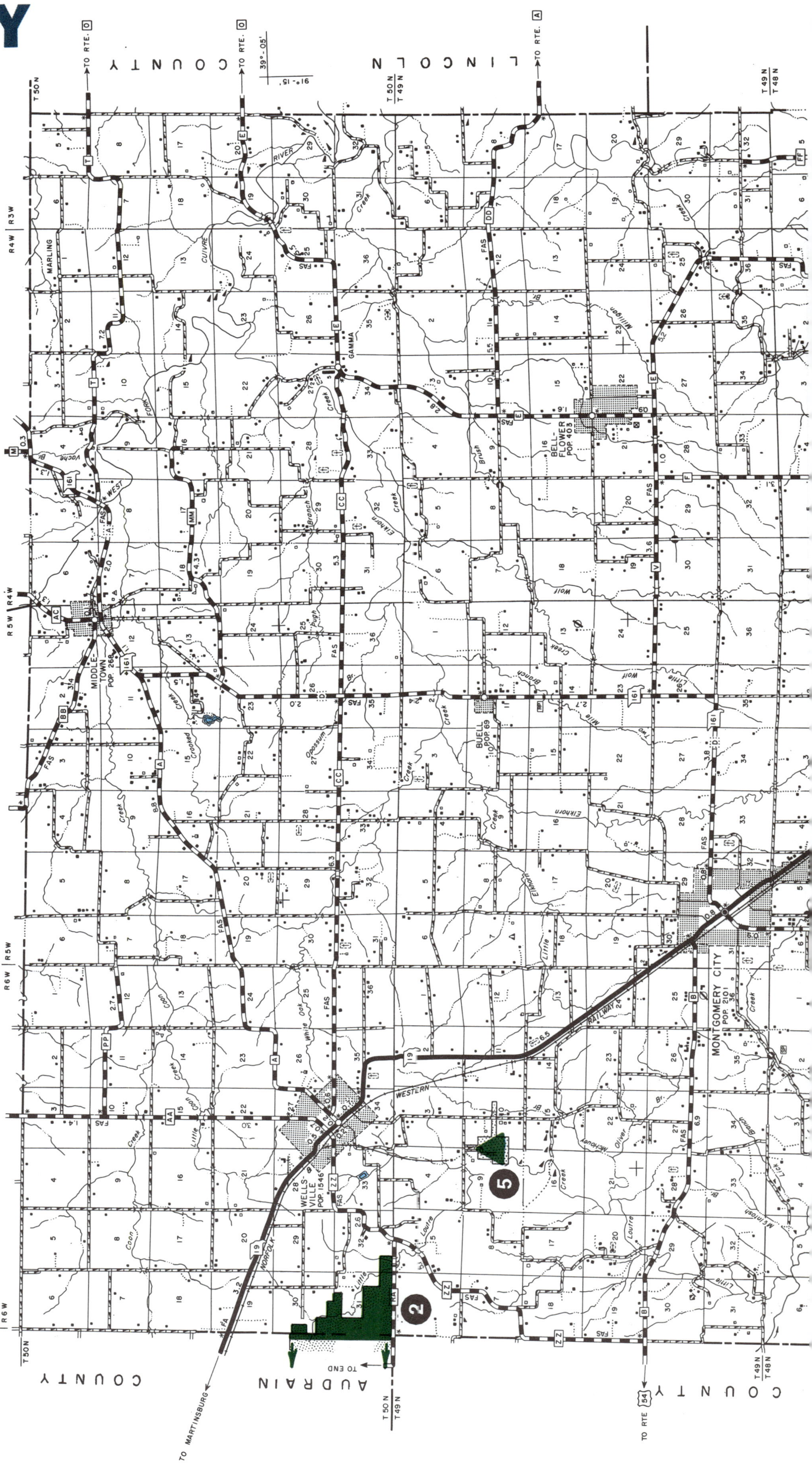

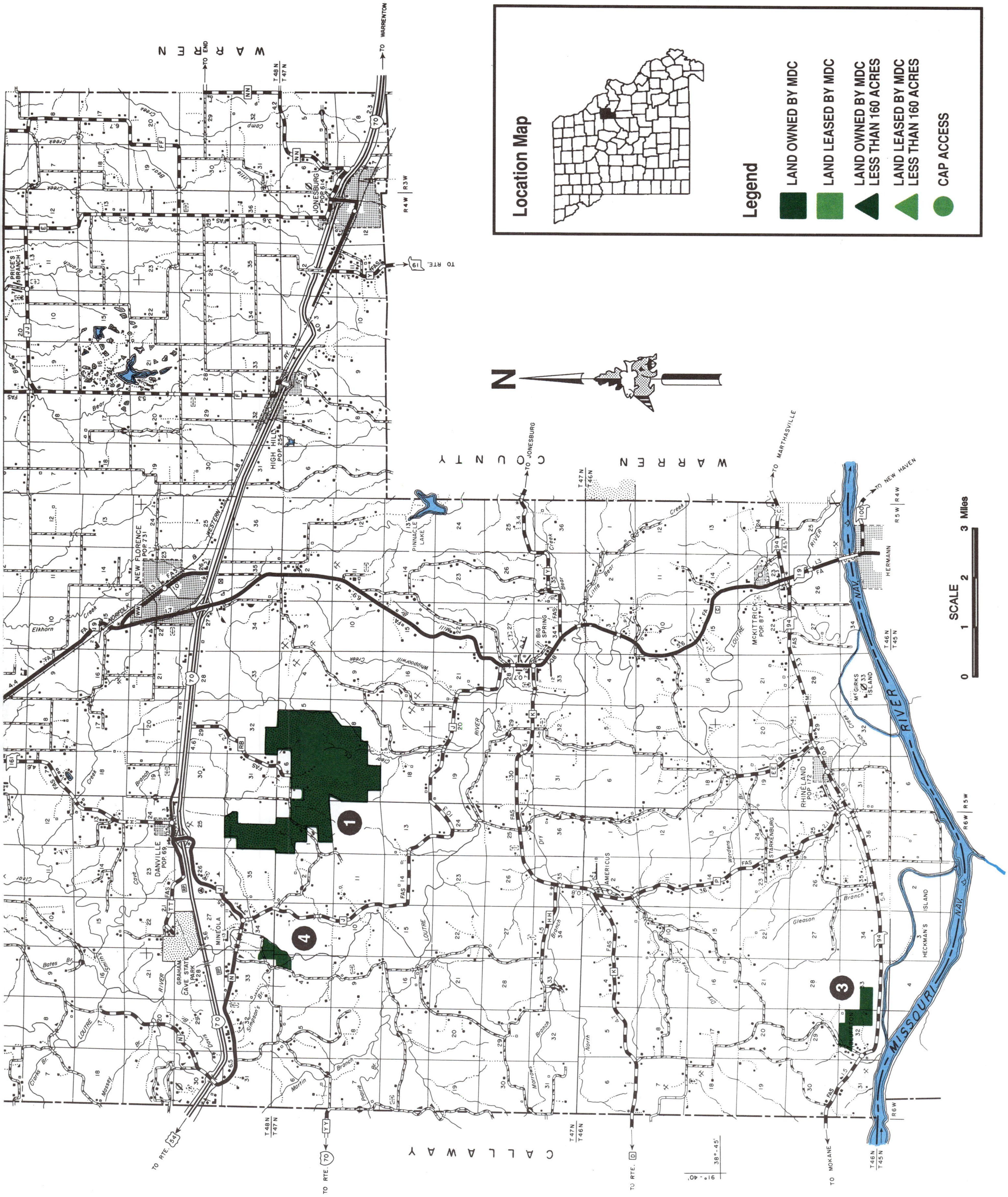
Location Map
Legend
LAND OWNED BY MDC
LAND LEASED BY MDC
LAND OWNED BY MDC LESS THAN 160 ACRES
LAND LEASED BY MDC LESS THAN 160 ACRES
CAP ACCESS
N
SCALE
0 1 2 3 Miles
WARREN
WARREN COUNTY
CALLAWAY
JONESBURG POP. 614
HIGH HILL POP. 254
NEW FLORENCE POP. 731
DANVILLE POP. 69
MINEOLA
GRAHAM CAVE STATE PARK
BIG SPRING
AMERICUS
McKITTRICK POP. 87
RHINELAND POP. 172
STARKENBURG
HERMANN
PINNACLE LAKE
McGIRKS ISLAND
HECKMAN'S ISLAND
MISSOURI RIVER
LOUTRE RIVER
TO WARRENTON
TO JONESBURG
TO MARTHASVILLE
TO NEW HAVEN
TO MOKANE
TO RTE. 54
TO RTE. 70
TO RTE. D
TO RTE. 19
TO END
38° 45'
91° 40'

MORGAN COUNTY

1 **Big Buffalo Creek Conservation Area**–see the Benton County map, page 20.

2 To reach **Carpenter Memorial Conservation Area,** take Highway 52 west of Versailles 1 mile, turn 7 miles left on State Road T, left on Highway 135 for 50 feet, then turn back (right) onto State Road T again. Follow State Road T until the pavement ends, and turn right on Gravel Road T-338 for about 2.5 miles. Watch for signs to the area. This 365-acre area is mainly forestland.

3 **Coffman Beach Access** is west of Eldon on Highway 52 to Highway Y. From Highway Y, turn on Lake Road Y-20 and watch for the cantilever sign marking this 8-acre area on Lake of the Ozarks. There is a boat ramp and wheelchair accessible dock.

4 **Gravois Mills Access** is marked by a cantilever sign on Highway 5 in the town of Gravois Mills. This is a 63-acre access on Lake of the Ozarks. It has a boat ramp, picnic area and disabled accessible fishing dock.

5 **Hite Prairie Conservation Area** is near the southwest edge of Versailles, .5 mile south of Highway 52. A diversity of flowers grows on this 103-acre prairie.

6 **Lamine River Conservation Area**–see the Cooper County map, page 58.

7 **Manito Lake Conservation Area**–see the Moniteau County map, page 144.

8 **Proctor Towersite** is south of Versailles on Highway 5, to the right on Highway J, then .75 mile right on Highway 135. This is 115 acres of forestland.

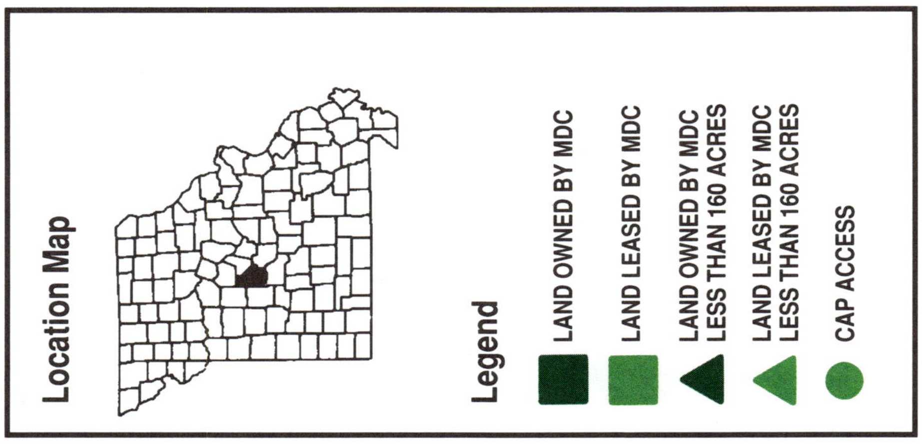

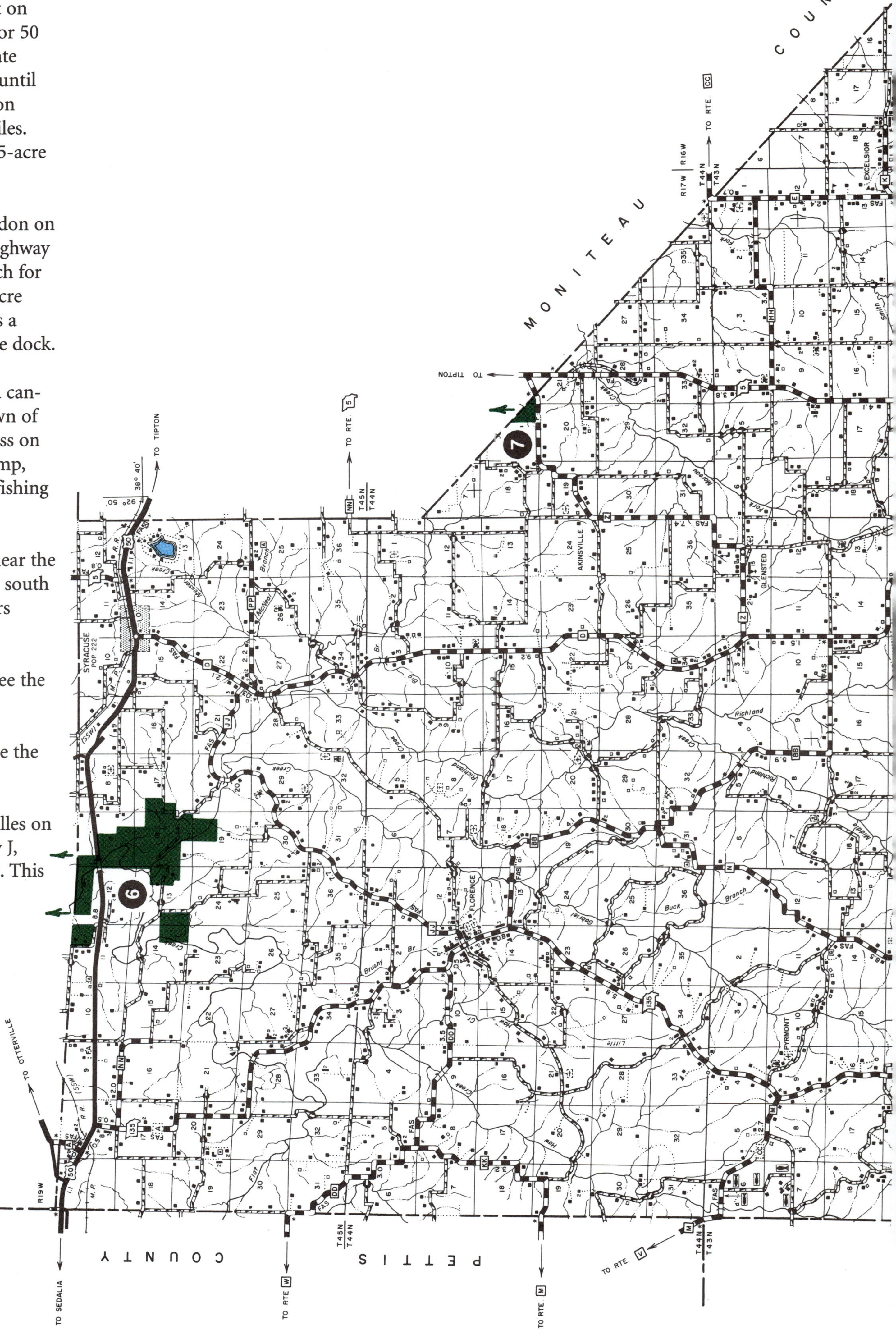

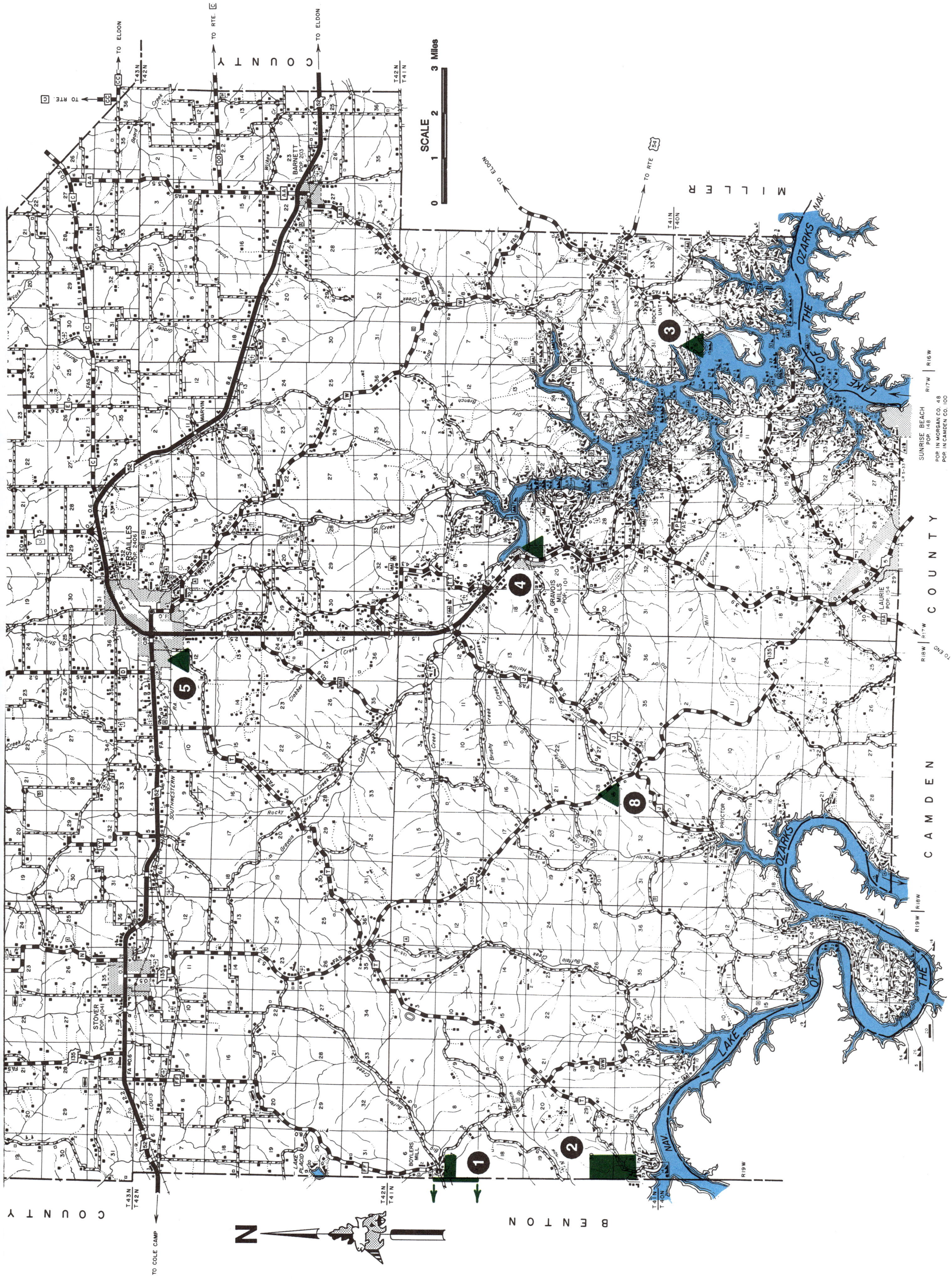
SCALE
0 1 2 3 Miles
TO ELDON
TO RTE. C
TO RTE. CC
TO RTE. 54
TO COLE CAMP
TO EDWARDS
MILLER
CAMDEN COUNTY
BENTON
VERSAILLES
BARNETT
MARVIN
STOVER
GRAVOIS MILLS
PROCTOR
LAURIE
BOYLERS MILL
LAKE PLACID
ROCKY MOUNT
SUNRISE BEACH
POP. 148
POP. IN MORGAN CO. 48
POP. IN CAMDEN CO. 100
LAKE OF THE OZARKS
NAV.
T43N
T42N
T41N
T40N
R19W
R18W
R17W
R16W
N

NEW MADRID COUNTY

1 **Donaldson Point Conservation Area** is east of New Madrid on Highway WW, then south on Highway AB to County Road 405. The area is across the levee on the right side of the road. This 5,785-acre area on the Mississippi River is mainly forestland and has large areas made up of bottomland hardwoods.

2 **John L. and Georgia Girvin Memorial Conservation Area** is east of Portageville on Highway 162, 1 mile south on an unnamed gravel road across the levee, then about 5 miles east on farm roads. This is 748 acres of forest along a 2-mile stretch of the Mississippi River.

3 **New Madrid Bend Access** is off Highway 61 in the city of New Madrid. This is 7 acres on the Mississippi River with a boat ramp.

4 **St. John's Bayou Access** is .5 mile east of New Madrid on Highway WW. This is 3.7 acres on the Mississippi River with no boat ramp.

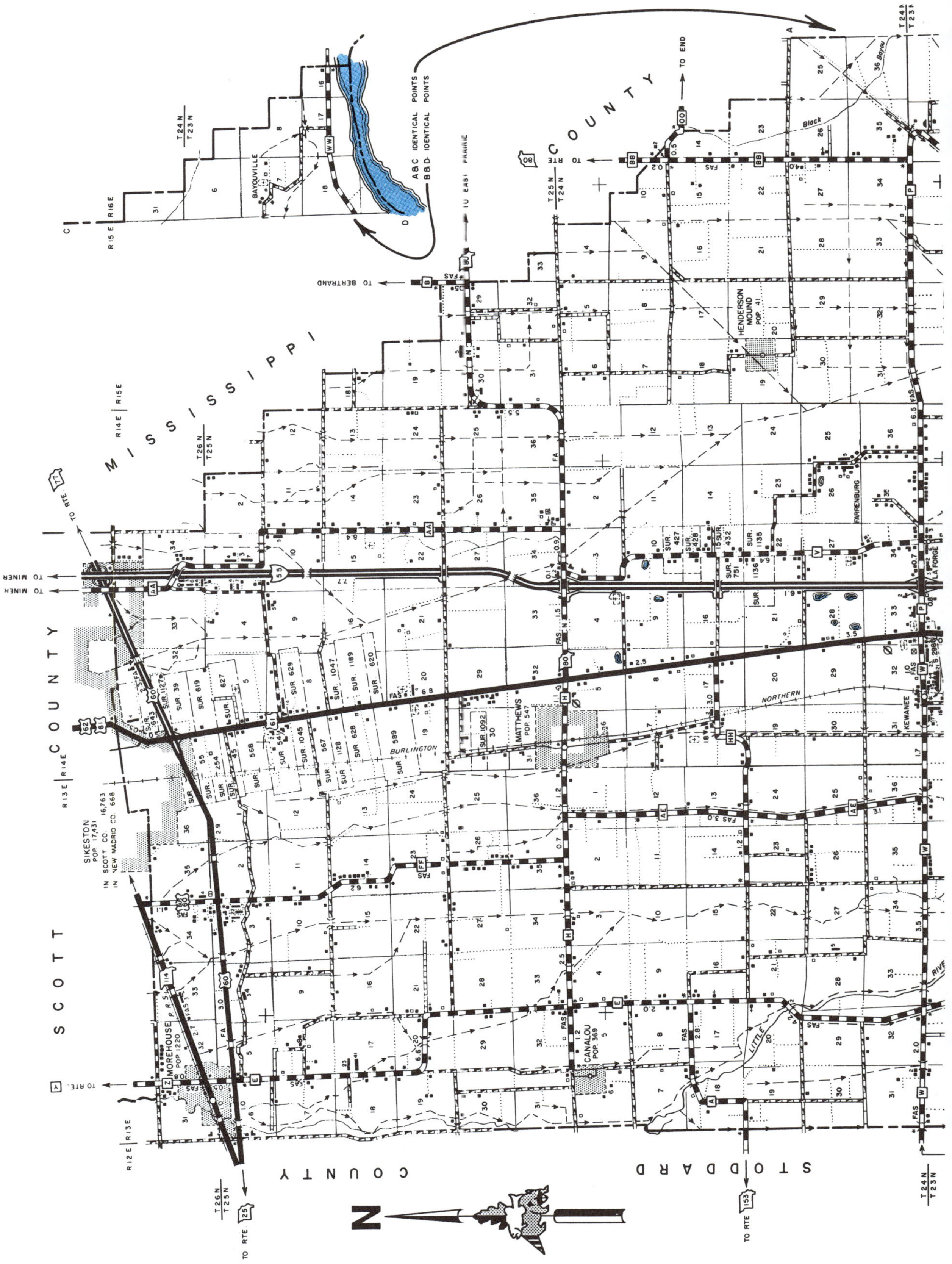

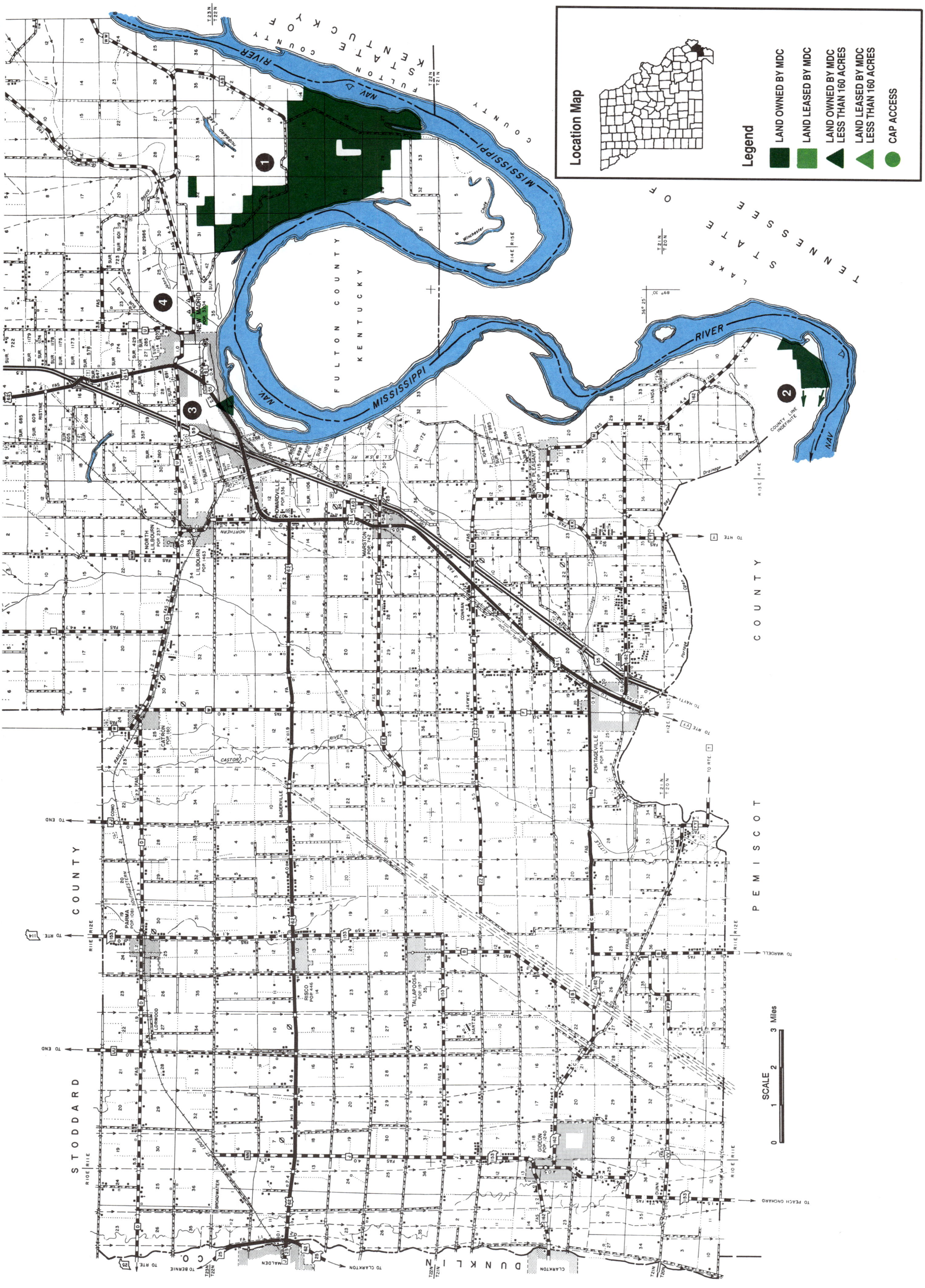
Location Map
Legend
LAND OWNED BY MDC
LAND LEASED BY MDC
LAND OWNED BY MDC LESS THAN 160 ACRES
LAND LEASED BY MDC LESS THAN 160 ACRES
CAP ACCESS
MISSISSIPPI RIVER
FULTON COUNTY KENTUCKY
STATE OF TENNESSEE
LAKE COUNTY
PEMISCOT COUNTY
STODDARD COUNTY
DUNKLIN CO.
NEW MADRID
SCALE
0 1 2 3 Miles

NEWTON COUNTY

1 **Allen Bridge Access** is 2 miles east of Neosho on Business Highway 60, then 1.5 miles north on Nighthawk Road. This is a 15-acre access to Shoal Creek with a boat ramp.

2 **Cherry Corner Access** is northeast of Neosho at the Alternate 71 crossing of Shoal Creek. At the junction of Highways MM and Alternate 71, turn east on Greenbriar, then right on Otter Drive. This is a 4-acre area with no boat ramp on Shoal Creek.

3 **Diamond Grove Prairie Conservation Area** is 4 miles west of Diamond on Highway V, and 1.25 miles north on a gravel road marked with an area sign. Most of this 611-acre area is a Designated Natural Area.

4 **Fort Crowder Conservation Area** is 3.5 miles east of Neosho on Highway HH, east of the junction of Alternate 71 and Highway HH. This is a 2,362-acre area with a 12-mile trail system open to horseback riders. There is also a wheelchair accessible archery range. This area is adjacent to land used by the National Guard, and access may be restricted at times.

5 **Goodman Towersite** is south of Neosho on Sorrel Road, 1 mile south of Highway AA. Turn right on Sorrel Road, and the 40-acre towersite is .75 mile down the road.

6 **Joplin Towersite** is south of the city of Joplin on Highway 86, 1 mile east on Highway P, then .4 mile south on Eland Road. This is a 10-acre forested area.

7 **Lime Kiln Access** is .5 mile east of Neosho on Business Highway 60, then .25 mile left on Lime Kiln Drive. This is an 11-acre area with two boat ramps on Shoal Creek.

8 **Neosho Towersite** is 3 miles east of Neosho on Business Highway 60. This is a 79-acre area.

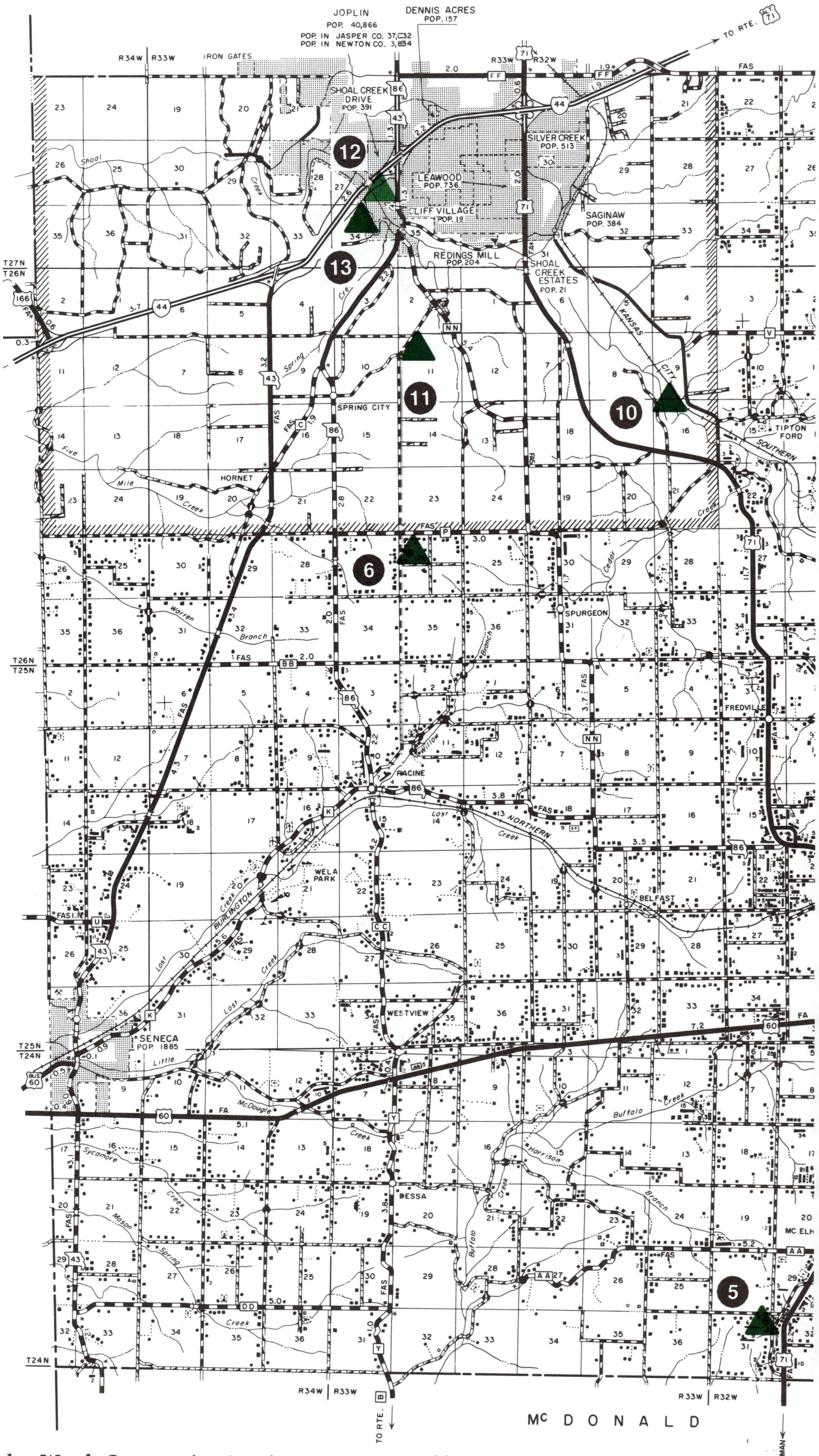

9 **Smack-out Access** is between Monett and Granby on Highway 60. Turn south on Ivy Drive .75 miles east of the Highway 60 bridge. This is a 1-acre area on Shoal Creek.

10 **Tipton Ford Access** is 6.5 miles west of Diamond on Highway V, then 1 mile south on Old Highway 71. This is a 90-acre area on Shoal Creek with no boat ramp.

11 **Walter Woods Conservation Area** is east of Redings Mill on Highway NN, 1.25 miles to Dutch Elm Road. Turn right on Dutch Elm, and go .75 mile to the area entrance on the left side of the road. Jacob's Branch flows through this 28-acre area (mainly forestland), and there is an asphalt interpretive trail (.7 mile). A lodge building with meeting rooms and a kitchen is available by reservation. The building is disabled accessible.

12 **Wildcat Access** is 1 mile south of I-44 on Highway 86 through the entrance to Wildcat Park. This is 1 acre on Shoal Creek (with a boat ramp) that lies adjacent to the park and its facilities.

13 To reach **Wildcat Glade Natural Area** (27 acres), go 1.1 miles south of Joplin on Route 86 to the Shoal Creek Bridge. Cross the bridge, and turn right immediately on an unnamed blacktop road. Go 1.5 miles.

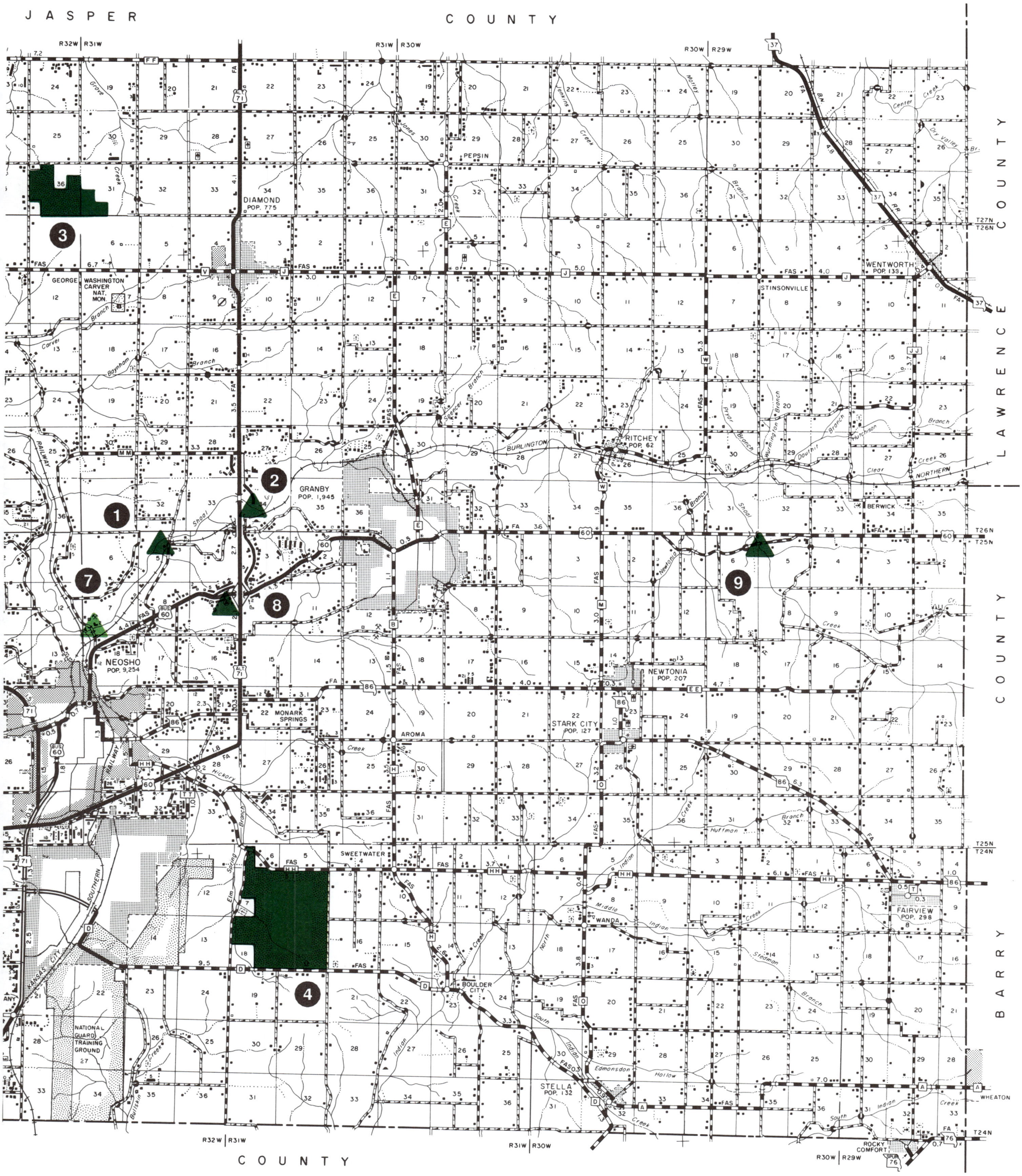
JASPER COUNTY
LAWRENCE COUNTY
BARRY COUNTY
COUNTY
R32W R31W
R31W R30W
R30W R29W
T27N T26N
T26N T25N
T25N T24N
T24N
DIAMOND POP. 775
PEPSIN
GEORGE WASHINGTON CARVER NAT. MON
GRANBY POP. 1,945
RITCHEY POP. 62
WENTWORTH POP. 139
STINSONVILLE
BERWICK
NEOSHO POP. 9,254
MONARK SPRINGS
NEWTONIA POP. 207
STARK CITY POP. 127
AROMA
SWEETWATER
FAIRVIEW POP. 298
WANDA
BOULDER CITY
STELLA POP. 132
NATIONAL GUARD TRAINING GROUND
ROCKY COMFORT
WHEATON
1
2
3
4
7
8
9

SCALE
0 1 2 3 Miles

N

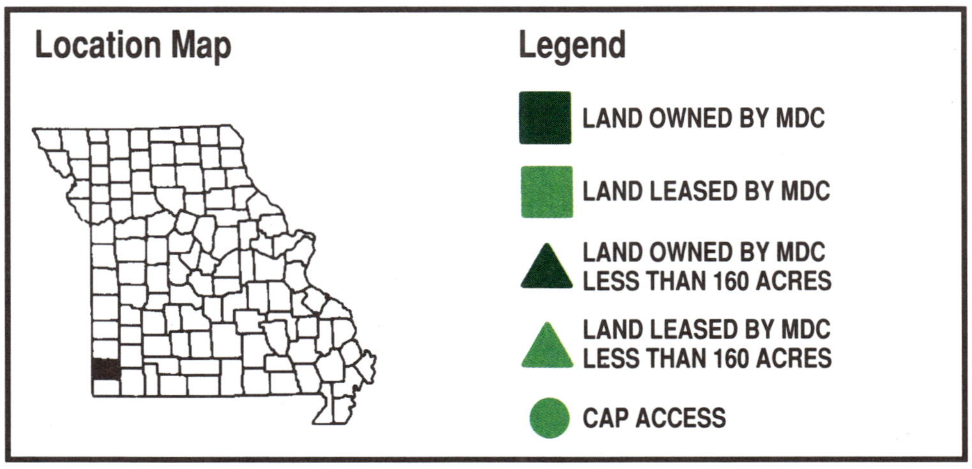
Location Map
Legend
LAND OWNED BY MDC
LAND LEASED BY MDC
LAND OWNED BY MDC LESS THAN 160 ACRES
LAND LEASED BY MDC LESS THAN 160 ACRES
CAP ACCESS

NODAWAY COUNTY

1 **Bilby Ranch Lake Conservation Area** is 14 miles west of Maryville on Highway 46. This 5,030-acre area has 8 ponds and Bilby Ranch Lake (120 acres for fishing). The lake has a boat ramp, a jetty that is accessible to wheelchairs and a picnic area.

2 **Bridgewater Access** is west of Highway 71 on Highway U. This is a 13-acre area.

3 **Bristle Ridge Access** is 1 mile west of Guilford where Highway M crosses the Platte River. This is a 1-acre area with no boat ramp.

4 **Keever Bridge Access** on the Platte River is south of Parnell off Highway 46. This is a 6-acre area with no boat ramp.

5 **Mozingo Lake** is east of Maryville on Highway 136. Two miles east of Highway EE, look for a blacktop road and go 1 mile north. Watch for signs. This is a 3,325-acre area with 1,000 acres of fishable water.

6 **Nodaway County Community Lake** is north of Maryville on Highway 148. This 73-acre lake has a boat ramp and wheelchair accessible fishing jetty. The land around the lake is mainly pasture and old croplands (237 acres) and has an archery range.

7 **Possum Walk Access** is 2 miles west of Clearmont on Highway C. This is a 5-acre area of forestland on the Nodaway River. There is no boat ramp.

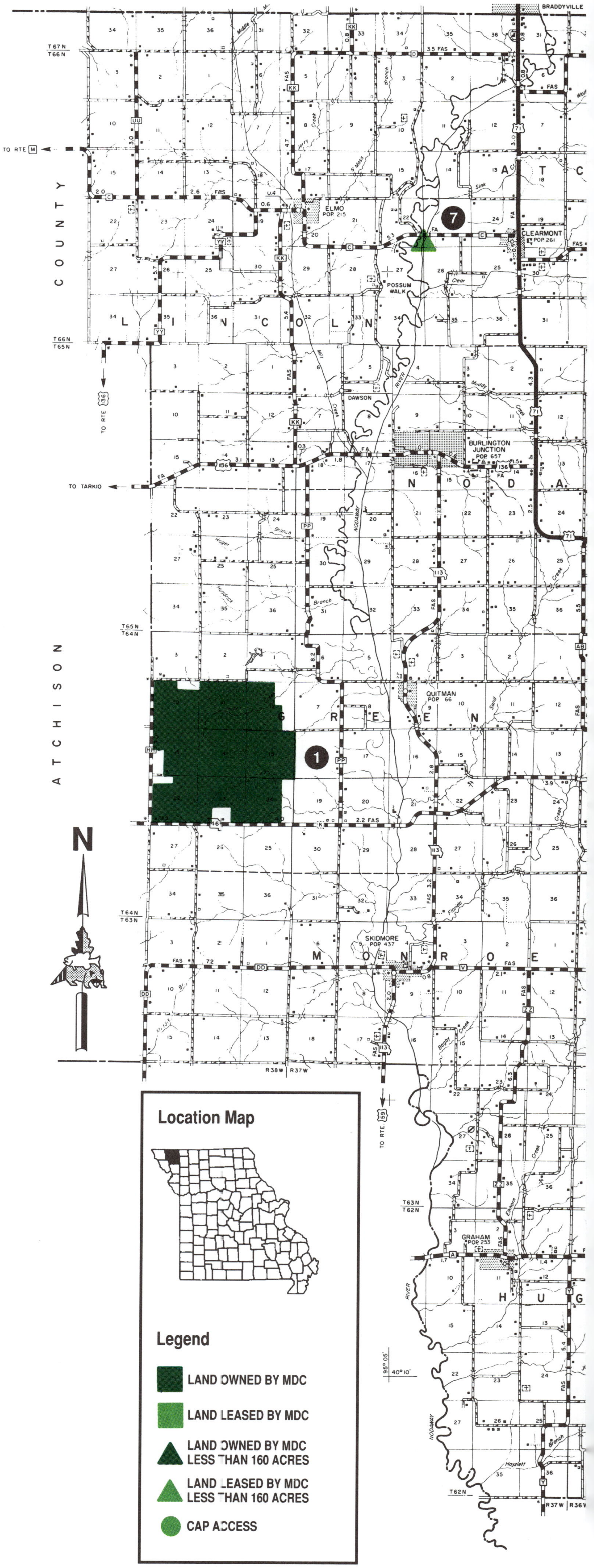

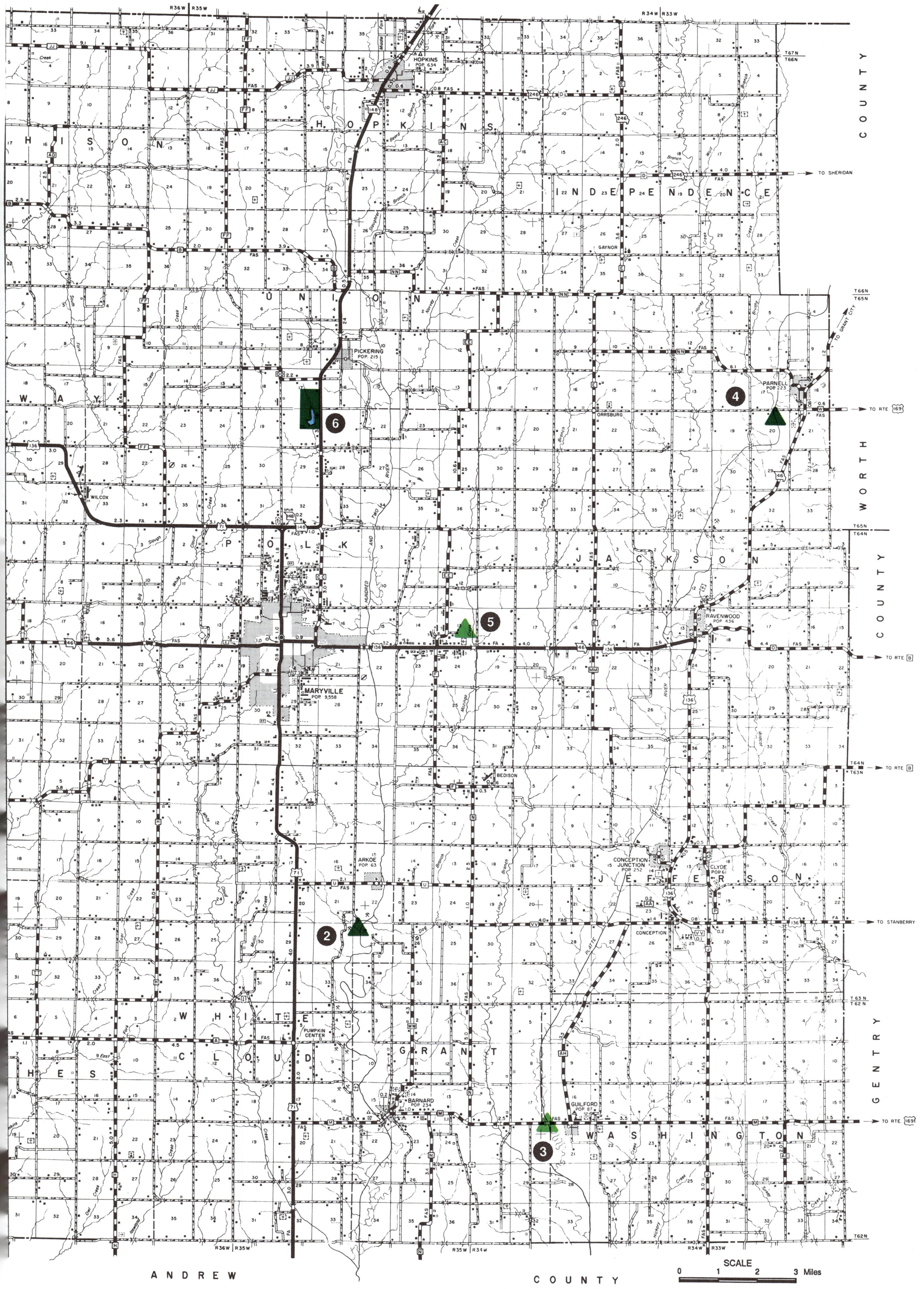
HOPKINS
POP 634
H O P K I N S
I N D E P E N D E N C E
TO SHERIDAN
GAYNOR
U N I O N
PICKERING
POP 215
PARNELL
POP 223
ORRSBURG
TO GRANT CITY
TO RTE 169
WILCOX
P O L K
J A C K S O N
W O R T H
C O U N T Y
RAVENWOOD
POP 436
MARYVILLE
POP 9,558
TO RTE B
BEDISON
ARKOE
POP 63
CONCEPTION
JUNCTION
POP 252
CLYDE
POP 61
J E F F E R S O N
TO STANBERRY
CONCEPTION
W H I T E
C L O U D
PUMPKIN
CENTER
G R A N T
BARNARD
POP 234
GUILFORD
POP 87
W A S H I N G T O N
TO RTE 169
G E N T R Y
C O U N T Y
A N D R E W
C O U N T Y
SCALE
0
1
2
3 Miles

OREGON COUNTY

1 **Cover Memorial Wildlife Area** is 5 miles west of Alton on Highway P, then 1 mile on County Road P-314. This is a 282-acre area composed of forestland (190 acres), old pastures (78 acres) and a 13-acre glade.

2 **Myrtle Access** is 15 miles east of Thayer on Highway 142, and 4 miles south on Highway H to the end of the pavement. Continue 2 miles on County Road H-278 to the area. This is a 29-acre area on the Eleven Point River with a boat ramp.

3 **Rose Hill Towersite** is 7 miles east of Thayer on Highway 142, 1.5 miles north on Highway EE, and 1 mile left on County Road 142-248. The tower sits on 8 acres of forestland.

4 **Warm Fork Conservation Area** is 5 miles east of Koshkonong on Highway Z, then north .1 mile on County Road P-323. This is a 185-acre area composed of open land (110 acres), forestland (15 acres) and a .75-mile stretch of Warm Fork of the Spring River.

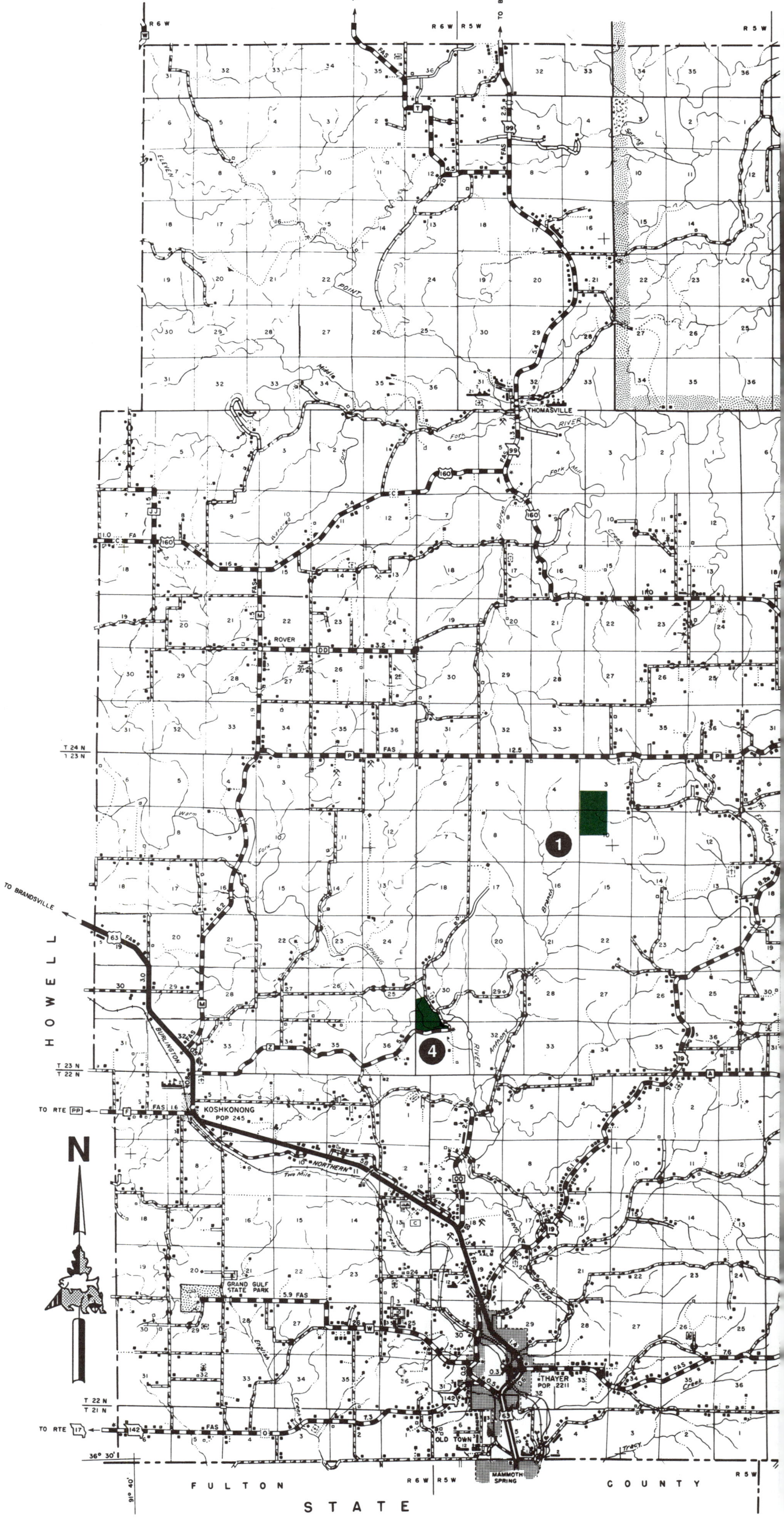

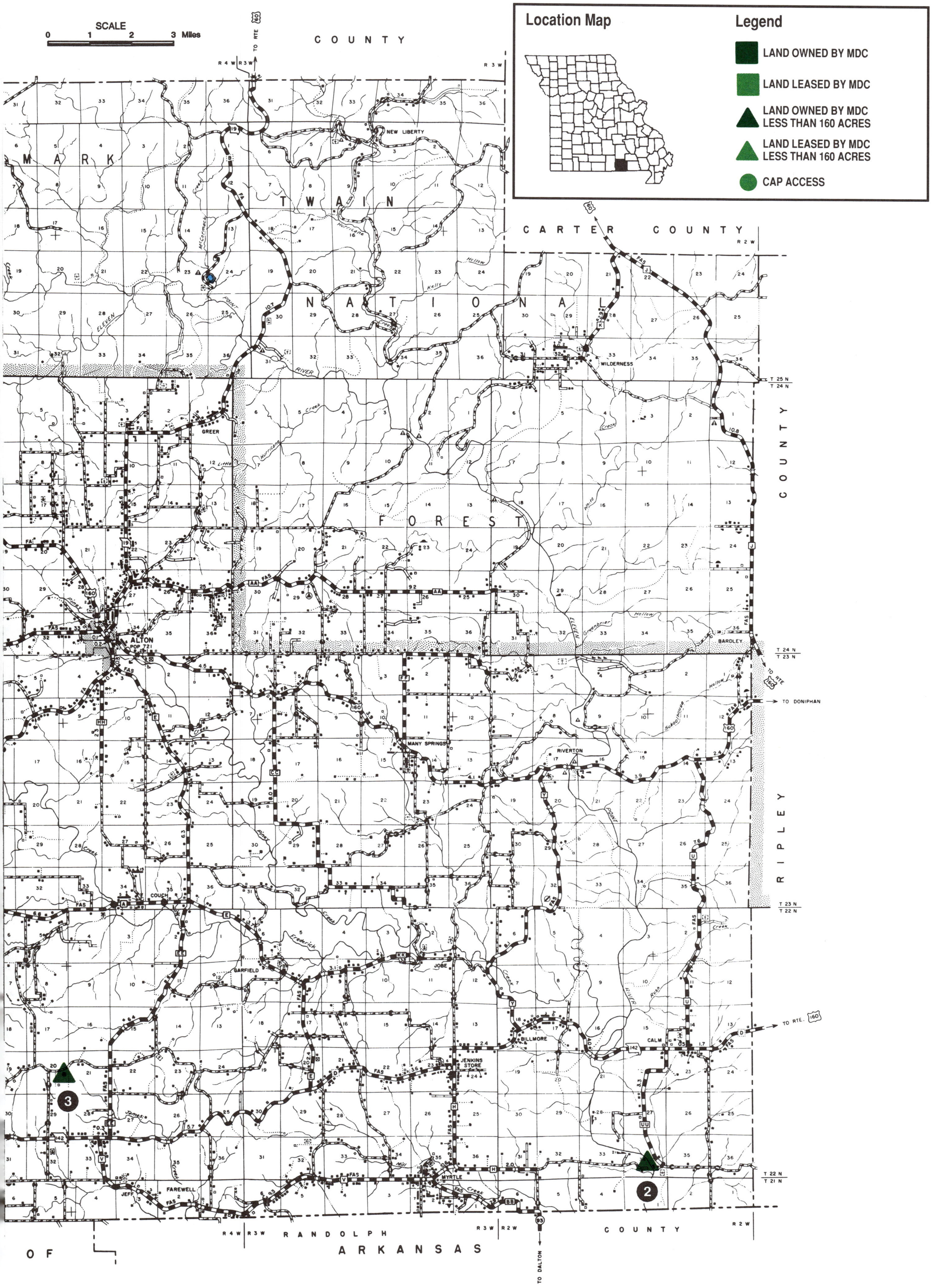
SCALE
0 1 2 3 Miles
Location Map
Legend
LAND OWNED BY MDC
LAND LEASED BY MDC
LAND OWNED BY MDC LESS THAN 160 ACRES
LAND LEASED BY MDC LESS THAN 160 ACRES
CAP ACCESS
MARK TWAIN NATIONAL FOREST
CARTER COUNTY
RIPLEY COUNTY
RANDOLPH COUNTY
OF ARKANSAS
NEW LIBERTY
WILDERNESS
GREER
ALTON POP 721
BARDLEY
TO DONIPHAN
MANY SPRINGS
RIVERTON
COUCH
GARFIELD
JOBE
BILLMORE
CALM
JENKINS STORE
MYRTLE
FAREWELL
JEFF
TO DALTON
TO RTE. 160
ELEVEN POINT RIVER
T 25 N
T 24 N
T 23 N
T 22 N
T 21 N
R 4 W
R 3 W
R 2 W
2
3

OSAGE COUNTY

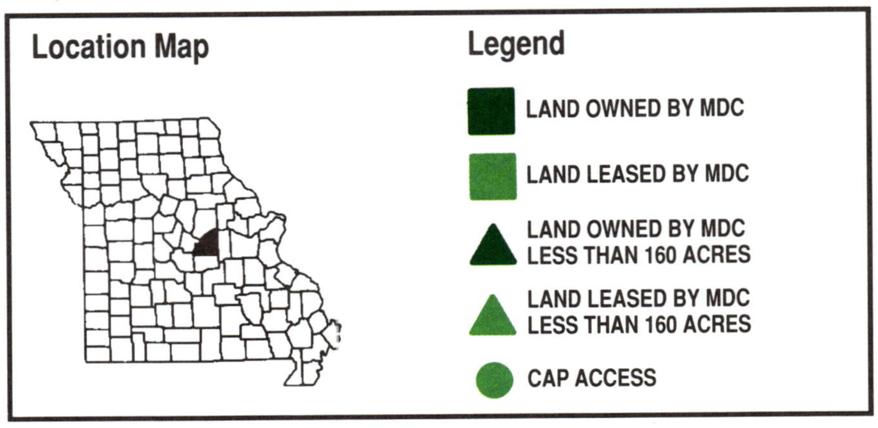

1 **Ben Branch Lake Conservation Area** is 10 miles northeast of Linn on Highway 89. This 44-acre lake has a boat ramp, picnic area and a fishing dock that is accessible to wheelchairs. It is surrounded by forestland (350 acres) and pastureland (120 acres).

2 **Bonnots Mill Access** is in the town of Bonnots Mill on Highway A. It is a 5-acre access with a boat ramp on the Osage River.

3 **Dr. Bernard Bruns Access** is north of Westphalia off Highway 63 (watch for the cantilever sign). This is a 77-acre area on the Maries River.

4 **Chamois Access** is in the town of Chamois off Highway 100. This is a 1-acre area with a boat ramp on the Missouri River.

5 To reach **Cooper Hill Conservation Area**, take Highway 50 to Mt. Sterling. Go 2 miles south on Highway A, then go 2 miles right on Highway D. This is a 214-acre area near the Gasconade River.

6 **Meta Towersite** (20 acres) is 2 miles north of the town of Meta on Highway 133.

7 **Painted Rock Conservation Area** is 7 miles southwest of Westphalia on Highway 133. This 1,480-acre area contains a stretch of the Osage River (no boat ramp) and a 5-acre lake suitable for fishing. Osage Bluff Hiking Trail winds 1.6 miles through the area.

8 **Pointers Creek Access** is 8 miles southeast of Linn on Highway CC, then east to the end of County Road RA. This is a 9-acre access with a boat ramp on the Gasconade River.

9 **Rollins Ferry Access** is south of Linn at the junction of highways E and 89. This is 16 acres with a boat ramp on the Gasconade River.

10 **Smoky Waters Conservation Area**–see the Cole County map, page 54.

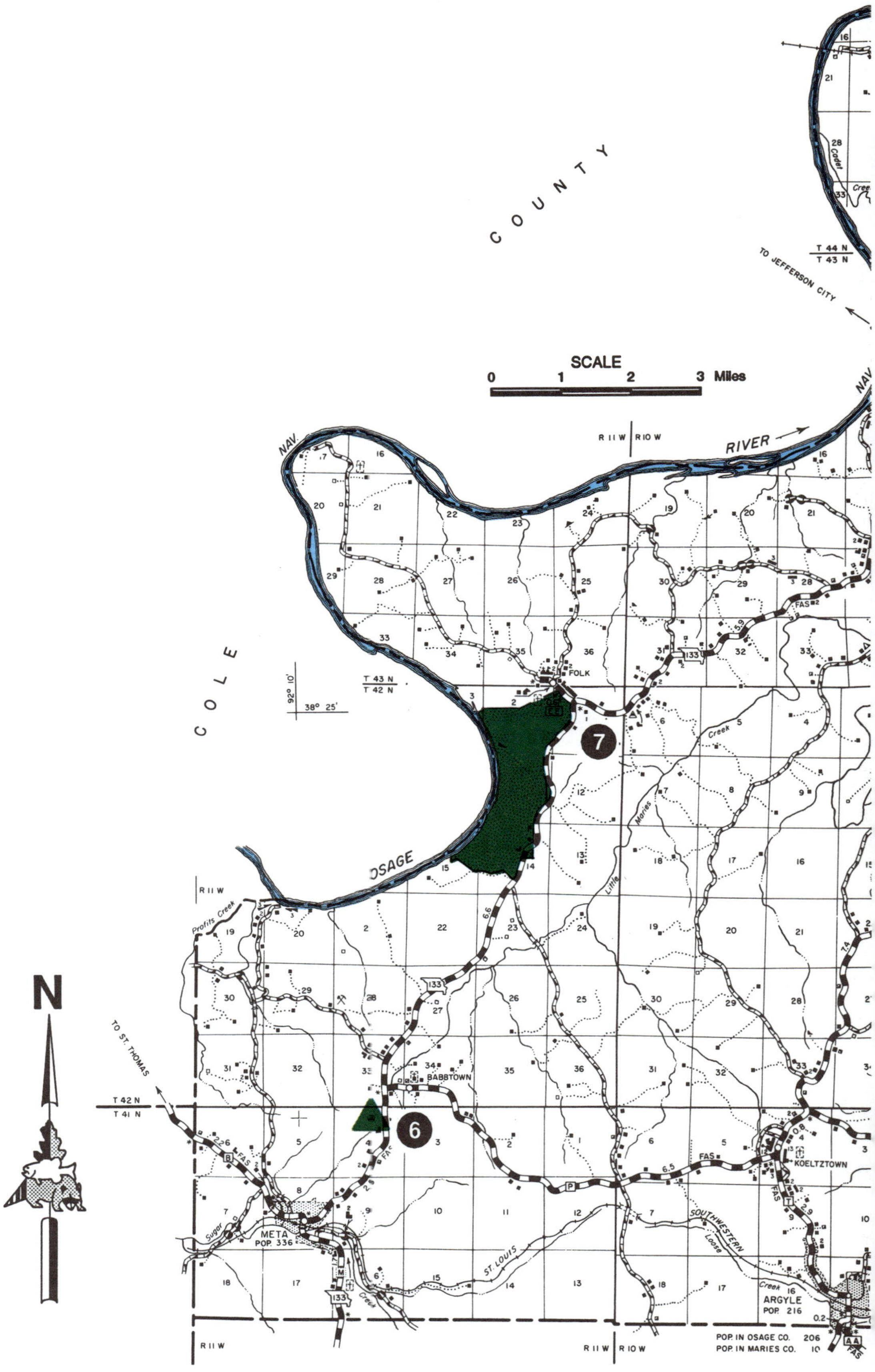

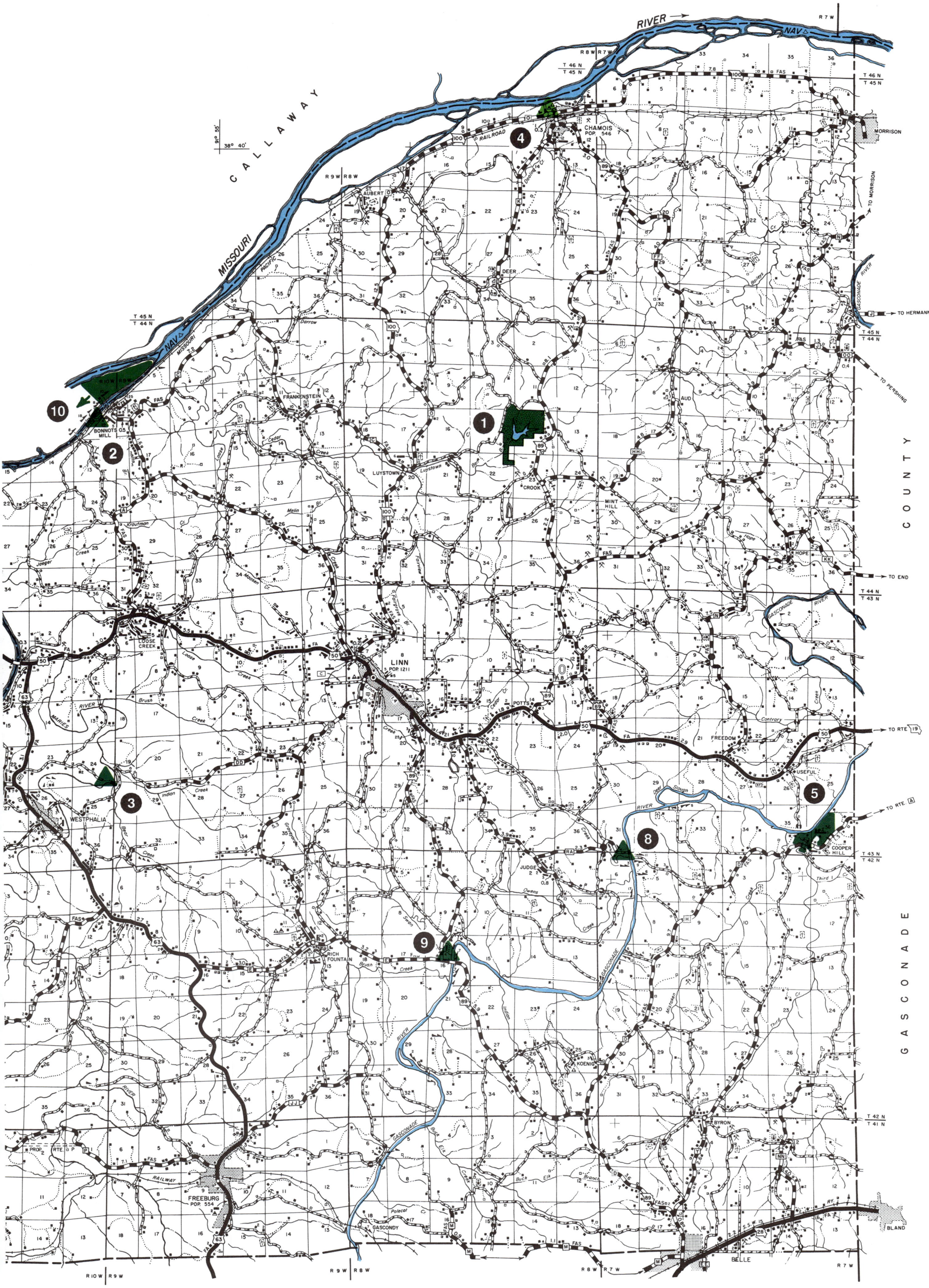
MISSOURI RIVER
CALLAWAY
GASCONADE COUNTY
GASCONADE RIVER
MARIES RIVER
CHAMOIS POP. 546
MORRISON
TO MORRISON
TO HERMANN
TO PERSHING
TO END
TO RTE 19
TO RTE A
ST. AUBERT
BONNOTS MILL
FRANKENSTEIN
LUYSTOWN
DEER
CROOK
MINT HILL
AUD
HOPE
LOOSE CREEK
LINN POP 1211
POTTS
FREEDOM
USEFUL
COOPER HILL
WESTPHALIA
JUDGE
RICH FOUNTAIN
KOENIG
BYRON
FREEBURG POP. 554
GASCONDY
BELLE
BLAND
1
2
3
4
5
8
9
10

OZARK COUNTY

1 **Blair Bridge Access** is 12 miles west of Pottersville on Highway KK to the end of the pavement, then .5 mile west on County Road H-354. This is 7 acres with a boat ramp on the North Fork of the White River.

2 **Bull Shoals Reservoir** is off Route 160. Several county and state roads (watch for signs) provide access to this 34,736-acre area. The area has 16,359 acres of fishable water, and is composed of Bull Shoals Lake, Little North Fork, Big Creek, Swan Creek and Beaver Creek. There are boat ramps, jetties and docks (accessible to wheelchairs), camping areas and picnic grounds. These facilities are managed by the Corps of Engineers. The area also extends into Taney County. For additional information, contact the Ozark District Office, Box 138, West Plains, MO 65775, 417/256-7161.

3 **Caney Mountain Conservation Area** is 5 miles north of Gainesville on Highway 181 (watch for the cantilever sign), then .5 mile west on an unnamed gravel entrance road. This 7,882-acre area contains dolomite glades, chert savannahs and numerous caves and springs. Caney Creek flows through the area, and 1,330 acres are designated Caney Mountain Natural Area. There is a horseback riding trail, two hiking trails, six picnic areas and four vistas. Parking, restrooms and an archery range on the area are accessible to wheelchairs. Interpretive nature programs are held at various times of the year.

4 **Florence C. Cook Access** is 3.5 miles east of Gainesville on Highway 160 and 3 miles north on County Road 308. This is a 4.7-acre area with 500 feet of frontage on Bryant Creek (upper end of Lake Norfork). There is a boat ramp.

5 **Norfork Lake** (1,000 acres) is about 20 miles southwest of West Plains, and may be reached from Highway O, T or 160. This 4,960-acre area contains the lake, boat ramps, picnic areas, fishing jetties and camper pads.

6 **Patrick Bridge Access** is on Highway H, 4 miles north of Highway 160. This is a 161-acre area of mainly forestland (135 acres) on the North Fork of the White River. There is a canoe launching area, campground, picnic area and short trail to Althea Spring. This is a trophy trout management area.

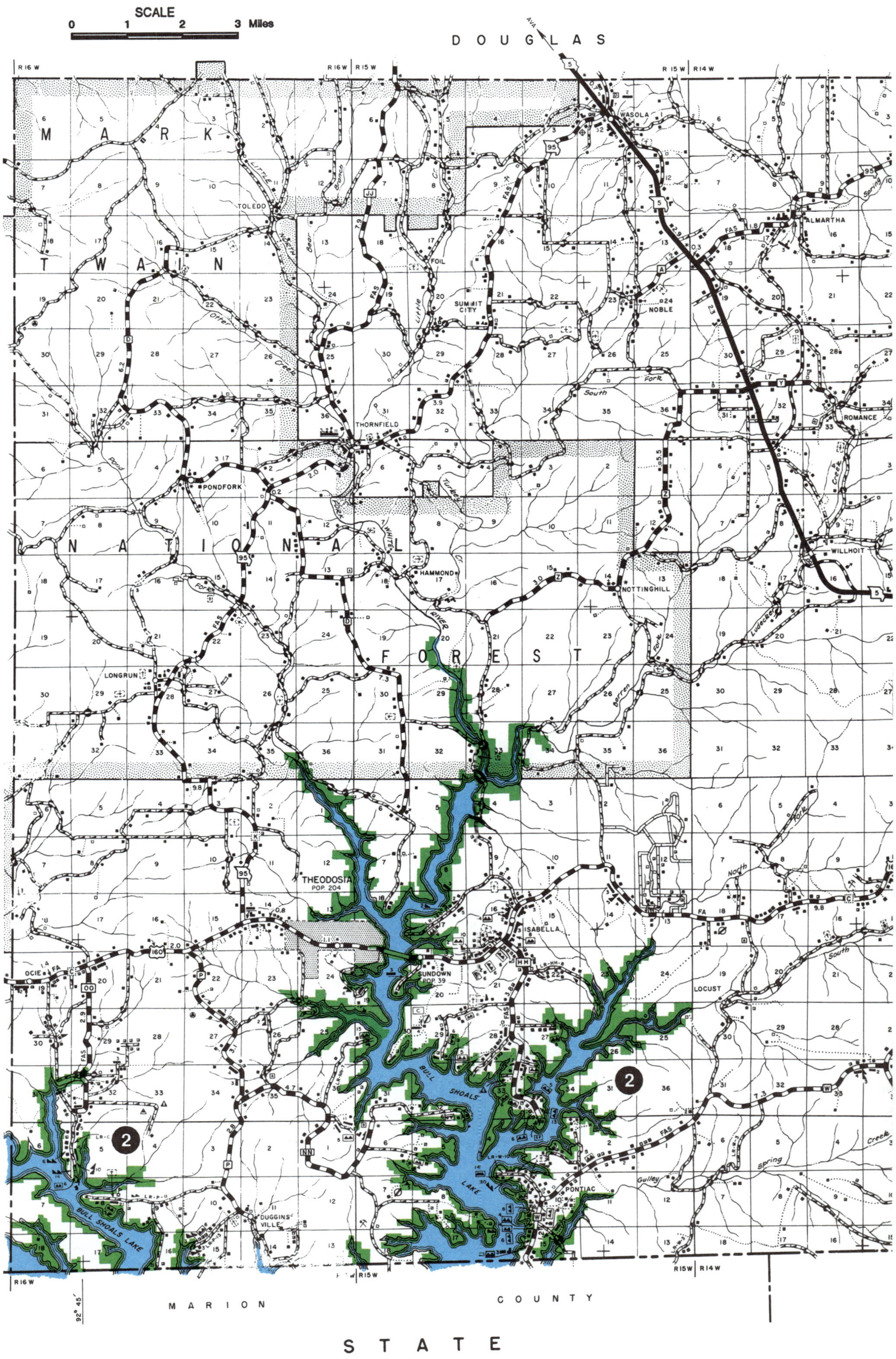

7 **Sycamore Access** is on Highway 181, 12 miles north of the town of Gainesville. This is a 16-acre access on Bryant Creek with no boat ramp.

8 **Tecumseh Towersite** (40 acres) is 10 miles east of Gainesville on Highway 160. The access road is closed to vehicles when the tower is not occupied.

9 **Timber Knob Towersite** is 2.25 miles south of Gainesville on Highway 5. This is a 40-acre area. The access road is closed to vehicles when the tower is not occupied.

10 **Warren Bridge Access** is 8 miles north of Tecumseh on Highway PP, and 2.5 miles west on County Road 328. This is 7 acres on Bryant Creek with no boat ramp.

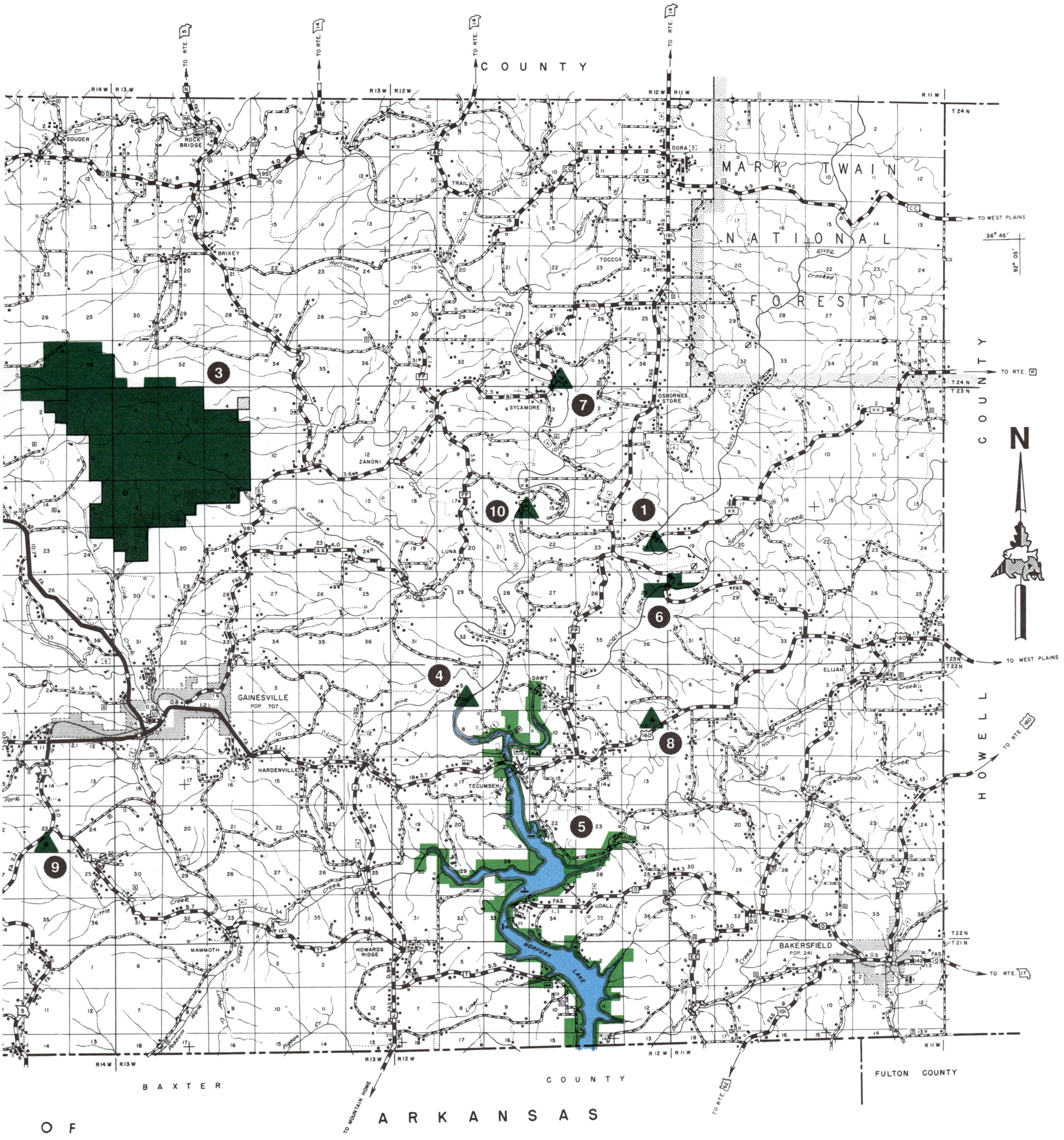

Location Map

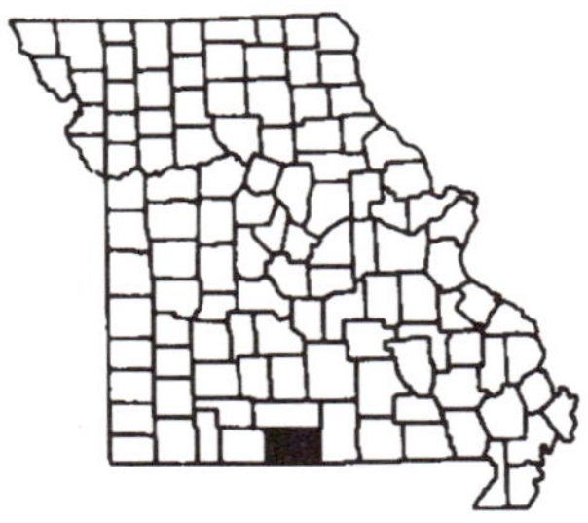

Legend

- LAND OWNED BY MDC
- LAND LEASED BY MDC
- LAND OWNED BY MDC LESS THAN 160 ACRES
- LAND LEASED BY MDC LESS THAN 160 ACRES
- CAP ACCESS

PEMISCOT COUNTY

1 **Caruthersville Rookery Conservation Area** is .25 mile south of the town of Caruthersville on Highway Y, .5 mile west on County Road 524, and .5 mile north on County Road 523 to the parking area. Much of this 43-acre area is in the form of an old borrow pit with shallow water and mud flats that attract many shore birds.

2 **Gayoso Bend Conservation Area** (968 acres) may be reached by boat only from the Mississippi River. Fishing, camping, hunting and trapping is permitted on the area.

3 **John L. and Georgia Girvin Memorial Conservation Area**–see the New Madrid County map, page 152.

4 **Little River Conservation Area and Jerry P. Combs Lake**–see the Dunklin County map, page 74.

5 **S. P. Reynolds Access** is at the east end of Ward Avenue in Caruthersville. This is a 1-acre area with a boat ramp on the Mississippi River within a city park. There is a picnic area.

6 **Twin Borrow Pits Conservation Area** is 1.5 miles east of Highway T on an unnamed county gravel road. Watch for signs. This is a 20-acre area 1 mile west of the Mississippi River.

7 To reach **Wolf Bayou Conservation Area** (276 acres), take the Wardel Exit off I-55, and follow the East Outer Road south to Highway BB. Follow BB over the levee until it ends. Wolf Bayou Natural Area is 200 acres of bottomland forest and oxbow lake. It is suitable for fishing, and there is a boat ramp and picnic area.

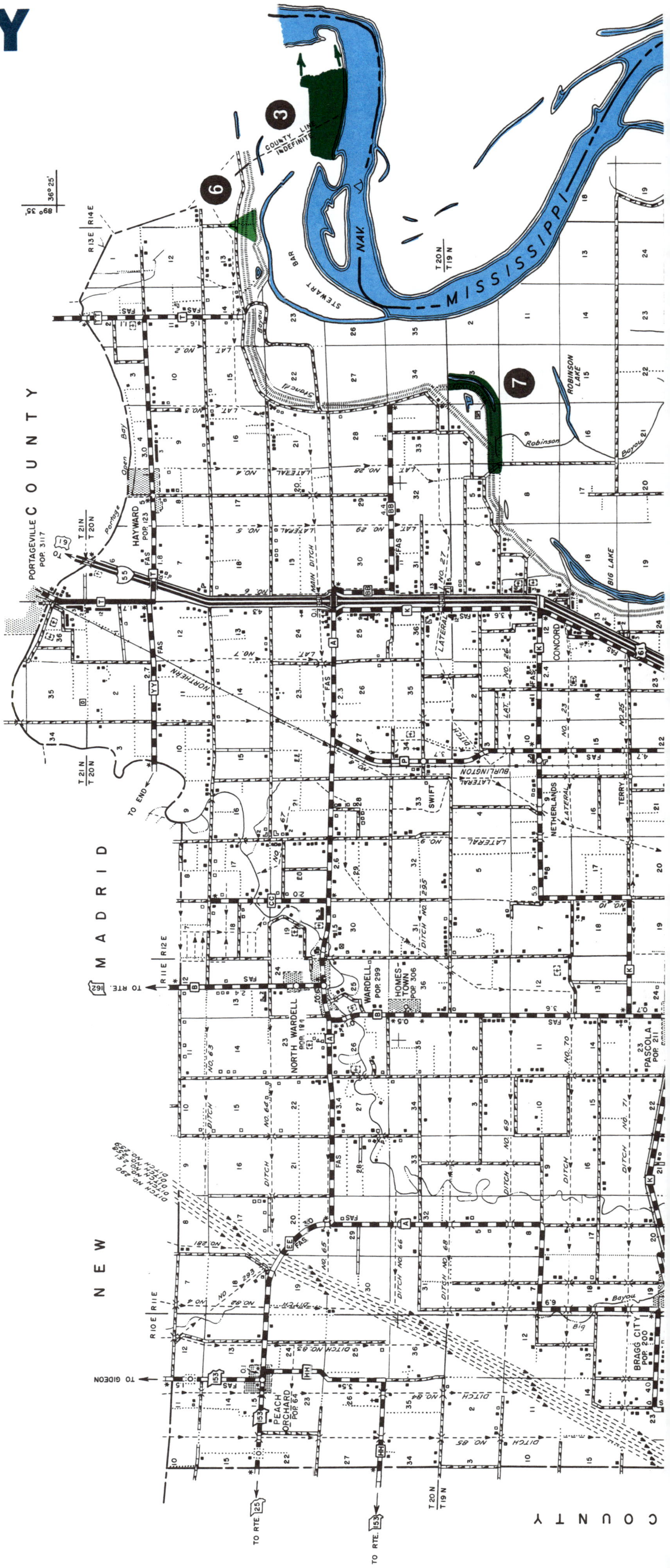

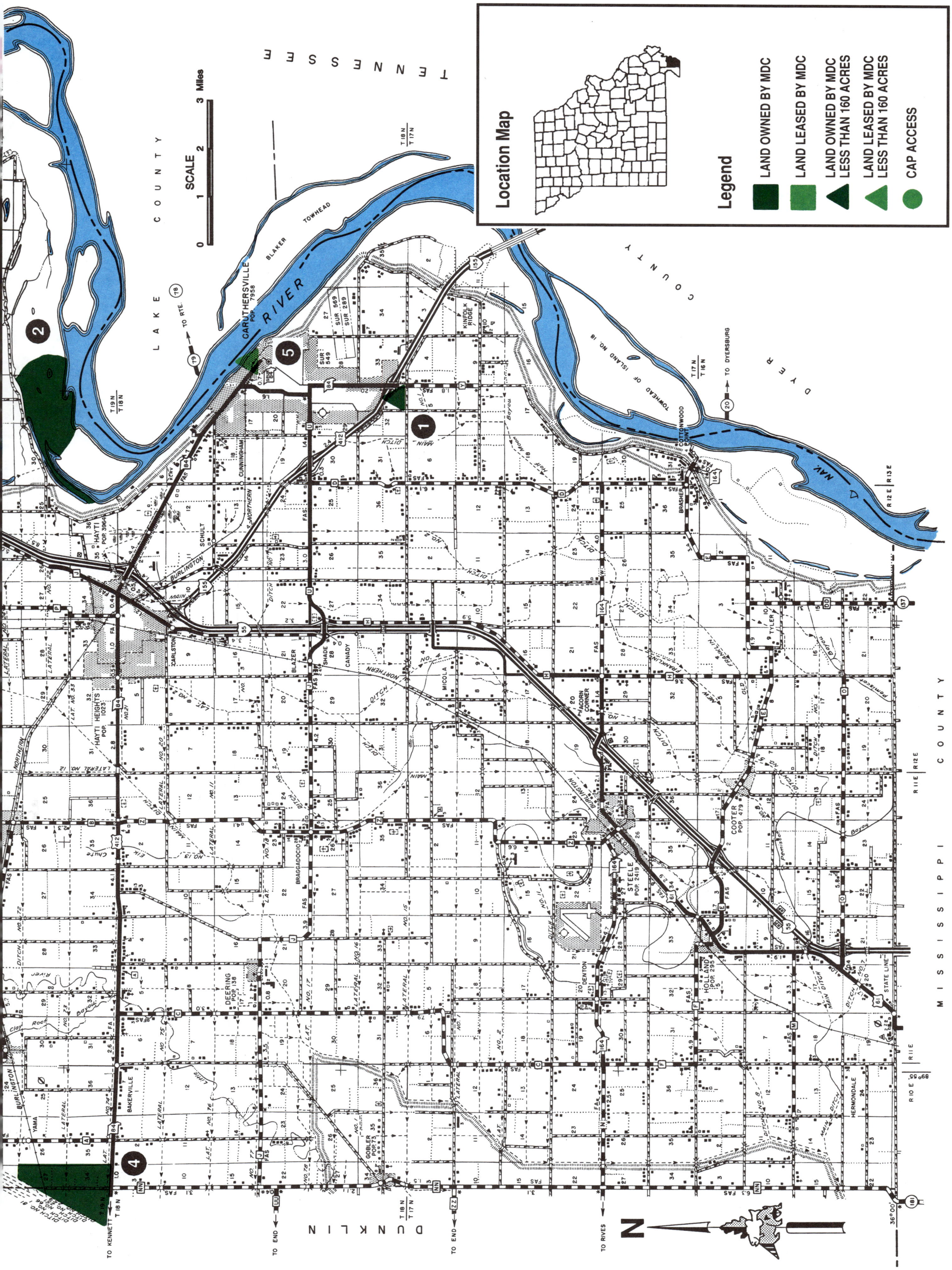

Location Map
Legend
LAND OWNED BY MDC
LAND LEASED BY MDC
LAND OWNED BY MDC LESS THAN 160 ACRES
LAND LEASED BY MDC LESS THAN 160 ACRES
CAP ACCESS
SCALE
0 1 2 3 Miles
TENNESSEE
LAKE COUNTY
DYER COUNTY
MISSISSIPPI COUNTY
DUNKLIN
RIVER
BLAKER TOWHEAD
TOWHEAD OF ISLAND NO. 18
CARUTHERSVILLE POP. 7958
HAYTI POP. 3964
HAYTI HEIGHTS POP. 1023
CARLSTON
CUNNINGHAM
SCHULT
BRAGGADOCIO
DEERING POP. 138
BAKERVILLE
GOBLER POP. 73
YAMA
SHADE
CANADY
MICOLA
ACORN CORNER
DENTON
STEELE POP. 2419
HOLLAND POP. 295
COOTER POP. 479
COTTONWOOD POINT
BRASHER
TYLER
KINFOLK RIDGE
HERMONDALE
STATE LINE
TO KENNETT
TO RIVES
TO DYERSBURG
TO END
TO RTE.

PERRY COUNTY

1 **Ball Mill Resurgence Natural Area** is north of Perryville. Take Highway 61, then Highway V. The pavement ends 3 miles down Highway V, so go 1.5 miles on County Road 916 and watch for the area sign. This 20-acre area is home to a sink at the base of a 50-foot limestone bluff. Rocks fall into the basin and are tumbled smooth by water that backs up into the sinkhole after heavy rain. There is a .25-mile trail to this Designated Natural Area.

2 **Perry County Community Lake** is .3 mile west of Perryville on County Road 700 off Highway T. The lake has two boat ramps, five jetties (wheelchair accessible) and two picnic areas. This is 110 acres surrounded by old pastures (140 acres) and forestland (60 acres).

3 **Perry Towersite** is 10 miles south of Perryville on Highway 51, and 3 miles west on Highway J. This is a 5-acre area.

4 **Red Rock Landing Conservation Area** (559 acres) is south of Perryville. Take Highway 61 south to Highway P. At the junction of highways P and C, take C to Crosstown. In Crosstown, take County Road 350 to the area.

5 To reach the largest tract of **Seventy-Six Conservation Area**, go 7 miles south of Perryville on Highway 61 to Longtown. At the south edge of Longtown, turn left on Highway D to Highway C. Turn right on Highway C and, in Brazeau, turn left on Highway D until it ends at a gravel road. The 818-acre forest is a short distance down this gravel road. The area has several sinkholes, three springs and good vantage points overlooking the Mississippi Valley.

6 **Tower Rock Natural Area** is off Highway A near Uniontown, 1 mile to the right on County Road 460. Use caution when crossing the railroad tracks. This is 31 acres on the Mississippi River and is home to a rock formation on the river known as Tower Rock (originally designated for a colony of cliff swallows). There is a picnic area and a .2-mile hiking trail overlooking the river.

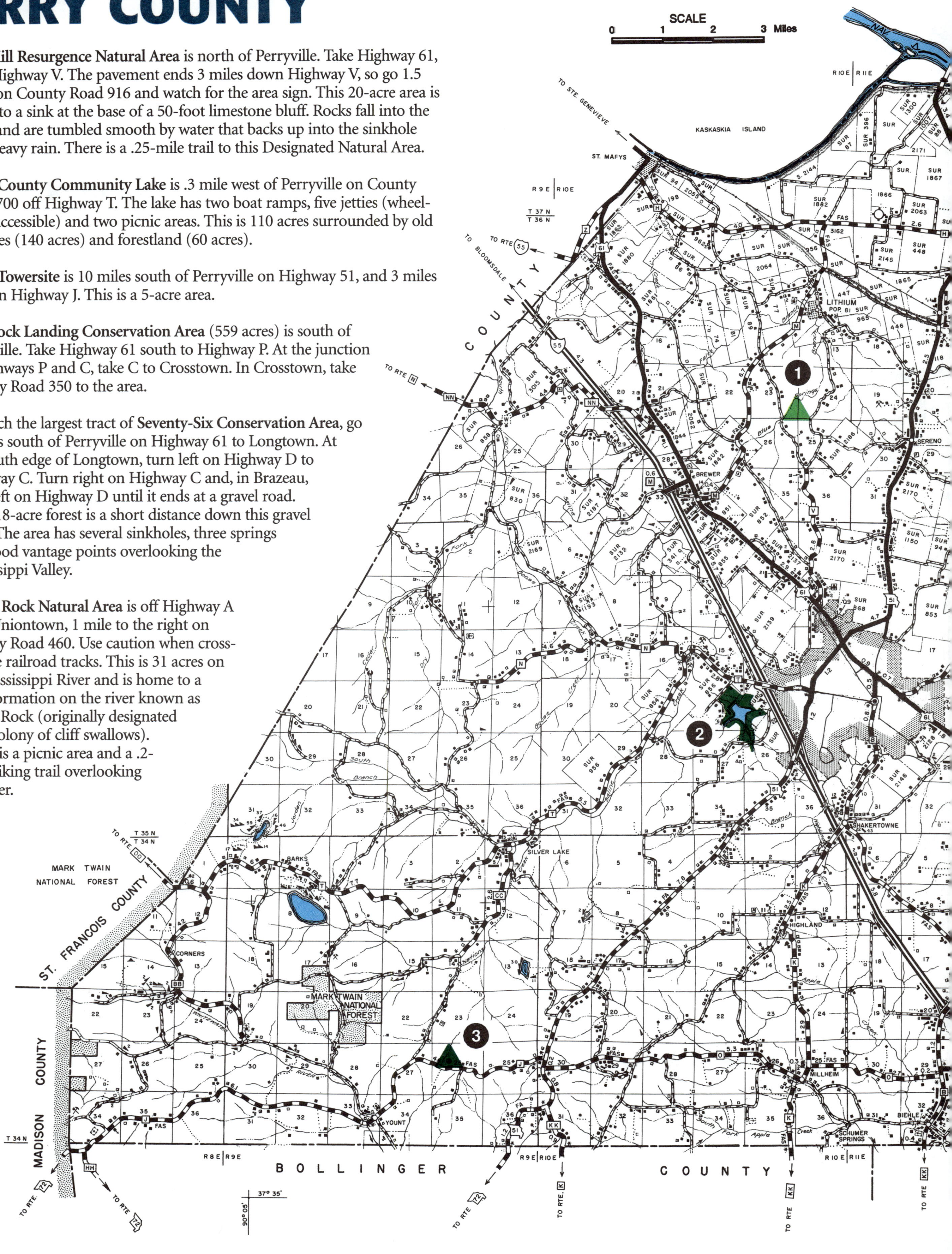

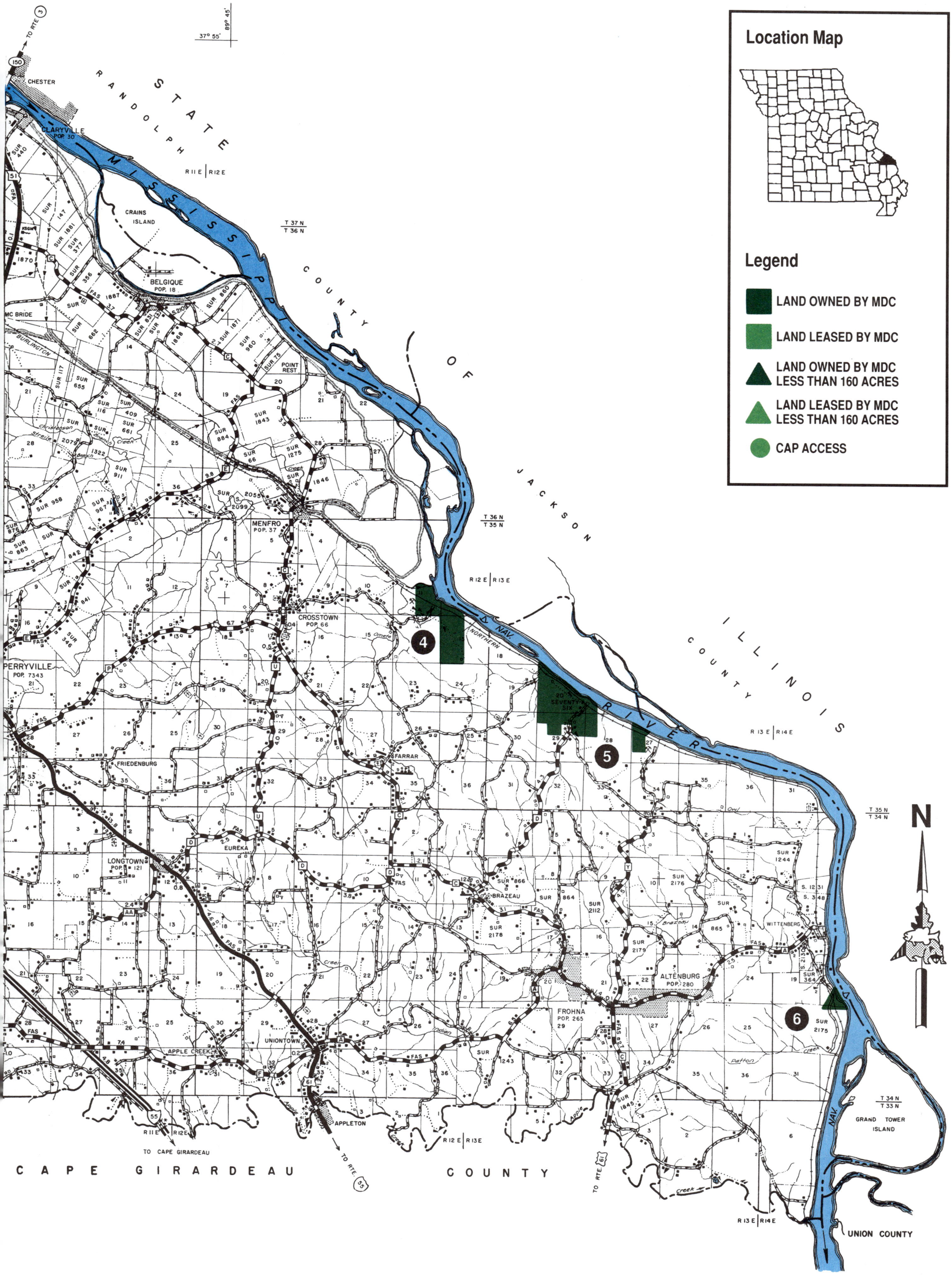
Location Map
Legend
LAND OWNED BY MDC
LAND LEASED BY MDC
LAND OWNED BY MDC LESS THAN 160 ACRES
LAND LEASED BY MDC LESS THAN 160 ACRES
CAP ACCESS
RANDOLPH COUNTY OF JACKSON
STATE
ILLINOIS
JACKSON COUNTY
MISSISSIPPI RIVER
CHESTER
CLARYVILLE POP. 30
CRAINS ISLAND
BELGIQUE POP. 18
MC BRIDE
POINT REST
MENFRO POP. 37
CROSSTOWN POP. 66
PERRYVILLE POP. 7343
FRIEDENBURG
EUREKA
LONGTOWN POP. 121
FARRAR
SEVENTY SIX
BRAZEAU
WITTENBERG
ALTENBURG POP. 280
FROHNA POP. 265
UNIONTOWN
APPLE CREEK
APPLETON
GRAND TOWER ISLAND
UNION COUNTY
CAPE GIRARDEAU COUNTY
TO CAPE GIRARDEAU
TO RTE. 3
TO RTE. 55
TO RTE. 61
T 37 N
T 36 N
T 35 N
T 34 N
T 33 N
R 11 E
R 12 E
R 13 E
R 14 E
37° 55'
89° 45'
N
4
5
6

PETTIS COUNTY

1 **Clover Dell Park Lake** is south off 32nd Street, 1 mile west of the Sedalia city limits. The lake is 12 acres and has a boat ramp. The park has an archery range and athletic fields.

2 **Drover's Prairie Conservation Area** is 10 miles south of Sedalia on Highway 65, and 1 mile west on Highway 52. This is an 80-acre native prairie.

3 **Farrington Park Lake**–see the Henry County map, page 90.

4 **Friendly Prairie Conservation Area** is 9 miles south of Sedalia on Highway 65, and 1.5 miles west on Manila Road. This is 40 acres of unplowed native prairie. More than 260 species of plants have been identified in this area.

5 **Grandfather Prairie Conservation Area** is 8 miles south of Sedalia on Highway 65, and 2 miles west on Mather Road. This unplowed, native prairie is 78 acres.

6 **Kahrs-Boger Park** is 11 miles southeast of Sedalia on Highway M. Lake Creek flows through this 108-acre area for .4 mile, and there is a pond (1.5 acres total for fishing).

7 **J. N. "Turkey" Kearn Memorial Wildlife Area**–see the Johnson County map, page 110.

8 **W. R. Kearn Memorial Conservation Area** is 3 miles west of Green Ridge on Route B, and 2 miles south on Hope Dale Road. The 120-acre area is on the west side of the road.

9 **Liberty Park Pond** is at the intersection of Third and Park streets, 1.5 miles north of the junction of highways 50 and 65 (off Highway 65). This is a 3-acre pond in a park equipped with fishing docks, public office and six picnic areas.

10 **Paint Brush Prairie Conservation Area** is 9 miles south of Sedalia on Highway 65, then left on Manila Road. Seventy-four acres of native prairie on the northwest end of this 314-acre area is a Designated Natural Area.

11 **Ralph and Martha Perry Memorial Conservation Area**–see the Johnson County map, page 110.

12 **Pinhook Access** is 1 mile north of Sedalia on Highway 65, 3 miles east on Highway HH, and 3.5 miles north on Highway EE. This is a 4-acre area on Muddy Creek with no boat ramp.

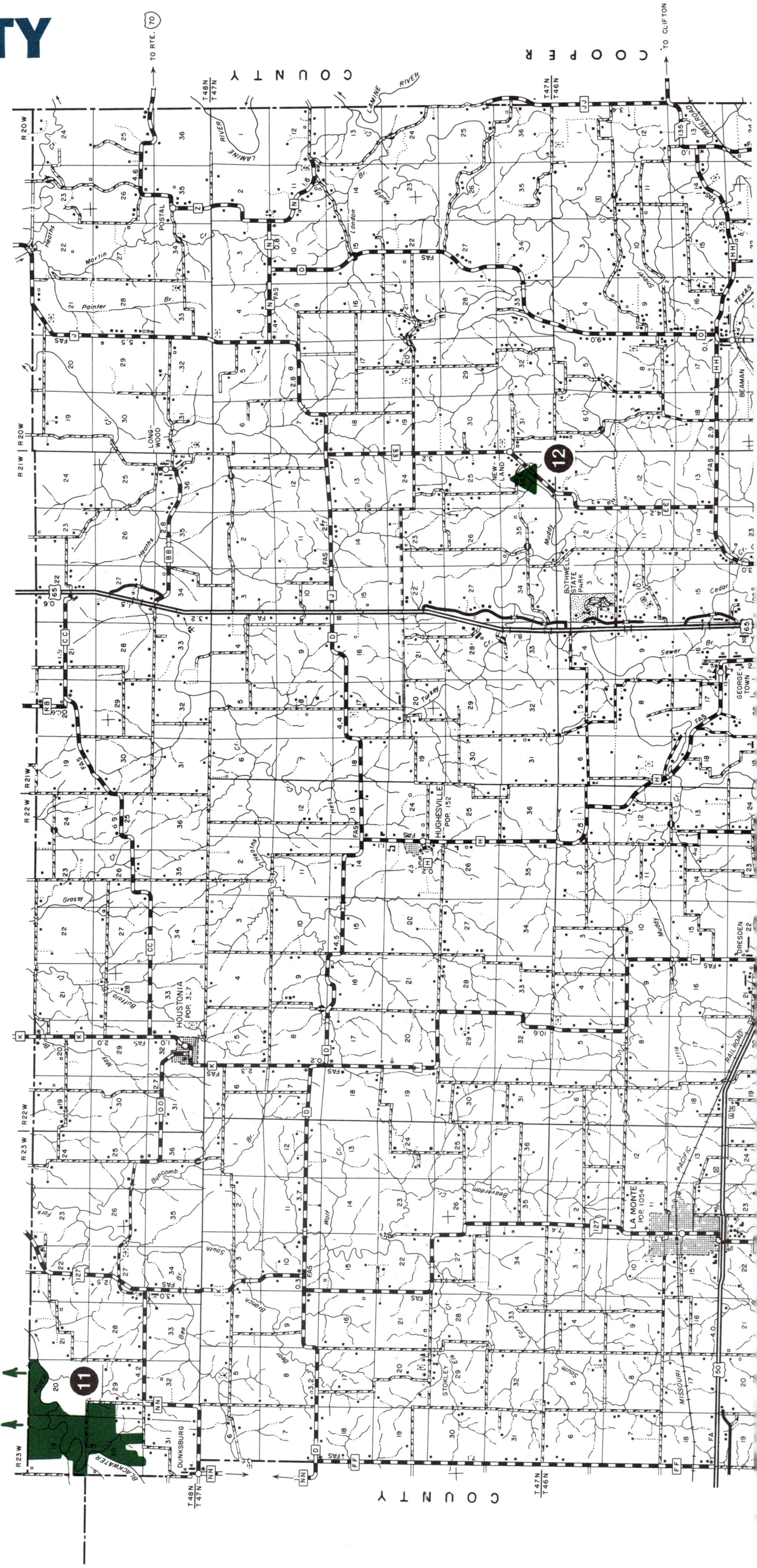

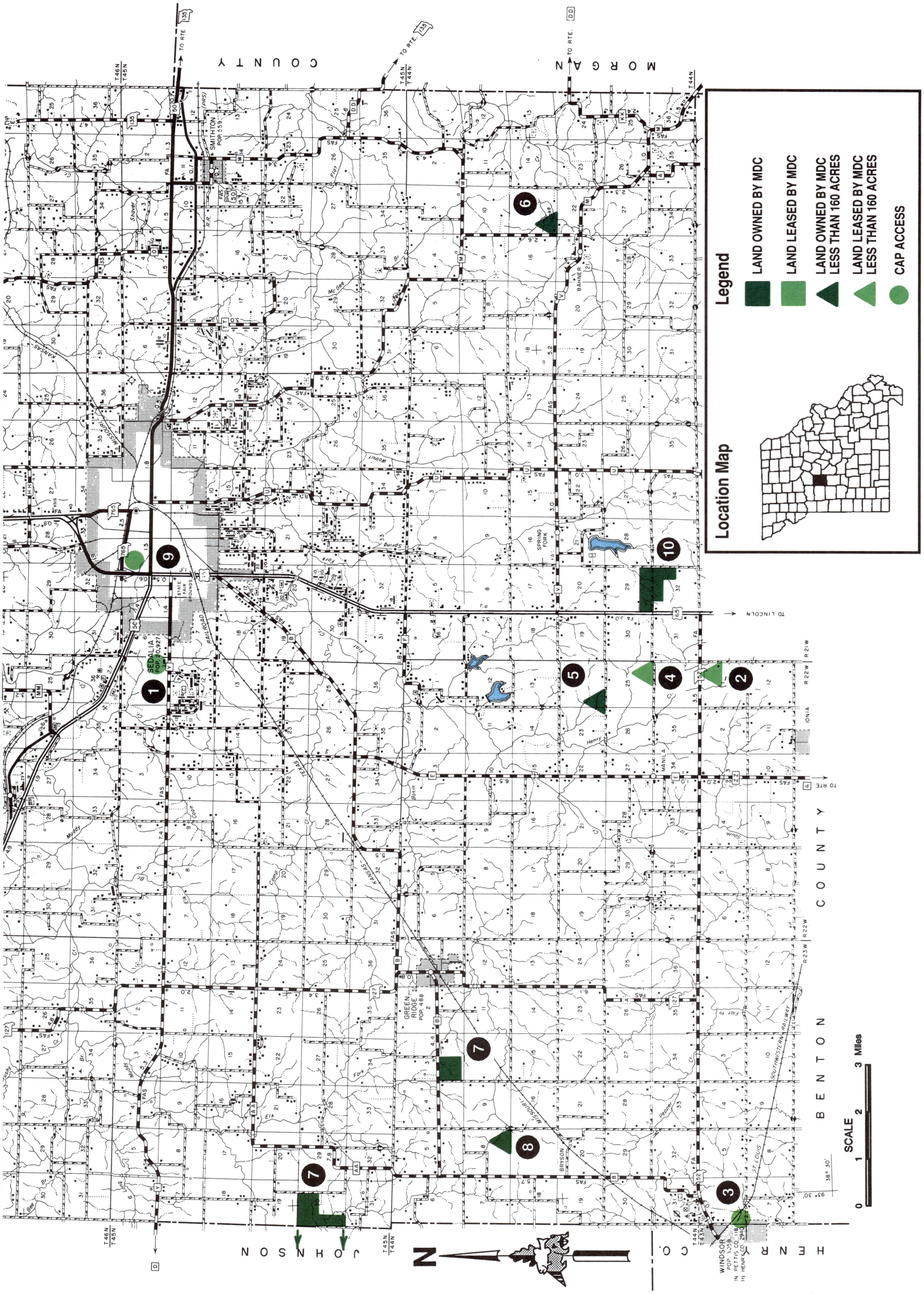
Legend
LAND OWNED BY MDC
LAND LEASED BY MDC
LAND OWNED BY MDC LESS THAN 160 ACRES
LAND LEASED BY MDC LESS THAN 160 ACRES
CAP ACCESS
Location Map
SCALE
0 1 2 3 Miles
MORGAN COUNTY
BENTON COUNTY
JOHNSON
HENRY CO.
SEDALIA POP. 20,927
SMITHTON POP. 559
GREEN RIDGE POP. 488
WINDSOR
STATE FAIR GROUNDS
SPRING FORK
IONIA
MANILA
BRYSON
TO LINCOLN

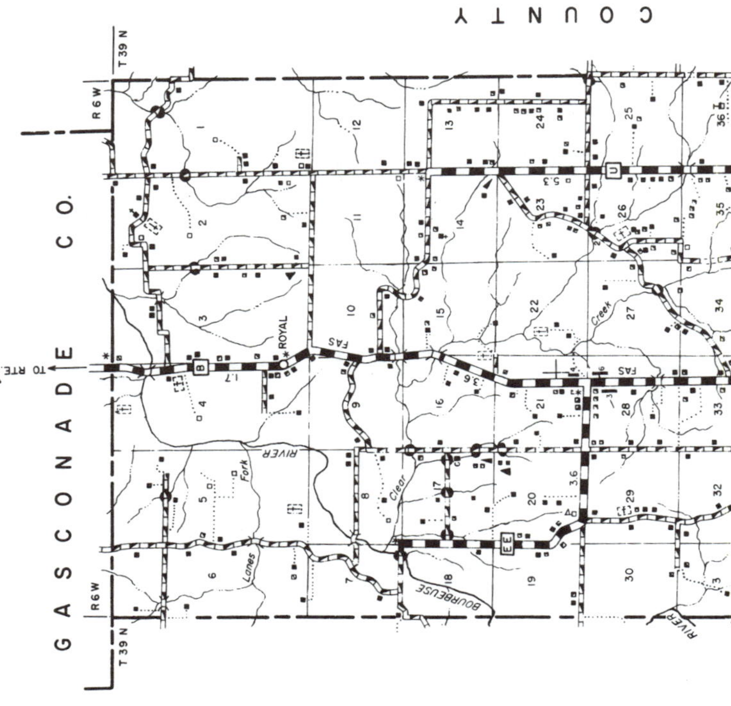

PHELPS COUNTY

1 **Beaver Creek Conservation Area** is 3 miles south of Rolla on Highway 63. The area is closed to vehicle traffic but may be accessed on foot. This 156-acre area contains a stretch of Beaver Creek. There is a picnic area, .25-mile hiking trail and an archery range.

2 **Jerome Access** is .1 mile off Highway D in the town of Jerome. This is a 10-acre area with a boat ramp on the Gasconade River.

3 **Little Prairie Community Lake** is on Route RA, 5 miles east of Rolla and off the north outer road of I-44. This 100-acre lake has a boat ramp, wheelchair accessible fishing jetties and 12 picnic areas. It is surrounded by about 343 acres of land.

4 **Maramec Spring Fish Hatchery and Trout Park** are inside Maramec Spring Park, operated by the James Foundation and one of the state's four trout parks. The park contains three museums and other facilities, such as volleyball and tennis courts, picnic areas, hiking trails and a restaurant.

5 **Pilot Knob Towersite** is 8 miles south of Rolla on Highway 63. This is a 4-acre area.

6 **Rosati Towersite** is .5 mile south of the town of Rosati on Highway KK. This 92-acre area is mainly forestland.

7 **Schuman Park Lake** (also known as Frisco Lake) is in Rolla, west on 18th Street then left on Cedar Street. This is a 5-acre lake (no boat ramp) in a 17-acre city park. There are two picnic areas.

8 **Scioto Lake** is a 5-acre body of water suitable for fishing inside St. James Park on James Lane, off Highway 8 within the St. James city limits. The park has a library, swimming pool, playground, tennis courts, ball fields, observation tower, nature trail and picnic areas.

9 **Woodson K. Woods Memorial Conservation Area**–see the Crawford County map, page 61.

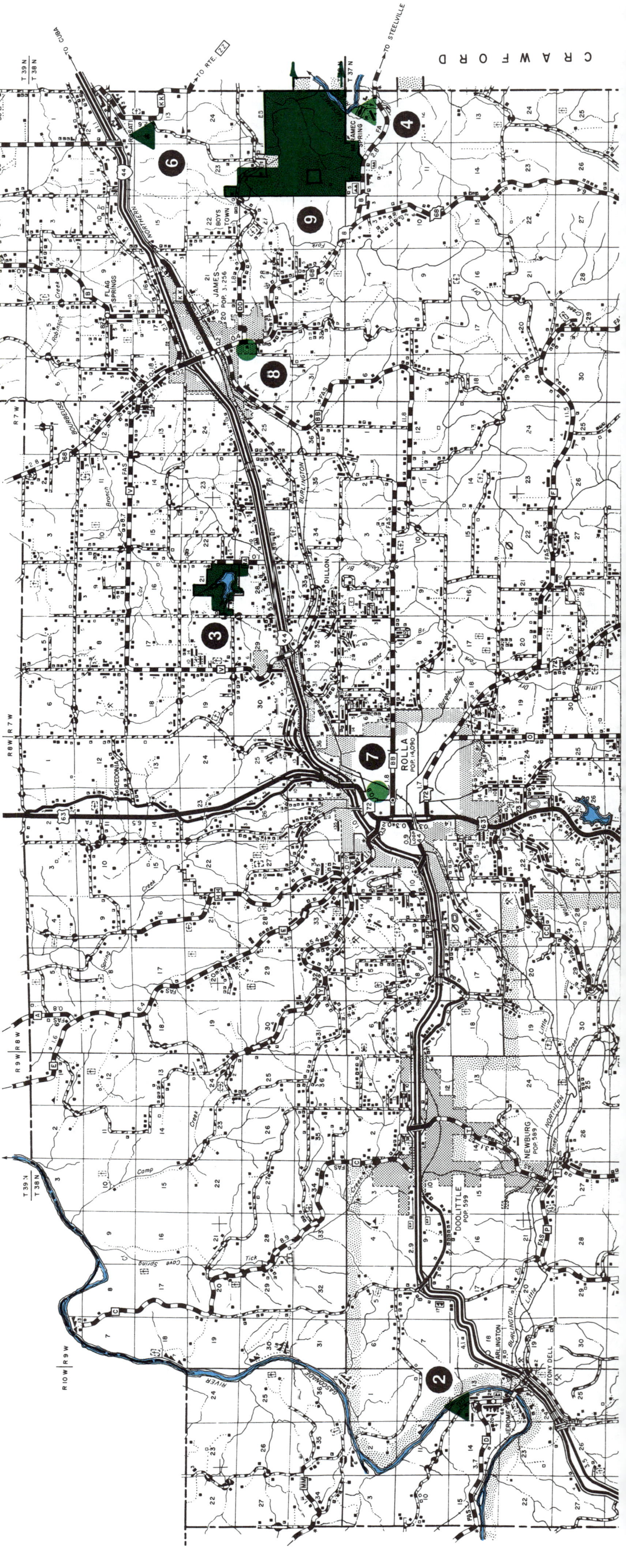

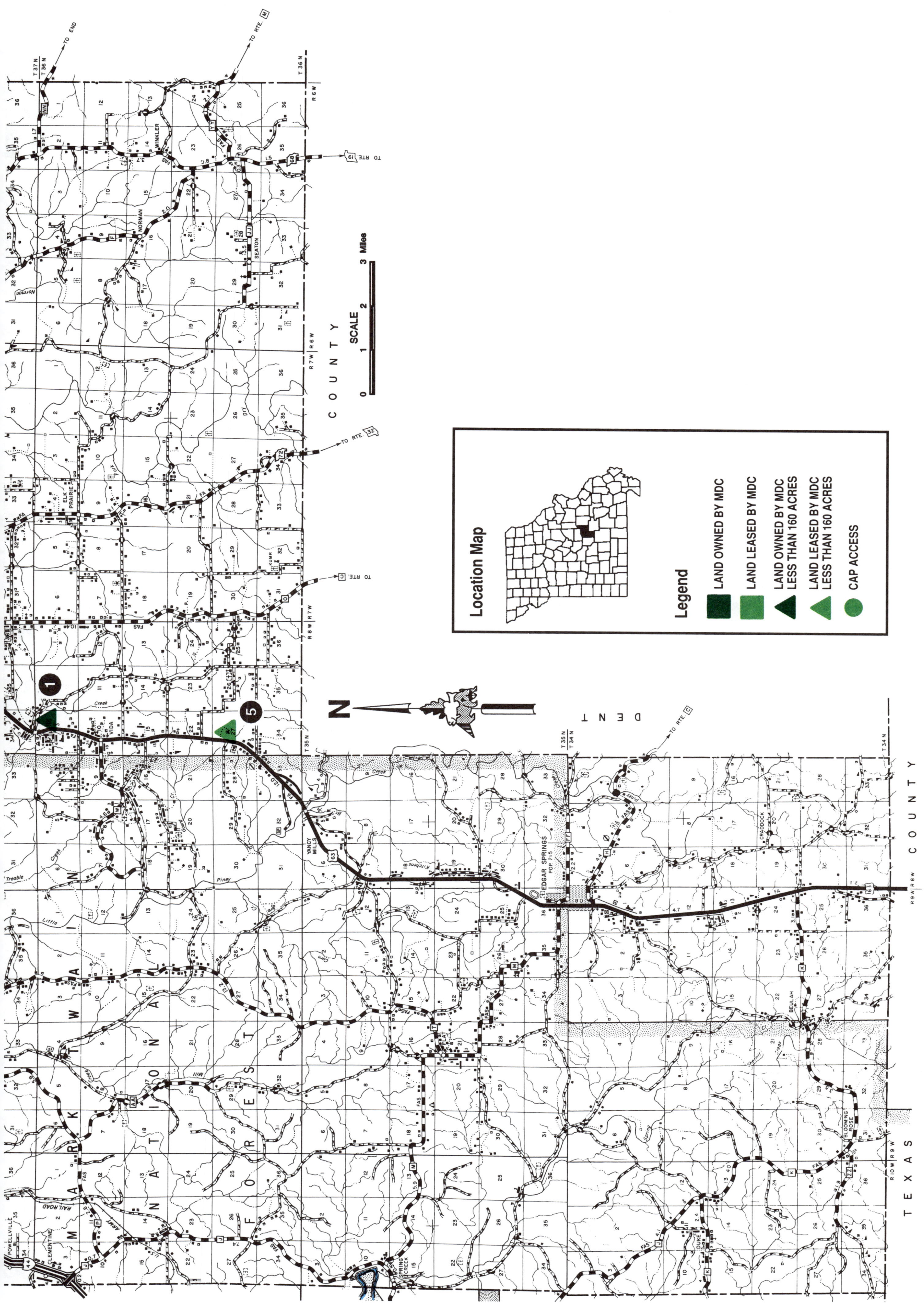
Location Map
Legend
LAND OWNED BY MDC
LAND LEASED BY MDC
LAND OWNED BY MDC LESS THAN 160 ACRES
LAND LEASED BY MDC LESS THAN 160 ACRES
CAP ACCESS
SCALE
0 1 2 3 Miles
MARK TWAIN NATIONAL FOREST
DENT COUNTY
TEXAS COUNTY
COUNTY
EDGAR SPRINGS
POP. 215
YANCY MILLS
ELK PRAIRIE
NORMAN
WINKLER
SEATON
VIDA
KUMA
CRADDOCK
BEULAH
BLOOMING ROSE
DUKE
SPRING CREEK
CLEMENTINE
POWELLVILLE

PIKE COUNTY

1 **Edward Anderson Conservation Area** is 11 miles south of Hannibal on Highway 79. The Mississippi River flows 2.25 miles along the edge of this 1,045-acre area.

2 **Calumet Creek Access** is 4 miles north of Clarksville on Highway 79. This is a 2.4-acre area on Calumet Creek, which accesses the Mississippi River. There is a boat ramp.

3 **DuPont Reservation Conservation Area** is 16 miles south of Hannibal on Highway 79 near Ashburn. A 1-mile stretch of the Mississippi River flows along the edge of this 1,320-acre area. DuPont Upland Forest Natural Area, a mixed upland hardwood forest, is an 80 acre Designated Natural Area. There is a .5-mile hiking trail and boat ramp. The area borders Ted Shanks Conservation Area.

4 **Hamberg Ferry Access** is 10 miles east on Highway P off Highway 79 near Elsberry. This is a 24-acre area on the Mississippi River with a boat ramp and wheelchair accessible parking and restrooms.

5 **Louisiana Access** is in the town of Louisiana, 6 blocks south of Highway 54 on Third Street, then 3 blocks east on South Carolina Street. This is 10 acres on the Mississippi River with a two-lane boat ramp and a wheelchair accessible dock.

6 **Ranacker Conservation Area** is 9 miles north of Bowling Green on Route RA off Highway 61. This 1,597-acre area is mainly forestland (1,190 acres) and contains a stretch of Peno Creek.

7 **Ted Shanks Conservation Area** is 17 miles south of Hannibal on Highway 79 and east on Route TT (area is marked by a sign). Route TT ends at the 4,025-acre area. Horseshoe Lake, Flag Lake, Rainbow Lake, Salt Pits and stretches of the Mississippi and Salt rivers provide anglers with about 575 acres of fishable water. There are two Designated Natural Areas, a 7-mile canoe trail on Salt River, area sloughs and canals, a wheelchair accessible trail, waterfowl blinds and group tours by appointment.

8 **Upper Mississippi Conservation Area**–see the Lewis County map, page 120.

9 **West Lake** is 2 miles east of Bowling Green on Route 54. At Vera, turn 2 miles south on an unnamed gravel road to reach the lake dam. The area is 45 acres of fishable water.

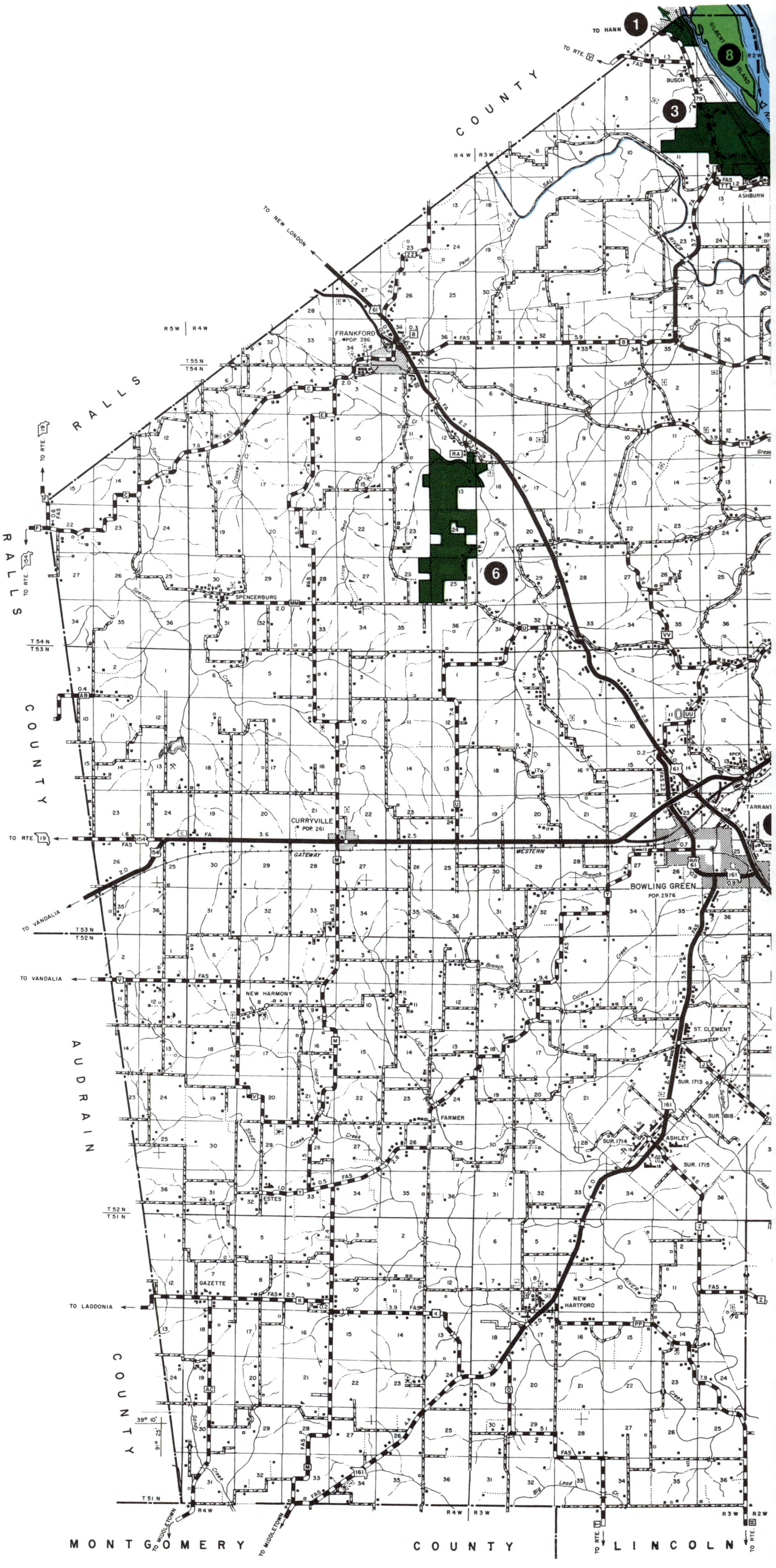

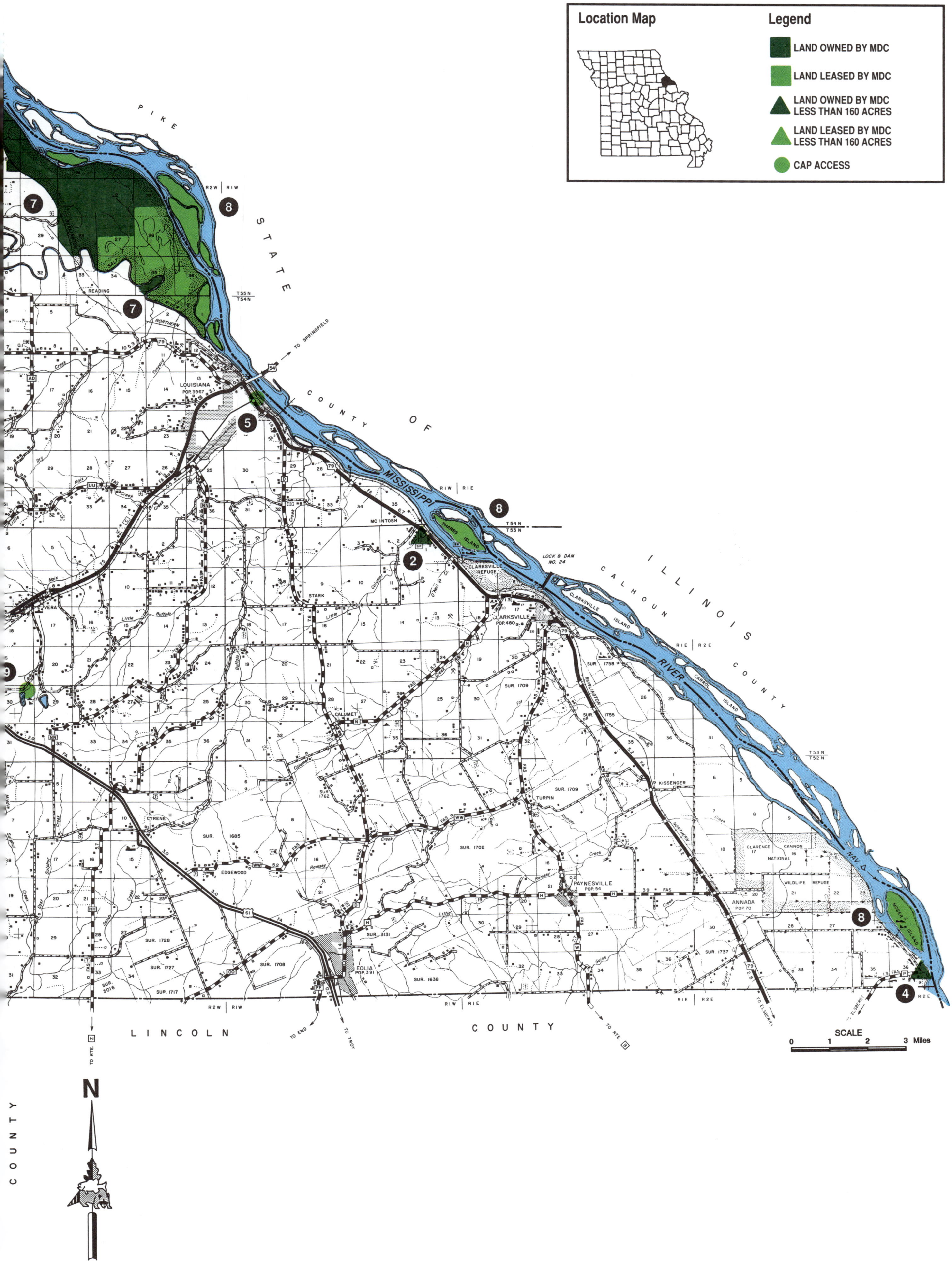
Location Map
Legend
LAND OWNED BY MDC
LAND LEASED BY MDC
LAND OWNED BY MDC LESS THAN 160 ACRES
LAND LEASED BY MDC LESS THAN 160 ACRES
CAP ACCESS
PIKE
STATE
OF
ILLINOIS
COUNTY
CALHOUN COUNTY
MISSISSIPPI RIVER
LINCOLN COUNTY
COUNTY
READING
LOUISIANA POP. 3967
TO SPRINGFIELD
MC INTOSH
PHARRS ISLAND
CLARKSVILLE REFUGE
LOCK & DAM NO. 24
CLARKSVILLE ISLAND
CLARKSVILLE POP. 480
STARK
VERA
CALUMET
CARROL ISLAND
KISSENGER
TURPIN
CYRENE
EDGEWOOD
PAYNESVILLE POP. 54
CLARENCE CANNON NATIONAL WILDLIFE REFUGE
ANNADA POP. 70
MOSIER ISLAND
EOLIA POP. 391
TO RTE. Z
TO END
TO TROY
TO RTE. B
TO ELSBERRY
ELSBERRY
R2W R1W
R1W R1E
R1E R2E
T55N T54N
T54N T53N
T53N T52N
SCALE
0 1 2 3 Miles
N

PLATTE COUNTY

1 **Humphrey Access** is southwest of Platte City off Highway N. This is a 12-acre area on the Platte River with a boat ramp.

2 **Anthony and Beatrice Kendzora Conservation Area**–see the Buchanan County map, page 26.

3 **Little Bean Marsh Conservation Area** is 30 miles north of Kansas City on Highway 45. Home to Bean Lake (200 acres suitable for fishing), this 427-acre area is mainly wetlands (306 acres). The area is visited by migrating white pelicans.

4 **Guy B. Park Conservation Area** is 3 miles north of Platte City on Highway 371. There is about 20 acres of fishable water on this 379-acre area.

5 To reach **Platte Falls Conservation Area,** take I-29 to Highway HH in Platte City, and go .5 mile north on the outer road (Knighton Avenue). Watch for an area sign. A 9-mile stretch of the Platte River flows through this 2,333-acre area, providing anglers with about 75 acres of fishable water. There are also ponds. The area has picnic areas, two hiking trails (totaling 2 miles), and shotgun and archery ranges.

6 **Ringgold Access** is 3 miles east of Camden Point on Highway E. This is a 22-acre area on the Platte River with no boat ramp.

7 **Schimmel City Access** is .5 mile off Highway 45, northeast of Farley. This is an 11-acre area on the Platte River with a boat ramp.

8 **Sharps Station Access** is on Interurban Road northeast of Platte City. This is 10 acres on the Platte River with a boat ramp. It is part of the 2,333-acre Platte Falls Conservation Area.

9 **Union Mill Access** is 2 miles west of Edgerton on Highway Z. This is 2 acres on the Platte River with no boat ramp.

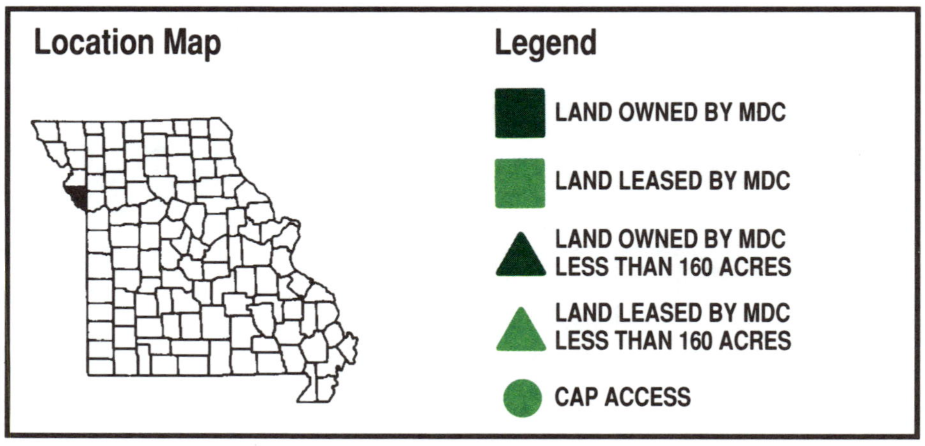

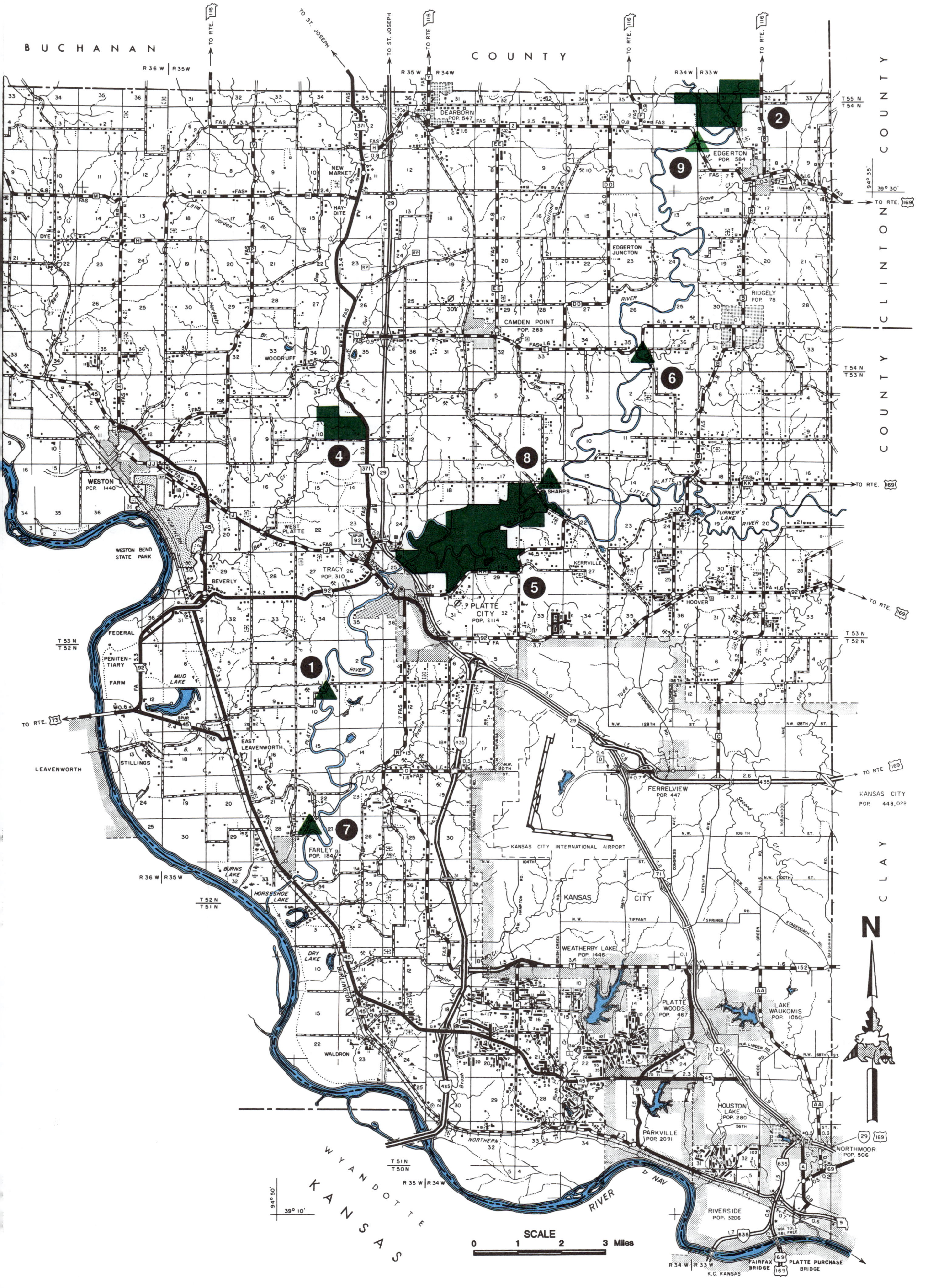
BUCHANAN COUNTY
CLINTON COUNTY
CLAY COUNTY
WYANDOTTE
KANSAS
TO ST. JOSEPH
TO RTE. 116
TO RTE. 169
TO RTE. 73
R 36 W R 35 W
R 35 W R 34 W
R 34 W R 33 W
T 55 N
T 54 N
T 53 N
T 52 N
T 51 N
T 50 N
94° 35'
39° 30'
94° 50'
39° 10'
DEARBORN
POP. 547
NEW MARKET
HAYDITE
DYE
EDGERTON
POP. 584
EDGERTON JUNCTION
RIDGELY
POP. 78
CAMDEN POINT
POP. 263
WOODRUFF
WESTON
POP. 1440
WESTON BEND STATE PARK
WEST PLATTE
TRACY
POP. 310
BEVERLY
SHARPS
KERRVILLE
PLATTE CITY
POP. 2114
HOOVER
TURNER'S LAKE
FEDERAL
PENITENTIARY
FARM
MUD LAKE
EAST LEAVENWORTH
STILLINGS
LEAVENWORTH
FARLEY
POP. 184
BURNS LAKE
HORSESHOE LAKE
DRY LAKE
WALDRON
FERRELVIEW
POP. 447
KANSAS CITY
POP. 448,028
KANSAS CITY INTERNATIONAL AIRPORT
KANSAS CITY
WEATHERBY LAKE
POP. 1446
PLATTE WOODS
POP. 467
LAKE WAUKOMIS
POP. 1050
HOUSTON LAKE
POP. 280
PARKVILLE
POP. 2091
NORTHMOOR
POP. 506
RIVERSIDE
POP. 3206
FAIRFAX BRIDGE
PLATTE PURCHASE BRIDGE
K.C. KANSAS
PLATTE RIVER
LITTLE PLATTE RIVER
RIVER
SCALE
0 1 2 3 Miles
N
1
2
4
5
6
7
8
9

POLK COUNTY

1 **Brush Creek Conservation Area** is 2 miles west of Humansville on Highway N, then .8 mile north on South 7th Road. It is a 158-acre forest.

2 **LaPetite Gemme Prairie Conservation Area** is 1 mile west on the first county road that lies west of the junction of Highway 13 and Business 13 (south of Bolivar). This is a 37-acre Designated Natural Area.

3 **Pleasant Hope Conservation Area** is 1.75 miles south of Pleasant Hope on Highway H, and 1.5 miles east on East 564th Road. Watch for the cantilever sign on Highway H. This 1,106-acre area is nearly all forestland (1,013 acres) and has a firearms range.

4 **Pomme de Terre Reservoir**–see the Hickory County map, page 92.

5 **Sentinel Conservation Area** is off Highway 64 at the town of Sentinel, 2.5 miles west and south on an unnamed county gravel road. It is a 160-acre forest.

6 **Stockton Reservoir**–see the Cedar County map, page 44.

7 **Sunset Park Access** is 4.5 miles east of the junction of highways 13 and Y on Highway Y. The pavement on Highway Y ends and the road forks, but continue on the left gravel fork .2 mile to the area. This is 21 acres with a boat ramp on the Pomme de Terre River.

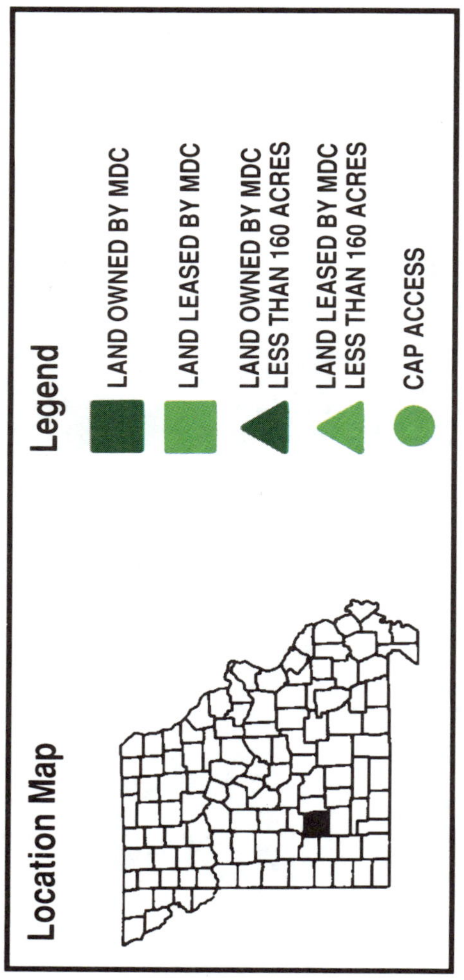

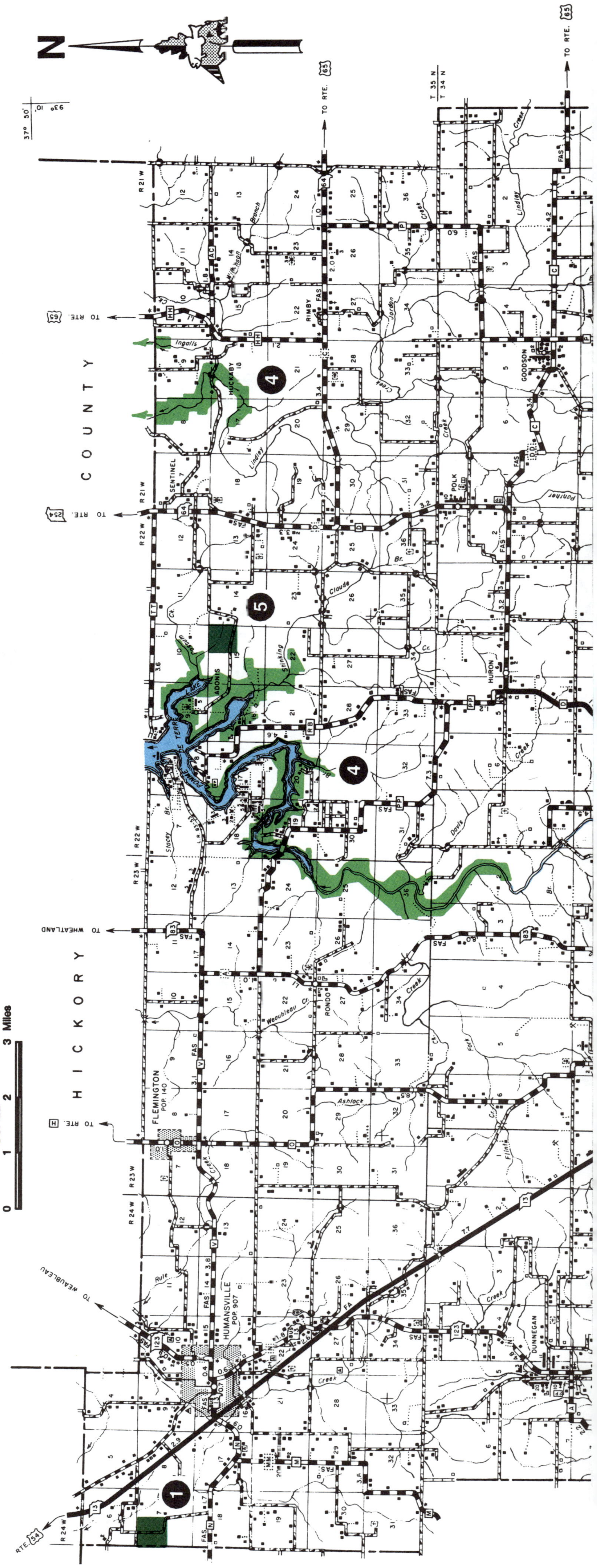

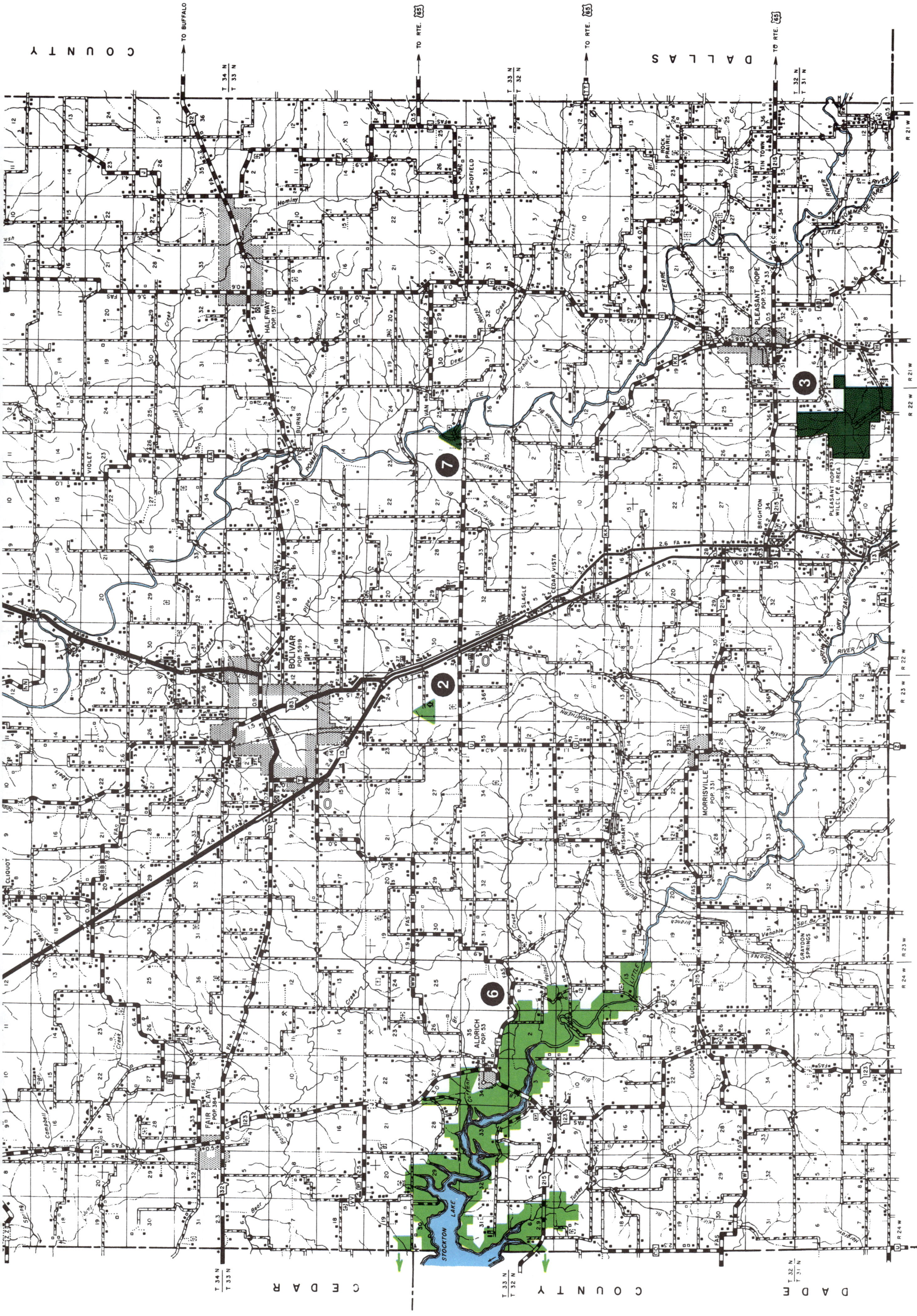

DALLAS COUNTY
GREENE COUNTY
CEDAR COUNTY
DADE
TO BUFFALO
TO RTE. 65
HALFWAY POP. 157
BOLIVAR POP. 5919
PLEASANT HOPE POP. 354
MORRISVILLE POP. 331
ALDRICH POP. 53
FAIR PLAY POP. 384
STOCKTON LAKE

PULASKI COUNTY

1 **Dixon Towersite** is 3 miles west of Dixon on Highway 133. The tower sits on 48 acres of forestland.

2 **Fort Leonard Wood Towersite** is 1 mile east of St. Robert on the South Outer Road of I-44. This is a 67-acre forested area.

3 **Gasconade Hills Conservation Area** is 10 miles south of Dixon on Highway 28, and west on County Road 28-462. The Gasconade River flows along this 370-acre area.

4 **Laughlin/Roubidoux Park** is downstream from Roubidoux Spring near the Highway 17 bridge over Roubidoux Creek. This 6.5-acre area has two picnic areas and a wheelchair accessible fishing dock.

5 **Riddle Bridge Access** is 6 miles north of St. Robert at the end of Highway Y. This is a 9-acre area with a boat ramp on the Gasconade River.

6 **Ross Access** is 2.5 miles west of Duke on County Road TT-820N. This is a 4.5-acre area on the Big Piney River with no boat ramp.

7 **Roubidoux Creek Conservation Area** is on Highway 17, 1 mile north of Waynesville. This is 244 acres on the Gasconade River and Roubidoux Creek with no boat ramps.

8 **Ryden Cave Conservation Area** is 1.5 miles west of Duke off Highway K. This is a 30-acre area. The cave is home to the Indiana bat, a federally endangered species. The cave is closed September 1 to April 30 to protect the bats.

9 **Schlicht Springs Access** is off Highway 133, 5 miles south of Crocker. This is 12 acres with a boat ramp on the Gasconade River.

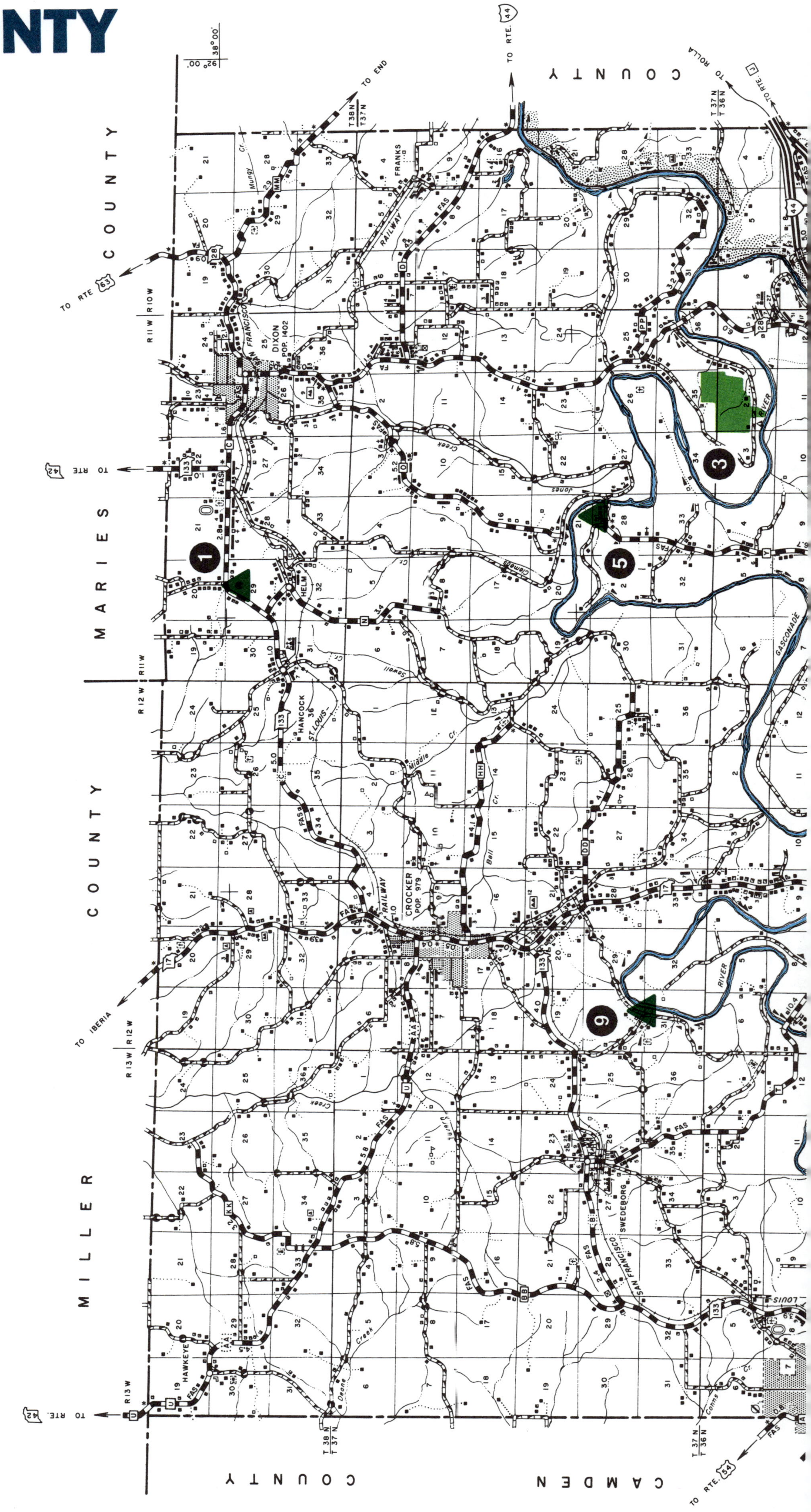

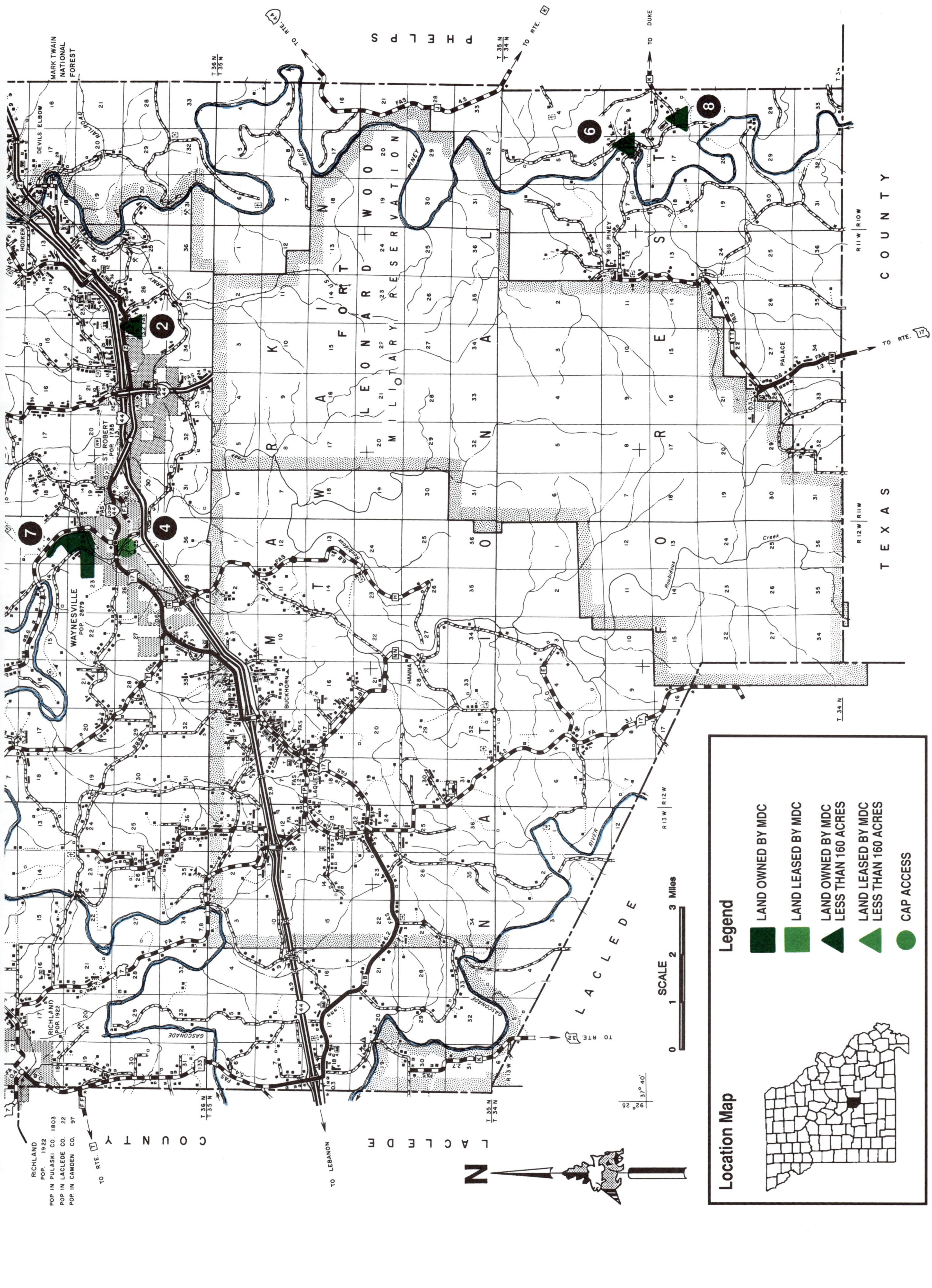

PHELPS
TEXAS COUNTY
LACLEDE
LACLEDE COUNTY
FORT LEONARD WOOD MILITARY RESERVATION
WAYNESVILLE POP. 2879
ST. ROBERT POP. 1735
RICHLAND POP. 1922
POP. IN PULASKI CO. 1803
POP. IN LACLEDE CO. 22
POP. IN CAMDEN CO. 97
DEVILS ELBOW
HOOKER
BIG PINEY
PALACE
HANNA
BUCKHORN
LAQUEY
MARK TWAIN NATIONAL FOREST
TO DUKE
TO LEBANON
SCALE
0 1 2 3 Miles
Legend
LAND OWNED BY MDC
LAND LEASED BY MDC
LAND OWNED BY MDC LESS THAN 160 ACRES
LAND LEASED BY MDC LESS THAN 160 ACRES
CAP ACCESS
Location Map

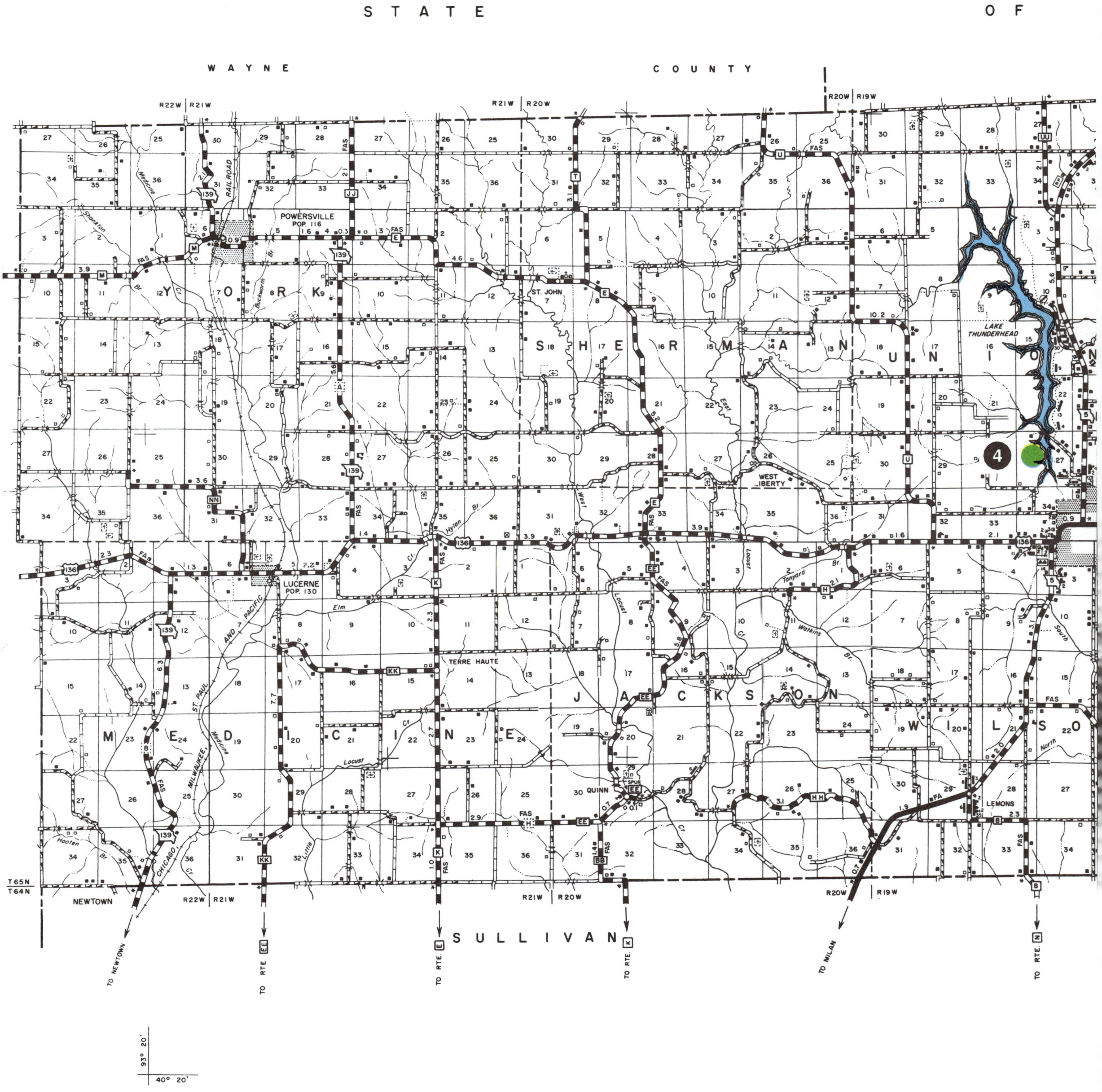

PUTNAM COUNTY

1 **Mineral Hills Conservation Area** is 3 miles south of Unionville on Highway 5, and 2.5 miles east on Highway F. Blackbird Creek flows through this 1,859-acre area that is mainly forestland (1,449 acres) and old pastures (410 acres).

2 **Rebel's Cove Conservation Area** is 2.5 miles north of Livonia on Highway N. Almost 10 miles of the Chariton River flows through the area, and the stretch is one of the last unchannelized portions of the river. The 4,176-acre area is composed of forestland, old fields and natural wetlands.

3 **Union Ridge Conservation Area**–see the Adair County map, page 6.

4 To reach **Unionville City Lake**, go 1 mile north of Unionville on Highway 5 to the first county road outside the city limits heading west. Take this road 1 mile, and cross the dam to the 85-acre lake. The lake has a boat ramp, and activities on the surrounding land are regulated by the city of Unionville.

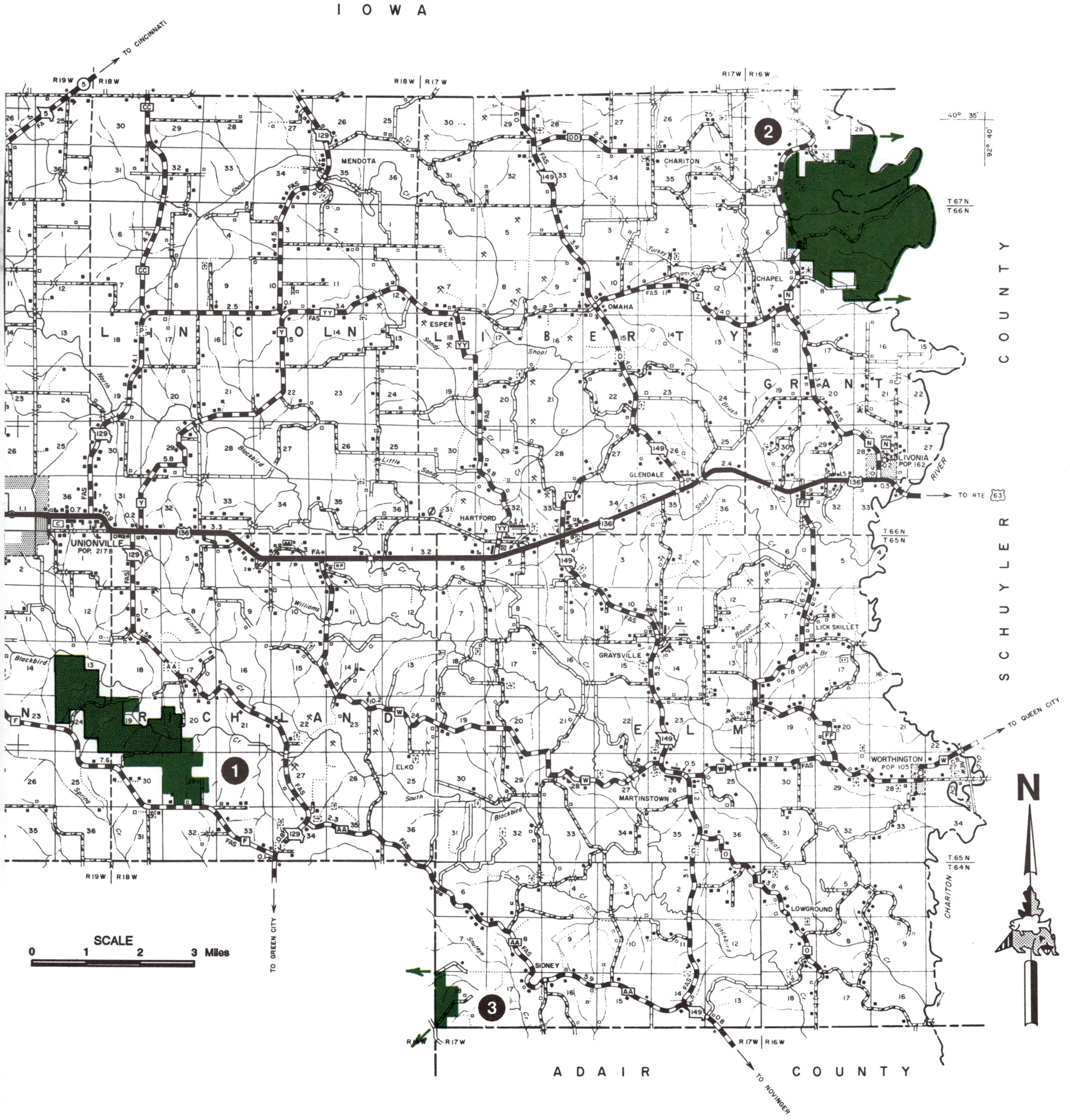

Location Map

Legend

LAND OWNED BY MDC

LAND LEASED BY MDC

LAND OWNED BY MDC LESS THAN 160 ACRES

LAND LEASED BY MDC LESS THAN 160 ACRES

CAP ACCESS

RALLS COUNTY

1 **Edward Anderson Conservation Area**–see the Pike County map, page 172.

2 **Indian Camp Access** is north of New London on Highway O (off Highway 61). This is a 10-acre area on the Salt River with a boat ramp.

3 **Robert H. Thompson Conservation Area** is an 11-acre area located 200 feet downstream from the Saverton Lock and Dam #22.

4 **Upper Mississippi Conservation Area**–see the Lewis County map, page 120.

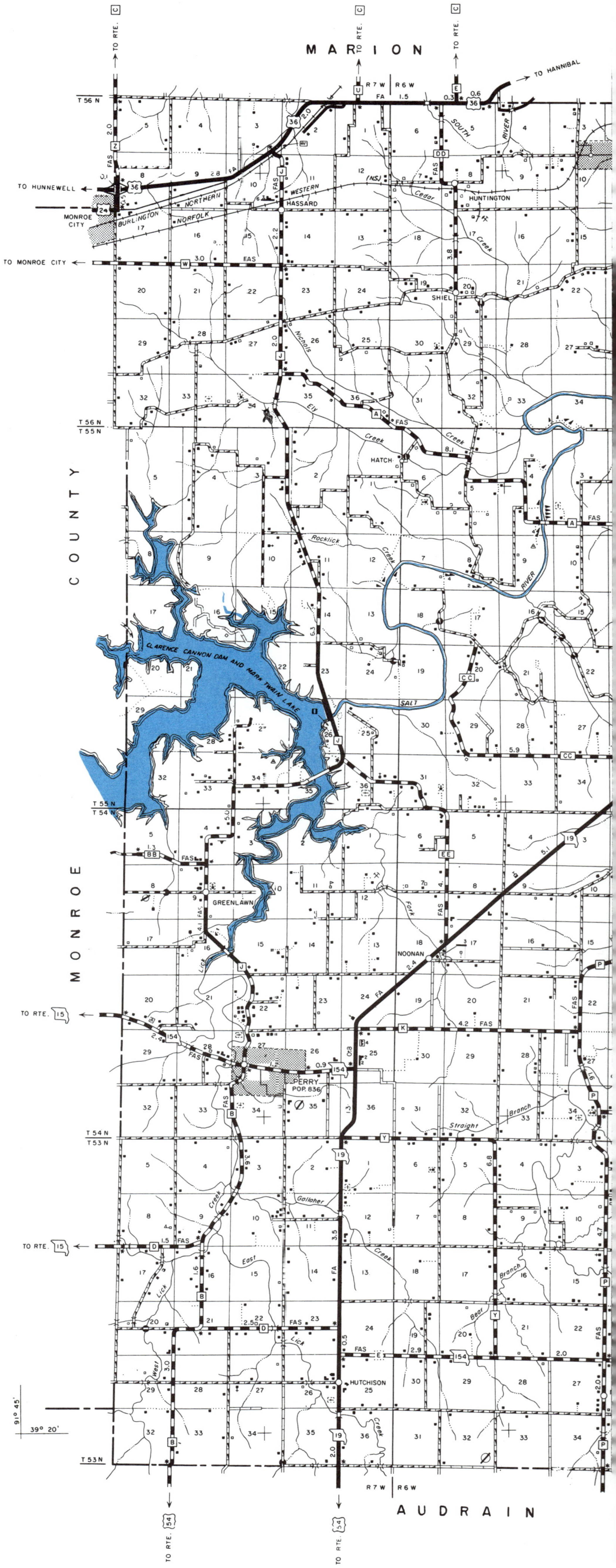

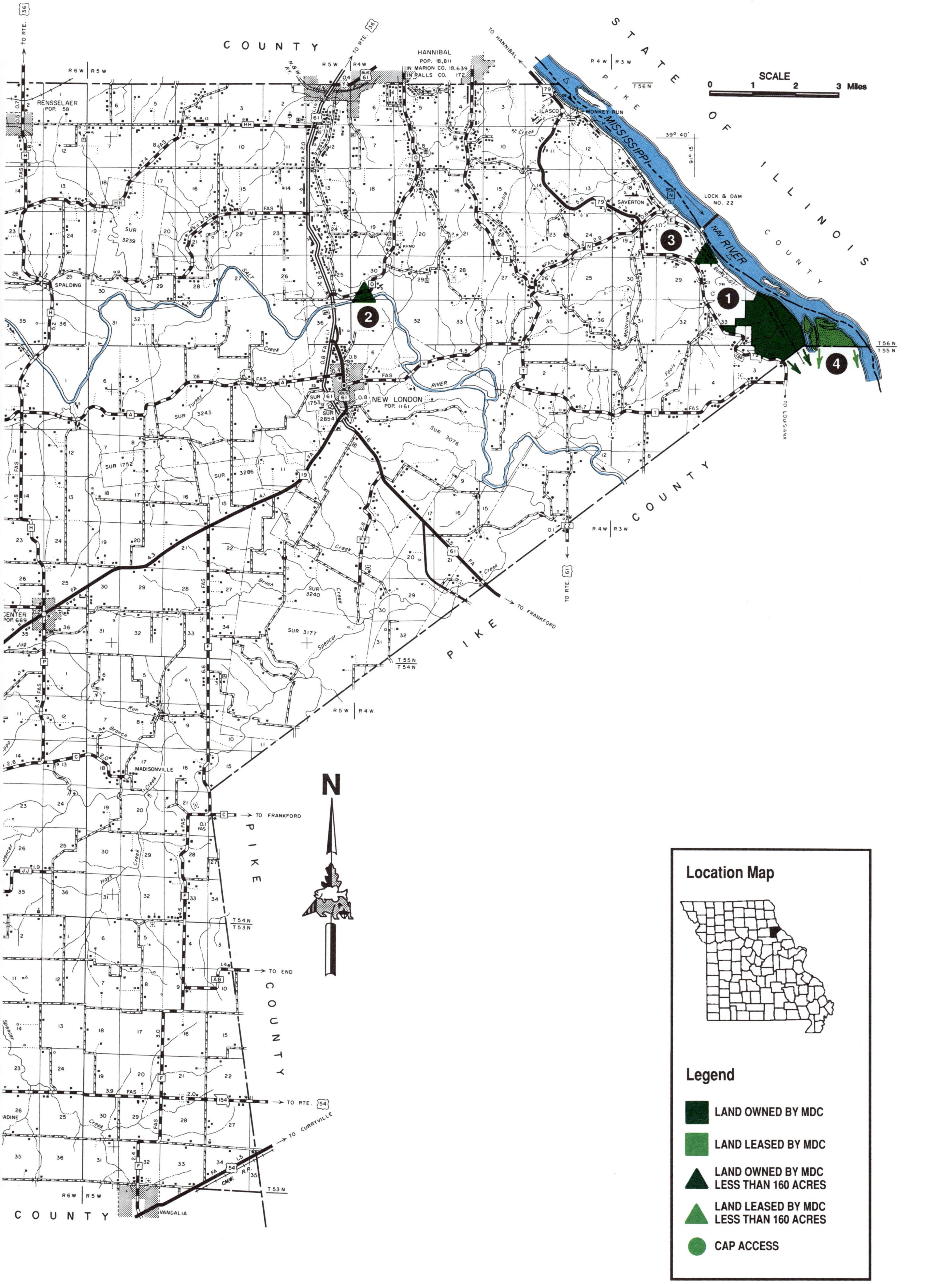
STATE OF ILLINOIS
PIKE COUNTY
COUNTY
SCALE
0 1 2 3 Miles
HANNIBAL
POP. 18,811
IN MARION CO. 18,639
IN RALLS CO. 172
RENSSELAER
POP. 58
ILASCO
MONKEY RUN
MISSISSIPPI
SAVERTON
LOCK & DAM
NO. 22
NAV RIVER
SPALDING
SALT
RIVER
NEW LONDON
POP. 1161
CENTER
POP. 669
MADISONVILLE
VANDALIA
TO HANNIBAL
TO LOUISIANA
TO FRANKFORD
TO END
TO RTE. 54
TO CURRYVILLE
TO RTE. 36
TO RTE. 61
PIKE COUNTY
PIKE
COUNTY
39° 40'
91° 15'
T56N
T55N
T54N
T53N
R6W
R5W
R4W
R3W
1
2
3
4
N
Location Map
Legend
LAND OWNED BY MDC
LAND LEASED BY MDC
LAND OWNED BY MDC LESS THAN 160 ACRES
LAND LEASED BY MDC LESS THAN 160 ACRES
CAP ACCESS

RANDOLPH COUNTY

1 **Bee Hollow Conservation Area**–see the Macon County map, page 130.

2 **Rudolf Bennitt Conservation Area**–see the Howard County map, page 96.

3 **Higbee City Lake** is 1 mile east on Division Street and County Road 228, then .5 mile north on County Road 261. The 15-acre lake is on the west side of the road. There is a boat ramp.

4 **Rothwell Park Lake** (24 acres) and **Water Works Lake** (24 acres) are 2.5 miles west of Moberly on Highway 24, and 1 mile south on Rothwell Park Road. Both lakes have boat ramps, and Water Works Lake is equipped with a wheelchair accessible fishing dock. There are ballfields, trails, tennis courts, an archery range and picnic areas inside the park.

5 **Thomas Hill Reservoir**–see the Macon County map, page 130.

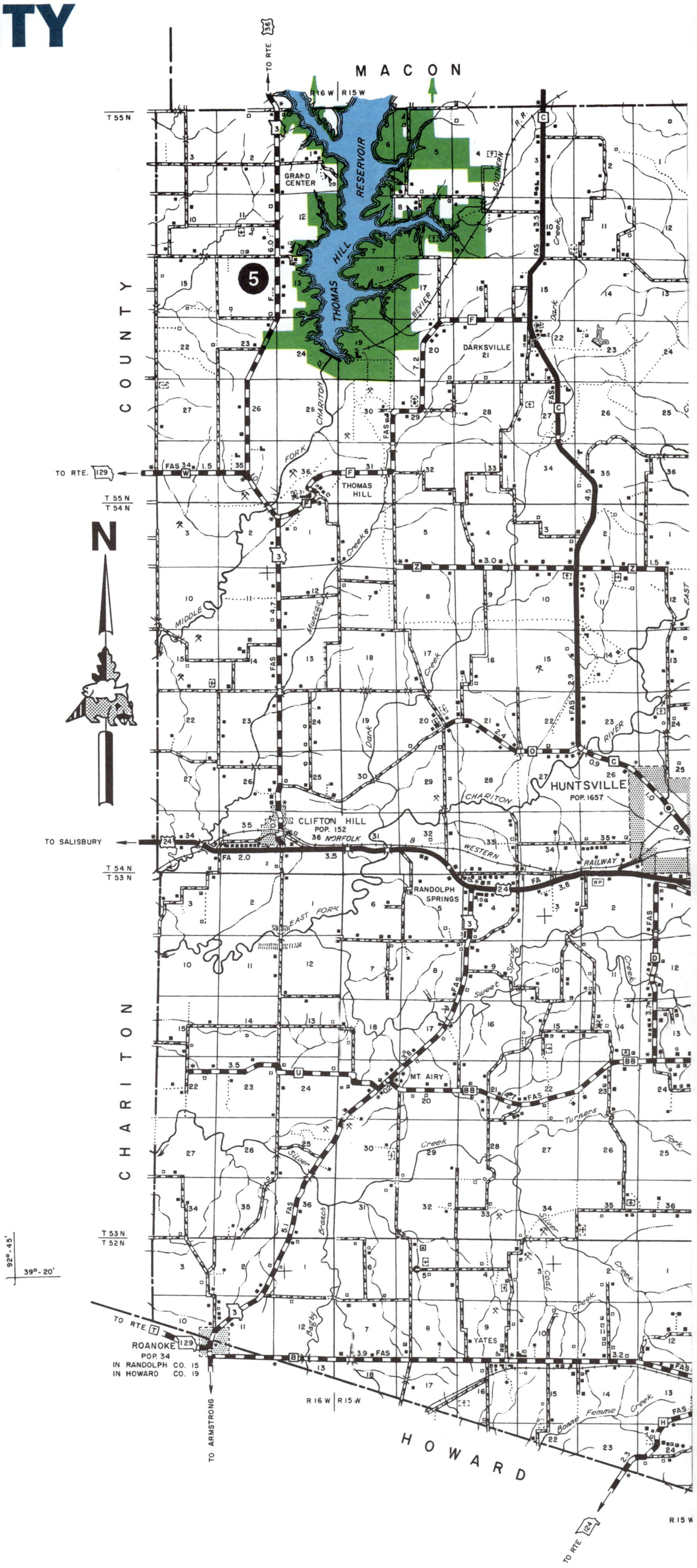

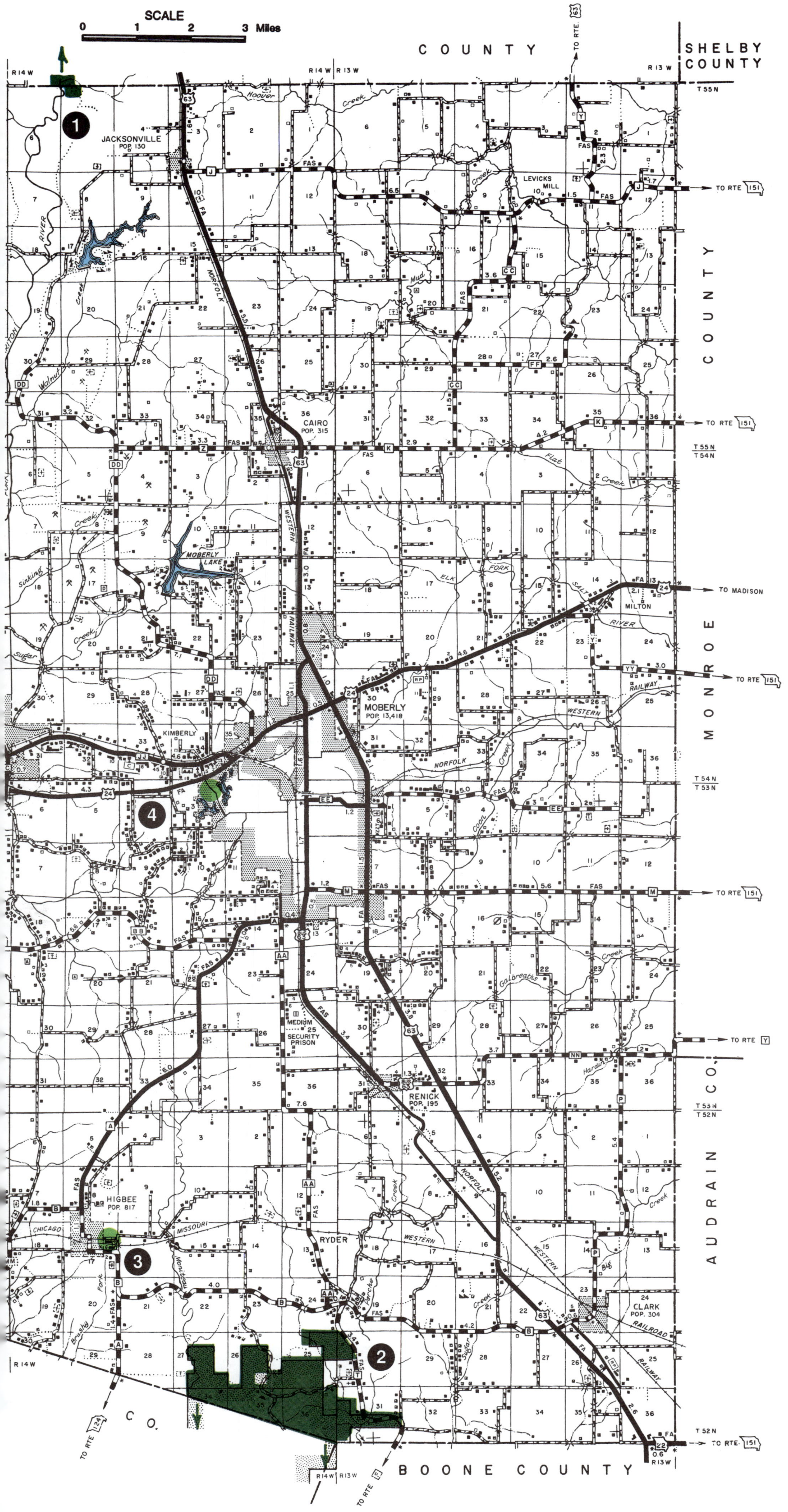

Location Map

Legend

- LAND OWNED BY MDC
- LAND LEASED BY MDC
- LAND OWNED BY MDC LESS THAN 160 ACRES
- LAND LEASED BY MDC LESS THAN 160 ACRES
- CAP ACCESS

RAY COUNTY

1 **Blue Jay Trail Access** is 5 miles east of Lawson on Route D and .5 mile north on Gravel Road 346. This is 30 acres on Crooked River with no boat ramp.

2 **Crooked River Conservation Area** is 4 miles west of Highway 13 on County Route FF. This is a 1,420-acre area with a 3-mile stretch of Crooked River. The area has nine ponds and a heron rookery.

3 **Foxglove Conservation Area** is .5 mile east of Vibbard on Gravel Road 174 and .5 mile north on Gravel Road 235. This area is 54 acres of mainly native prairie (45 acres).

4 **Hardin Conservation Area** is .5 mile south of Hardin on Highway J. This is 238 acres of mostly forestland and a 40-acre blue hole created in the flood of '93.

5 **James Bridge Access** is 3 miles west on Highway FF off Highway 13 and .25 mile west on Gravel Road 167. This is 22 acres on Crooked River with no boat ramp.

6 **Lawson City Lake** is 1 mile north of Highway D in Lawson on Raum Street. The reservoir is 25 acres of fishable water and has a boat ramp, fishing jetty (accessible to wheelchairs), 1.7-mile gravel hiking trail and a picnic area.

7 **Morton Bridge Access** is 3 miles east of Richmond on Highway 10 and 1 mile north on Gravel Road 255. This is 24 acres on Crooked River (.25-mile stretch) with no boat ramp.

8 **Pigg's Landing Access** is marked by a sign on the south side of Highway 210, 1 mile west of Orrick. This is a 9-acre area with a boat ramp on Fishing River.

9 **Ray County Community Lake** is 2 miles north of Richmond on Highway 13, 1 mile west on Gravel Road 228, then 1 mile south on Gravel Road 229. The lake is 25 acres of fishable water with a boat ramp and several fishing jetties (some are accessible to wheelchairs).

10 **Frank E. Wagner Conservation Area** is 2.5 miles north of Richmond on Highway 13, 4 miles east on Route B, 2 miles east on Gravel Road 156, and .75 miles north on Gravel Road 155. This is a 130-acre area composed mainly of cropland (71 acres) and forestland (49 acres).

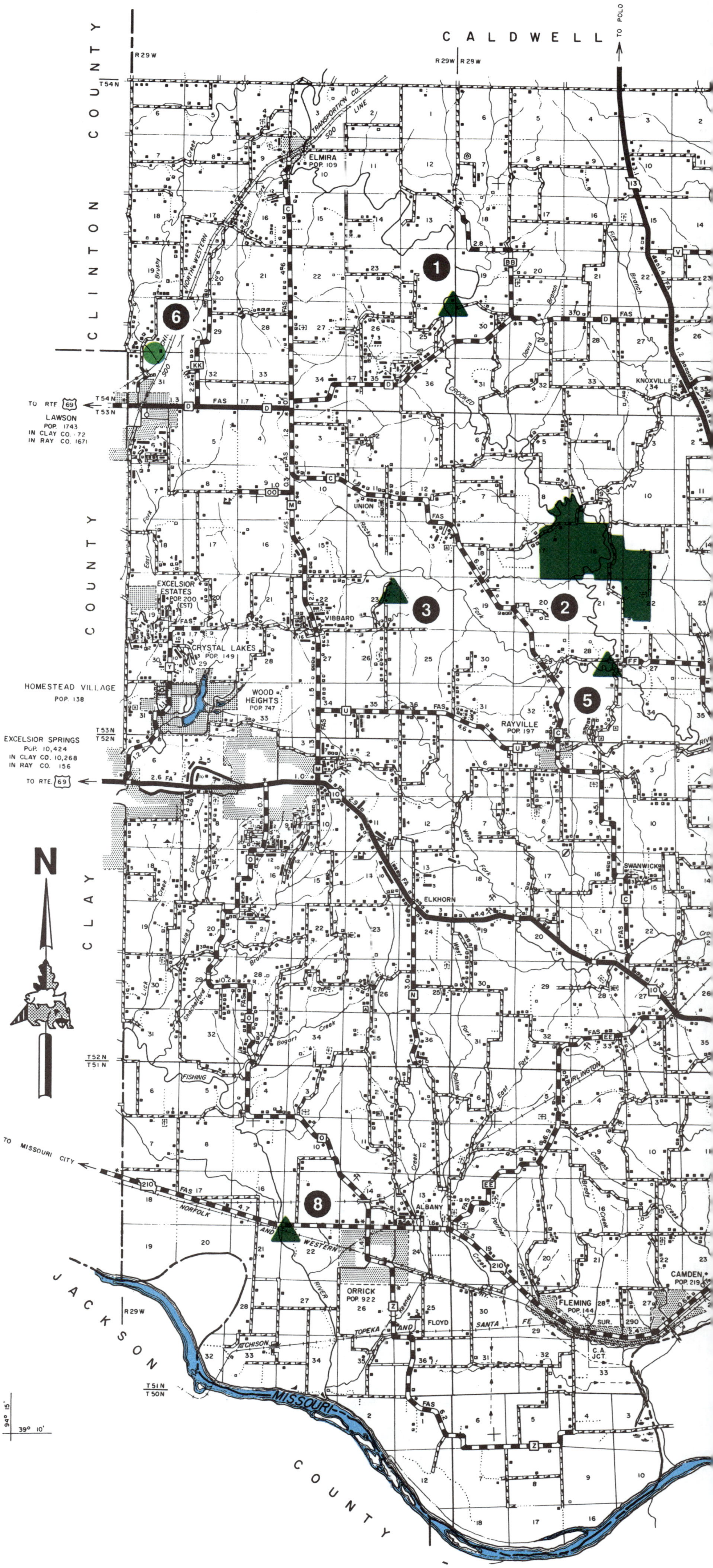

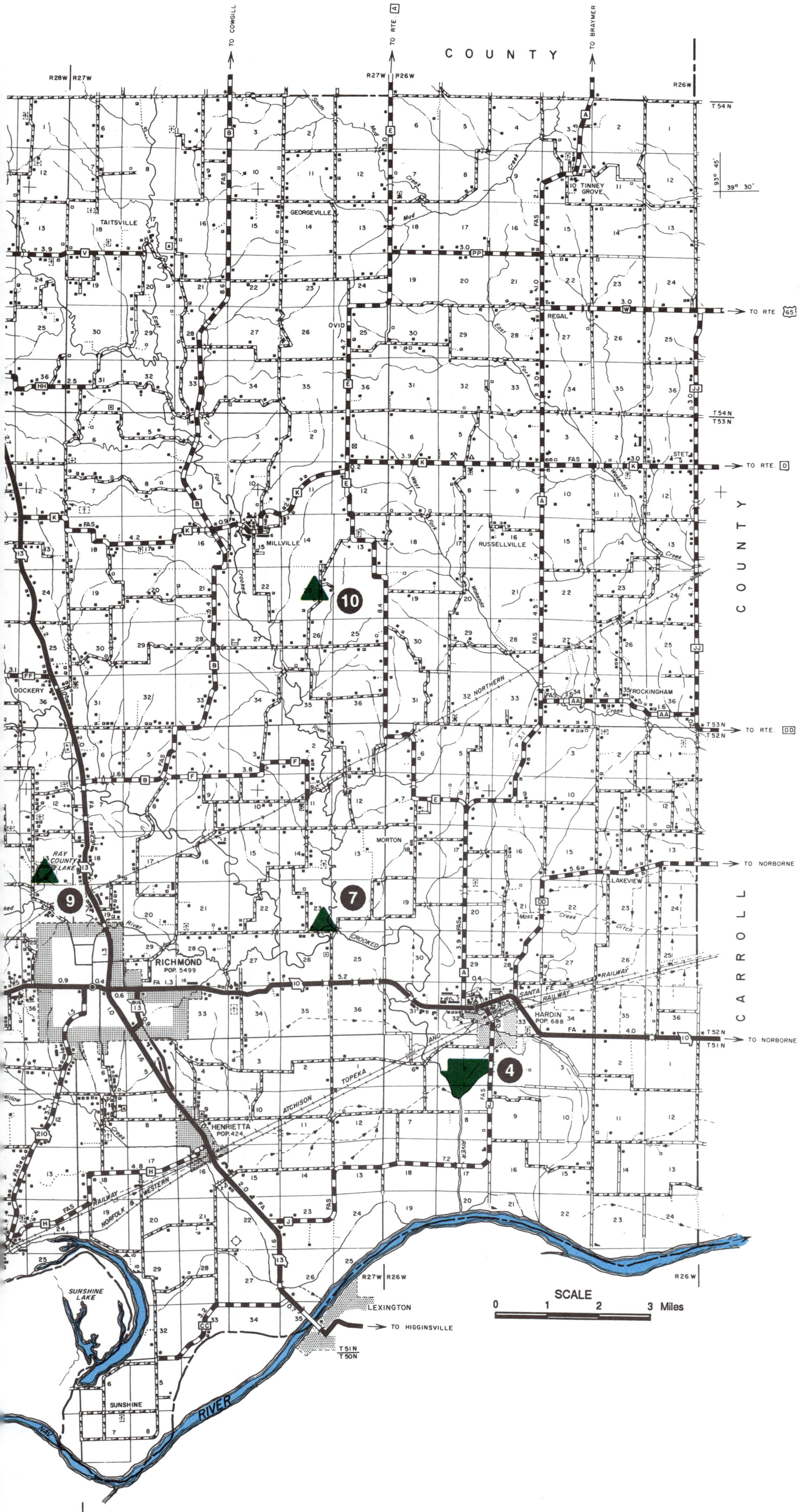
TO COWGILL
TO RTE A
TO BRAYMER
COUNTY
CARROLL COUNTY
TAITSVILLE
GEORGEVILLE
TINNEY GROVE
OVID
REGAL
MILLVILLE
RUSSELLVILLE
DOCKERY
ROCKINGHAM
STET
MORTON
LAKEVIEW
RAY COUNTY LAKE
RICHMOND
POP. 5499
HARDIN
POP. 688
HENRIETTA
POP. 424
LEXINGTON
SUNSHINE LAKE
SUNSHINE
RIVER
Crooked
Wakenda
Mud Creek
East Fork
West Fork
Moss Creek
NORTHERN
SANTA FE RAILWAY
ATCHISON TOPEKA AND
NORFOLK WESTERN RAILWAY
TO RTE 65
TO RTE D
TO RTE DD
TO NORBORNE
TO HIGGINSVILLE
T54N
T53N
T52N
T51N
T50N
R28W
R27W
R26W
93° 45'
39° 30'
10
9
7
4
SCALE
0 1 2 3 Miles

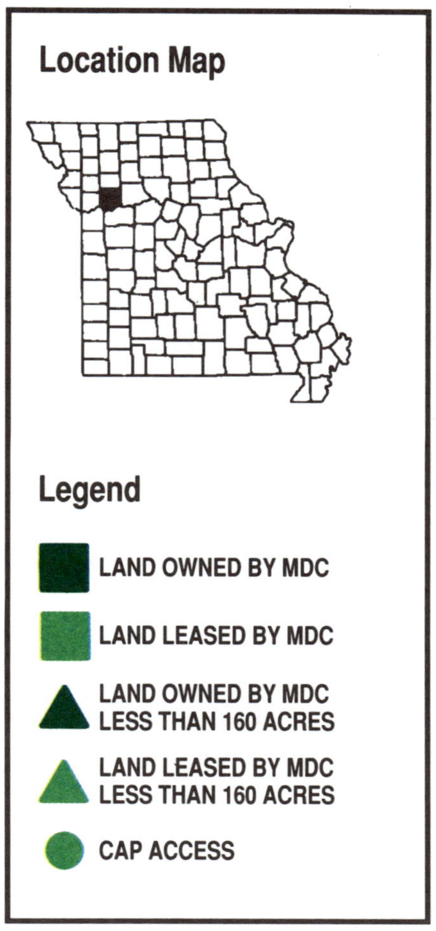
Location Map
Legend
LAND OWNED BY MDC
LAND LEASED BY MDC
LAND OWNED BY MDC LESS THAN 160 ACRES
LAND LEASED BY MDC LESS THAN 160 ACRES
CAP ACCESS

REYNOLDS COUNTY

1 **Centerville Access** is north of the town of Centerville on Highway 21. This 55-acre area has .33 mile of frontage on the West Fork of Black River. There is no boat ramp.

2 **Clearwater Conservation Area** consists of several tracts. The main tract is located near the Highway HH Spur on Highway HH (off Highway 34). Webb Creek flows through this 6,148-acre forest, and Clearwater Lake is adjacent to the area. Natural features of interest include two fens, a rhyolite knob, a sinkhole pond and a dry sink.

3 **Clearwater Lake** (adjacent to Clearwater Conservation Area) is 5 miles west of Piedmont on Highway HH. Clearwater Lake is 1,630 acres, and the entire area is 12,659 acres. There are three boat ramps, two fishing docks (accessible to wheelchairs), three viewing blinds, a .5-mile trail, camper pads and picnic areas.

4 **Current River Conservation Area**–see the Carter County map, page 40.

5 **Ketcherside Conservation Area**–see the Iron County map, page 100.

6 To reach **Lesterville Access**, take Highway 21 to the north side of Lesterville, then take Peola Road to County Road 2. This is a 53-acre area on the Black River.

7 **Logan Creek Conservation Area** consists of four tracts. The main tract is 3 miles north of Highway 106 on Highway B. This is a 4,783-acre area with several natural features. They include a .5-acre pond shrub swamp, gasconade dolomite sinkhole, two roubidoux dolomite sink holes and a roubidoux dolomite glade.

8 To reach **Lower Taum Sauk Lake**, take Highway U north of Lesterville to the junction of highways U and A. Turn left on Highway A to the 1,340-acre area. The lake is 200 acres of fishable water and has a boat ramp.

9 **Riverside Conservation Area**–see the Iron County map, page 100.

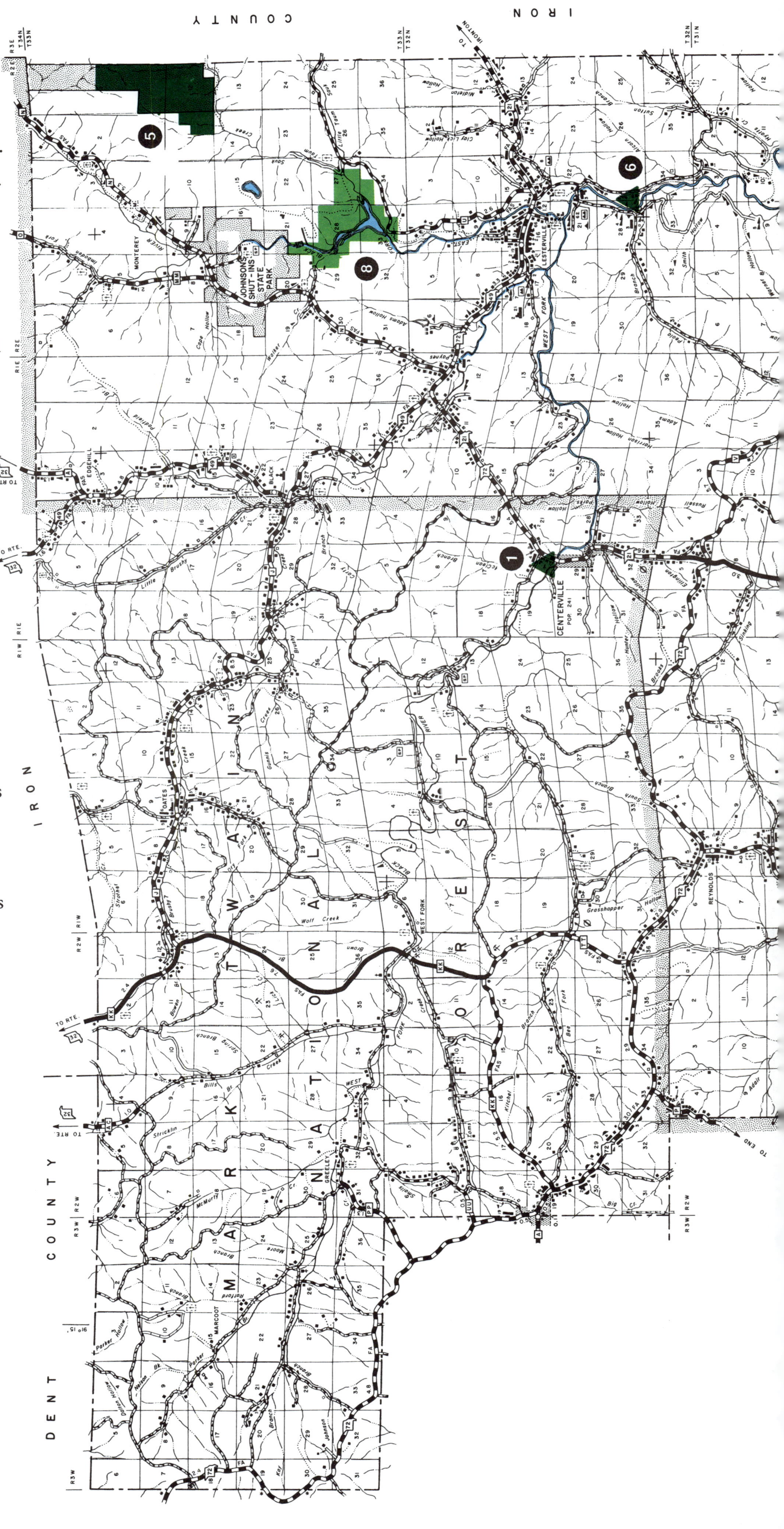

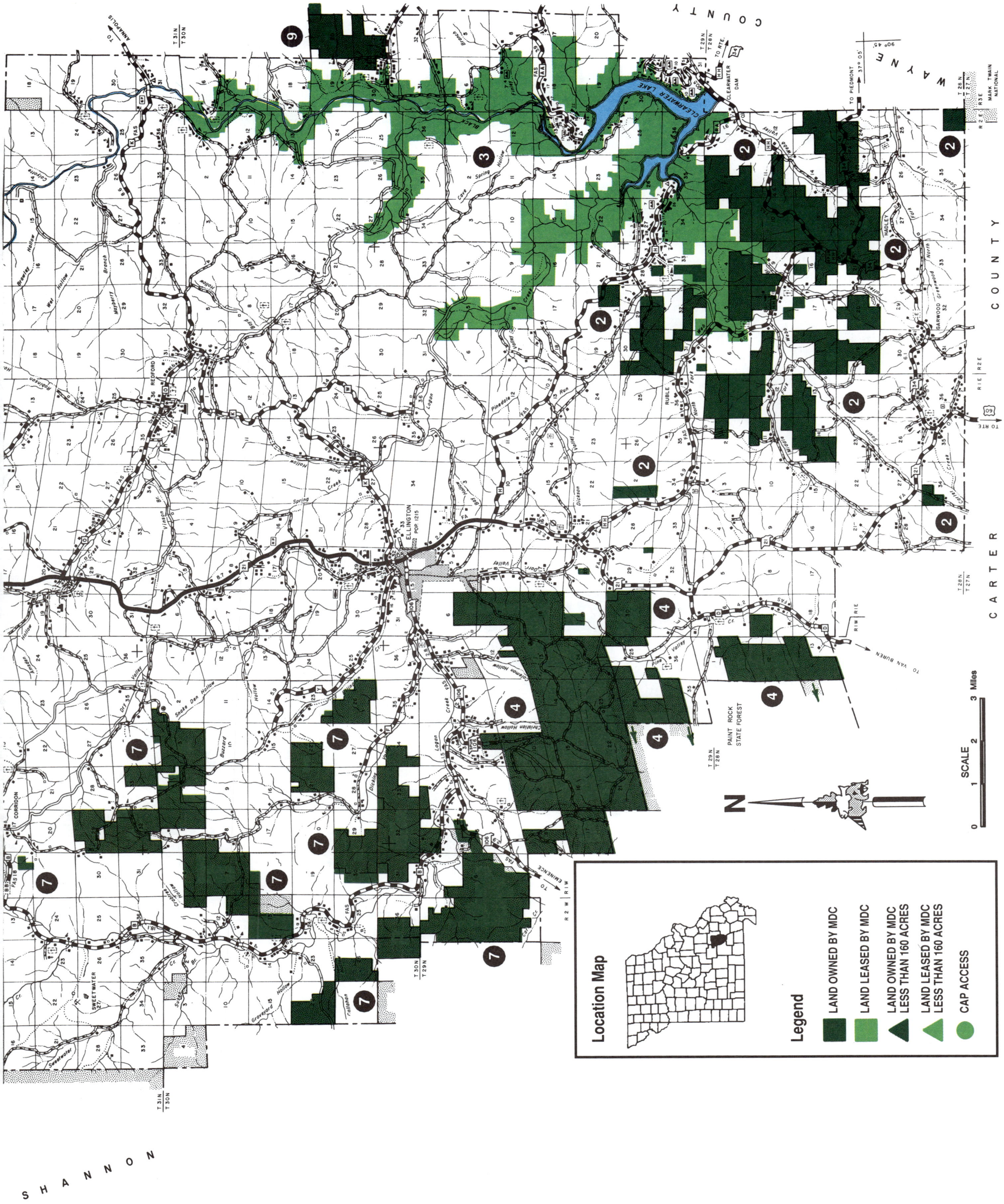
Location Map
Legend
LAND OWNED BY MDC
LAND LEASED BY MDC
LAND OWNED BY MDC LESS THAN 160 ACRES
LAND LEASED BY MDC LESS THAN 160 ACRES
CAP ACCESS
SCALE
Miles
CLEARWATER LAKE
CLEARWATER DAM
ELLINGTON POP. 1215
REDFORD
RUBLE
HADLEY
GARWOOD
CORRIDON
SWEETWATER
PAINT ROCK STATE FOREST
MARK TWAIN NATIONAL
TO PIEDMONT
TO ANNAPOLIS
TO VAN BUREN
TO EMINENCE
W A Y N E
C O U N T Y
C A R T E R
C O U N T Y
S H A N N O N

RIPLEY COUNTY

1 **Doniphan Towersite** (10 acres) is 1.5 miles north of the junction of highways 21 and 160 on Highway 21, then .25 mile west on Gravel Road 21-N-2. Watch for the area sign.

2 **Fourche Creek Conservation Area** consists of five tracts. The main tract is on Highway P, 6 miles from the junction of highways P and 142. Fourche Creek flows .5 mile through this mainly forested, 3,776-acre area.

3 **Grandin Towersite** is .5 mile north of the town of Grandin on Highway 21, 6 miles east on Highway B, then 5 miles south on Highway K. The tower sits on 160 acres of forestland.

4 **Greenville Ford Access** is 4 miles north of the junction of highways 21 and 106 on Highway 21, then 4 miles on Highway K to the bridge over the Little Black River. This 3-acre area is adjacent to the bridge and has a boat ramp.

5 **Hemenway Conservation Area** is .5 mile south of Naylor on Highway W and 2.5 miles west on Highway H. This 172-acre area is mainly old pastureland.

6 **Little Black Conservation Area** (2,322 acres) is on Highway NN, .25 mile from the junction of highways 21 and NN. The area contains a stretch of the Little Black River and about 5 acres of deep muck and forested fens. There are nine ponds and, together with the river, they provide anglers with 15 acres of fishable water.

7 **Mudpuppy Conservation Area** is 6 miles east of Doniphan on Highway 160, then 4 miles to the end of the pavement on Highway BB. This 1,378-acre area contains a 3.5-mile stretch of the Little Black River (with no boat ramp) and is open to horseback riders.

8 **Sand Ponds Conservation Area** (and the **N. B. Altvater Addition** to the area) is .5 mile south of Naylor on Highway 142, 3.75 miles on Highway W, then left on County Road W-3. This 302-acre area includes Sand Ponds Natural Area–68 acres of forest on unusual dune and swale topography.

9 **T. L. Wright Memorial Access** is on the Current River in the T. L. Wright Memorial Park in downtown Doniphan. This is an 8-acre area with a boat ramp and picnic area.

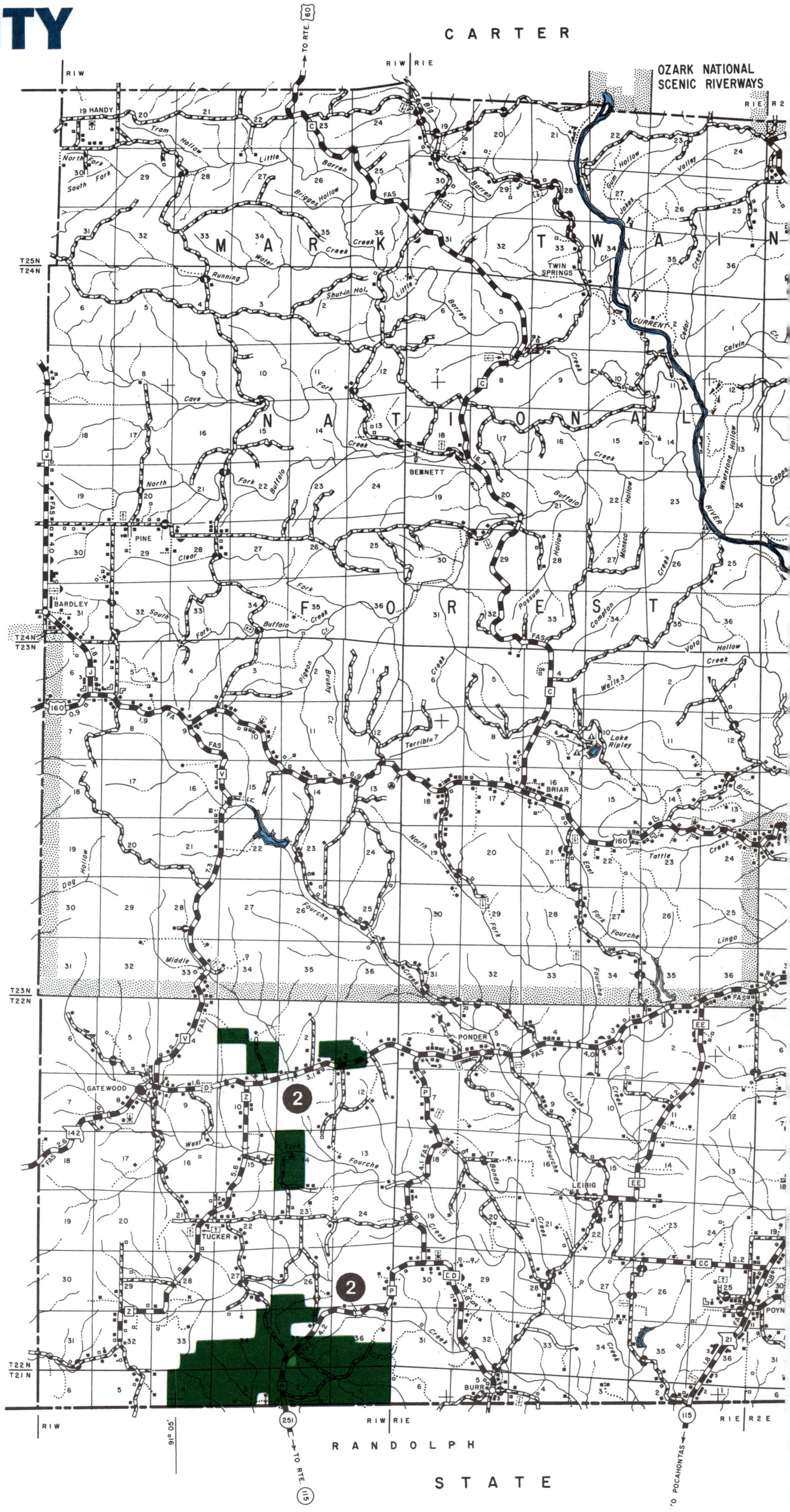

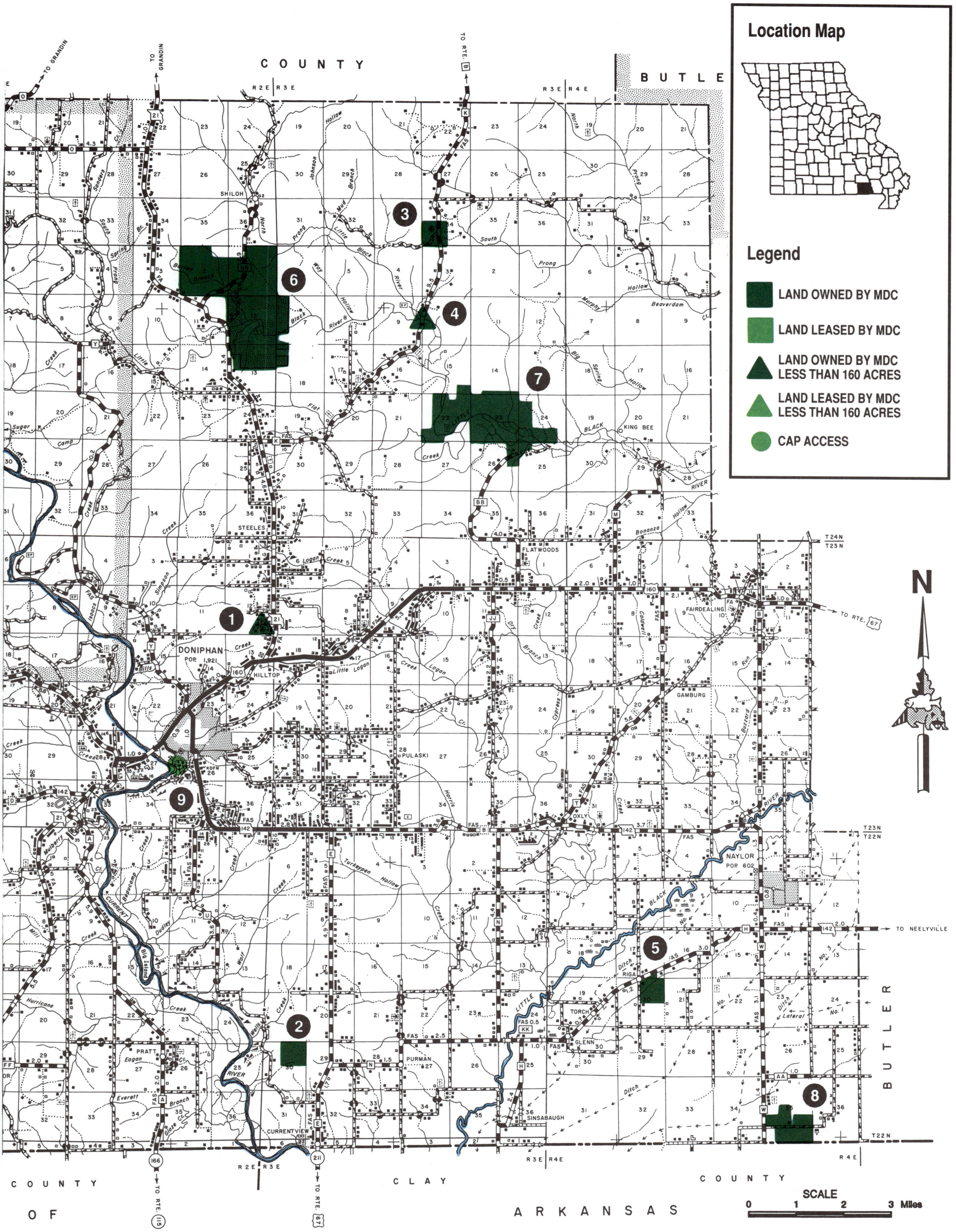

Location Map
Legend
LAND OWNED BY MDC
LAND LEASED BY MDC
LAND OWNED BY MDC LESS THAN 160 ACRES
LAND LEASED BY MDC LESS THAN 160 ACRES
CAP ACCESS
COUNTY
BUTLE
BUTLER
COUNTY OF
CLAY
ARKANSAS
COUNTY
TO GRANDIN
TO RTE. B
TO RTE. 67
TO NEELYVILLE
TO RTE. 115
TO RTE. 67
SHILOH
STEELES
FLATWOODS
KING BEE
FAIRDEALING
DONIPHAN
POP. 1,921
HILLTOP
GAMBURG
PULASKI
OXLY
NAYLOR
POP. 602
RIGA
TORCH
GLENN
PURMAN
PRATT
SINSABAUGH
CURRENTVIEW
CURRENT RIVER
BLACK RIVER
LITTLE BLACK RIVER
SCALE
0 1 2 3 Miles
N

ST. CHARLES COUNTY

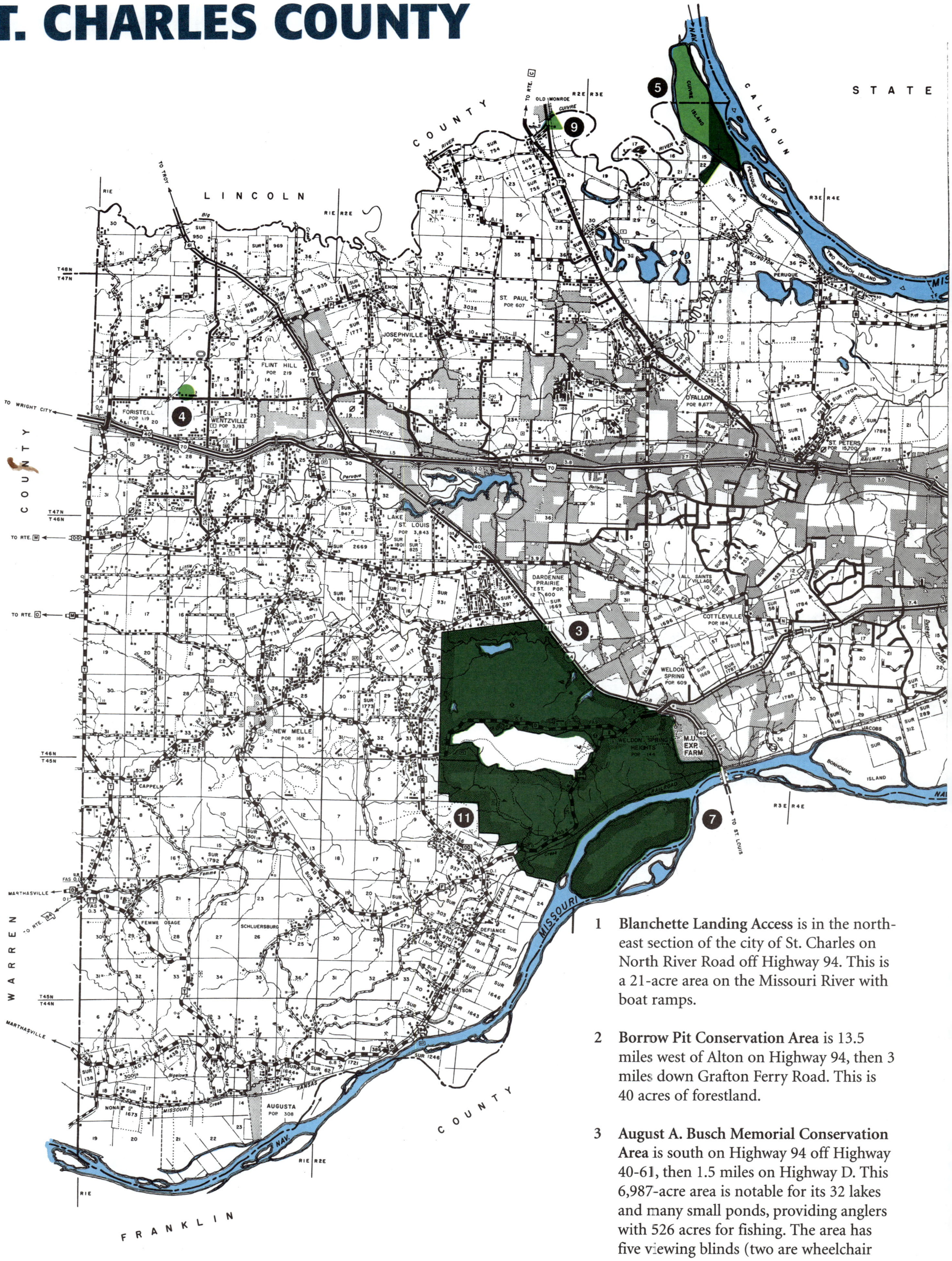

1 **Blanchette Landing Access** is in the northeast section of the city of St. Charles on North River Road off Highway 94. This is a 21-acre area on the Missouri River with boat ramps.

2 **Borrow Pit Conservation Area** is 13.5 miles west of Alton on Highway 94, then 3 miles down Grafton Ferry Road. This is 40 acres of forestland.

3 **August A. Busch Memorial Conservation Area** is south on Highway 94 off Highway 40-61, then 1.5 miles on Highway D. This 6,987-acre area is notable for its 32 lakes and many small ponds, providing anglers with 526 acres for fishing. The area has five viewing blinds (two are wheelchair accessible), seven hiking trails, picnic areas, a staffed firearms range, fishing jetties and floating docks that are wheelchair accessible. The area has interpretive programs, exhibits and demonstration sites, and is used by bicyclists.

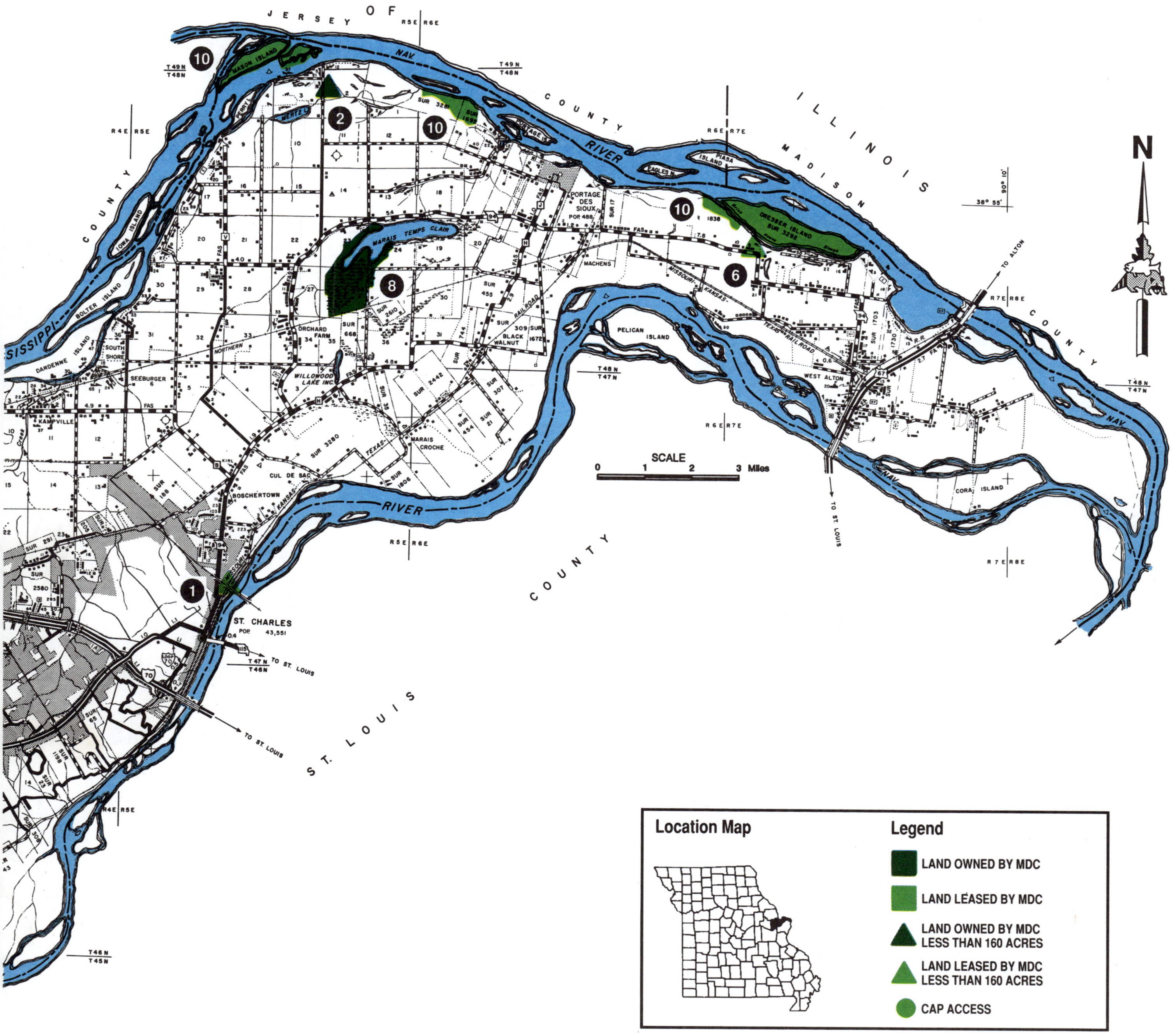

4 **Community Club Lake** is north of I-70 on Pearce Boulevard, then 3 miles west on Meyer Road. This 5-acre lake is in a city park. The area has picnic areas and sponsors fishing clinics for kids. Boats are not allowed on the lake.

5 **Cuivre Island Conservation Area**–see the Lincoln County map, page 122.

6 **Dresser Island Access** is about 4 miles west of West Alton on Highway 94. The area is adjacent to part of the Upper Mississippi Wildlife Conservation Area and has two boat ramps and a fishing dock. Ten miles of levees are open to hiking.

7 **Howell Island Conservation Area** is on Olive Street/Eatherton Road, 3 miles off Highway 40-61. This 2,548-acre area is mostly forestland and is surrounded by part of the Missouri River and Centaur Chute (10 miles). The island is reached by a causeway over Centaur Chute.

8 To reach **Marais Temps Clair Conservation Area**, go north on Highway 94, then go 3 miles on Highway H to Island Road. This 918-acre area is 1.5 miles down Island Road. It is composed mainly of wetlands (598 acres) and cropland (300 acres) and has 10 pools and several drainage channels (20 acres suitable for fishing). There are nine waterfowl blinds and 10 miles of levees suitable for hiking.

9 **Old Monroe Access** (1 acre) is on an unnamed county road, just east of the junction of Highway 79 and the Cuivre River in Old Monroe. There is no boat ramp.

10 **Upper Mississippi Conservation Area**–see the Lewis County map, page 120.

11 Access to **Weldon Spring Conservation Area** is from several parking areas along Highway 94, south of Highway 40-61. This is a 7,356-acre area with four lakes totaling 47 acres and an access to the Missouri River. One lake has a disabled accessible floating dock. Weldon Spring Hollow Natural Area is 385 acres of upland and bottomland forest and rugged riverbreaks topography. Other natural features include limestone cliffs and bluff escarpments. There are four hiking trails, 5 to 8 miles in length, two of which are suitable for biking (Katy Trail and Lost Valley Trail).

ST. CLAIR COUNTY

1 **Birdsong Conservation Area** is 15 miles east of El Dorado Springs on Highway 54, then 2.5 miles south on Route J. The area is 430 acres of mainly forestland and contains Brush Creek Natural Area (26 acres) that is habitat for the federally endangered Niangua darter. Brush Creek flows through the area.

2 To reach **Blackjack Access**, go west of the junction of highways 54 and 13 on Highway 54. Turn 3 miles south on Highway J, then 2 miles west on Highway W. Brush Creek and Sac River may be accessed here, and there is a boat ramp.

3 **Dave Rock Conservation Area** is 1 mile south of Lowry City on Highway 13, then 1 mile west on County Road NE 700. Watch for the area cantilever sign on Highway 13. The area includes a 44-acre Designated Natural Area that features sandstone glade, savannah and forest natural communities.

4 **Kings Prairie Access** is 2.5 miles east of Vista on Highway T and along County Road SE 200. Two miles of Weaubleau Creek flow through this 281-acre area (10 acres for fishing).

5 **Lichen Glade Conservation Area** (26 acres) is 6 miles west of Osceola on Highway B. Lichen Glade Natural Area is an example of a xeric habitat, populated by stunted trees and grasses adapted to living in rocky, dry soil.

6 The main entrance to **Schell-Osage Conservation Area** is in Vernon County, 20 miles east of Nevada on Highway 54, 12 miles north on County Road AA, then 1 mile down County Road RA. This is 8,633 acres on Truman Reservoir (Osage River arm) and also home to Atkinson Lake, Schell Lake and Barber Lake for a total of 900 acres of fishable water. The area contains 45.8 acres of native prairies (a Designated Natural Area), two oxbows, 440-acre Greentree Reservoir, a bald eagle restoration project, 22 waterfowl blinds (some are wheelchair accessible), picnic areas and a 1-mile hiking trail. Exhibits and interpretative programs about eagles are available, and visitors may arrange for slide shows and tours.

7 **Taberville Access** is .25 mile south of the town of Taberville on County Route H. This is a 5-acre area on Truman Reservoir (Osage River arm). There is a boat ramp.

8 **Taberville Prairie Conservation Area** (1,680 acres) is .5 mile east of Appleton City on Highway 52, 2 miles south on Highway A, then 7 miles south on Highway H. It is a registered National Natural Landmark and Missouri's largest prairie natural area (1,360 acres). There are seven ponds providing anglers with 3.5 acres of fishable water and a 2-mile hiking trail.

9 **Harry S Truman Reservoir Wildlife Management Lands**–see the Benton County map, page 20.

10 **Wah-Kon-Tah Prairie** is 2.5 miles northeast of El Dorado Springs on Highway 82. This is a 1,040-acre native prairie.

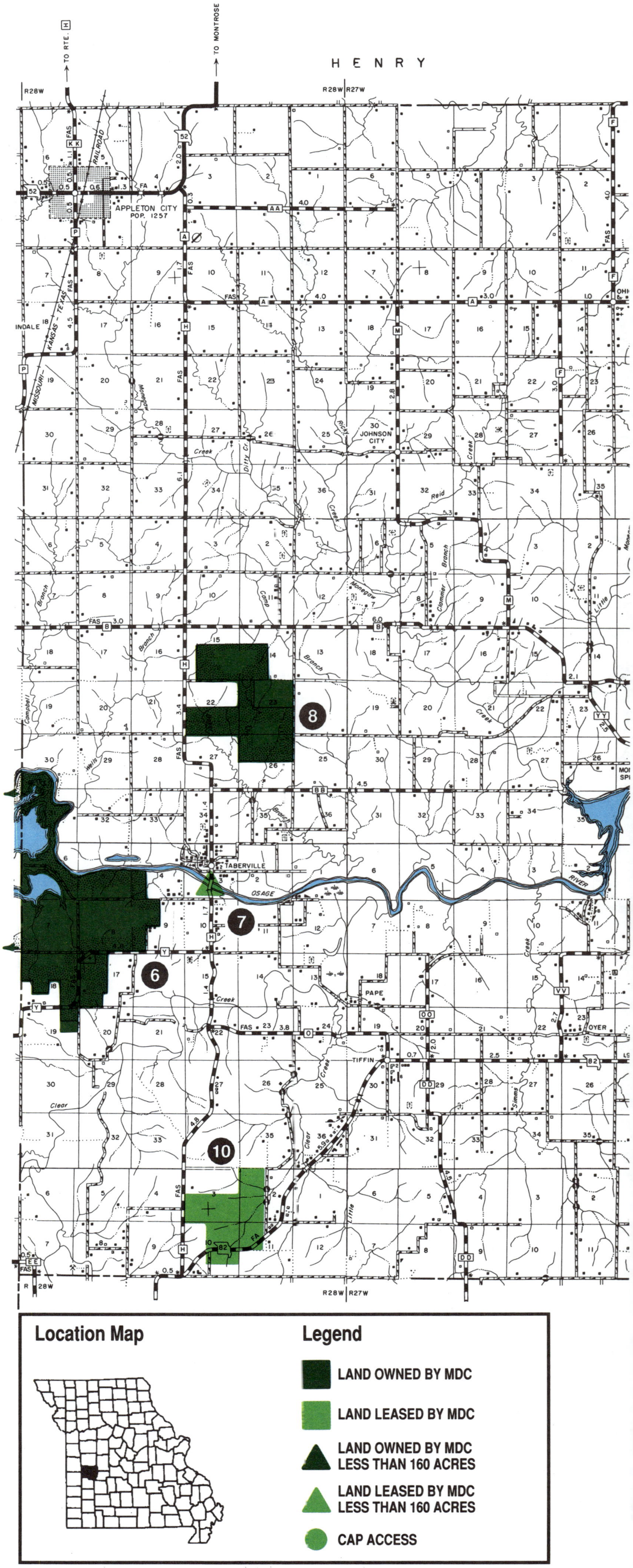

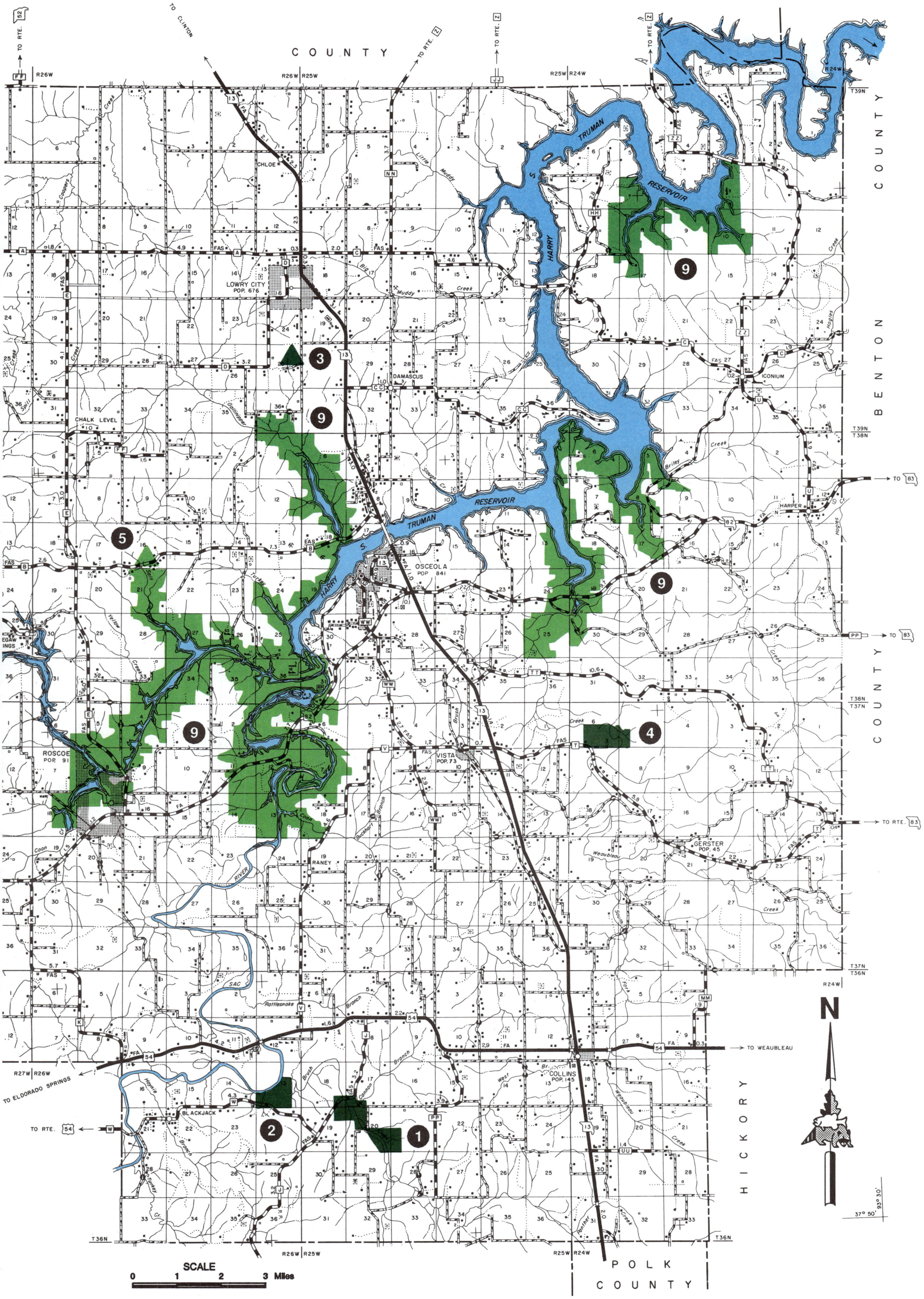
COUNTY
TO CLINTON
TO RTE. 52
TO RTE. Z
TRUMAN
RESERVOIR
HARRY S. TRUMAN RESERVOIR
CHLOE
LOWRY CITY POP. 676
DAMASCUS
CHALK LEVEL
ICONIUM
HARPER
OSCEOLA POP. 841
ROSCOE POP. 91
VISTA POP. 73
GERSTER POP. 45
RANEY
BLACKJACK
COLLINS POP. 145
TO 83
TO RTE. 83
TO WEAUBLEAU
TO ELDORADO SPRINGS
TO RTE. 54
BENTON COUNTY
HICKORY
POLK COUNTY
SAC RIVER
SCALE
0 1 2 3 Miles
N
R26W
R25W
R24W
R27W
T39N
T38N
T37N
T36N

ST. FRANCOIS COUNTY

1 **Bismarck Lake Conservation Area**–see the Iron County map, page 100.

2 **Farmington City Lake** is near the Industrial Park in the city of Farmington. From Industrial Park, turn right on Doubet Road, then left on Progress Road. The parking area for the 5-acre lake is on Progress Road. There is a fishing jetty accessible to wheelchairs.

3 **Gruner Ford Access** is 3 miles south of Farmington on Highway H. This is a 4-acre area on the St. Francis River with no boat ramp.

4 **Iron Mountain City Lake** is west of Farmington on Highway W to Parkway Boulevard, then left on Gallatin. This 100-acre lake has a boat ramp and picnic area.

5 **Knob Lick Towersite** is 10 miles south of Farmington on Highway 67, then right on Canterberry County Road. This is an 80-acre area with a 5-acre glade overlooking a large valley.

6 **Leadwood Access** is on Highway 8 near the town of Leadwood. This is 6.3 acres on Big River with no boat ramp.

7 **Quarry Pond** is on the Mineral Area Community College campus just east of Highway 67. This is a 2-acre pond.

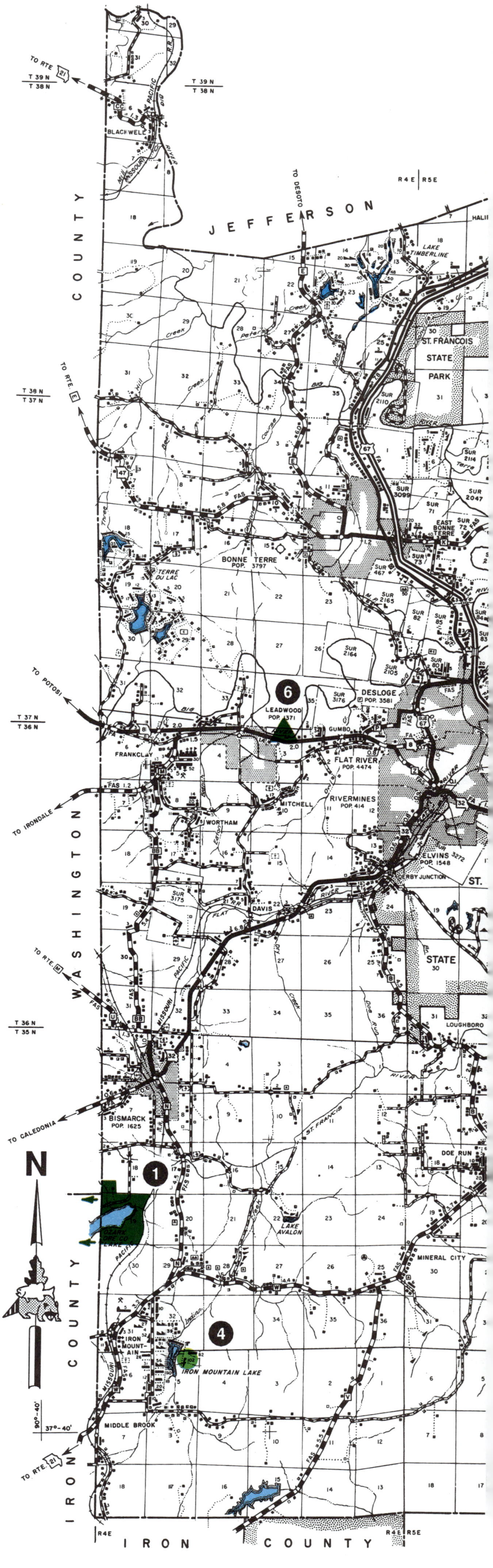

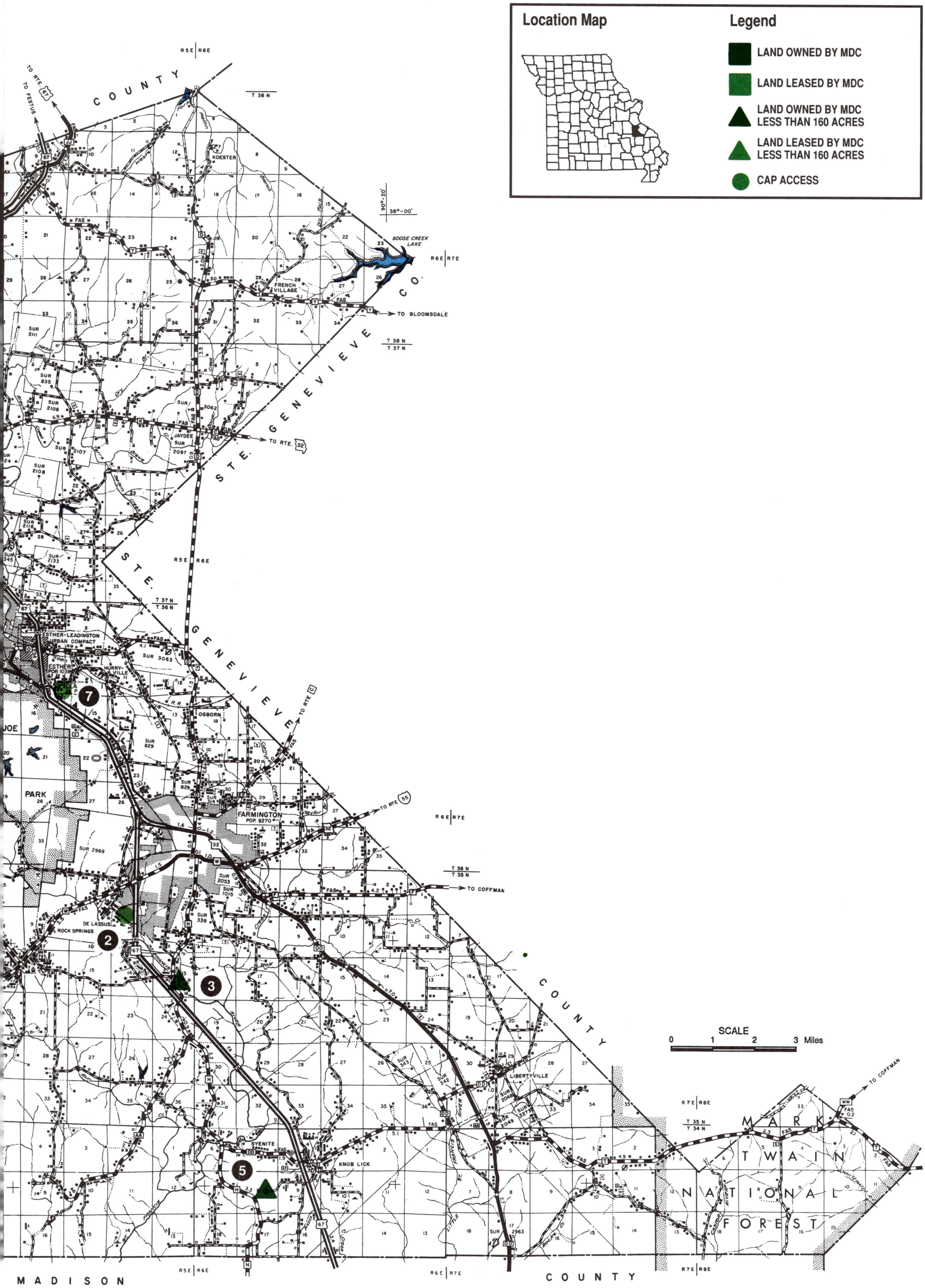
Location Map
Legend
LAND OWNED BY MDC
LAND LEASED BY MDC
LAND OWNED BY MDC LESS THAN 160 ACRES
LAND LEASED BY MDC LESS THAN 160 ACRES
CAP ACCESS
SCALE
0 1 2 3 Miles
COUNTY
STE. GENEVIEVE CO.
STE. GENEVIEVE COUNTY
GOOSE CREEK LAKE
TO BLOOMSDALE
TO FESTUS
TO RTE. 67
TO RTE. 32
TO RTE. 55
TO RTE C
TO COFFMAN
KOESTER
FRENCH VILLAGE
JAYDEE
ESTHER-LEADINGTON URBAN COMPACT
ESTHER
HURRYVILLE
OGBORN
FARMINGTON POP. 8270
PARK
DE LASSUS
ROCK SPRINGS
LIBERTYVILLE
SYENITE
KNOB LICK
MARK TWAIN NATIONAL FOREST
MADISON COUNTY
R5E R6E
R6E R7E
R7E R8E
T 38 N
T 37 N
T 36 N
T 35 N
T 34 N
90°-20'
38°-00'
7
2
3
5

STE. GENEVIEVE COUNTY

1 **Coffman Towersite** is southwest of the junction of highways 55 and 32 near Ste. Genevieve. From the junction, head south on Highway B for 10 miles, then turn west on Highway CC. This 80-acre forested area is .5 mile down Highway CC.

2 **Magnolia Hollow Conservation Area** is south of Bloomsdale on Highway 61, 1 mile east on Highway V, then north on White Sands County Gravel Road. Establishment Creek flows 2 miles through this 1,751-acre forested area.

3 **Pickle Springs Natural Area** is east of Farmington on Highway 32, east on Highway AA, then north on Dorlac Road. This is a 256-acre Designated Natural Area and contains sandstone knobs, arches, canyons and cliffs. There is a 2-mile hiking trail.

4 **Saint Mary's City Access** is in the middle of the town of Saint Mary. Watch for an area sign on Highway 61. This is 10 acres with a boat ramp on Old River.

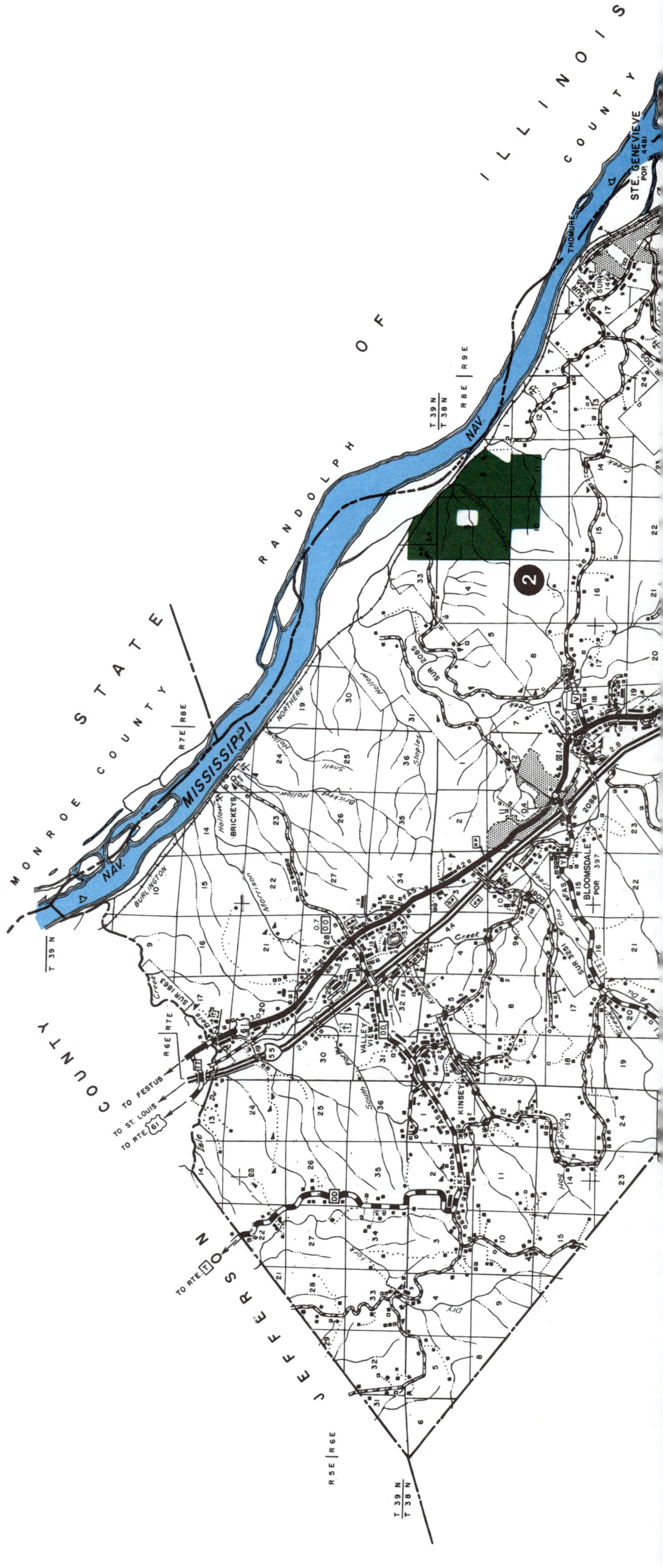

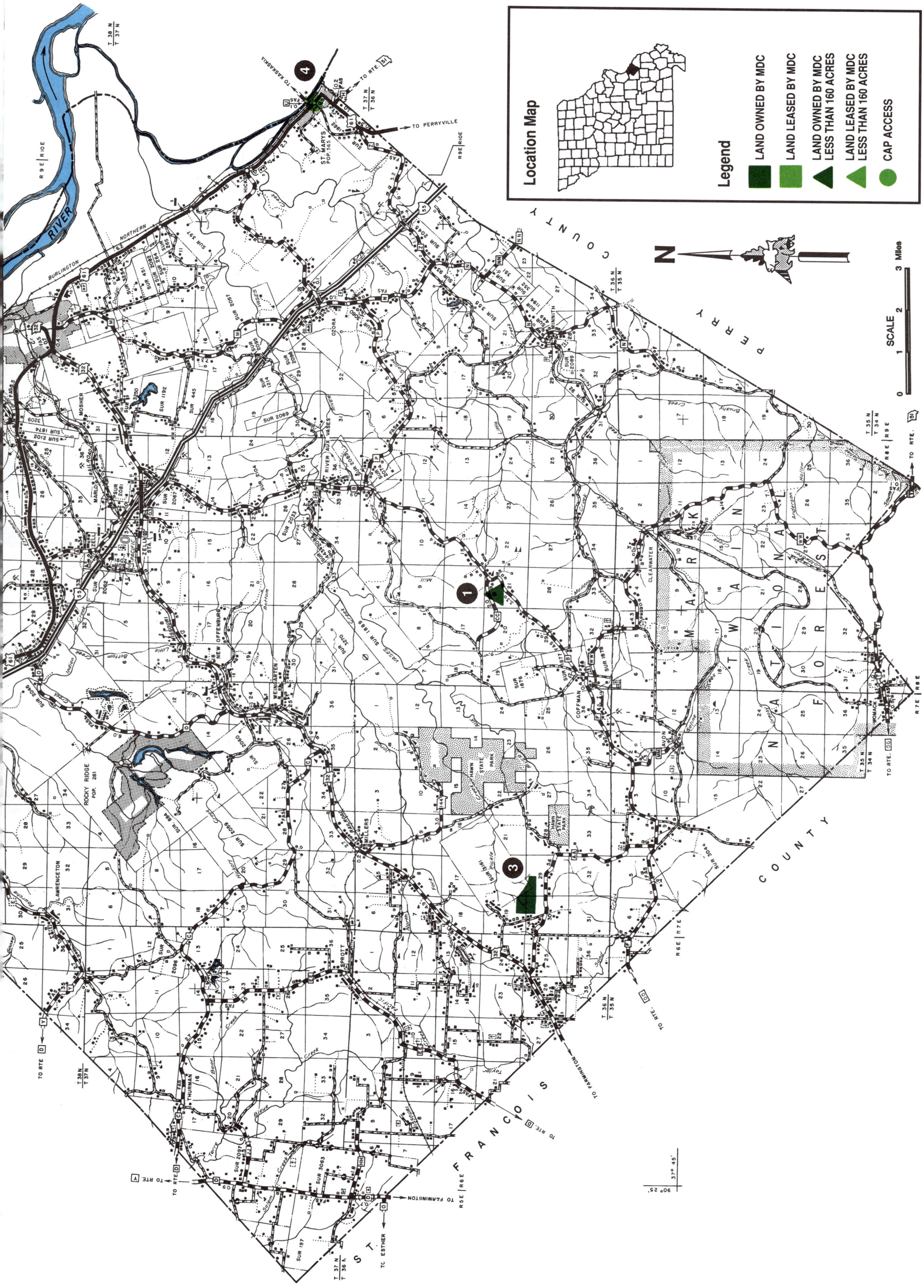
Location Map
Legend
LAND OWNED BY MDC
LAND LEASED BY MDC
LAND OWNED BY MDC LESS THAN 160 ACRES
LAND LEASED BY MDC LESS THAN 160 ACRES
CAP ACCESS
N
SCALE
0 1 2 3 Miles
PERRY COUNTY
ST. FRANCOIS COUNTY
MARK TWAIN NATIONAL FOREST
RIVER
ST. MARYS POP. 565
ROCKY RIDGE POP. 281
HAWN STATE PARK
TO PERRYVILLE
TO KASKASKIA
TO FARMINGTON
TO ESTHER

ST. LOUIS COUNTY

1 To reach **Allenton Access**, exit Highway 44 at Allenton Six Flags Road and go south. Turn east on Main Street, then south on Huntersford Road to the entrance. This is 10 acres on the Meramec River with a boat ramp.

2 **Bee Tree Lake** is south on Telegraph Road off I-270, left on Becker, then down Finestown Avenue. Bee Tree Lake (9 acres) sits on this forested and developed 198-acre area. There is a museum, playground, memorial garden, pavilion, picnic areas and five hiking trails through the natural areas.

3 **Bellefontaine Park Lake** is south off I-270 on Bellefontaine Road. This is a 2-acre lake suitable for fishing.

4 **Benton Park Lake** is west off I-55 on Arsenal. This is a 1-acre lake suitable for fishing.

5 **Boathouse Lake** and **Horseshoe Lake** are in Carondelet Park on Loughborough Avenue off I-55. Boathouse Lake is 4.4-acres and has a wheelchair accessible fishing dock. Horseshoe Lake is 2.4 acres. Both have nearby picnic areas.

6 **Clifton Heights Park Lake** is south on Hampton Avenue off I-44, west on Elizabeth, then left on Ester. Clifton Heights Lake is .7 acres and suitable for fishing.

7 **Creve Coeur Lake** is west on Dorsett Road off I-270, then right on Marine. This is a 1,141-acre park home to Creve Coeur Lake (320 acres of fishable water with two boat ramps). The park also has tennis courts, ball fields, ice-skating facilities and many picnic areas.

8 **Emmenegger Nature Park**–see page 203.

9 To reach **Fairgrounds Park Lake**, take I-70 to West Florissant. Take West Florissant to North Florissant and go left. From North Florissant, turn right on Prairie Avenue to the park. Fairgrounds Lake is 8.8 acres of fishable water.

10 **Forest 44 Conservation Area** is one block south off I-44 on Highway 141, .9 mile west on Meramec Station Road, then .4 mile left on Hillsboro Road. This is a 958-acre area of mainly forestland (841 acres). Williams Creek flows through the area, and there are 10 hiking and horseback riding trails ranging .5 to 3.5 miles in length.

11 To reach **Gravois Creek Conservation Area** (14 acres), take Reavis Barracks to Avenue H. Access to the area is through the Gateway Trailnet Hiking and Biking Trail.

12 **Greentree Park Access** is on Marshall Road between Highway 141 and I-270. This is 10 acres with a boat ramp on the Meramec River.

13 **Hickory Woods Conservation Area** is north of I-70 on Lindbergh, west on Long Road and north on Fee Fee road. This is also the entrance to Gentry Park. This is a 10-acre area with a 1.5-mile hiking trail.

14 **Howell Island Conservation Area**–see the St. Charles County map, page 193.

15 **Hyde Park Lake** is on Salisbury Street, west off I-70. Hyde Park Lake is .1 acre and suitable for fishing.

16 **January-Wabash Park Lake** is south on New Florissant Road off I-270, then right on January Avenue. This is a 5-acre lake suitable for fishing.

17 **Jefferson Lake, Lake #7 and Lake #8** are all located in Forest Park, a large urban park with museums, zoo, picnic areas, playing fields and many other facilities. These three lakes are suitable for fishing and have wheelchair accessible docks.

18 **Kiwanis Lake** is off I-70, right on St. Charles Road to Fee Fee Road, then right on Oak Street. This is an 8-acre area with one trail.

19 **Roger Klamberg Woods Conservation Area** is south of Ellisville on Kiefer Creek Road off State Road 100. The woods is reached by using the entrance to Bluebird Park. It is a 67-acre, mainly forested area with two trails totaling about 1.5 miles in length.

20 **Lafayette Park Lake** is north on Jefferson Avenue off I-44, then right on Lafayette. This is a 1-acre lake suitable for fishing.

21 **New Ballwin Lake** is west off I-270 on Manchester Road, then left on New Ballwin Road. This is a 1-acre area.

22 **North Riverfront Park Access** is 2 miles south of Highway 270 on Riverview Drive. This is a 5-acre area with 1 mile of frontage along the Mississippi River. There are picnic areas and a 3-mile asphalt biking trail.

23 **North Riverfront Park Lake** is north on Broadway off I-70, then right on Riverview Boulevard. This is an 8.5-acre lake suitable for fishing.

24 **O'Fallon Park Lake** is south on West Florissant Avenue off I-70. This is a 5.3-acre lake suitable for fishing.

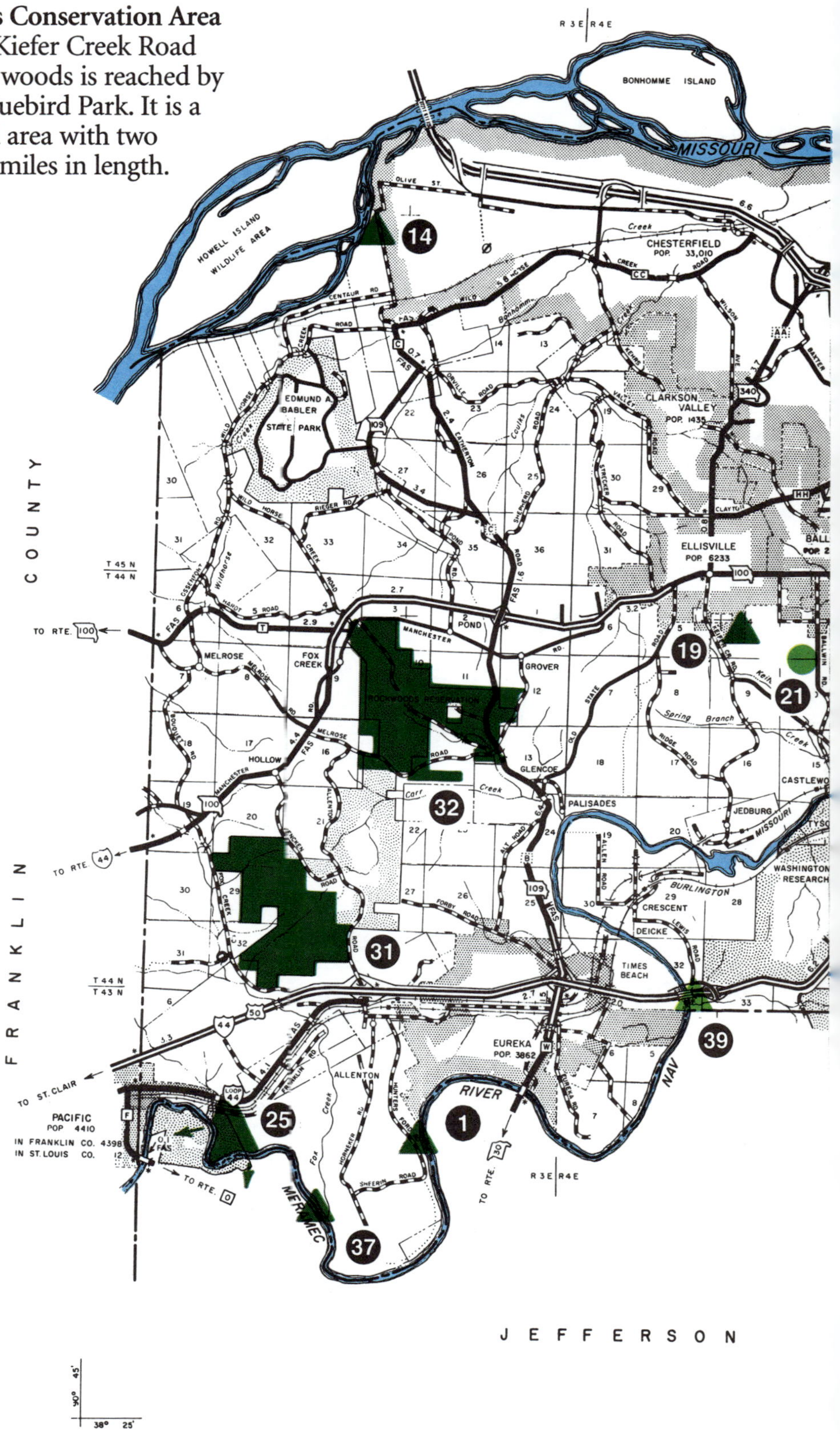

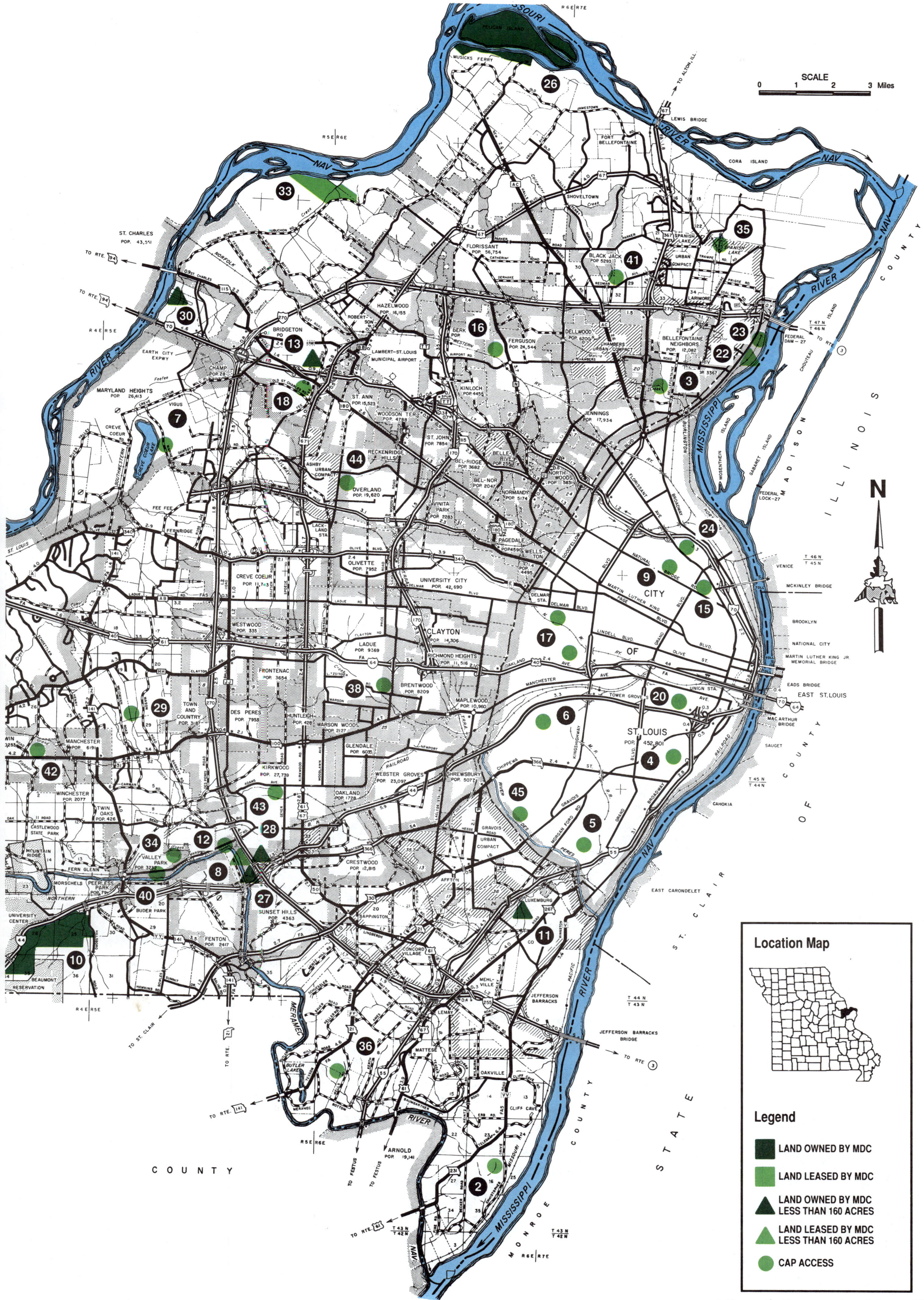
SCALE
0 1 2 3 Miles
N
Location Map
Legend
LAND OWNED BY MDC
LAND LEASED BY MDC
LAND OWNED BY MDC LESS THAN 160 ACRES
LAND LEASED BY MDC LESS THAN 160 ACRES
CAP ACCESS
MISSOURI RIVER
MISSISSIPPI RIVER
MERAMEC RIVER
PELICAN ISLAND
MUSICKS FERRY
TO ALTON, ILL.
LEWIS BRIDGE
CORA ISLAND
FORT BELLEFONTAINE
SHOVELTOWN
FLORISSANT POP. 56,754
BLACK JACK POP. 5293
SPANISH LAKE
HAZELWOOD POP. 16,155
BRIDGETON POP.
LAMBERT-ST. LOUIS MUNICIPAL AIRPORT
FERGUSON POP. 24,544
DELLWOOD POP. 6200
BELLEFONTAINE NEIGHBORS POP. 12,082
ST. CHARLES POP. 43,551
MARYLAND HEIGHTS POP. 26,413
CREVE COEUR
JENNINGS POP. 17,934
ST. ANN POP. 15,523
OVERLAND POP. 19,620
NORMANDY POP. 5174
OLIVETTE POP. 7952
UNIVERSITY CITY POP. 42,690
CREVE COEUR POP. 11,743
CLAYTON POP. 14,306
LADUE POP. 9369
RICHMOND HEIGHTS POP. 11,516
FRONTENAC POP. 3654
BRENTWOOD POP. 8209
MAPLEWOOD POP. 10,960
TOWN AND COUNTRY POP. 3187
DES PERES POP. 7953
MANCHESTER POP. 6191
WINCHESTER POP. 2077
KIRKWOOD POP. 27,739
WEBSTER GROVES POP. 23,097
SHREWSBURY POP. 5077
GLENDALE POP. 6035
CRESTWOOD POP. 12,815
SUNSET HILLS POP. 4363
FENTON POP. 2417
VALLEY PARK POP. 3232
ST. LOUIS POP. 452,801
CITY OF ST. LOUIS
EAST ST. LOUIS
MADISON COUNTY
ILLINOIS
ST. CLAIR COUNTY
MONROE COUNTY
STATE OF ILLINOIS
JEFFERSON BARRACKS
JEFFERSON BARRACKS BRIDGE
OAKVILLE
MEHLVILLE
LEMAY
ARNOLD POP. 19,141
BUTLER LAKE
BEAUMONT RESERVATION
UNIVERSITY CENTER
CASTLEWOOD STATE PARK
EADS BRIDGE
MCKINLEY BRIDGE
MARTIN LUTHER KING JR. MEMORIAL BRIDGE
MAC ARTHUR BRIDGE
CAHOKIA
VENICE
BROOKLYN
NATIONAL CITY
TO FESTUS

25 **Pacific Palisades Conservation Area**–see the Jefferson County map, page 108.

26 **Pelican Island Natural Area** is on the Missouri River. Access is by boat or through Sioux Passage Park in St. Louis County. This 2,260-acre area contains a 5.8-mile stretch of the Missouri River and 3.8 miles of the Car of Commerce Chute.

27 **Possum Woods Conservation Area** is on Cragwold Road, .3 mile past the entrance to the Powder Valley Nature Center. This is a 15-acre forested area.

28 **Powder Valley Conservation Nature Center** (and neighboring Emmenegger Nature Park) is .5 mile south on Lindbergh Road (off I-44), .5 mile west on Watson Road (also called Route 366), then north on Geyer Road. From Geyer, it is just 200 yards to Cragwold Road. Turn west on Cragwold and watch for signs to the nature center. This is a 112-acre forested area with three hiking trails. The nature center has two floors of exhibits. See page 203 for more information.

29 **Queeny Park Lake** is west on Manchester off I-270, then right on Weidmann Road. This is a 1-acre area suitable for fishing. Jarville Lake (2 acres) is nearby.

30 **Riverwoods Conservation Area** is at the end of Old St. Charles Rock Road Bridge across from St. Charles. This is a 123-acre mostly forested area.

31 **Rockwoods Range** is on Fox Creek Road, west of the intersection of Highway 44 and Allenton Six Flags Road. The 1,388-acre area is on the south side of Fox Creek Road. This is nearly all forestland and has two horseback riding trails. It is home to the A.P. Greensfelder Round House Memorial.

32 **Rockwoods Reservation** is 4.1 miles north on Highway 109 off Highway 44, then 1.5 miles north on Glencoe Road. This 1,844-acre area contains three caves, three springs and a stretch of Hamilton Creek. The area is nearly all forestland, but has a Conservation Education Center where school groups and scouts attend nature and conservation programs. There are eight hiking trails, interpretative programs, displays and several picnic areas.

33 **St. Stanislaus County Park North** is off I-70 on the Earth City Expressway, which turns into Aubuchon Road. Continue north on Aubuchon about 3.5 miles to the area. This is a 937-acre area on the Missouri River.

34 **Simpson Park Lake** is north of I-44 on Route 141 (Meramec Station Road), and 1 mile right on Marshall Road. This 72-acre lake has four picnic areas and a wheelchair accessible fishing dock.

35 **Spanish Lake** is north on Bellefontaine Road off I-270, then 1 mile to the right on Spanish Pond Road. This 230-acre area is home to Spanish Lake and **Sunfish Lake**, which together, provide anglers with about 65 acres of fishable water. There are two boat ramps and jetties, two picnic areas and two hiking trails.

36 **Suson Park Lakes** are south off I-270 on Tesson Ferry Road (also called Highway 21), then left on Wells Road. The park contains three ponds totaling 8 acres of fishable water.

37 **Swiftwater Bend Conservation Area** is south on Highway W, south on Highway FF, west on Highway F, then down McNamee School Road. This 69-acre area contains a .5-mile stretch of the Meramec River. The river is accessible by a .25-mile hiking trail.

38 **Tilles Park Lake** is on McKnight Road, south of I-40. The lake is 2 acres and suitable for fishing.

39 **Times Beach Access** is just under the I-44 bridge at the Lewis Road Exit. This is 3 acres on the Meramec River with a boat ramp.

40 **Valley Park Access** is on Marshall Road. This is 5 acres on the Meramec River with a boat ramp.

41 **Veteran's Memorial Park Lake** is north on Highway 367 off I-270, then left on Redman Road. The lake is 1 acre and suitable for fishing.

42 **Vlasis Park Lake** is west on Manchester Road off I-270, then right on Seven Trails Drive. Vlasis Lake is .5 acre and suitable for fishing.

43 **Walker Lake** is north on Highway 61-67, and left on Adams. This is a 2.5-acre lake, with two picnic areas and a wheelchair accessible fishing dock.

44 **Wild Acres Park** is north on Ashby from Page Avenue, then right on Midland Blvd. This is a 2.5-acre lake suitable for fishing.

45 **Willmore Park North and South Lakes** are in Willmore Park, which is west on Loughborough off I-55, left on Gravois, then right on Hampton Avenue. Together, these lakes provide anglers with about 8 acres of fishable water.

Jim Rathert

While in St. Louis County, visit

Powder Valley Conservation Nature Center

and Emmenegger Nature Park

Powder Valley Conservation Nature Center is on Cragwold Road in Kirkwood. From I-44 Eastbound, take the Watson Road exit and turn north on Geyer Road. Follow Geyer 200 yards to Cragwold Road. Go 1 mile west on Cragwold. Watch for signs to the nature center on both Geyer and Cragwold roads. Emmenegger Nature Park is located .33 mile west of the nature center entrance. Exhibits, trails and nature center programs are just a few of the attractions awaiting drop-by visitors or school groups, scouts and other organizations that call ahead.

Outside

- more than 110 acres of forestland at Powder Valley and 95 acres of forestland at Emmenegger Nature Park
- three hiking trails at Powder Valley, one of which is accessible to people who use wheelchairs
- more trails at Emmenegger; one overlooks the Meramec River

Inside

- two floors of hands-on nature exhibits
- indoor wildlife viewing area
- resource library, four classrooms and a 250-seat auditorium
- regularly scheduled public programs and guided walks announced in the center's monthly newsletter, *Making Tracks*
- a 3,000-gallon aquarium home to native fish and turtles
- sales of books, calendars and fishing and hunting permits

Building Hours

Open year-round, seven days a week, 8 a.m. to 5 p.m. Closed Thanksgiving, Christmas and New Year's Day.

Area Hours

Open daily, 8 a.m. to 8 p.m. during daylight savings; and daily, 8 a.m. to 6 p.m. standard time. Closed Thanksgiving, Christmas and New Year's Day.

Pets, horses, hunting, fishing, collecting, littering, fires and alcoholic beverages are not permitted. Bicycling, jogging, in-line skating and motorized vehicles are not permitted on trails. Call the Nature Center for regulations applying to Emmenegger Nature Park.

Powder Valley Conservation Nature Center
Missouri Department of Conservation
11715 Craigwold Road
Kirkwood, MO 63122
314/301-1500

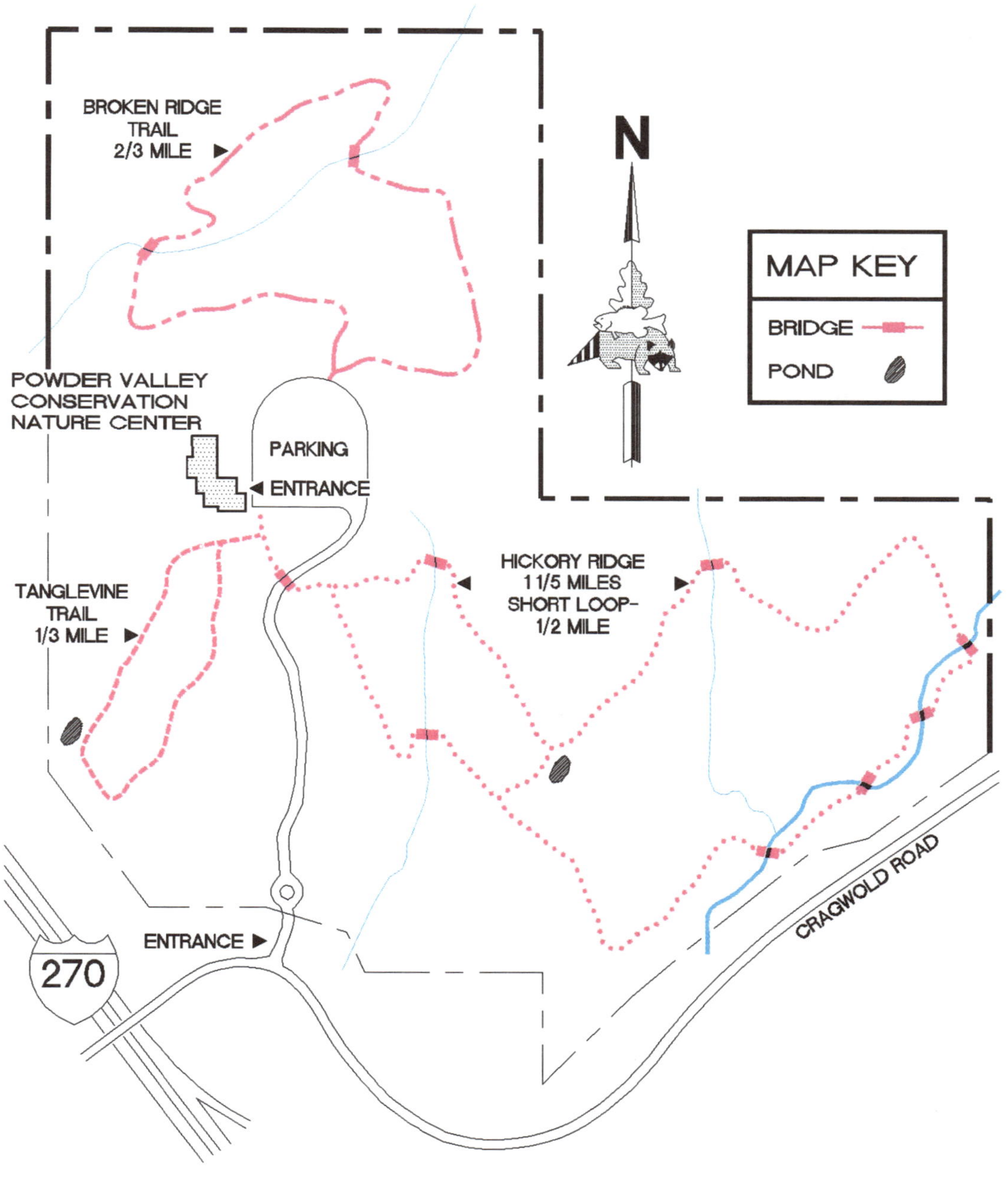

More than 110 forested acres surround the Powder Valley Nature Center (opposite page), which is open to visitors free of charge.

SALINE COUNTY

1 **Blind Pony Lake Conservation Area** consists of three tracts. The largest is on Route ZZ, 6.5 miles northeast of the junction of Highway 127 and I-70 in Sweet Springs. This is a 2,207-acre area with 210 acres of fishable water (four ponds and a lake). There are several picnic areas, a boat ramp, and tours of the hatchery are available.

2 **Blue Lick Conservation Area** is 3.5 miles north of I-70 on Highway 65, then 1.5 miles west on an unnamed county road. This 390-acre area features several saline and mineral springs.

3 **Grand Pass Conservation Area** is 8 miles west of Marshall on Highway 65 and 5 miles north on Highway N. This 5,296-acre area is on the Missouri River and contains Teteseau Lake (7 acres). The surrounding land is mainly cropland (1,500 acres) and wetlands (2,000 acres). An observation tower overlooks the lake, and there is a 1-mile hiking trail.

4 **McAllister Springs Access** is on the Blackwater River off Highway K, east of Sweet Springs. There is a boat ramp on the river and about 4 acres of surrounding forestland.

5 **Malta Bend Community Lake** is 3 miles east of Grand Pass on Highway 65, and 1.25 miles south on Highway 127. The lake is 4 acres and is surrounded by about 20 acres of forestland and old fields. There is no boat ramp.

6 **Marshall Habilitation Center Lake** is at the intersection of Highway 41 and Lincoln Street at the north city limits of Marshall. This is a 12-acre lake suitable for fishing.

7 **Marshall Junction Conservation Area** is 2 miles south of the junction of I-70 and Highway 65, then 3 miles west on Highway CC, and 1 mile north on County Road RB. The Blackwater River flows 2.5 miles through this 774-acre area, which is composed of forestland (464 acres), old fields (155 acres) and cropland (155 acres).

8 **Miami Access** is on Highway 41, .25 mile south of the Missouri River bridge. This is 13 acres with a boat ramp on the Missouri River.

9 **Ralph and Martha Perry Memorial Conservation Area**–see the Johnson County map, page 110.

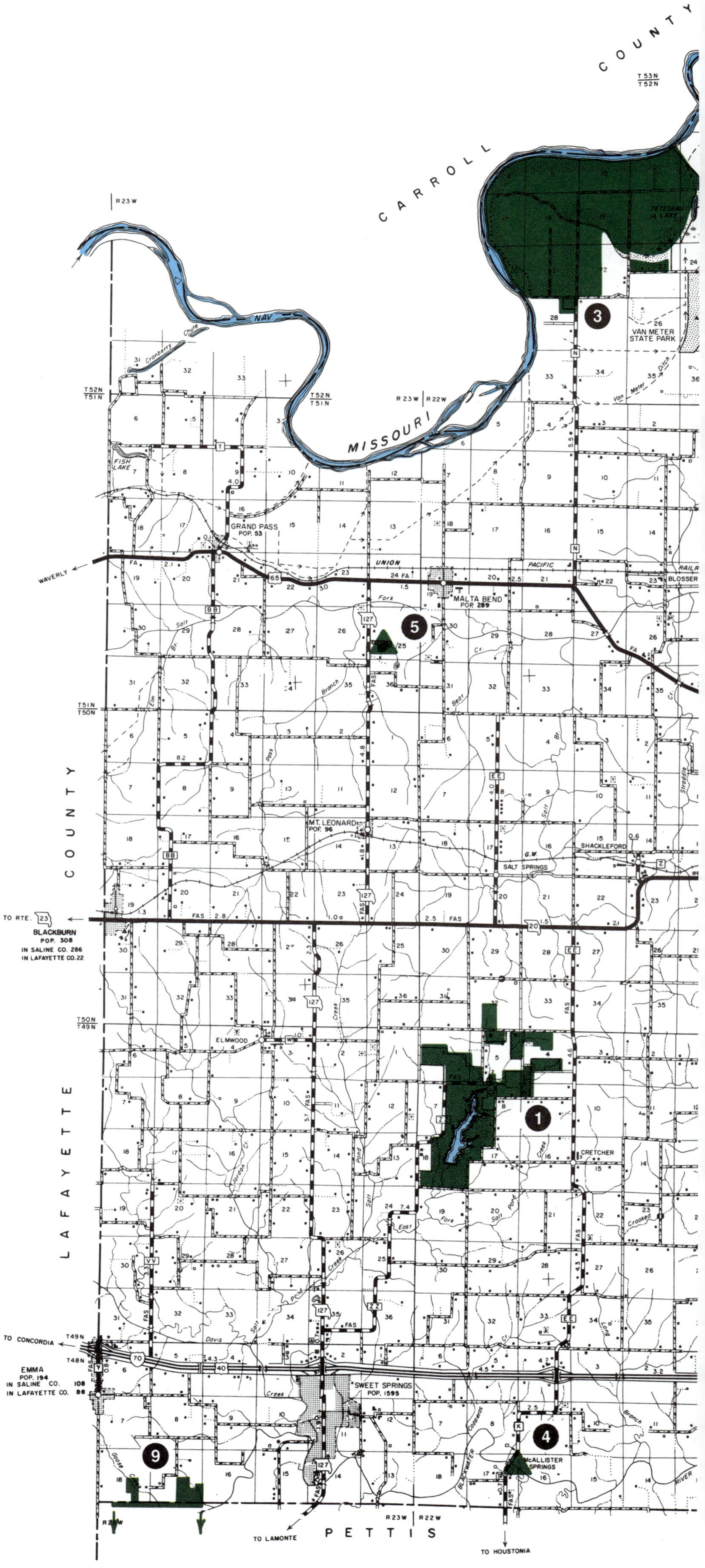

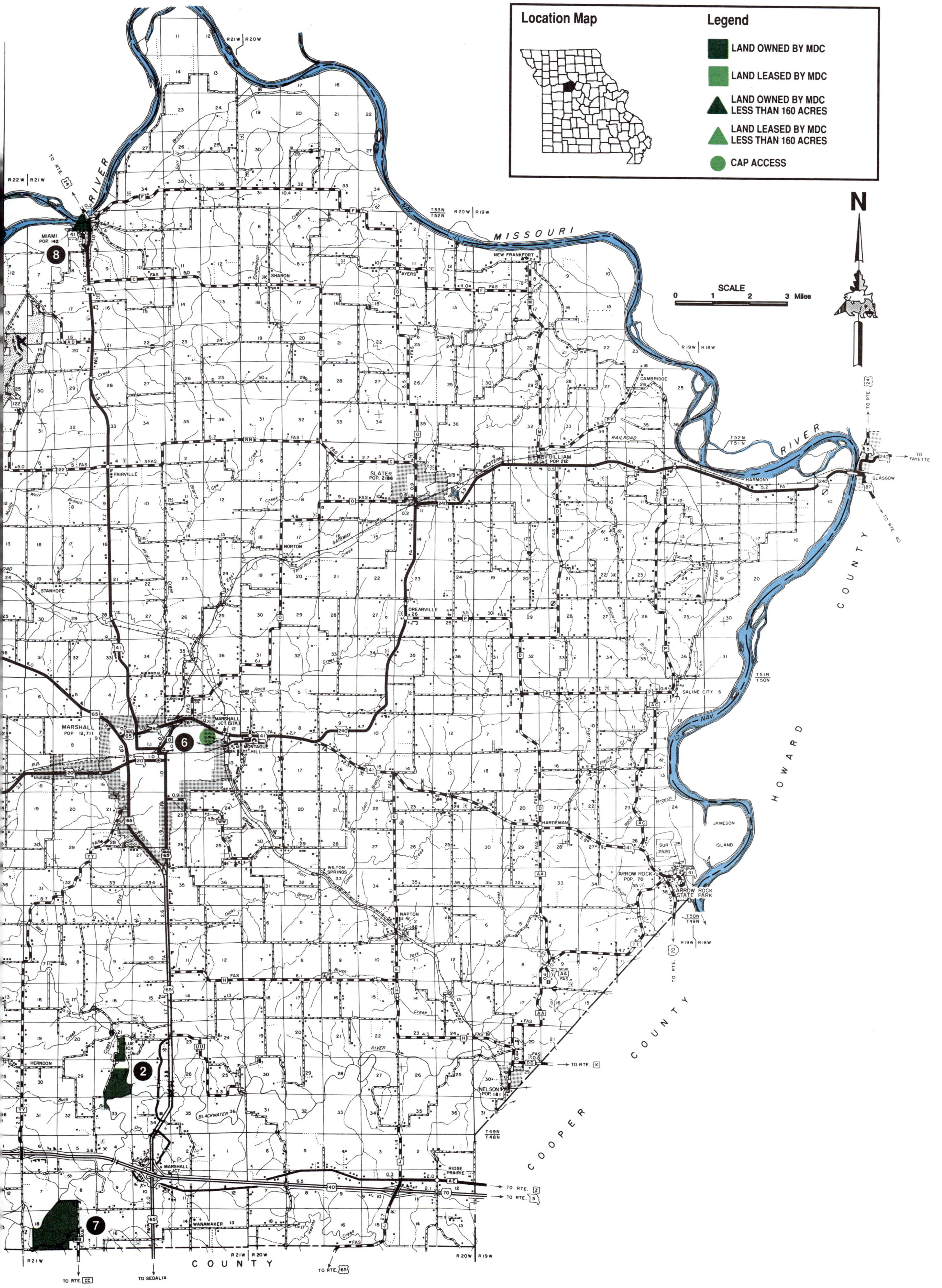
Location Map
Legend
LAND OWNED BY MDC
LAND LEASED BY MDC
LAND OWNED BY MDC LESS THAN 160 ACRES
LAND LEASED BY MDC LESS THAN 160 ACRES
CAP ACCESS
N
SCALE
0 1 2 3 Miles
MISSOURI RIVER
MIAMI POP. 142
SHARON
AYERS
NEW FRANKFORT
CAMBRIDGE
FAIRVILLE
SLATER POP. 2186
GILLIAM POP. 212
HARMONY
GLASGOW
TO FAYETTE
NORTON
STANHOPE
OREARVILLE
SALINE CITY
MARSHALL POP. 12,711
MARSHALL JCT. (STA.)
MONTAGUE HILL
HARDEMAN
JAMESON ISLAND
WILTON SPRINGS
ARROW ROCK POP. 70
ARROW ROCK STATE PARK
NAPTON
HERNDON
NELSON POP. 181
MARSHALL JCT.
RIDGE PRAIRIE
WANAMAKER
HOWARD COUNTY
COOPER COUNTY
TO SEDALIA
TO RTE. CC
TO RTE. K
TO RTE. Z
TO RTE. 5
TO RTE. 65
TO RTE. 70
TO RTE. 40
TO RTE. 24

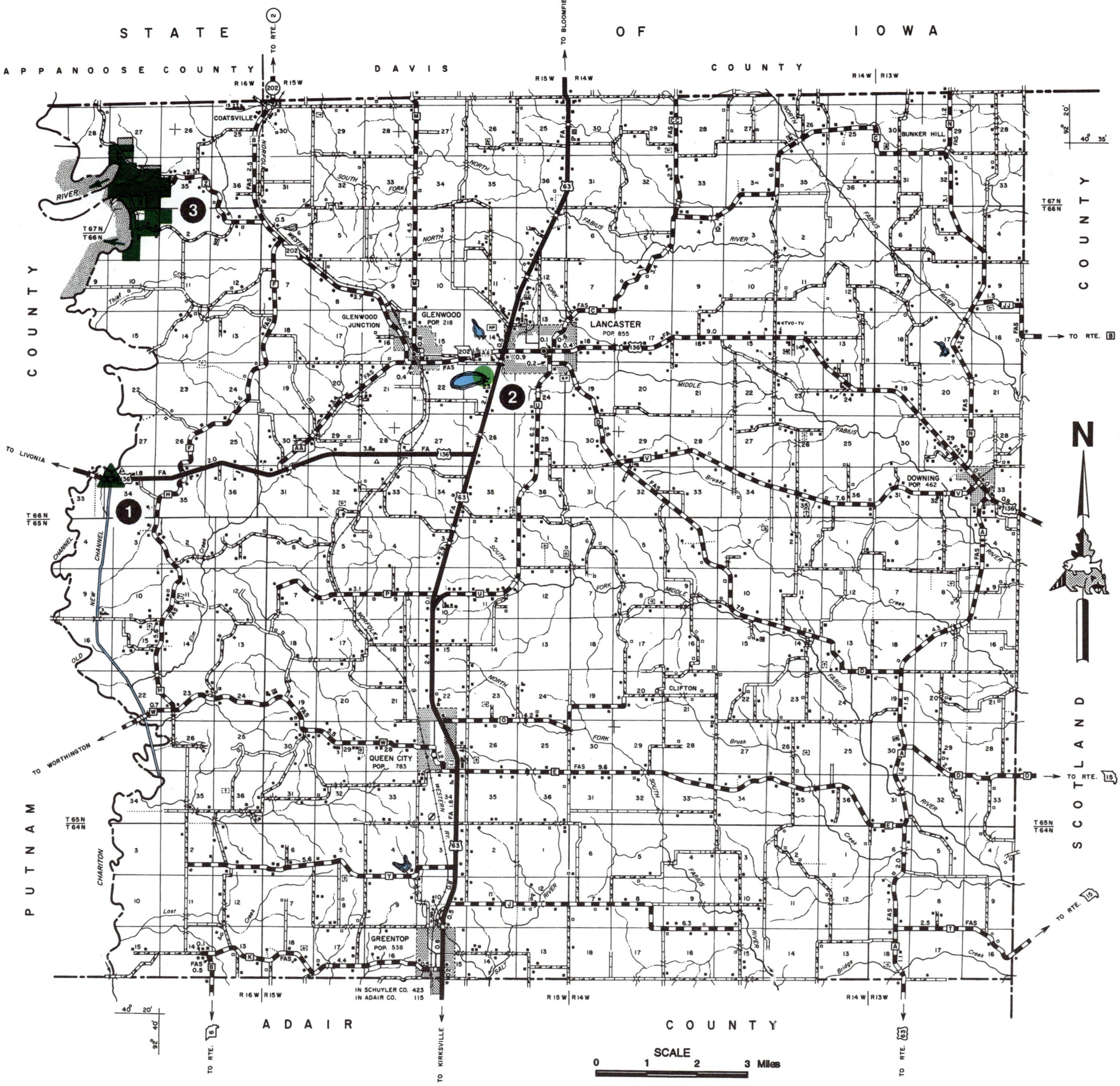

SCHUYLER COUNTY

1 **Archangel Access** is 1 mile east of Livonia on Highway 136. This is 10 acres with no boat ramp on the Chariton River.

2 **Lancaster City Lake** is marked by a cantilever sign on Highway 63, 1 mile north of the junction of highways 63 and 136. The lake is 47 acres and has a boat ramp.

3 **Rebel's Cove Conservation Area**–see the Putnam County map, page 180.

Location Map

Legend

- LAND OWNED BY MDC
- LAND LEASED BY MDC
- LAND OWNED BY MDC LESS THAN 160 ACRES
- LAND LEASED BY MDC LESS THAN 160 ACRES
- CAP ACCESS

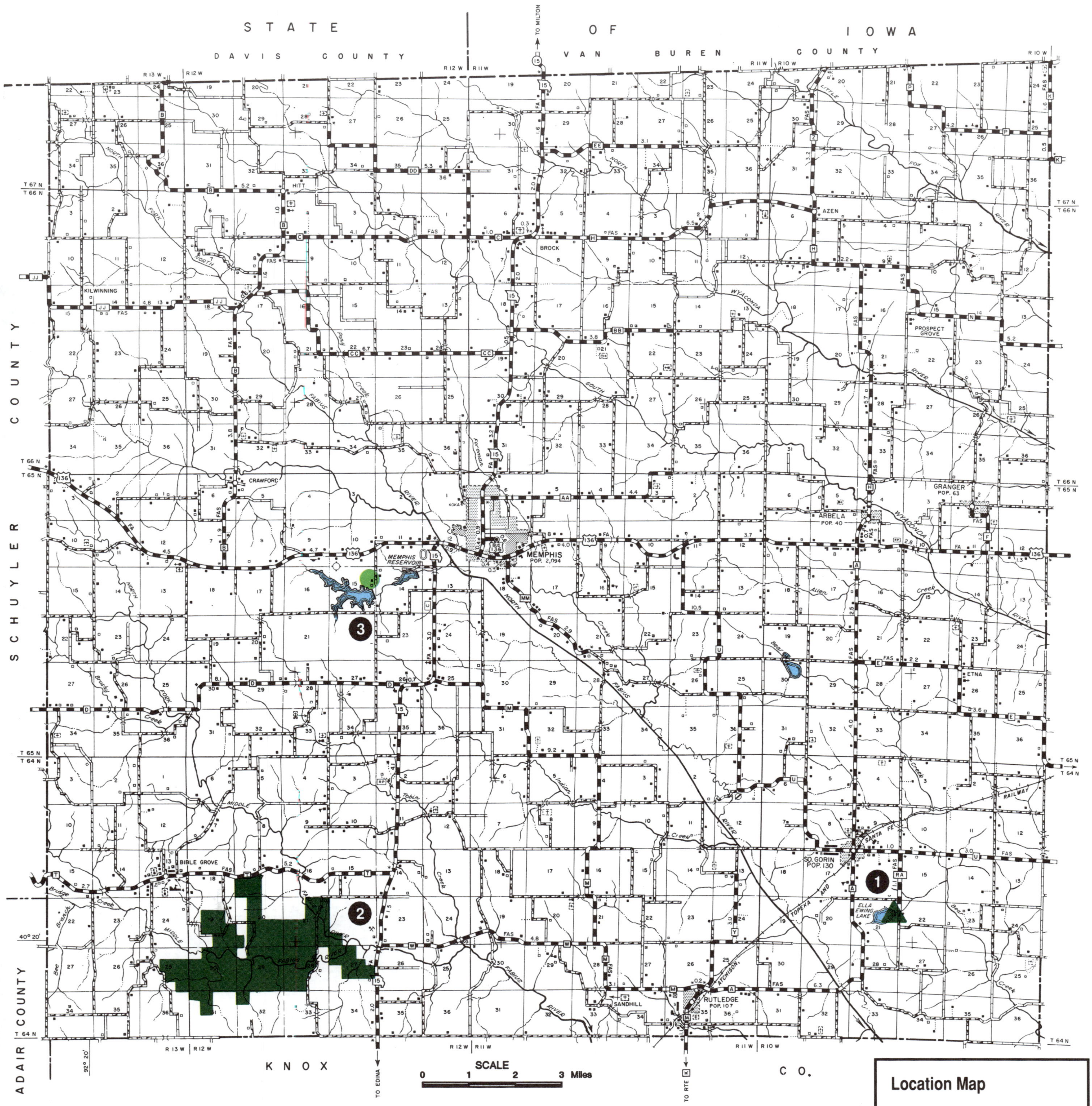

SCOTLAND COUNTY

1 **Ella Ewing Lake Conservation Area** is 1 mile east of Gorin on Route U and 1 mile south on Route RA. This is a 15-acre lake with a boat ramp surrounded by 45 acres of forestland and old fields.

2 **Indian Hills Conservation Area** is 8 miles south of Memphis on Highway 15, 3.5 miles west on Highway T, then south 1 mile at the area sign. This 3,691-acre area contains stretches of Middle Fabius River and North Fork of the Middle Fabius. There is a 30-nest heron rookery and horseback riding is allowed with a permit.

3 **Lake Showme** is 2 miles west of Memphis on Highway 136, then 1 mile south on the first road named Lake Showme Drive. The lake is 225 acres and is equipped with boat ramps and one wheelchair accessible fishing dock. Activities and facilities on the surrounding land are regulated by the city of Memphis.

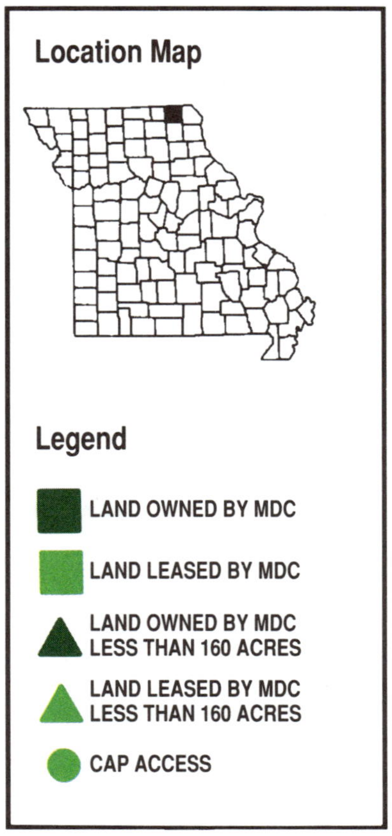

SCOTT COUNTY

1 **General Watkins Conservation Area** is off Highway 61 on an unnamed gravel road, 1.5 miles south of Benton. This 1,101-acre area has three lakes and five ponds, providing anglers with 36 acres of fishable water. There is a 1-mile hiking trail and picnic area.

2 **Tywappity Community Lake** is reached by heading east on Highway A from the center of Chaffee. At the city limits, go 2 miles on Highway RA. The lake (37 acres with a boat ramp) is surrounded by 79 acres of mainly forestland. **Tywappity Towersite** is .5 mile from the parking area at Tywappity Lake. There is no trail to the tower.

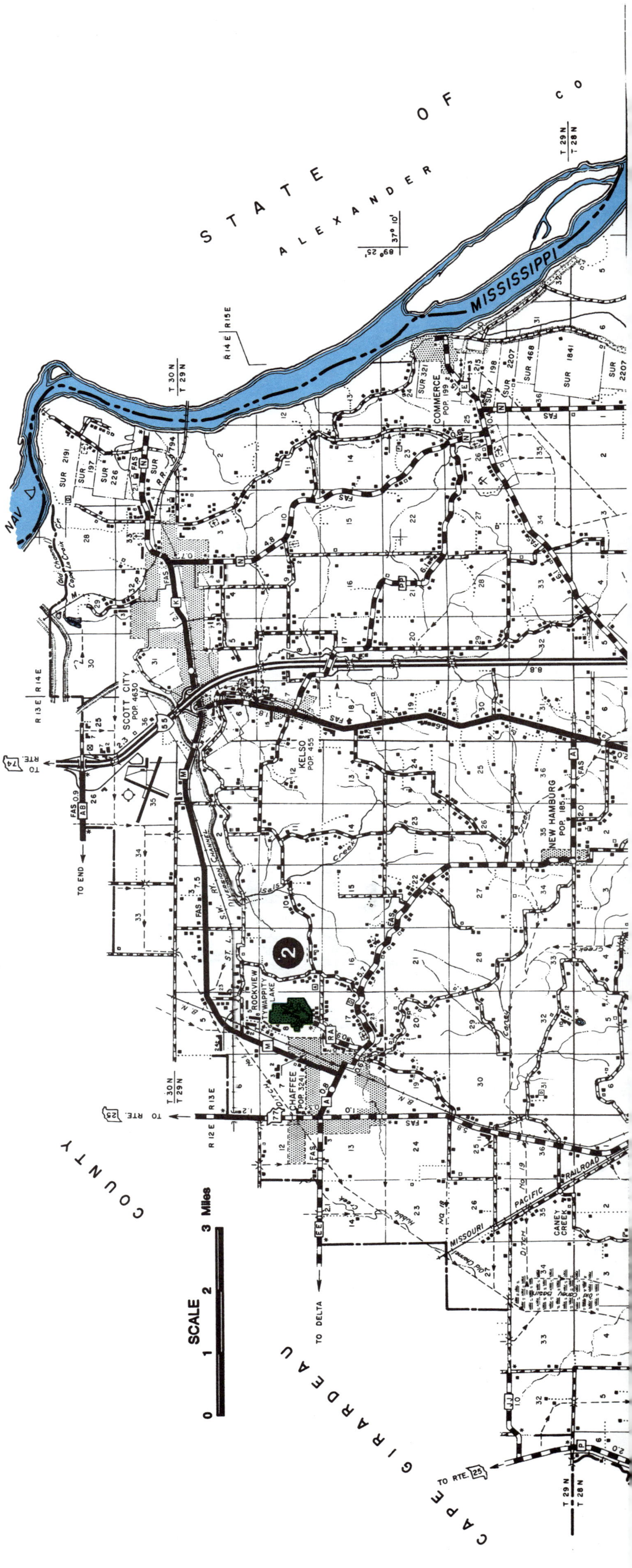

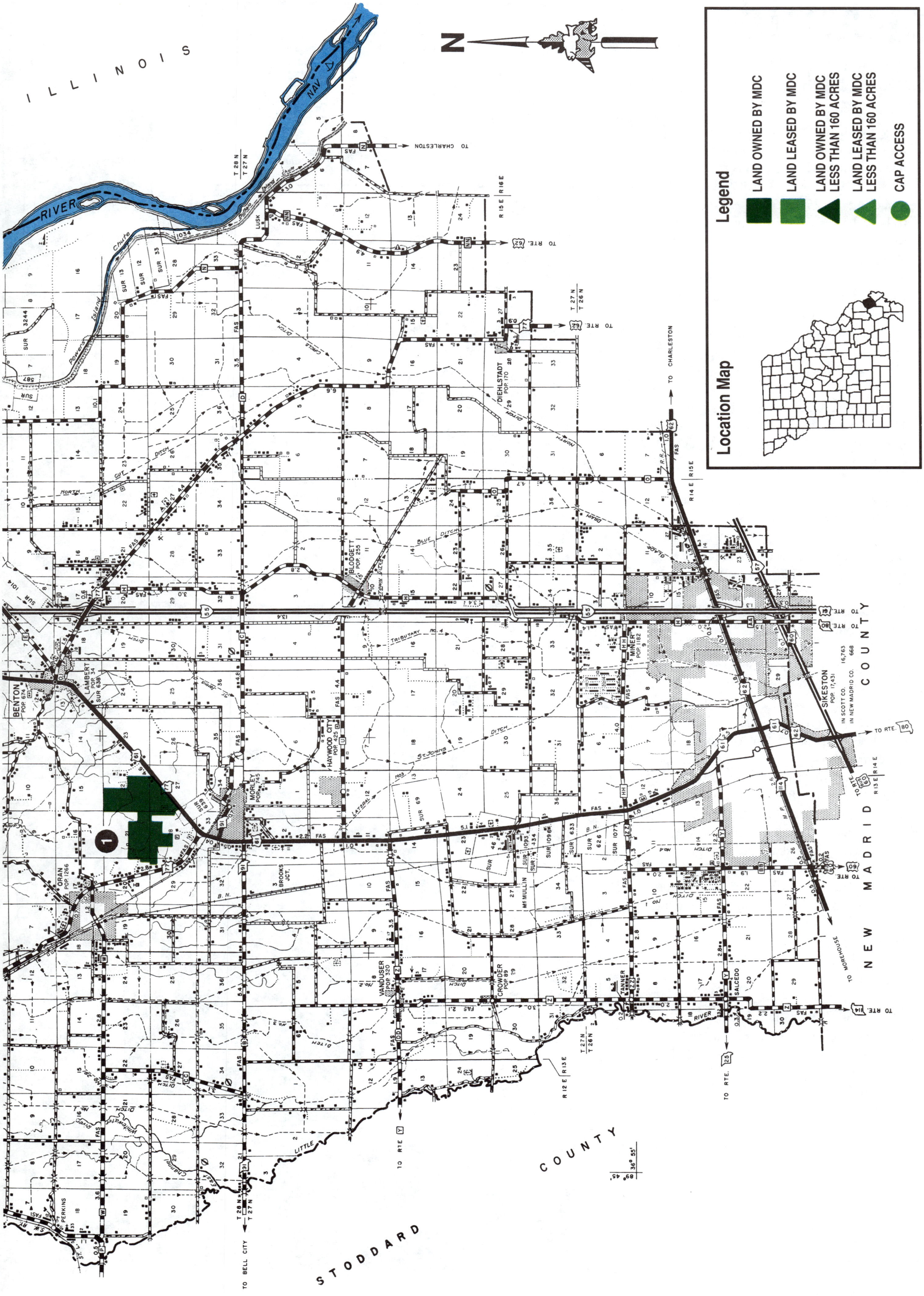
Legend
LAND OWNED BY MDC
LAND LEASED BY MDC
LAND OWNED BY MDC LESS THAN 160 ACRES
LAND LEASED BY MDC LESS THAN 160 ACRES
CAP ACCESS
Location Map
N
ILLINOIS
RIVER
NAV
STODDARD COUNTY
NEW MADRID COUNTY
BENTON POP 674
LAMBERT POP 354
ORAN POP 1266
MORLEY POP 745
HAYWOOD CITY POP 425
BLODGETT POP 255
DIEHLSTADT POP 170
MINER POP 1102
SIKESTON POP 17,431 IN SCOTT CO. 16,763 IN NEW MADRID CO. 668
VANDUSER POP 320
CROWDER POP 89
BROOKS JCT.
McMULLIN
PERKINS
SALCEDO
TANNER
LUSK
TO CHARLESTON
TO BELL CITY
TO MOREHOUSE
LITTLE RIVER
ST. JOHNS DITCH
BLUE DITCH
TRIBUTARY NO. 4
T 28 N
T 27 N
T 26 N
R 12 E
R 13 E
R 14 E
R 15 E
R 16 E
1

SHANNON COUNTY

1 **Angeline Conservation Area** is north of Eminence and extends 15 miles west to near Summersville. Access is from highways 19, 106 and D. This 36,929-acre has a hiking trail, 25 miles of field roads open to horseback riders and 1-mile stretches of the Current and Jacks Fork rivers. There is also a firearms range.

2 **Birch Creek Conservation Area** (5,645 acres) is 6 miles south of Birch Tree on Highway 99. This area is mostly forested and has 3 miles of field roads open to horseback riders.

3 **Buttin Rock Access** is near the Highway 19 bridge over the Jacks Fork River on the north edge of Eminence. It is 9 acres and suitable for launching canoes.

4 **Current River Conservation Area**–see the Carter County map, page 40.

5 **Logan Creek Conservation Area**–see the Reynolds County Map, page 188.

6 **Peck Ranch Conservation Area**–see the Carter County map, page 40.

7 **Rocky Creek Conservation Area** consists of several tracts. The main tract is 2 miles south of Eminence on Highway 19. This is a 39,486-acre forested area. It is home to two designated natural areas, Powder Mill Cave and Mill Mountain. The Current River, Jacks Fork River, Rocky Creek, Indian Creek and Bloom Creek all flow through the area, and there is a 7-mile hiking trail and about 15 miles of field roads open to horseback riders.

8 **Sunklands Conservation Area** is 13 miles north of Eminence on Highway 19, then 1 mile west on County Road 19-324. This 35,789-acre area is mostly forested. The Current River and Barren Fork Creek flow through the area.

9 **Thomasville Towersite** (40 acres) is 3 miles east of Mountain View on Highway 60, 7 miles south on Highway T, then 1 mile west on Highway TT. The .25-mile road off TT leading to the tower is closed to vehicle traffic, but visitors may walk in.

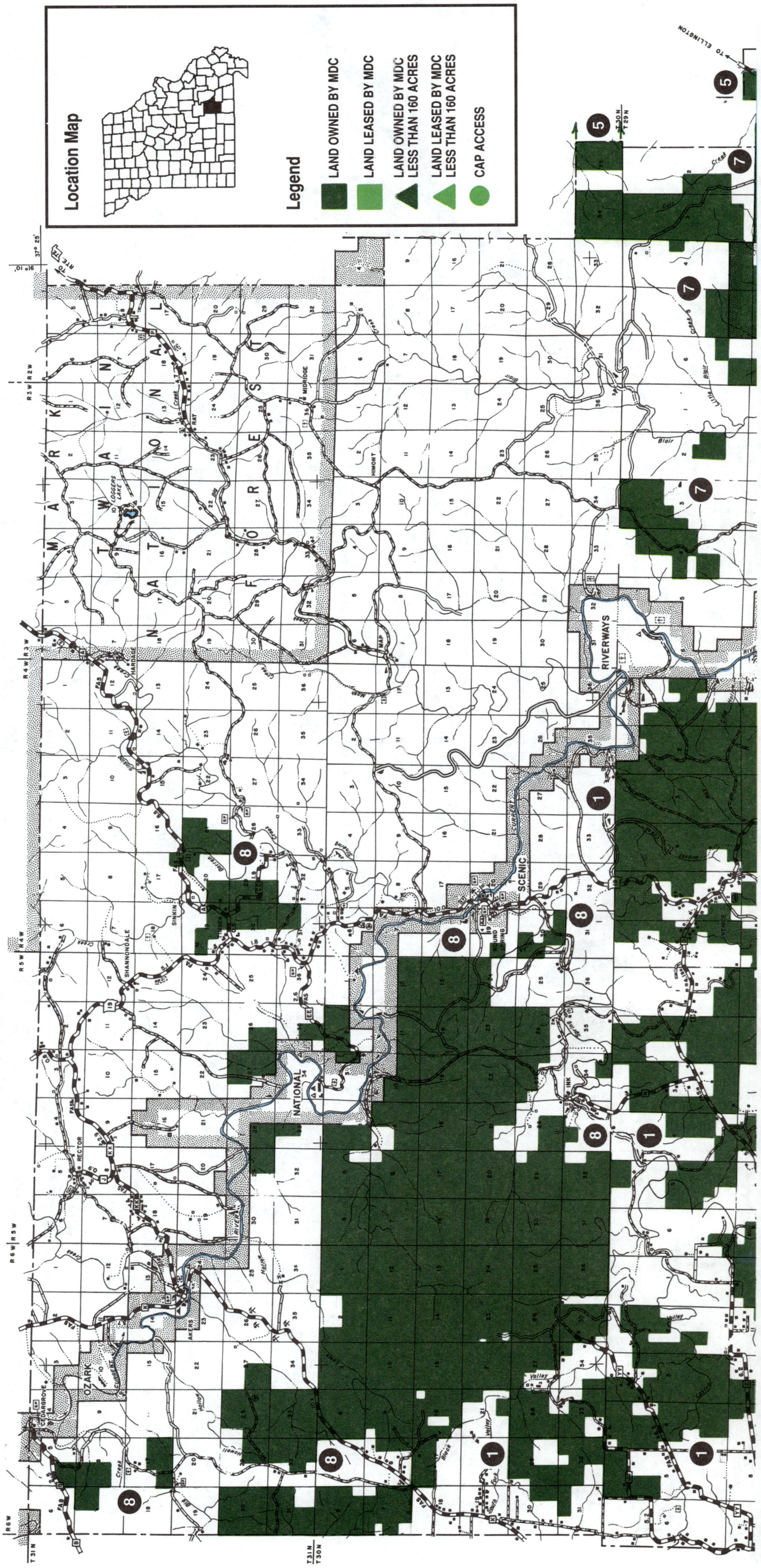

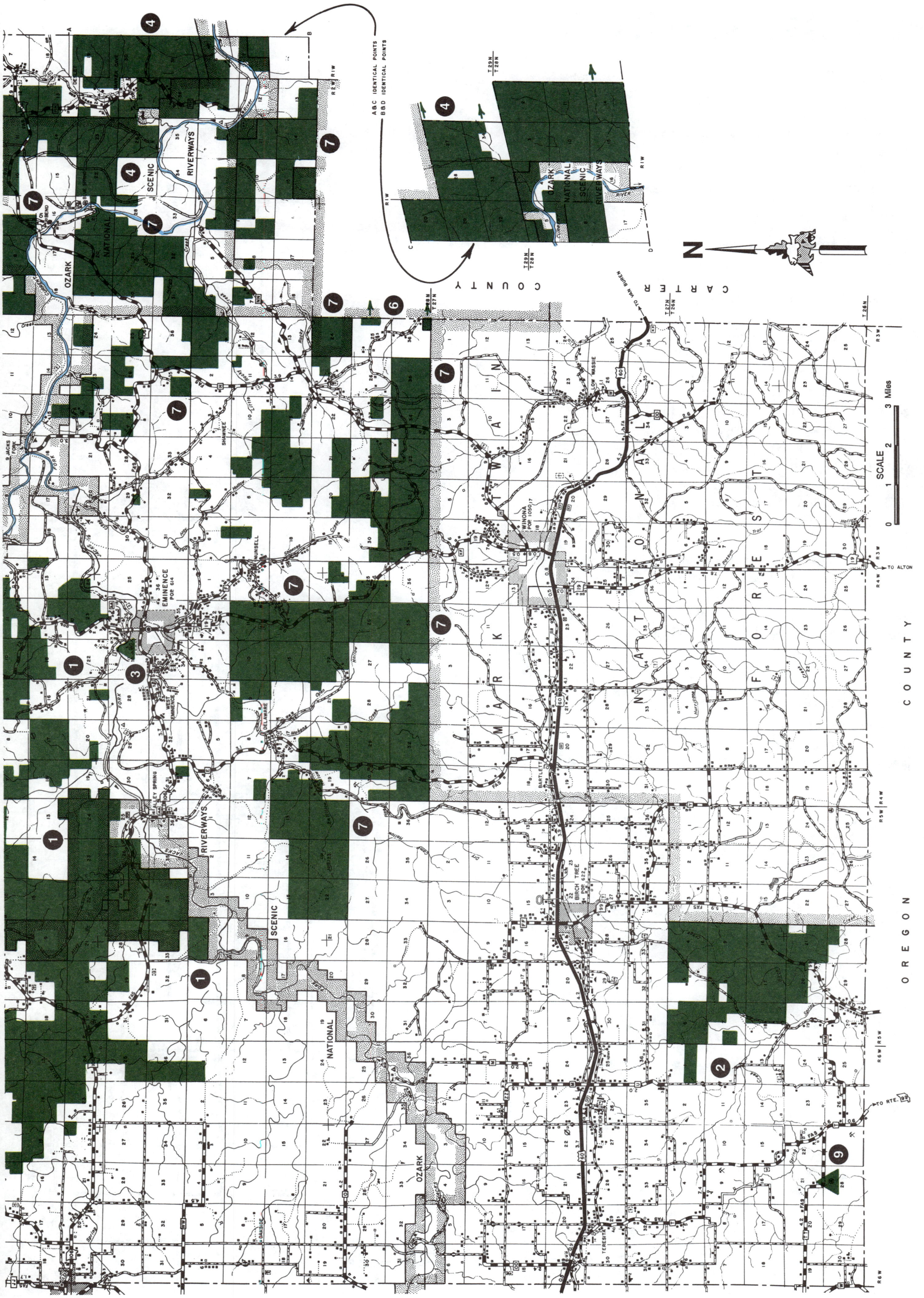
A&C IDENTICAL POINTS
B&D IDENTICAL POINTS
OZARK NATIONAL SCENIC RIVERWAYS
CARTER COUNTY
TO VAN BUREN
MARK TWAIN NATIONAL FOREST
WINONA POP. 1050
EMINENCE POP. 614
WEST EMINENCE
SHAWNEE
MUNSELL
DELAWARE
ALLEY SPRING
LOW WASSIE
BARTLETT
BIRCH TREE POP. 622
MONTIER
TERESITA
OREGON COUNTY
TO ALTON
TO RTE. 99
SCALE
0 1 2 3 Miles
N

SHELBY COUNTY

1 **Arrow-Wood Conservation Area** is 3 miles north of Shelbina on Highway 15 near the North Fork of the Salt River crossing. This 153-acre area is one of the few places in the state where southern arrow-wood grows. A 1-mile stretch of the North Fork of the Salt River flows through the area, and there is a boat ramp.

2 **Hunnewell Access** is 1 mile west of the town of Hunnewell near the Highway 36 crossing of the North Fork of the Salt River. This is an 18-acre area with no boat ramp.

3 **Hunnewell Lake Conservation Area** is 3 miles north of the town of Hunnewell on County Road Z. This 1,714-acre area is home to Hunnewell Lake (228 acres), 75 acres of mature white oak forest, and 16 acres of warm water hatchery ponds (tours available on request). The hatchery consists of 24 ponds and provides visitors with opportunities to view migratory birds and a variety of amphibian and fish-rearing demonstrations.

4 **Mound View Access** is 3 miles east of Shelbina. Take Gravel Road 462 east 2.5 miles, go 1 mile north on Gravel Road 453, then go .5 mile east on Gravel Road 470. This is 41 acres on the North Fork of the Salt River.

5 **Pin Oak Conservation Area** is 6 miles west of Shelbina on Highway 36, and 4 miles north of Lentner on Highway N. This is a 65-acre area composed mainly of forest-land (50 acres) and wetlands (10 acres).

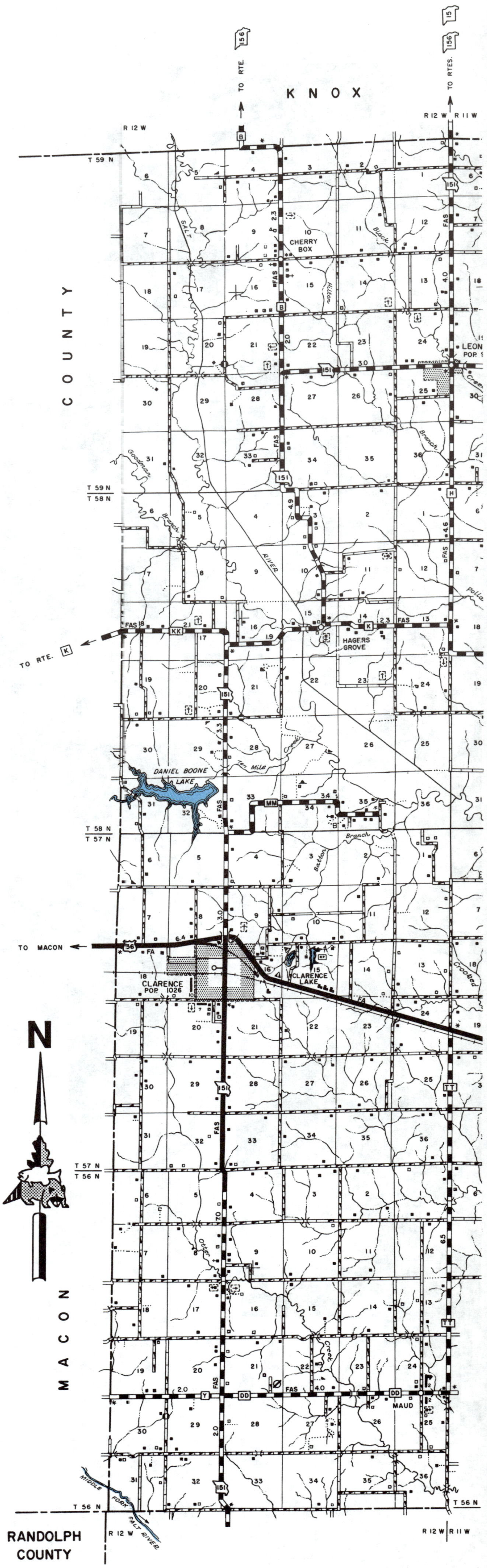

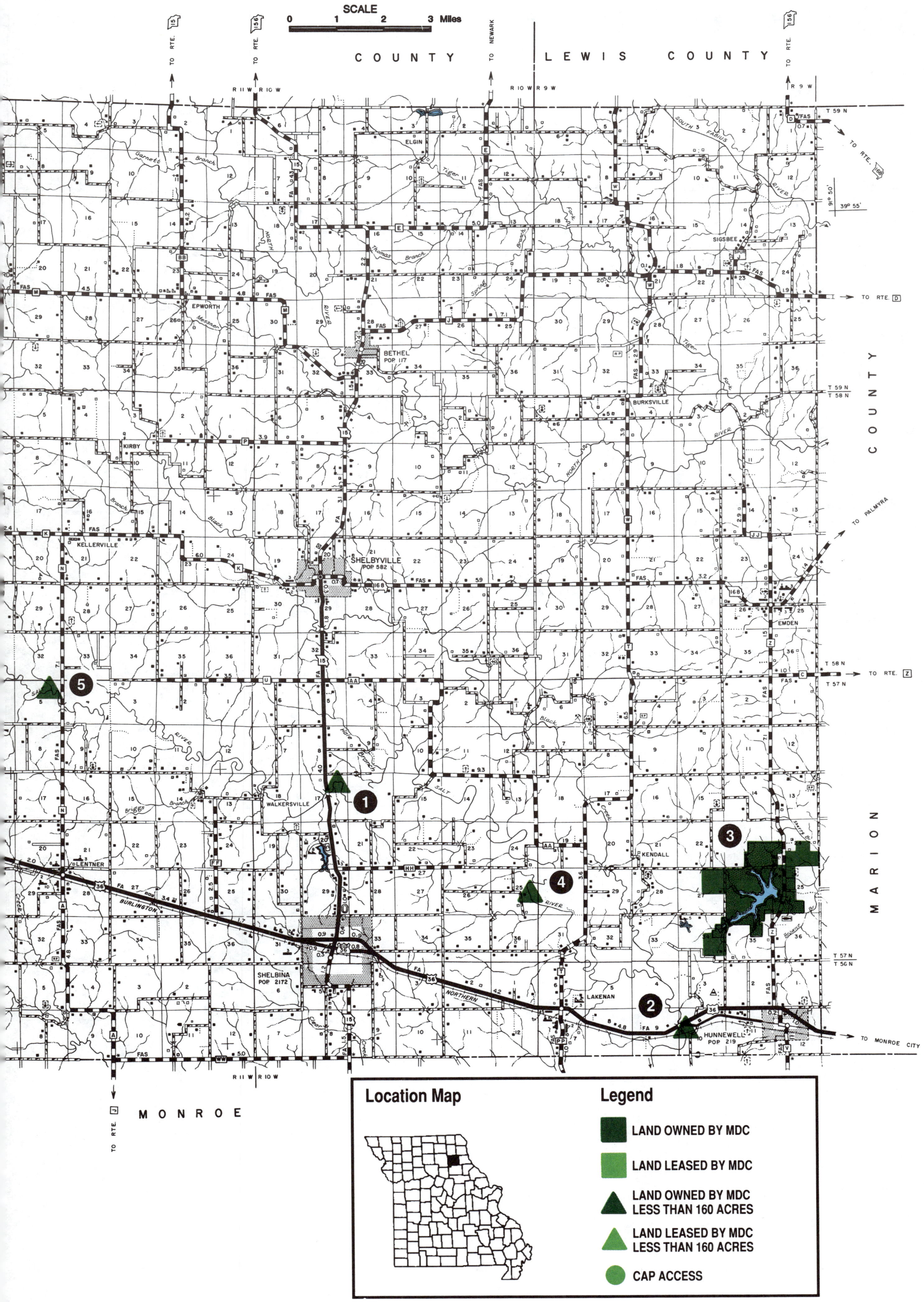
SCALE
0 1 2 3 Miles
TO RTE. 15
TO RTE. 156
TO NEWARK
TO RTE. 156
COUNTY
LEWIS COUNTY
R 11 W R 10 W
R 10 W R 9 W
R 9 W
T 59 N
TO RTE. 168
91° 50'
39° 55'
SOUTH FABIUS RIVER
ELGIN
Garnett Branch
Tiger Fork
NORTH
Thomas Branch
Spring Branch
SIGSBEE
EPWORTH
Mesner
TO RTE. D
BETHEL POP 117
Tiger Fork
BURKSVILLE
T 59 N
T 58 N
COUNTY
KIRBY
RIVER
NORTH
Black Branch
TO PALMYRA
KELLERVILLE
SHELBYVILLE POP 582
EMDEN
T 58 N
T 57 N
TO RTE. Z
SALT
RIVER
Hays Branch
SALT
Black
Briggs Branch
WALKERSVILLE
MARION
LENTNER
KENDALL
RIVER
BURLINGTON
T 57 N
T 56 N
SHELBINA POP 2172
NORTHERN
LAKENAN
HUNNEWELL POP 219
TO MONROE CITY
R 11 W R 10 W
TO RTE. J
MONROE
Location Map
Legend
LAND OWNED BY MDC
LAND LEASED BY MDC
LAND OWNED BY MDC LESS THAN 160 ACRES
LAND LEASED BY MDC LESS THAN 160 ACRES
CAP ACCESS

STODDARD COUNTY

1 **Aquilla Access** is 1 mile west of Aquilla on Highway M, then north on County Road 223 to the bridge over the Castor River. This is a 92-acre area along a .5-mile stretch of the river with no boat ramp.

2 **Crowley's Ridge Conservation Area** is 10 miles east of Puxico on Highway PP. This 1,878-acre area contains six ponds and Hall's Lake (50 acres of fishable water). The land around the lake is composed of old pastures (978 acres), forestland (470 acres) and cropland (377 acres).

3 **Dexter City Lake** is 1 mile south of Highway 60 on One Mile Road in Dexter. This is 11 acres of fishable water with no boat ramp.

4 **Duck Creek Conservation Area**–see the Bollinger County map, page 22.

5 **Holly Ridge Conservation Area** is east of Bloomfield on Highway E, then 2 miles south on County Road 517. This 1,000-acre area is home to Holly Ridge and Beech Springs natural areas, which feature sand forests and acid seeps and springs. There is an archery range and a 4.5-mile horse and hiking trail.

6 **Oak Ridge Conservation Area** is 1 mile west of Dudley on Old Highway 60. This is a 243-acre forested area.

7 **Otter Slough Conservation Area**–see the Butler County map, page 28.

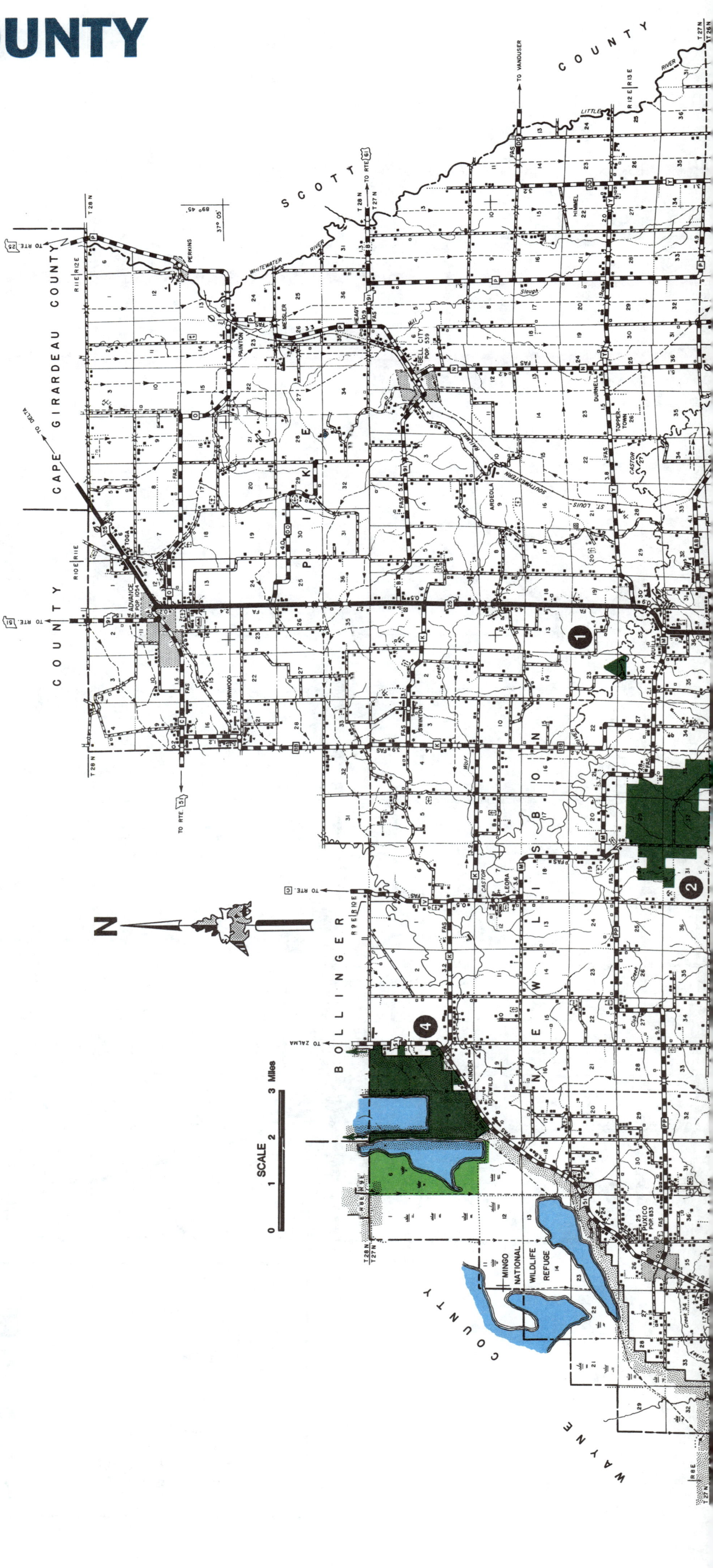

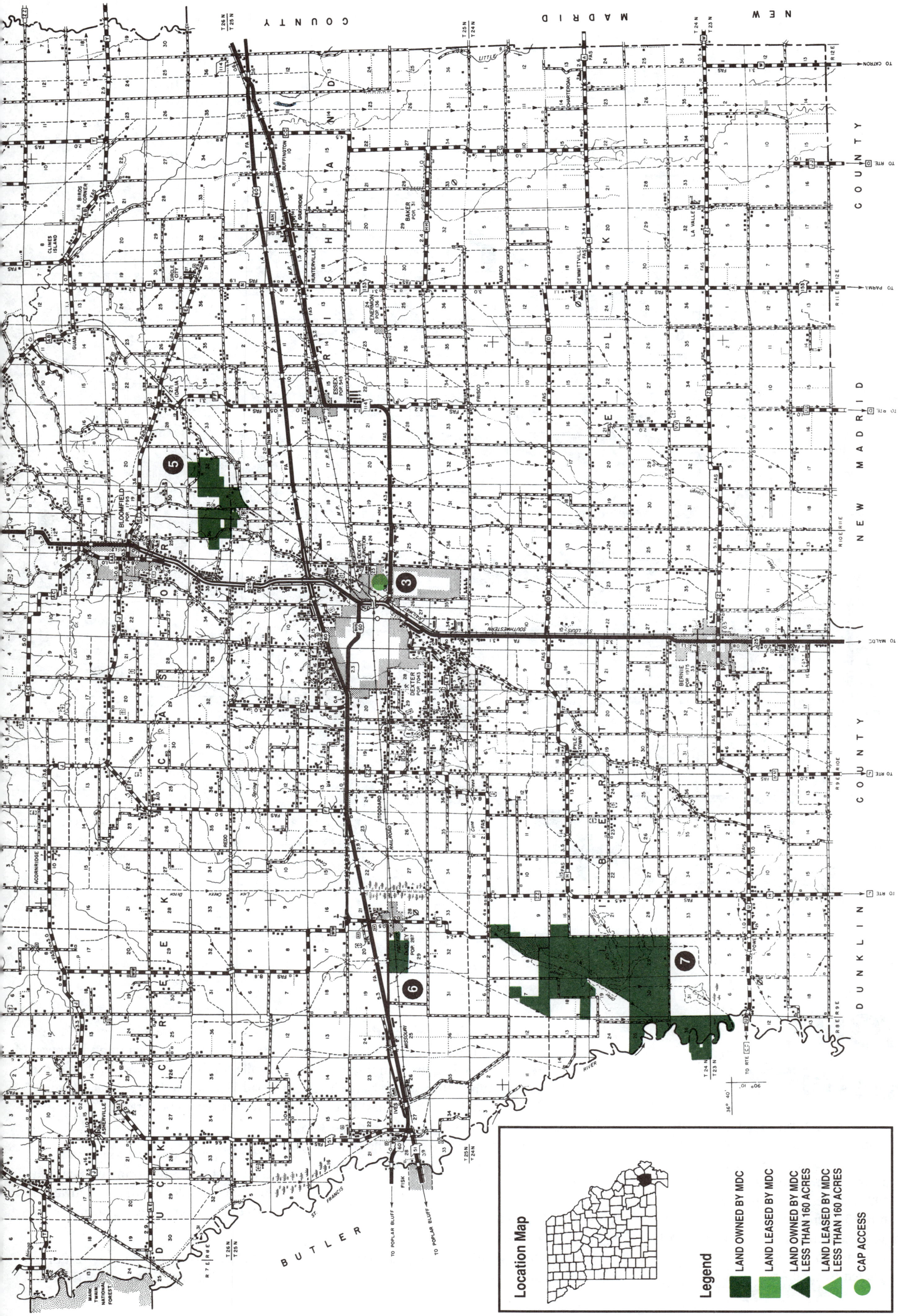
Location Map
Legend
LAND OWNED BY MDC
LAND LEASED BY MDC
LAND OWNED BY MDC LESS THAN 160 ACRES
LAND LEASED BY MDC LESS THAN 160 ACRES
CAP ACCESS
NEW MADRID COUNTY
DUNKLIN COUNTY
BUTLER
DEXTER POP 7043
BLOOMFIELD POP 1795
BERNIE POP 1975
ESSEX POP 545
TO POPLAR BLUFF
TO MALDEN
TO PARMA
TO CATRON

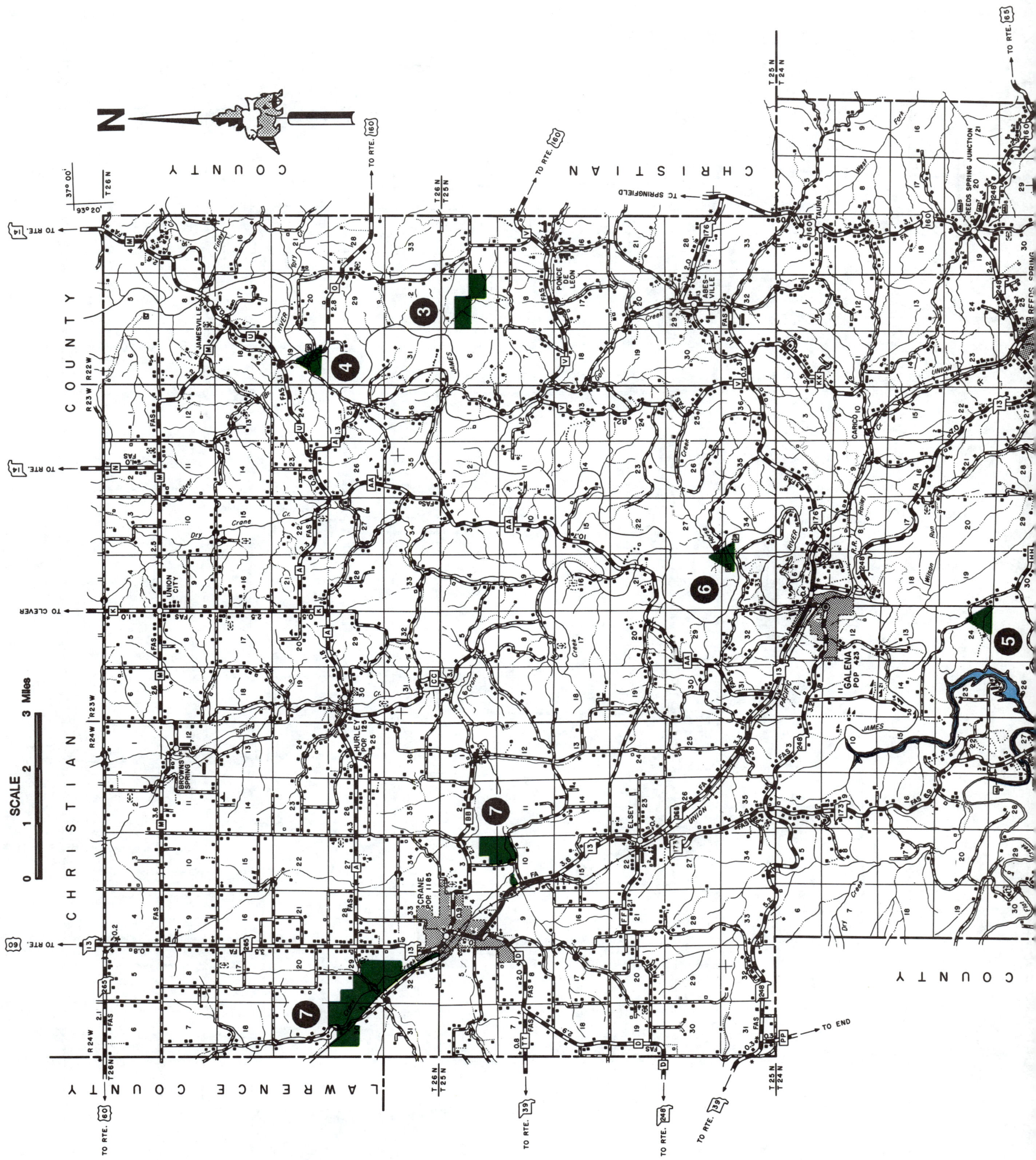

STONE COUNTY

1 **Ashe Juniper Natural Area** is south of Branson on Highway 65, 7 miles west on Highway 86, and 4 miles north on Highway JJ to an unnamed gravel road. Go 100 yards left on the gravel road to reach the 31-acre area. This area is an almost pure stand of ashe juniper (Ozark white cedar) on a glade and is a Designated Natural Area.

2 **Ruth and Paul Henning Conservation Area** is west of Branson on Highway 76. Roark Creek flows .5 mile through this 1,534-acre area. White River Balds Natural Area (dolomite and limestone glades) is a 362-acre Designated Natural Area and can be explored along a 1.75-mile hiking trail. There is a naturalist on duty April through October.

3 **Holland State Forest** is 1 mile north of Ponce de Leon on Road V-20, then .5 mile west on Road V-20C. This is a 160-acre forested area.

4 **Hooten Town Access** is west of Nixa on Highway 14, 7 miles south on Highway M, 1 mile south on Highway U, then east on Road V-20 to the James River. This is a 5-acre area with a boat ramp.

5 **Jessie Hollow Conservation Area** is 3 miles west of Reeds Spring on Highway 76, then 2.5 miles north on Highway HH. This is a 120-acre forested area.

6 **H. L. Kerr Access** is 5.5 miles west of Highway 160 on Highway 176, then 2

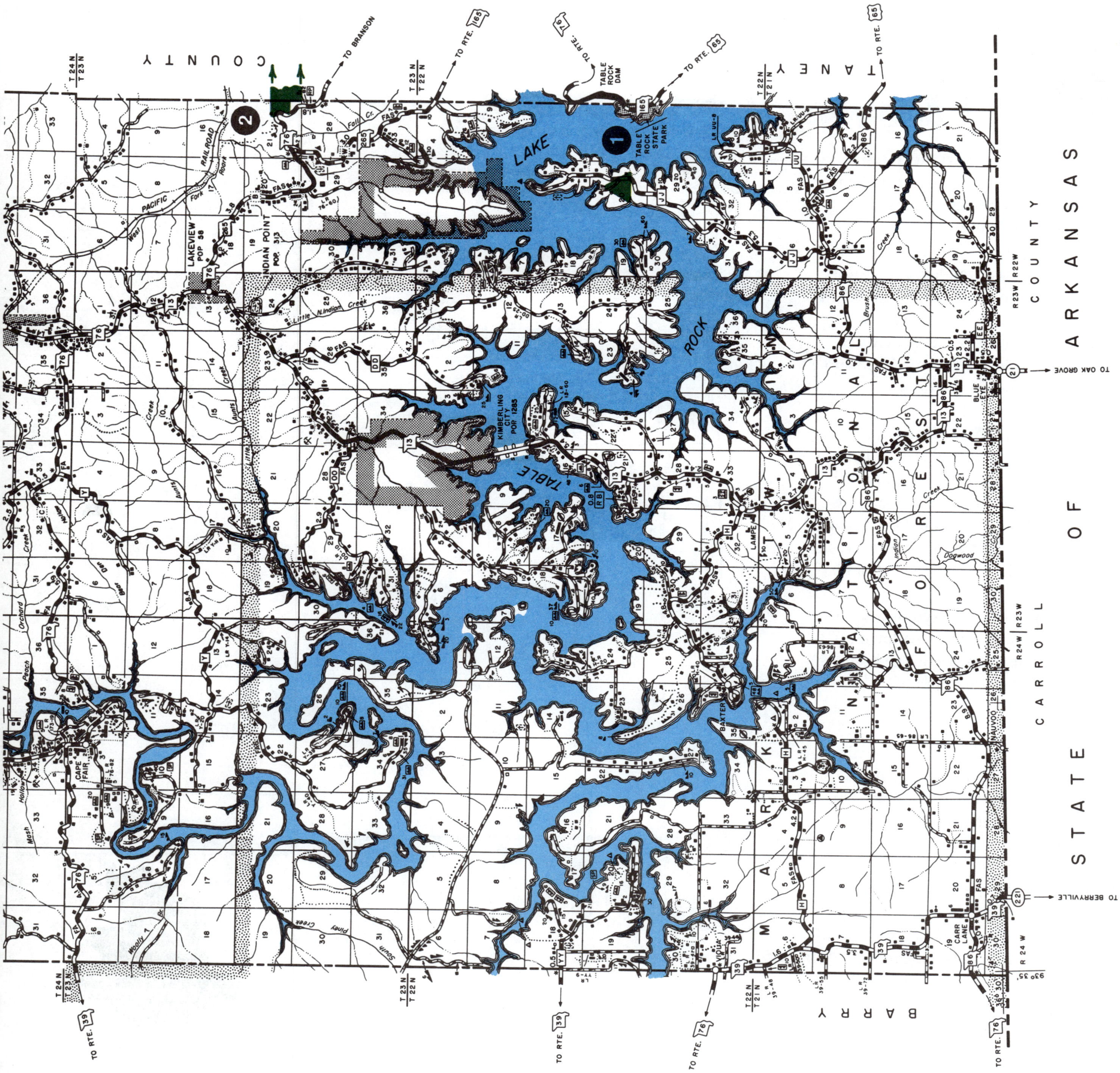

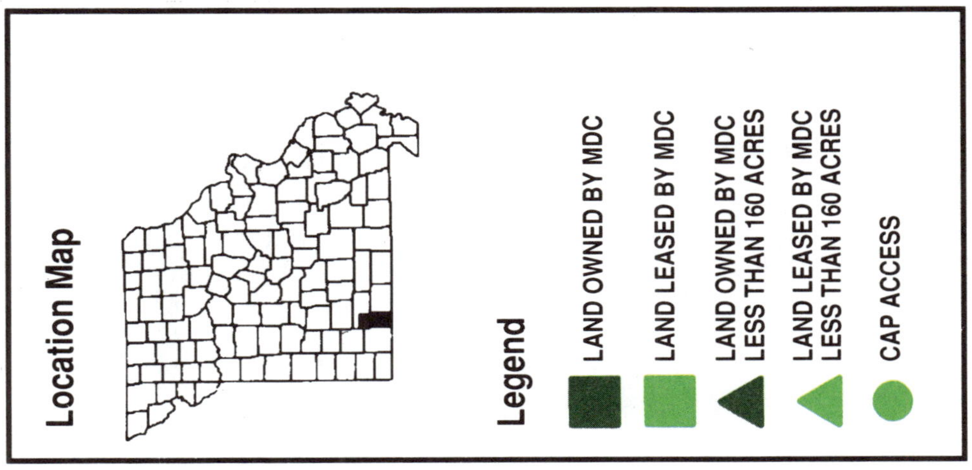

miles north on Horse Creek Road (176-40). This is 39 acres on the James River with a boat ramp.

7 **Wire Road Conservation Area** consists of two tracts. The main tract is north of Crane on Highway 13, then 1 mile west on Road 13-190. Crane Creek flows 3.2 miles through this 730-acre area and is open for catch-and-release trout fishing.

SULLIVAN COUNTY

1 To reach **Dark Hollow Natural Area,** go 2 miles north of Green City on Highway 129. When 129 curves left, continue straight (north) onto a county gravel road for about .75 mile. The road surface becomes very muddy, and visitors usually must hike the last .5 mile to the area. The entire 293-acre area is a Designated Natural Area made up of mesic upland forest, dry-mesic upland forest and remnant prairie.

2 **Elmwood Lake** is near Milan. From Highway 6 on the east edge of Milan, go north on Highway 5. After about a mile, turn west on the first county road just past the first bridge. This road ends in a T-intersection. Turn south (left) to reach the 180-acre lake. There is a boat ramp.

3 **Green City Lake** is marked by a cantilever sign 2 miles east of the town of Green City on Highway 6. The lake is 60 acres of fishable water and has a boat ramp.

4 **Henry's Mill Access** is 5 miles north of Milan on Highway 5, then 2.5 miles west on an unnamed gravel road. This is a 34-acre area with no boat ramp on Locust Creek.

5 **Locust Creek Conservation Area** is 2 miles southwest of Milan on Highway 5. One of the few unchanneled sections of Locust Creek (5.7 miles) meanders through this 3,162-acre area.

6 **Rocky Ford Access** is 3 miles north of Browning off Highway 5. This is 4 acres on Locust Creek with no boat ramp.

7 **Sears Community Lake** is 3 miles northeast of Milan on Highway 6, north on Highway Y, then down Access Route RA. The lake is 19 acres and is surrounded mainly by old pastures (55 acres) and forestland (9 acres). There is no boat ramp.

8 **Union Ridge Conservation Area**–see the Adair County map, page 6.

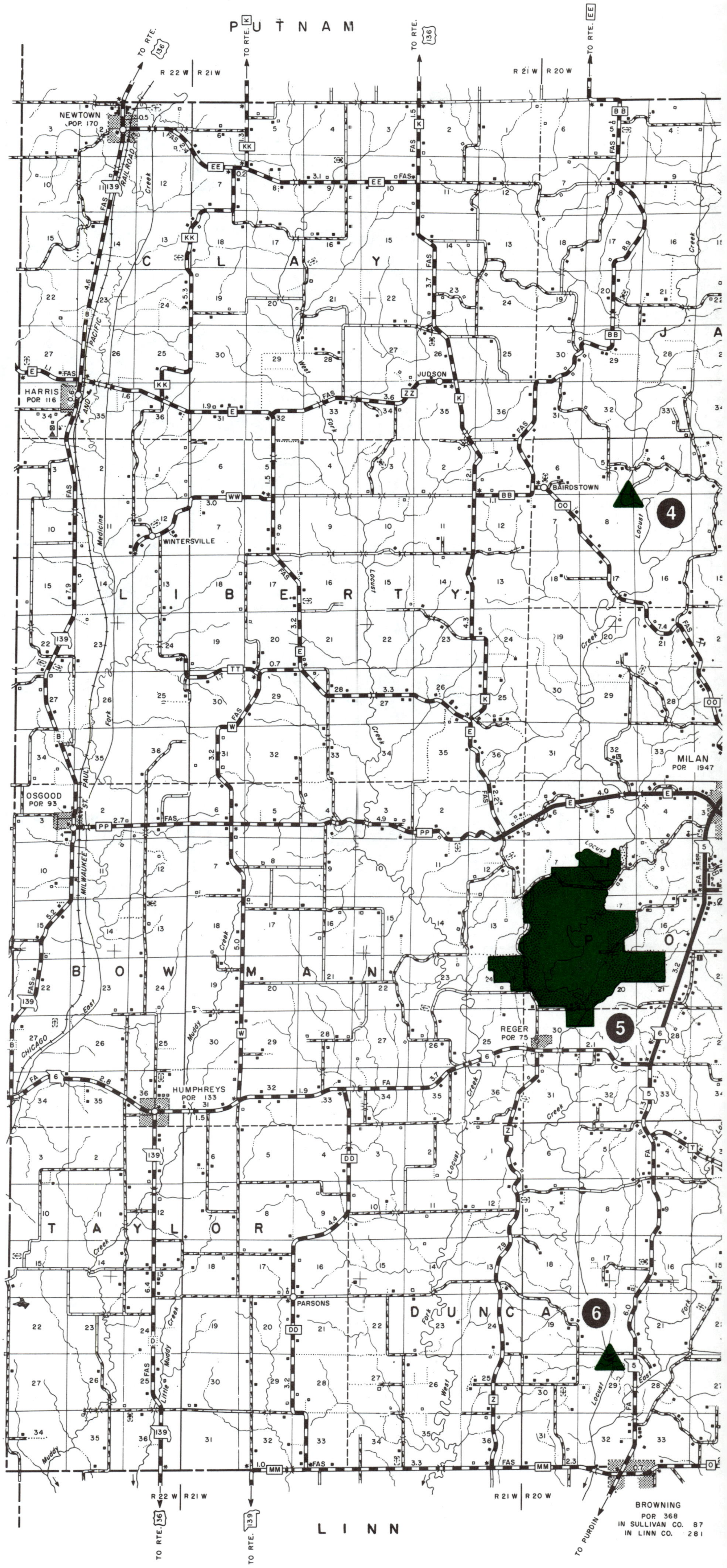

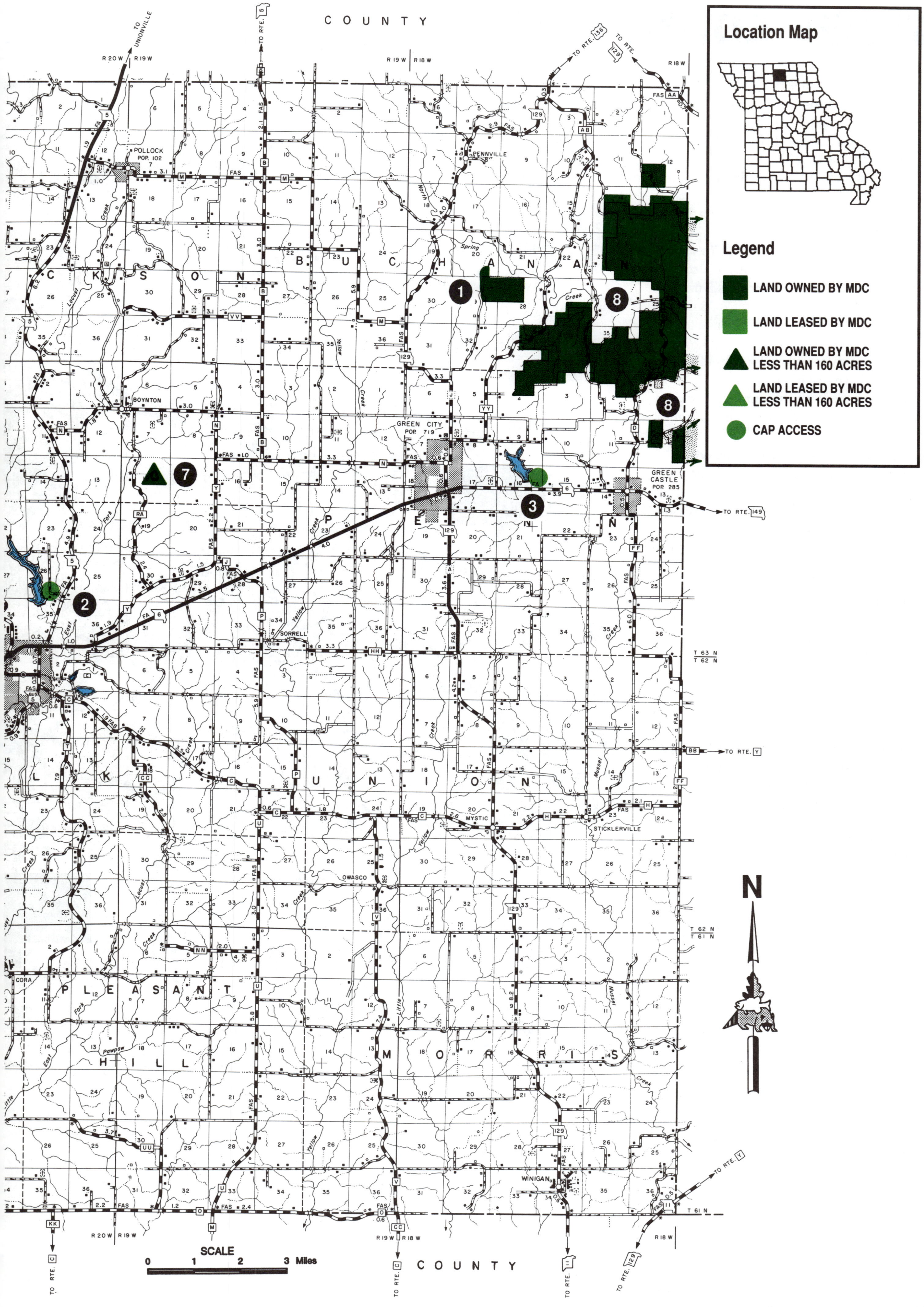
COUNTY
Location Map
Legend
LAND OWNED BY MDC
LAND LEASED BY MDC
LAND OWNED BY MDC LESS THAN 160 ACRES
LAND LEASED BY MDC LESS THAN 160 ACRES
CAP ACCESS
TO UNIONVILLE
TO RTE. 5
TO RTE. 136
TO RTE. 129
POLLOCK POP. 102
PENNVILLE
BOYNTON
GREEN CITY POP. 719
GREEN CASTLE POP. 285
TO RTE. 149
SORRELL
TO RTE. Y
MYSTIC
STICKLERVILLE
OWASCO
CORA
WINIGAN
BUCHANAN
JACKSON
PEPEN
UNION
PLEASANT HILL
MORRIS
T 63 N
T 62 N
T 61 N
R 20 W
R 19 W
R 18 W
SCALE
0 1 2 3 Miles
N
COUNTY
TO RTE. C
TO RTE. 11
TO RTE. 129

TANEY COUNTY

1 **Boston Ferry Conservation Area** is 3 miles north of Branson on Highway 65, 1 mile east on Road 65-70, then 1 mile north (left) on Road 65-70B. This 180-acre area is adjacent to Lake Taneycomo. There are picnic areas and a 1-mile hiking trail.

2 **Bull Shoals Reservoir**–see the Ozark County map, page 162.

3 **Drury-Mincy Conservation Area** consists of two tracts. The main tract is 5 miles east of Branson on Highway 76, 6 miles south on Highway J, then .75 mile to the right on County Road J40. This 5,699-acre area (mainly forestland) borders Bull Shoals Lake and has many acres (1,200) of dolomite glade and savannah, one cave, a sinkhole pond and many stands of cane.

4 **Ruth and Paul Henning Conservation Area**–see the Stone County map, page 216.

5 **Hilltop Towersite** is north of Forsyth on Highway 160, then 4.5 miles north on Highway H. The tower and 3-acre area are on the east side of the road.

6 **Hollister Towersite** is 6 miles south of Branson on Highway 65, .75 mile west on Highway 265, then .25 mile south on Road 265-30. This is 180 acres of mainly forestland (160 acres) with two small ponds and a sinkhole. There is also a picnic area.

7 **Shepherd of the Hills Fish Hatchery** is 6 miles southwest of Branson off Highway 165 near Table Rock Dam. This 211-acre area is adjacent to Lake Taneycomo. Exhibits and tours of the fish hatchery are available.

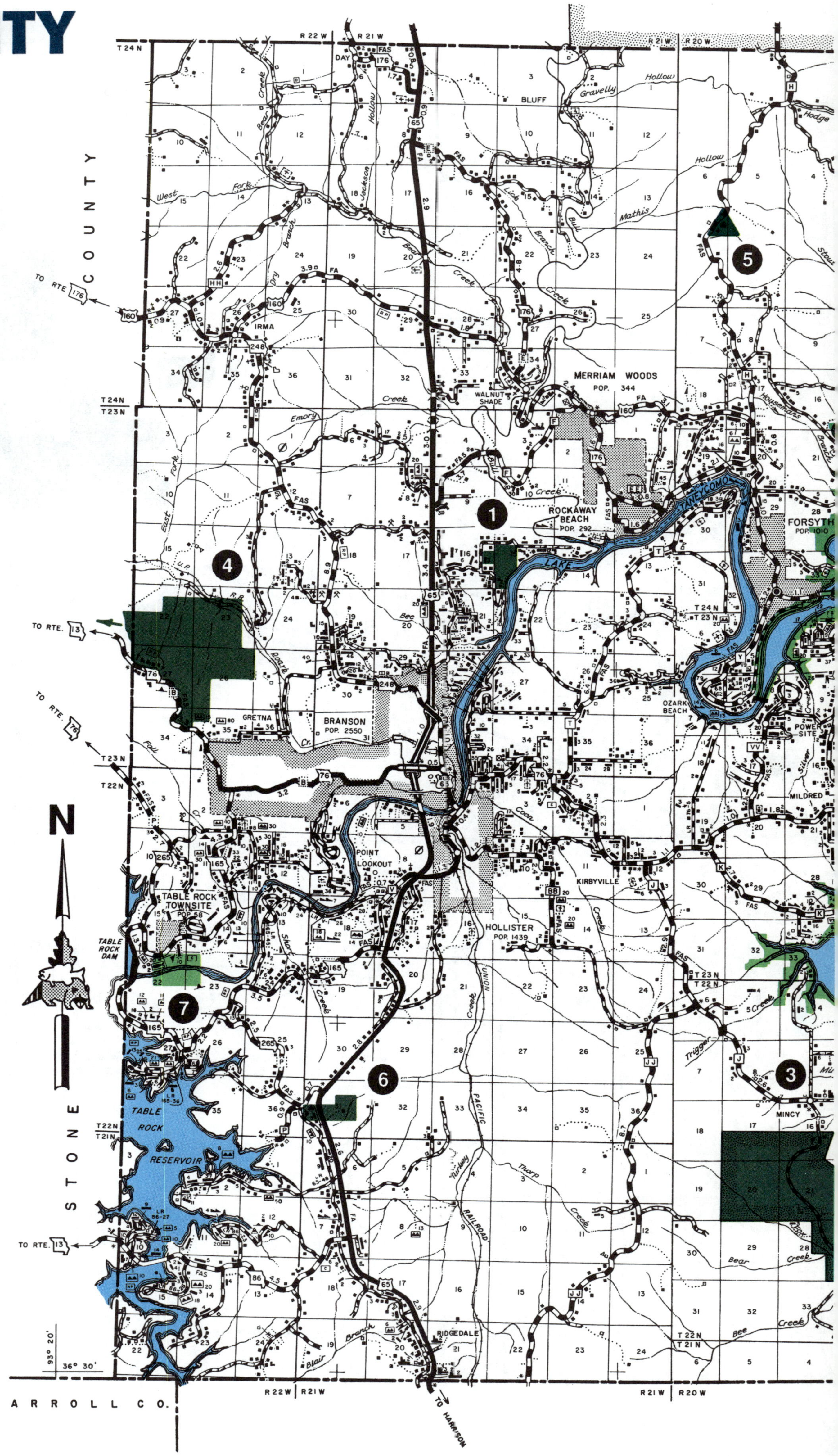

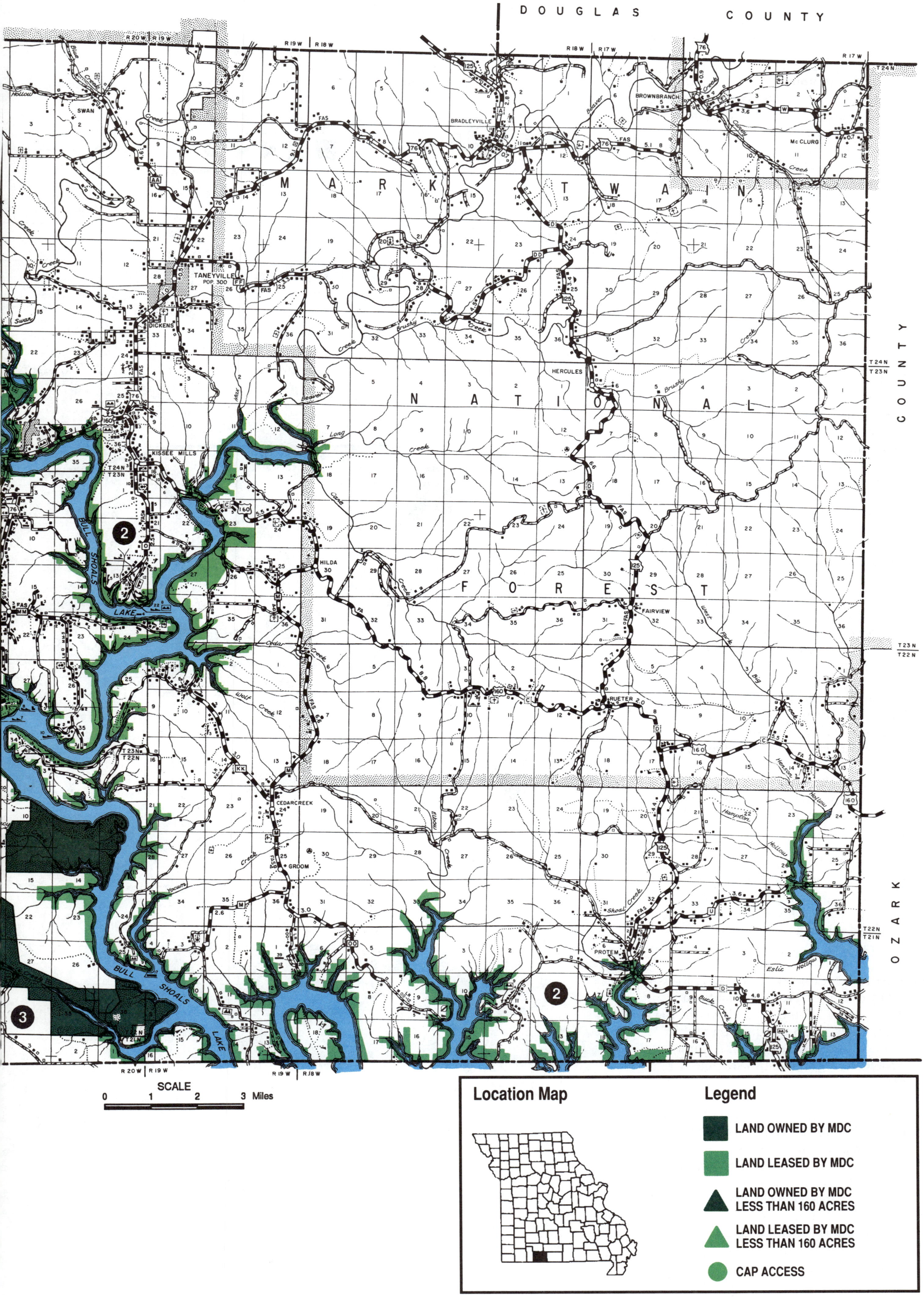
DOUGLAS COUNTY
OZARK COUNTY
MARK TWAIN
NATIONAL
FOREST
SWAN
TANEYVILLE
POP. 300
DICKENS
KISSEE MILLS
BRADLEYVILLE
BROWNBRANCH
Mc CLURG
HERCULES
HILDA
FAIRVIEW
RUETER
CEDARCREEK
GROOM
PROTEM
BULL SHOALS LAKE
T24N
T23N
T22N
T21N
R 20 W
R 19 W
R 18 W
R 17 W
SCALE
0
1
2
3 Miles
Location Map
Legend
LAND OWNED BY MDC
LAND LEASED BY MDC
LAND OWNED BY MDC LESS THAN 160 ACRES
LAND LEASED BY MDC LESS THAN 160 ACRES
CAP ACCESS

TEXAS COUNTY

1 **Austin Community Lake** is 1.5 miles north of Highway 60 on Highway MM, then east on Greenwood Road. This 22-acre lake has a boat ramp and is surrounded by about 29 acres of forestland. There is a picnic area.

2 **Baptist Camp Access** is 6 miles south of Houston on Highway 63, and 1 mile west on Highway RA. This is a 5-acre access with no boat ramp on the Big Piney River. There is a picnic area.

3 **Barn Hollow Natural Area** is 1 mile north of Mountain View on Highway Y, right on County Road 353 for 1.25 miles, then right (east) on County Road 394 for 1 mile. This 252-acre area is on the left side of the road. There is a .5-mile hiking trail.

4 **P. F. Barnes Conservation Area** is west of Houston on Highway 17, 3.25 miles south on Highway 38, and 2 miles south on Barnes Road. This is 120 acres of mainly forestland.

5 **Lois Arlene Boesel Outdoor Education Area** (8.5 acres) is .25 mile west of Houston on Cleveland Road (off Highway 63 near the south edge of town), then .25 mile down Skyview Drive.

6 **Boiling Spring Access** is 7 miles west of Highway 63 on Highway BB. This is 9 acres on the Big Piney River. There is a boat ramp.

7 **Cabool Towersite** is 2 miles north of Cabool on Highway M. The tower sits on 15.5 acres. When the tower is not occupied, the access road is closed to vehicles.

8 **Dog's Bluff Access** is on Highway 17, 3 miles west of Highway 63. This is a 3.3-acre area on the Big Piney River with no boat ramp.

9 **Dripping Springs Natural Area** is accessible by river only. From Baptist Camp Access, it is about a 1-mile float down the Big Piney River. This is an 8.5-acre wet bluff and spring and is a Designated Natural Area.

10 **Peter A. Eck Conservation Area** is accessible by river only. From the Slabtown Access, it is a 1.5-mile float down the Big Piney River. A 256-acre portion of this 316-acre area is a Designated Natural Area featuring an old growth, pine-oak forest.

11 **Horseshoe Bend Natural Area** (215 acres) is 3.5 miles west of Highway 63 at the Houston city limits. Turn west on Oak Hill Drive, west on Forest Drive, north on Mineral Drive, then west on Horseshoe Drive. The area has a 69-acre Designated Natural Area. It contains Blankenship Cave, which is the largest of a series of small caves in the limestone bluffs above the river.

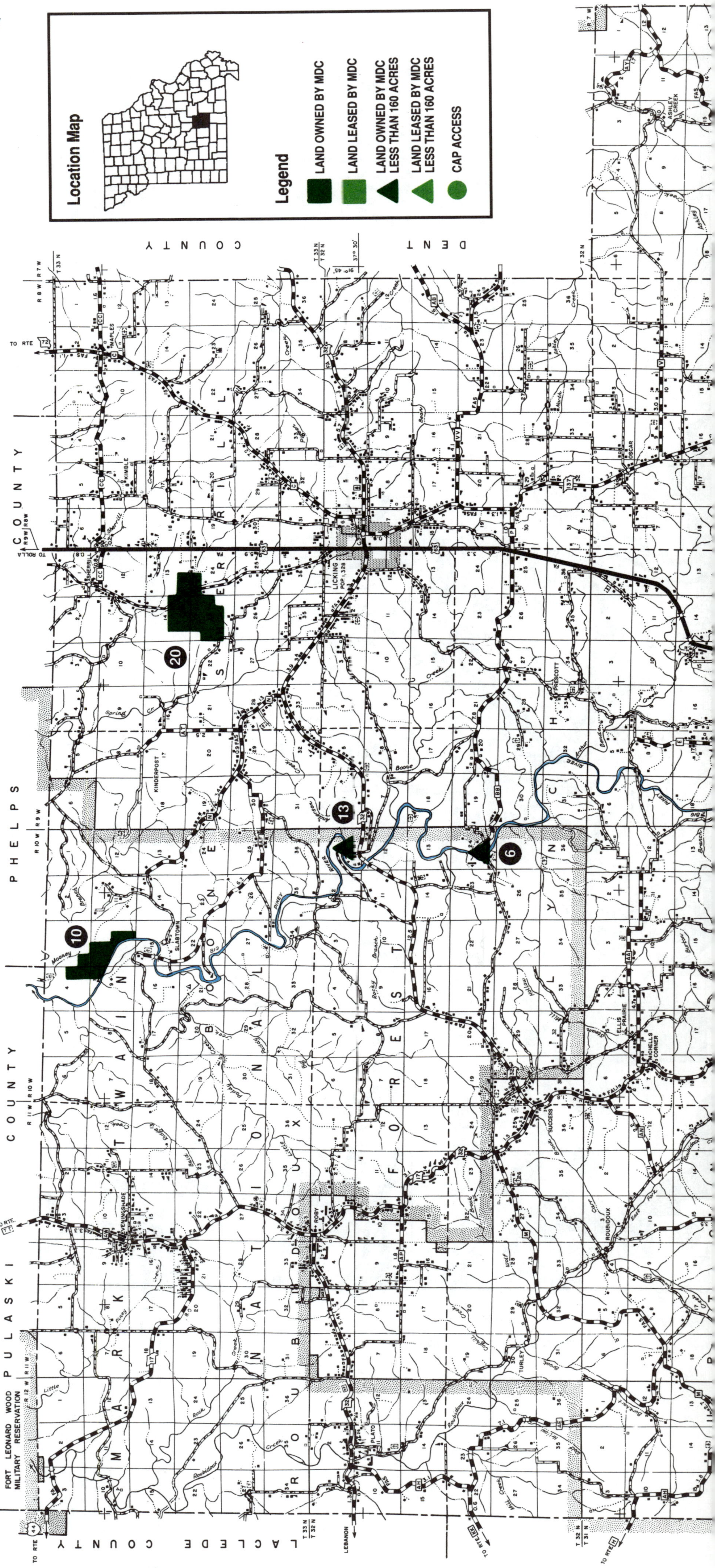

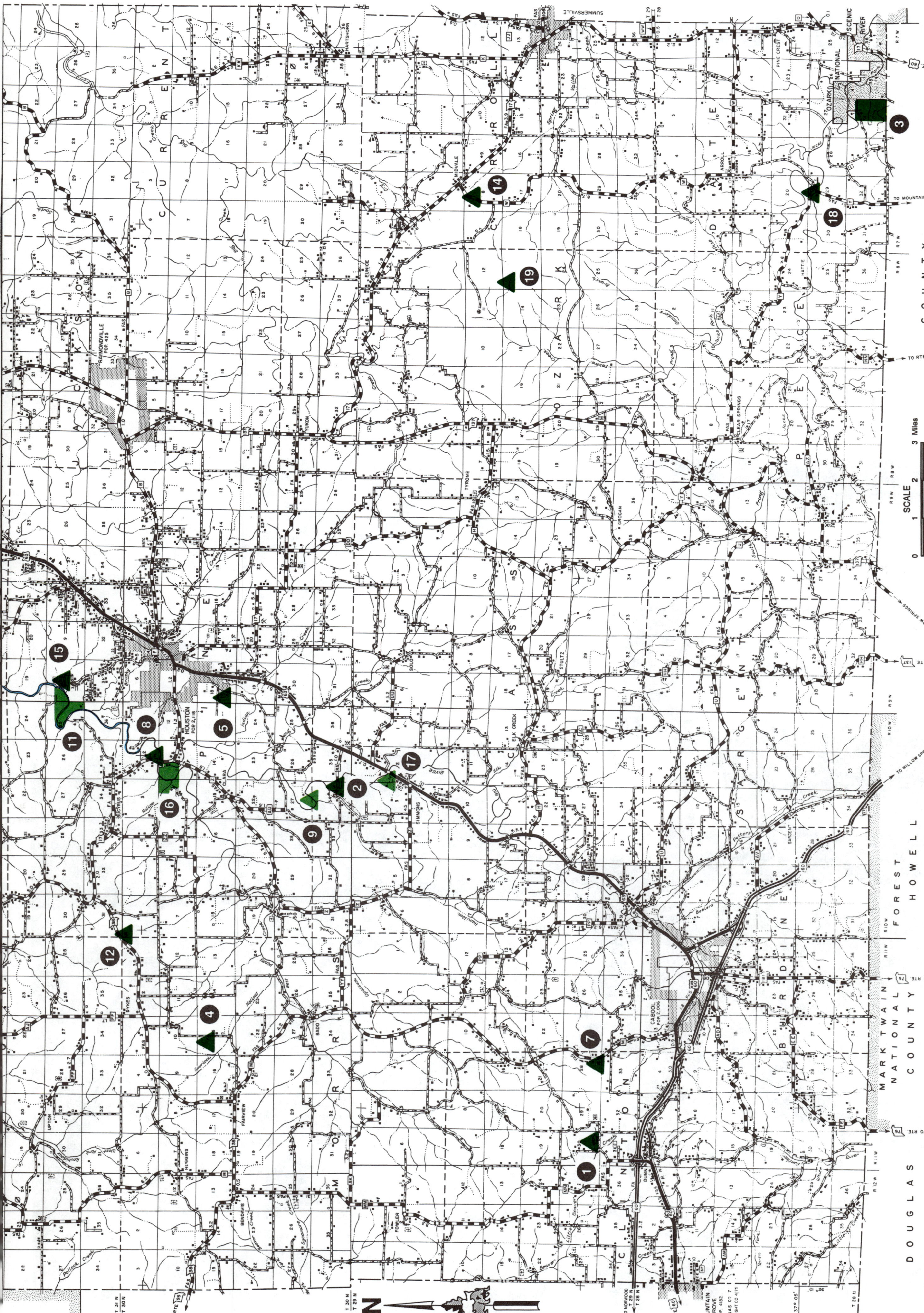
SUMMERSVILLE
OZARK NATIONAL SCENIC RIVER
TO MOUNTAIN VIEW
RAYMONDVILLE POP 425
HOUSTON POP 2,118
CABOOL POP 2,006
MOUNTAIN GROVE
TO WILLOW SPRINGS
SCALE
3 Miles
MARK TWAIN NATIONAL FOREST
HOWELL COUNTY
DOUGLAS COUNTY

TEXAS COUNTY continued...

12 **Houston Towersite** is 6 miles west of Houston on Highway 17 and 1 mile south on Highway 38. This is a 17-acre area.

13 **Mason Bridge Access** is 7 miles west of Licking on Highway 32, then north on Mason Road. This is a 7-acre area on the Big Piney River with a boat ramp.

14 **Midvale Conservation Area** is .25 mile south of Highway 17 on Highway W. This is an 80-acre forested area.

15 **Mineral Springs Access** is 3.5 miles west of Highway 63 at the north Houston city limits. Turn west on Oak Hill Drive, west on Forest Drive, then north on Mineral Drive. This is 6.65 acres on the Big Piney.

16 **Piney River Narrows Natural Area** is on Highway 17, 1.75 miles west of Houston. This 258-acre area contains a 50-acre Designated Natural Area composed of dolomite pinnacles along a ridge between the Big Piney River and Piney Creek.

17 **Simmons Ford Access** is 7.5 miles south of Houston on Highway 63. This is 2.75 acres on the Big Piney River with a gravel boat launching area.

18 **South Prong Access** is 6 miles north of Mountain View on Highway Y. This is 3.2 acres with no boat ramp on the South Prong of the Jack's Fork River.

19 **Summersville Towersite** is 1 mile south on Highway W (off Highway 17), 1.5 miles west on Wallen Road, then 1.75 miles west on Tower Road. This is a 26-acre area.

20 **George O. White State Forest Nursery** is north of Licking on U.S. 63, 1 mile west on Highway CC, and 2 miles south on Shafer Road. This 749-acre area has a 100-acre nursery and, with prior arrangements, it may be toured by visitors. It also contains Quercus Flatwoods Natural Area, which is made up of post oak flatwoods (48 acres). There is a picnic area and 3-acre lake suitable for fishing. See page 225.

Paul Childress

Mark Sullivan

Paul Childress

Paul Childress

George O. White State Forest Nursery

The George O. White State Forest Nursery has been producing trees since the mid-1930s, when the U.S. Forest Service acquired it to grow shortleaf pine for restoring the national forests. The Conservation Department took over the nursery in 1942 and named it after Missouri's first state forester, George O. White, on the occasion of his retirement in 1960.

One of White's lasting contributions to the state's forests was his conviction that private landowners would better protect the health of the land if they had an active stake in it. He promoted tree plantings to convince landowners to help in the prevention of forest fires—then the most destructive threat to Missouri's forests.

Today, the nursery produces between 5 and 6 million trees, which are distributed free or at cost to about 14,000 people. Fifty-five to 60 different species of trees are grown at the nursery. Most are distributed to individuals, but some are used on Department lands for projects such as streambank stabilization and reforestation.

Applications for tree and shrub planting stock are available in mid-November at Natural Resources Conservation Service and University of Missouri Extension offices or at Missouri Department of Conservation service centers.

VERNON COUNTY

1 **Balltown Access** is about 9 miles north of Nevada off Highway 71. Take the Horton Exit and travel 1.5 miles west and .5 mile north on an unnamed gravel road. This is a 194-acre area on the Little Osage River.

2 **Big Drywood Creek Conservation Area** is 6 miles west of Nevada on Highway 54. This is a 12.5-acre area on Big Drywood Creek.

3 **Bristow Conservation Area** (158 acres) is 3 miles east of the junction of highways 71 and E, then .5 mile south at Camp Creek on an unnamed gravel road.

4 **Bushwhacker Lake Conservation Area**–see the Barton County map, page 16.

5 To reach **Cephas Ford Access**, go 7 miles north of Nevada on Highway 71. Watch for a sign marking the area near a gravel road heading east. When the gravel road ends in a T-intersection, turn south and go .25 mile to the entrance. This is 107 acres with a boat ramp on the Marmaton River.

6 **Clear Creek Conservation Area**–see the Barton County map, page 16.

7 **Douglas Branch Conservation Area** is about 6 miles north of Nevada off Highway 71. Watch for the cantilever sign. This is a 511-acre area with prairie (195 acres), old pastures (188 acres) and forestland. It frequently floods in spring and fall.

8 **Flight Lake Conservation Area** (159 acres) is 7 miles north of Nevada on Highway 71, 2.5 miles west on Highway D, and 4 miles south on Highway O. Watch for the area sign on Highway O, and take the gravel road near the sign 1 mile south. Turn west on the next gravel road marked with a cantilever sign. Natural features include Flight Lake (a 42-acre marsh) and 54 acres of native prairie.

9 **Four Rivers Conservation Area**–see the Bates County map, page 18.

10 **Gama Grass Prairie Conservation Area** is about 12 miles north of Nevada off Highway 71. Watch for the highway sign on 71, and go 1 mile west on a gravel road to the parking lot. Fifty-three acres of this 80-acre area is native prairie with a high percentage of gama grass growing on it.

11 **Gay Feather Prairie Conservation Area** is 7.5 miles east of Highway 71 on Route E, then 2 miles south on an unnamed county road (watch for the area sign). Most of the area is prairie (111 acres), but there is a small stand of upland hardwoods (blackjack).

12 **Little Osage Prairie** is 6 miles south of Nevada on Highway 71, then 1.5 miles west on an unnamed gravel road (watch for the area sign). This is an 80-acre Designated Natural Area with a diversity of native plants.

13 **Osage Prairie Conservation Area** is 6 miles south of Nevada on Highway 71, 1.5 miles west on an unnamed gravel road (marked with a sign), then .5 mile south on another unnamed road. This 1,466-acre area conatains a 615-acre Designated Natural Area of native grass and timber sites.

14 **Schell-Osage Conservation Area**–see the St. Clair County map, page 194.

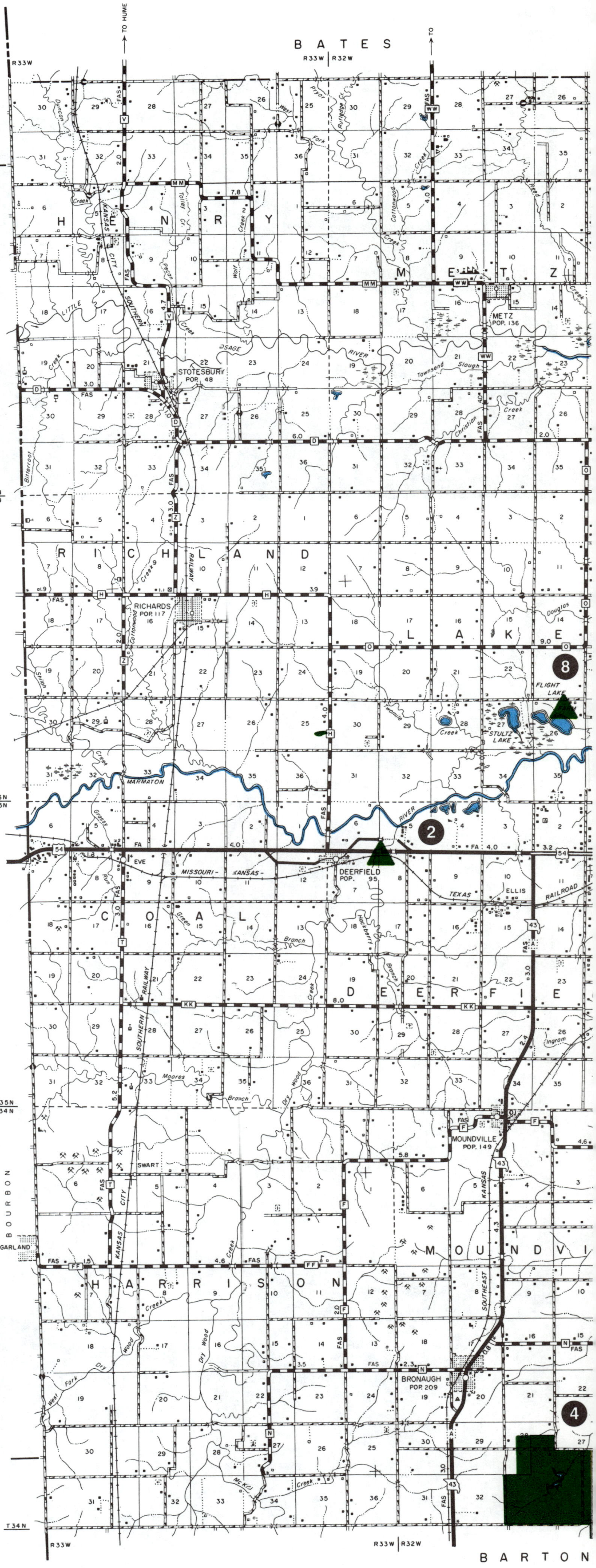

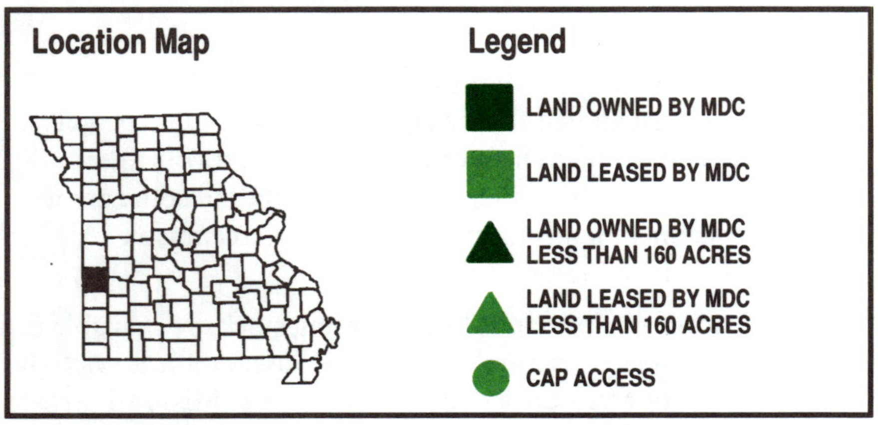

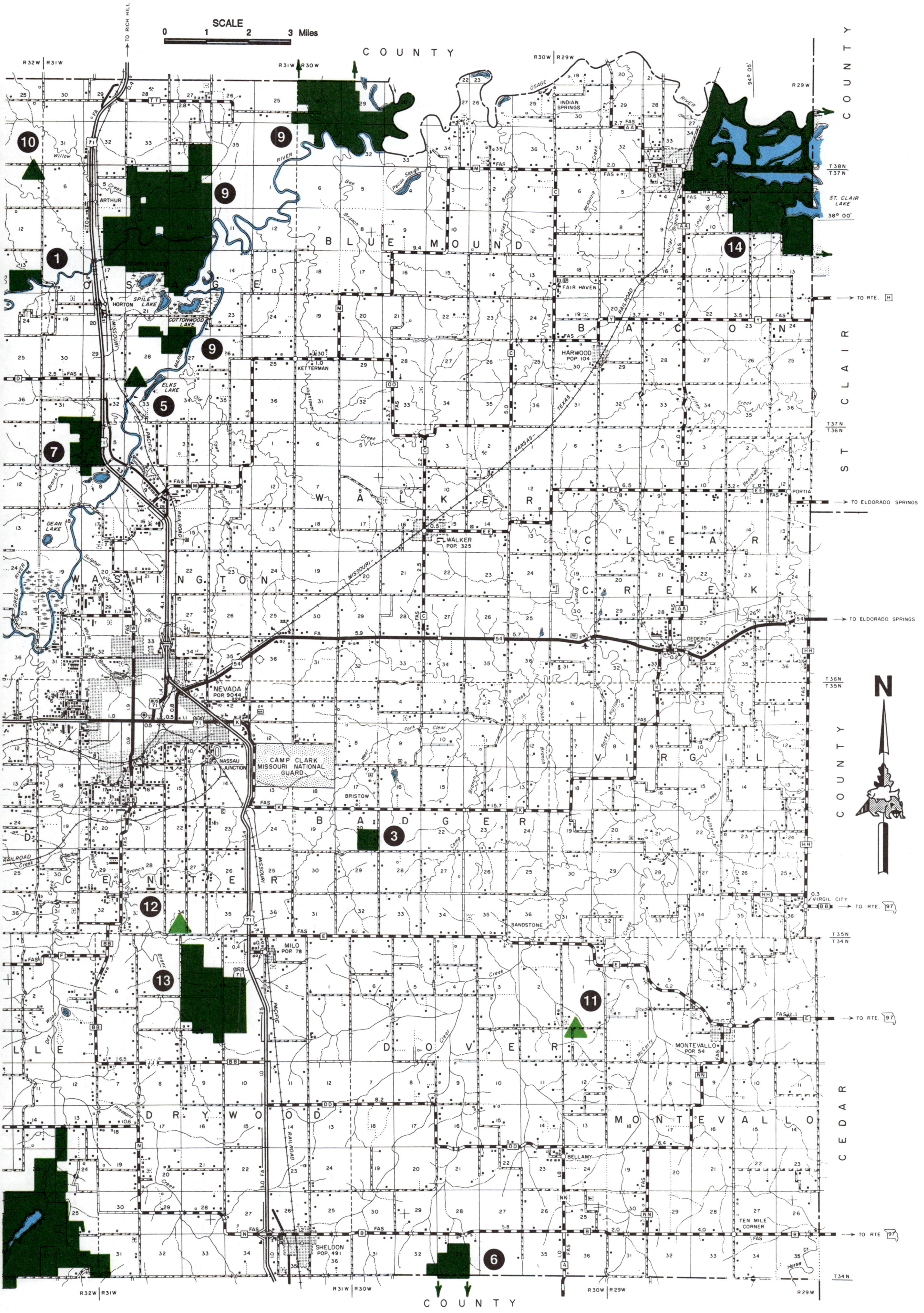
SCALE
0 1 2 3 Miles
TO RICH HILL
COUNTY
ST. CLAIR COUNTY
CEDAR COUNTY
N
OSAGE
BLUE MOUND
BACON
WALKER
CLEAR CREEK
WASHINGTON
VIRGIL
BADGER
CENTER
DOVER
DRYWOOD
MONTEVALLO
OSAGE RIVER
ST. CLAIR LAKE
INDIAN SPRINGS
ARTHUR
HORTON
SPILLE LAKE
COTTONWOOD LAKE
ELKS LAKE
DEAN LAKE
KETTERMAN
FAIR HAVEN
HARWOOD POP. 104
WALKER POP. 325
PORTIA
DEDERICK
NEVADA POP. 9044
NASSAU JUNCTION
CAMP CLARK MISSOURI NATIONAL GUARD
BRISTOW
SANDSTONE
MILO POP. 78
VIRGIL CITY
MONTEVALLO POP. 54
BELLAMY
TEN MILE CORNER
SHELDON POP. 491
MISSOURI-KANSAS-TEXAS RAILROAD
MISSOURI PACIFIC RAILROAD
TO RTE. H
TO ELDORADO SPRINGS
TO ELDORADO SPRINGS
TO RTE. 97
TO RTE. 97
TO RTE. 97
T38N
T37N
T36N
T35N
T34N
R32W
R31W
R30W
R29W
94° 05'
38° 00'

WARREN COUNTY

1 **Daniel Boone Conservation Area** is 6 miles west of Jonesburg on Highway Y. Near the Oak Grove Church, a cantilever sign directs visitors 3 miles southwest on Tower Road (a county gravel road). This 3,520-acre area is almost all forestland. Five ponds provide anglers with 3 acres of fishable water, and there are three picnic areas.

2 **Little Lost Creek Conservation Area** is 5 miles south of Pendelton on Highway B. This 2,899-acre area lies between Highway B on the west and Highway EE on the east. The area is nearly all forestland (2,698 acres), and Little Lost Creek has water in it seasonally.

3 **Frank Reifsnider State Forest** is 2 miles east of Warrenton on Highway M, and 3 miles south on Schuetzen Ground Road. This 1,465-acre area contains part of North Fork Charrette Creek, 22 acres of mature white oak forest (a Designated Natural Area), and numerous clay pits that were dug in the 1890s to obtain clay for brick-making. There are picnic areas and a .5-mile hiking trail.

4 **Warrenton Towersite** is 3 miles southwest of Warrenton on Highway U. This is a 10-acre forested area.

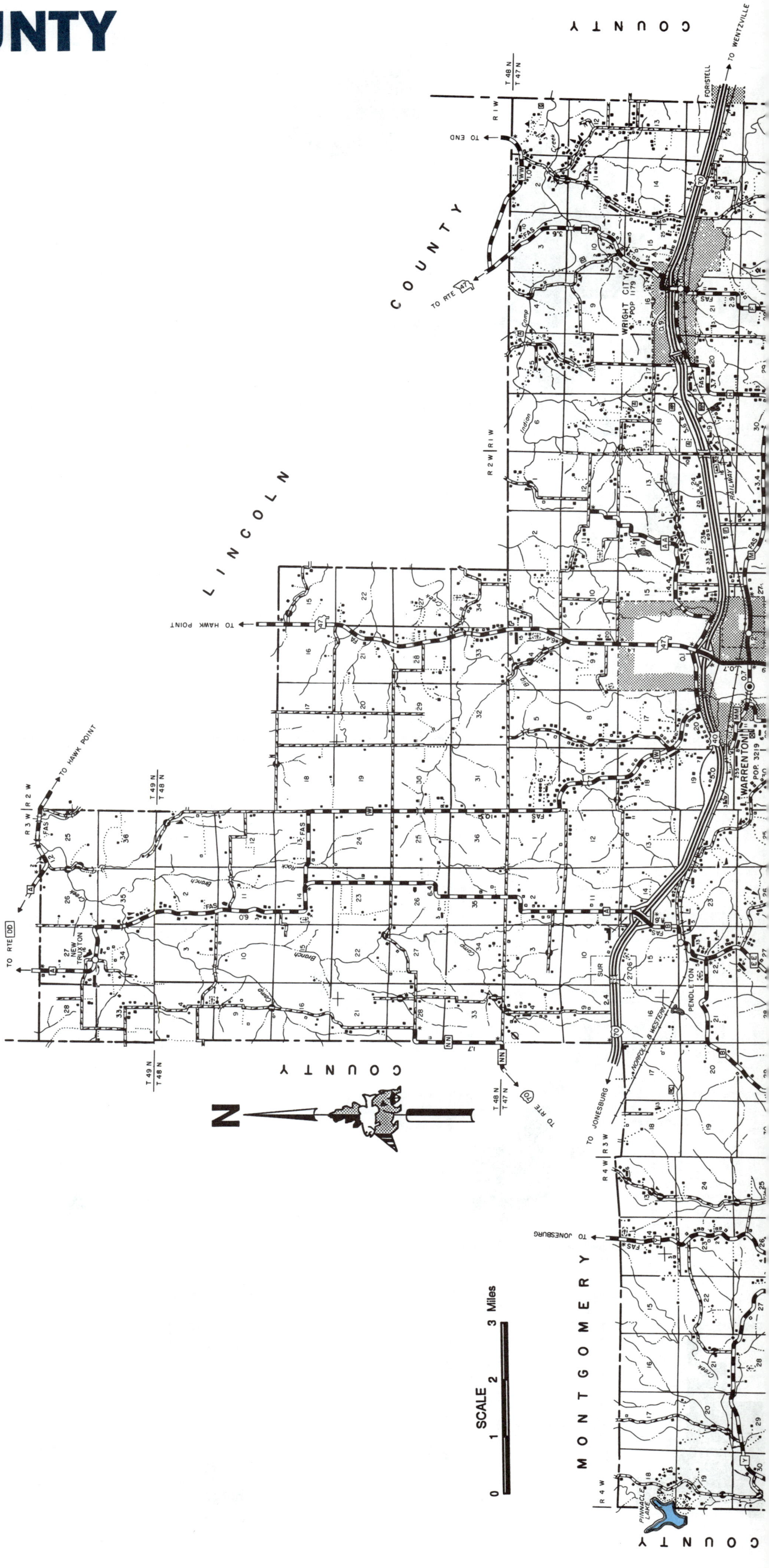

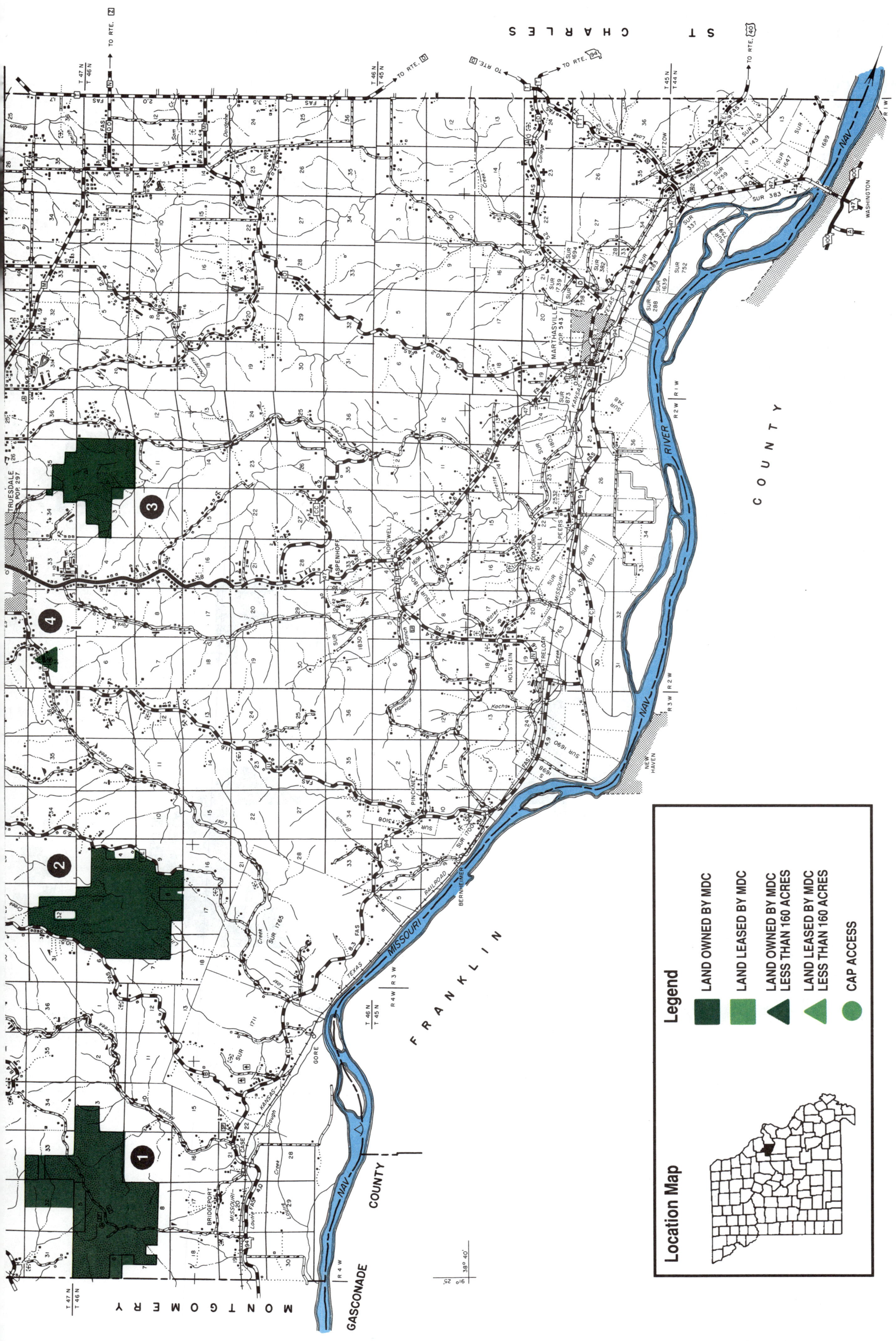
Legend
LAND OWNED BY MDC
LAND LEASED BY MDC
LAND OWNED BY MDC LESS THAN 160 ACRES
LAND LEASED BY MDC LESS THAN 160 ACRES
CAP ACCESS
Location Map
MONTGOMERY
ST CHARLES
FRANKLIN
COUNTY
GASCONADE COUNTY
MISSOURI RIVER
NAV
TRUESDALE POP 297
MARTHASVILLE POP 543
DUTZOW
WASHINGTON
NEW HAVEN
BERNHEIMER
PINCKNEY
HOPEWELL
ASPENHOFF
CONCORD HILL
PEERS
TRELOAR
HOLSTEIN
BRIDGEPORT
GORE
T 47 N
T 46 N
T 45 N
T 44 N
R 4 W
R 3 W
R 2 W
R 1 W
38° 40'
91° 25'

WASHINGTON COUNTY

1 **Bismarck Lake Conservation Area**–see the Iron County map, page 100.

2 **Bootleg Access** is marked with a cantilever sign 10 miles south of Potosi on Highway 21. This is 303 acres on the Big River with no boat ramp.

3 **Buford Mountain Conservation Area**–see the Iron County map, page 100.

4 **Hughes Mountain Natural Area** is 11 miles south of Potosi on Highway 21, 5 miles left on Highway M, then .25 mile right on Cedar Creek Road. This is a 462-acre area composed of forestland (241 acres) and glades (210 acres). Lost Creek and Cedar Creek flow through the area, and there is a 2-mile hiking trail.

5 **Kingston Access** is on County Road 317, 1 mile west of Washington State Park off Highway 21. This is 58 acres on Mineral Fork with no boat ramp.

6 **Little Indian Creek Conservation Area**–see the Franklin County map, page 76.

7 **Pea Ridge Conservation Area** consists of several tracts and is marked with a sign on Highway 185, between Meramec State Park and Potosi. This 8,269-acre area is mainly forestland (8,052 acres) and contains 4 miles of Indian Creek. The area has four caves and three springs.

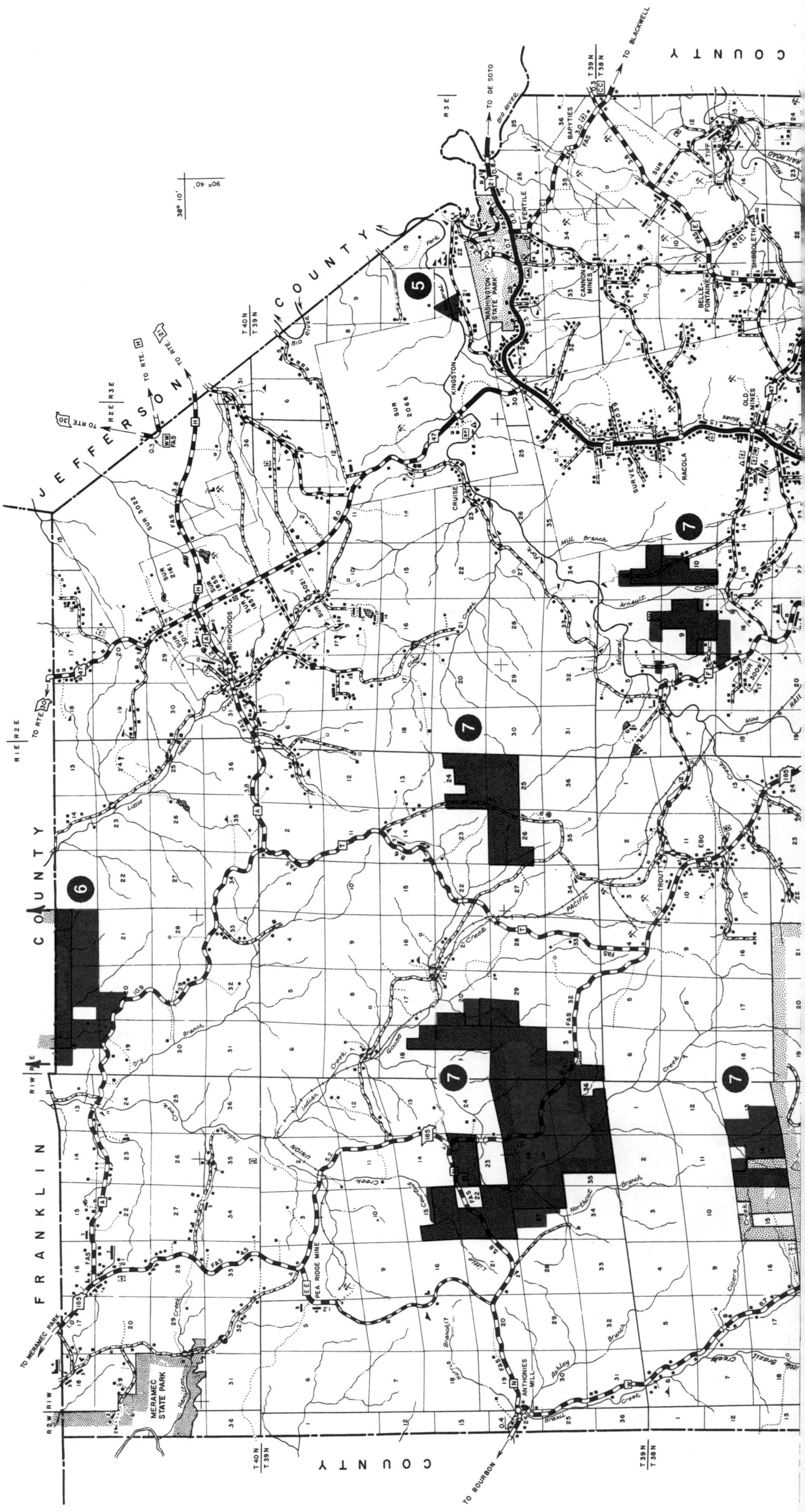

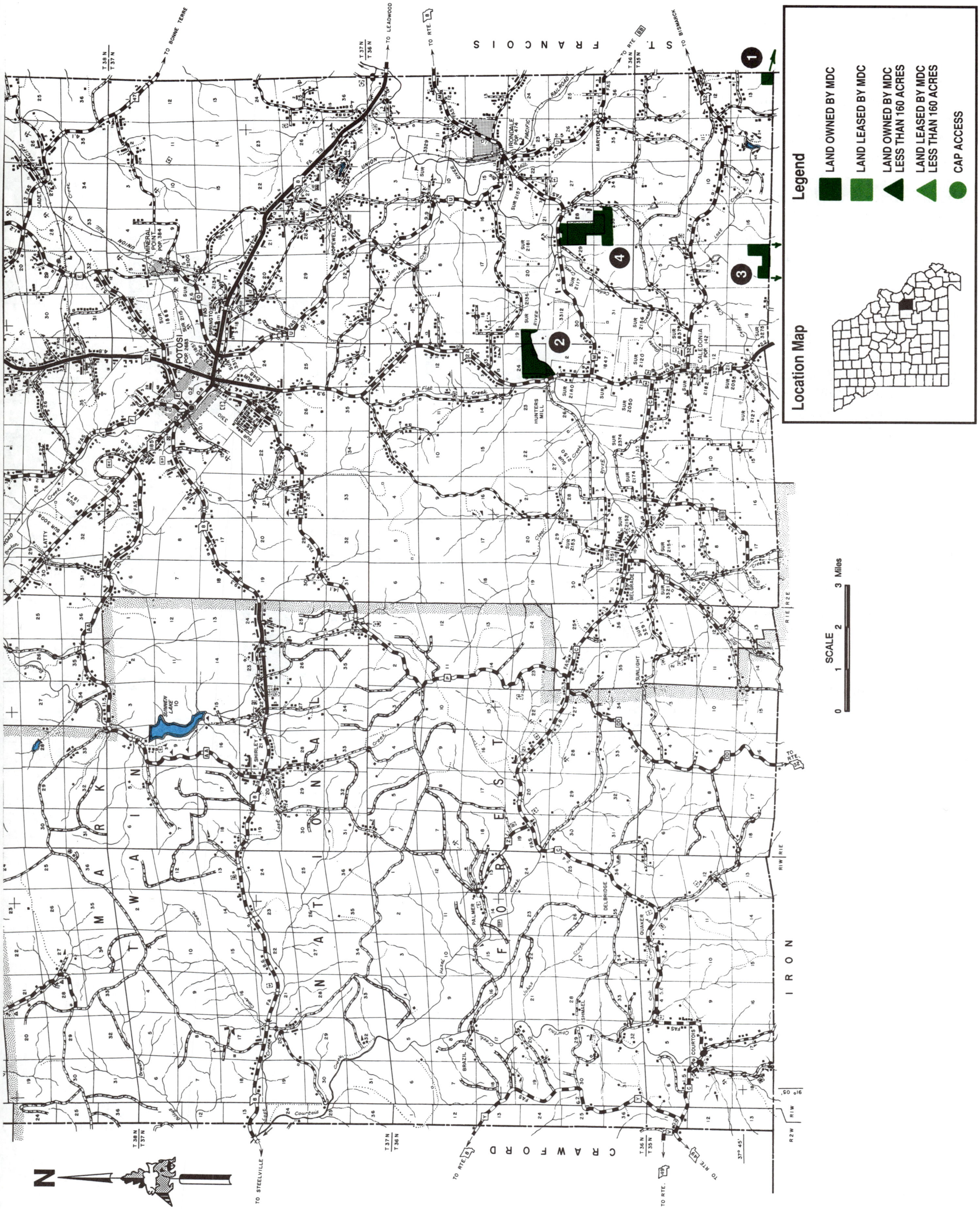
Legend
LAND OWNED BY MDC
LAND LEASED BY MDC
LAND OWNED BY MDC LESS THAN 160 ACRES
LAND LEASED BY MDC LESS THAN 160 ACRES
CAP ACCESS
Location Map
SCALE
0 1 2 3 Miles
ST. FRANCOIS
CRAWFORD
IRON
POTOSI POP. 2683
MINERAL POINT POP. 384
IRONDALE POP. 474
CALEDONIA POP. 142
TO STEELVILLE
TO BONNE TERRE
TO LEADWOOD
TO BISMARCK

WAYNE COUNTY

1 **Coldwater Access** is 9 miles north of Highway 34 on Highway 67, then 5 miles west on County Road 302. This is a 78-acre access on the St. Francis River. There is no boat ramp.

2 The main tract of **Coldwater Conservation Area** is marked with a sign on Highway 67, north of Greenville. This is a 9,470-acre forested area.

3 **Duck Creek Conservation Area**–see the Bollinger County map, page 22.

4 **Flatwoods Conservation Area** is on Highway B, about 4 miles east of Piedmont. This is a 909-acre forest with an archery and firearms range (firearms range is accessible to wheelchairs).

5 **Graves Mountain Conservation Area**–see the Iron County map, page 100.

6 **Iron Bridge Access** is south of Greenville on Highway E, then north on County Road 223. A spring, which flows into Bear Creek, runs through the west portion of this 70-acre area. There is no boat ramp, but the area is used by swimmers.

7 **Lon Sanders Canyon Conservation Area** is east of Piedmont on Highway 34, and .5 mile north on Canyon Road. This is a 130-acre forest with a 1-mile hiking trail. Natural features include shut-ins in McKenzie Creek.

8 **Riverside Conservation Area**–see the Iron County map, page 100.

9 **University Forest Conservation Area**–see the Butler County map, page 28.

10 **Wappapello Reservoir** is on Highway 34, 2.5 miles west of Highway 67. This 1,880-acre area is home to the reservoir, St. Francis River and Logan Creek, providing anglers with about 45 acres of fishable water. A 4.2-mile stretch of the Ozark Trail runs through the area.

11 **Yokum School Conservation Area** is 13 miles south of Greenville on Highway D. This is a 160-acre forest with one small pond (not suitable for fishing).

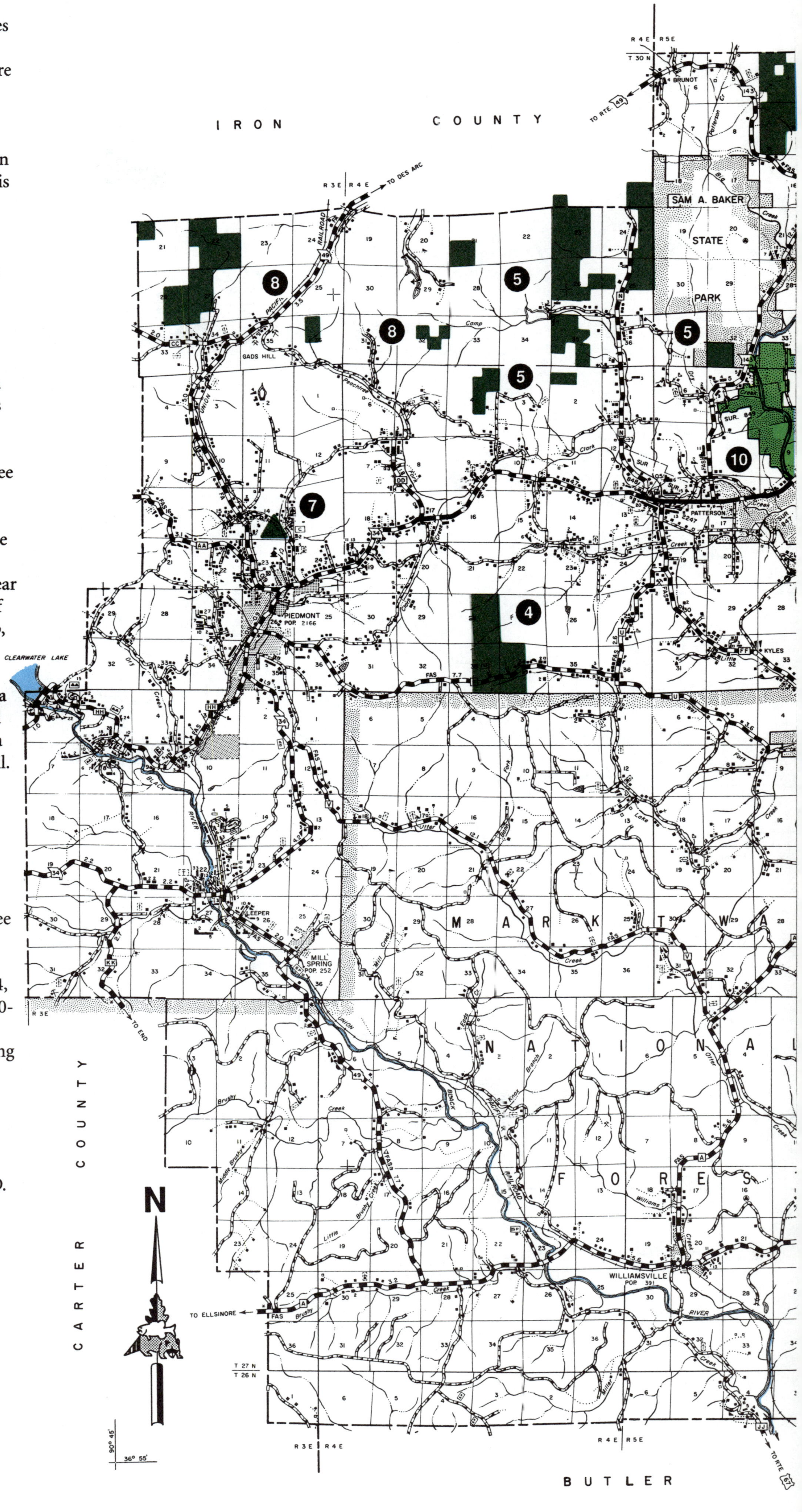

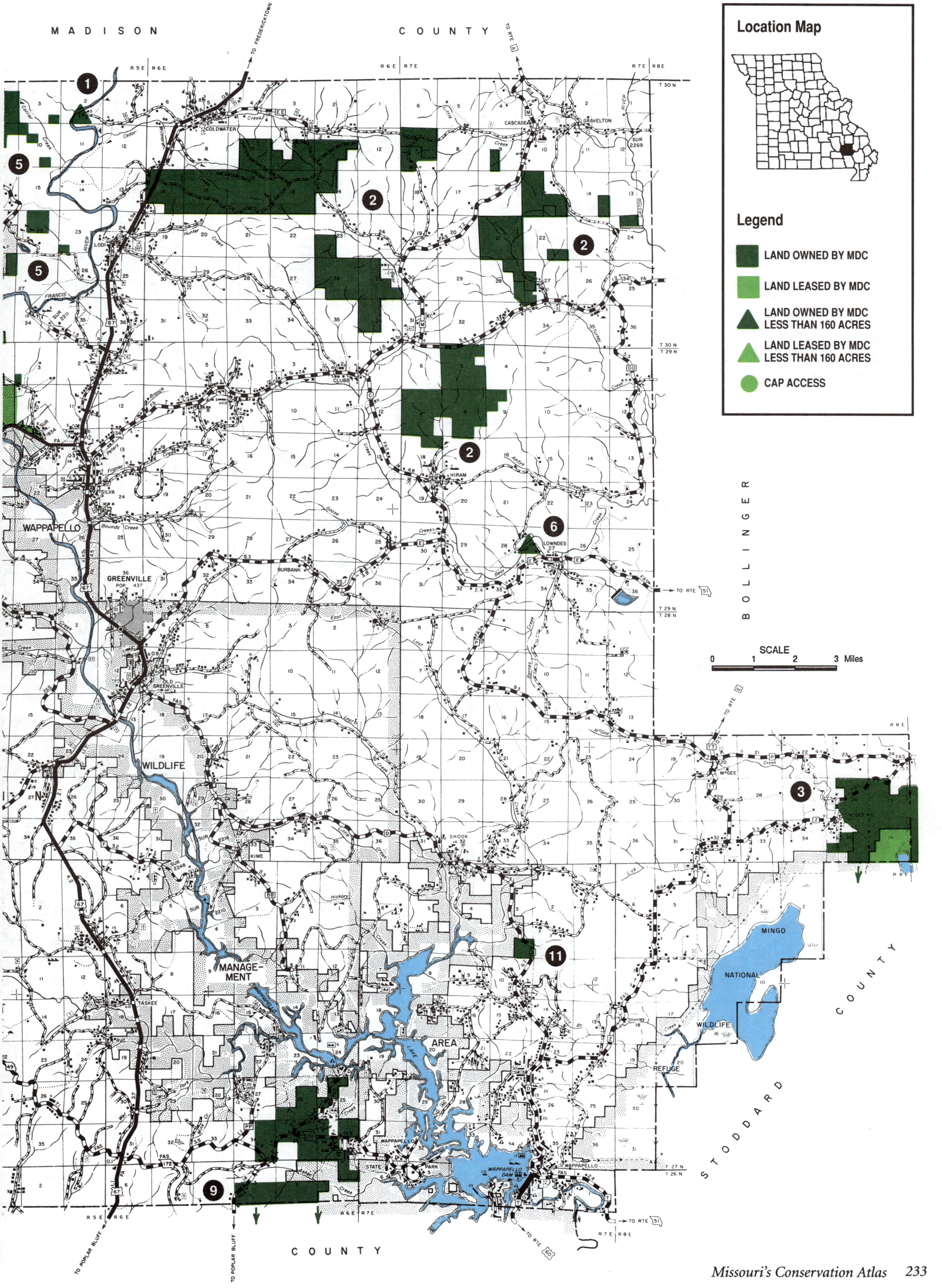
MADISON COUNTY
BOLLINGER
STODDARD COUNTY
COUNTY
TO FREDERICKTOWN
TO RTE A
TO RTE 51
TO RTE E
TO RTE 60
TO POPLAR BLUFF
COLDWATER
CASCADE
GRAVELTON
LODI
CLUBB
HIRAM
SILVA
WAPPAPELLO
GREENVILLE
POP. 437
OLD GREENVILLE
BURBANK
LOWNDES
MCGEE
SHOOK
KIME
TASKEE
WILDLIFE
MANAGE-
MENT
AREA
MINGO
NATIONAL
WILDLIFE
REFUGE
WAPPAPELLO STATE PARK
WAPPAPELLO DAM
T 30 N
T 29 N
T 28 N
T 27 N
T 26 N
R 5 E
R 6 E
R 7 E
R 8 E
Location Map
Legend
LAND OWNED BY MDC
LAND LEASED BY MDC
LAND OWNED BY MDC LESS THAN 160 ACRES
LAND LEASED BY MDC LESS THAN 160 ACRES
CAP ACCESS
SCALE
0 1 2 3 Miles

WEBSTER COUNTY

1 **Compton Hollow Conservation Area** is 5 miles south of I-44 on Highway B, then .8 mile east on a gravel road with a cantilever sign marking the area. There is a 1-mile hiking trail on this 840-acre area.

2 **Niangua Conservation Area** is made up of three separate tracts. The main tract is reached by taking I-44 to the first exit east of Marshfield, turning 1.6 miles west on County Road 606, then .2 mile southeast on County Road 615. (The tract located west of Highway CC is not accessible.) This 837-acre area is almost all forestland. There is a 1-mile hiking trail.

3 **Rader Access** is 7 miles east of Niangua on Highway F, then 3.5 miles north on Highway ZZ. The 62-acre area is on the left side of the road. One-half mile of the Osage Fork of the Gasconade River can be accessed from the area, but there is no boat ramp or parking lot.

4 **Seymour Towersite** (5 acres) is on the north side of Highway P, .7 mile east of the junction of highways C and P.

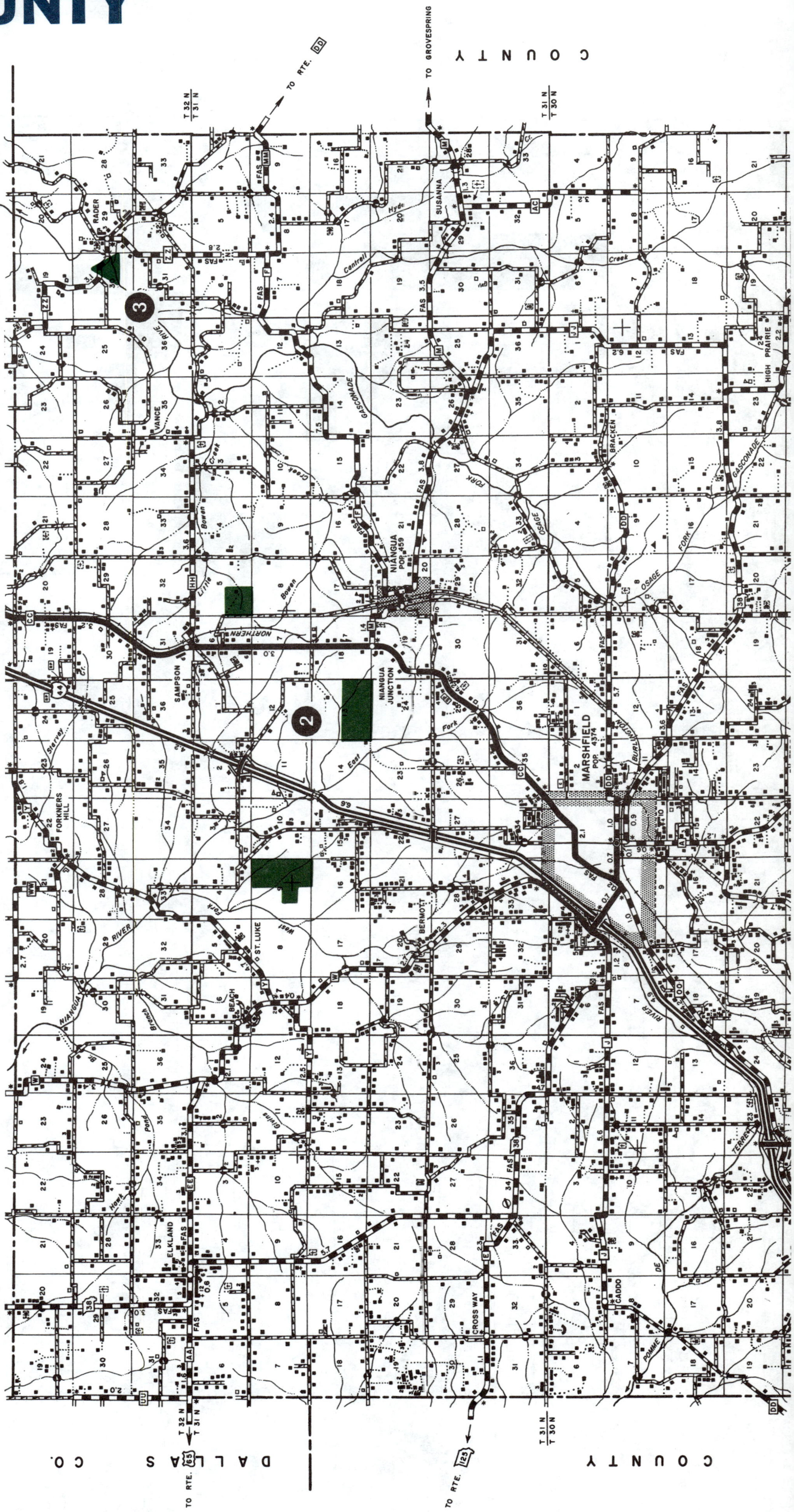

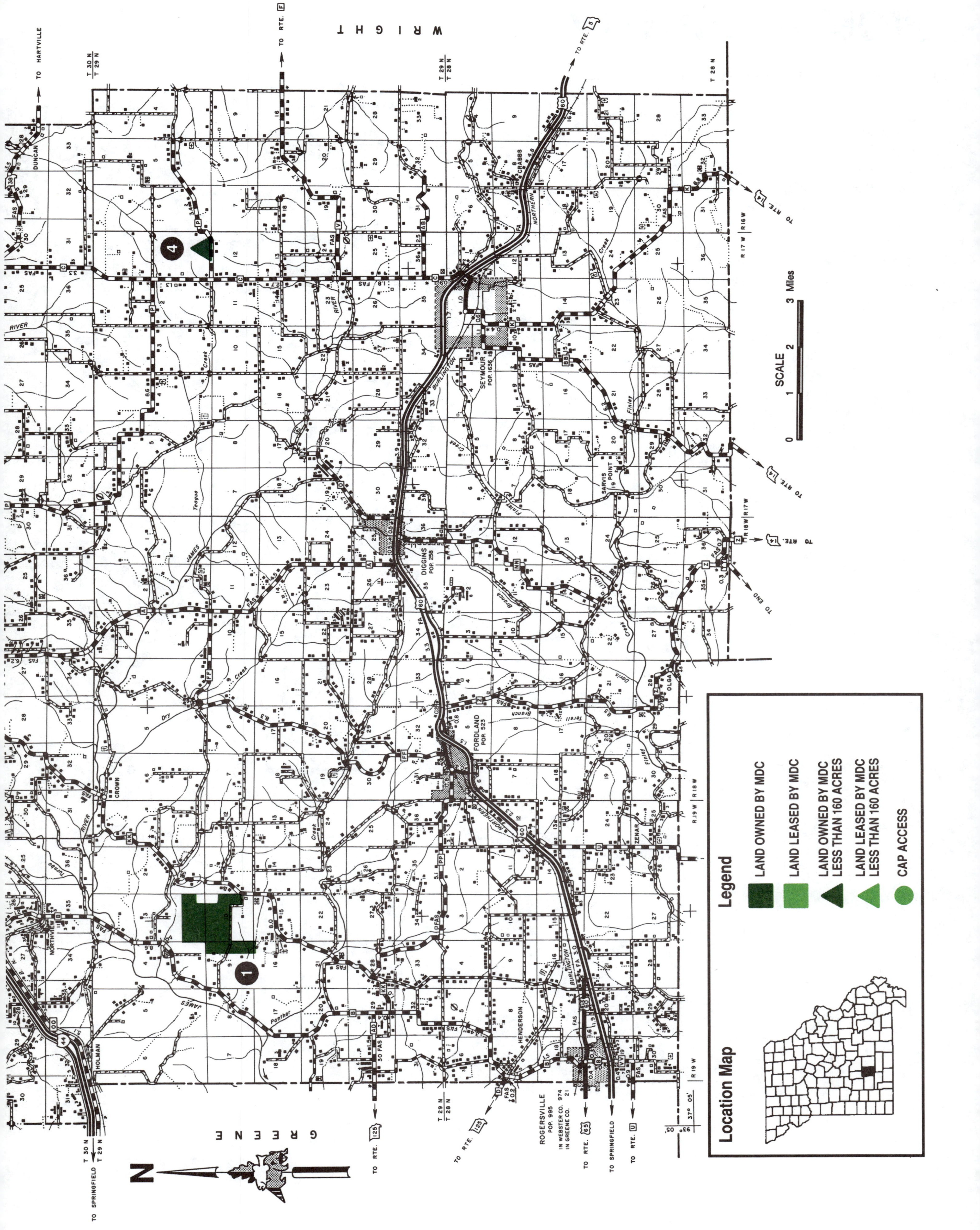

Legend
LAND OWNED BY MDC
LAND LEASED BY MDC
LAND OWNED BY MDC LESS THAN 160 ACRES
LAND LEASED BY MDC LESS THAN 160 ACRES
CAP ACCESS
Location Map
SCALE
0 1 2 3 Miles
WRIGHT
GREENE
SEYMOUR POP. 1636
DIGGINS POP. 258
FORDLAND POP. 523
ROGERSVILLE POP. 995 IN WEBSTER CO. 974 IN GREENE CO. 21
NORTHVIEW
DUNCAN
CRABBS
HENDERSON
HOLMAN
OLGA
ZENAR
SARVIS POINT
CROWN
TO HARTVILLE
TO SPRINGFIELD

WORTH COUNTY

1 **Denver Access** is .5 mile east and .5 mile south of the town of Denver on unnamed gravel roads. This is a 55-acre area on the East Fork of the Grand River with a boat ramp.

2 **Emmett and Leah Seat Memorial Conservation Area**–see the Gentry County map, page 82.

3 **Sheridan Access** is east of the town of Sheridan on Highway 246. This is 1 acre with no boat ramp on the Platte River.

4 **Worth County Community Lake** is 2 miles south of the town of Oxford off Highway U, then 1.5 miles east on a gravel road. This 20-acre lake (suitable for fishing) has a boat ramp and is surrounded by about 35 acres of prairie. There is a .5-mile hiking trail.

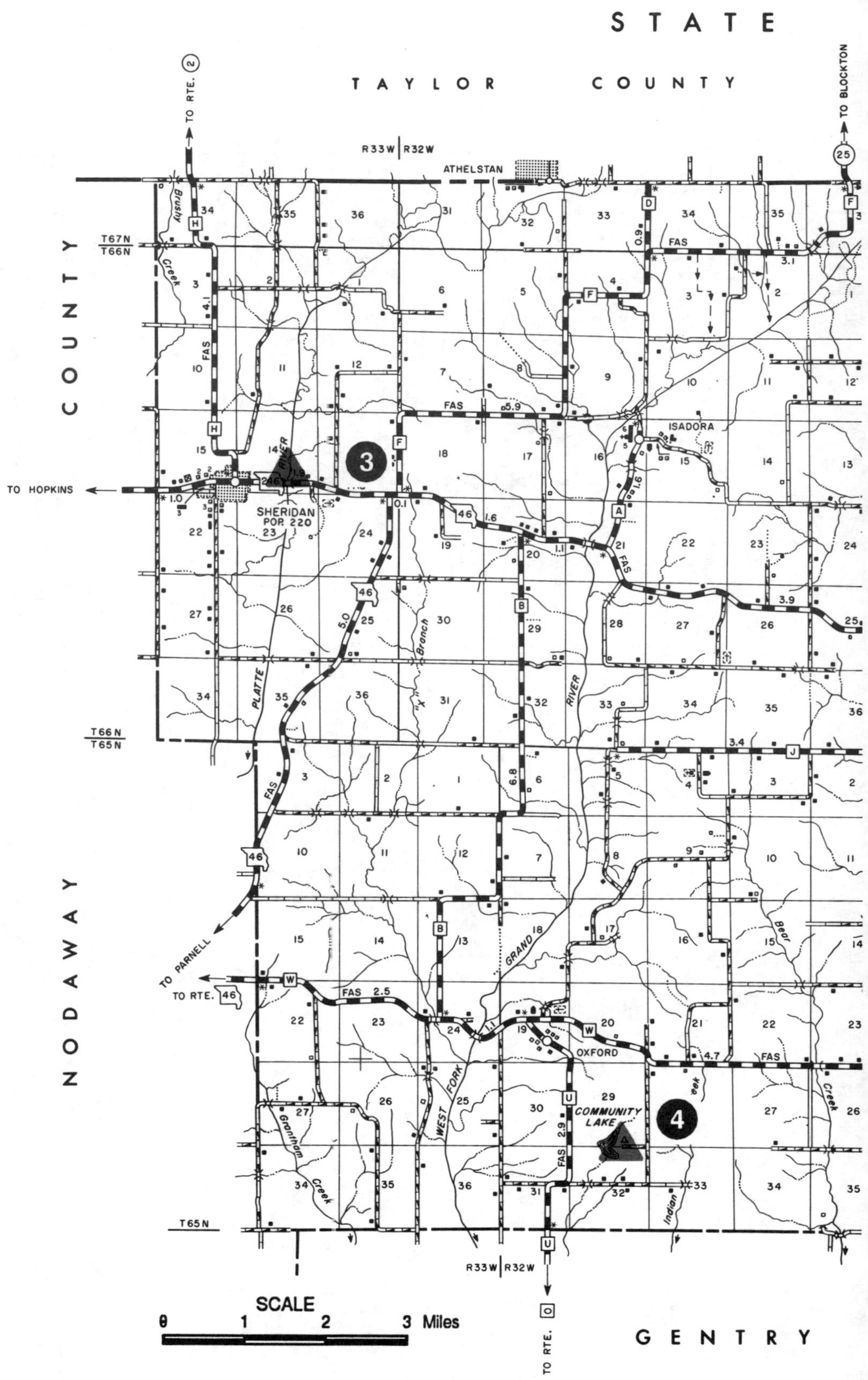

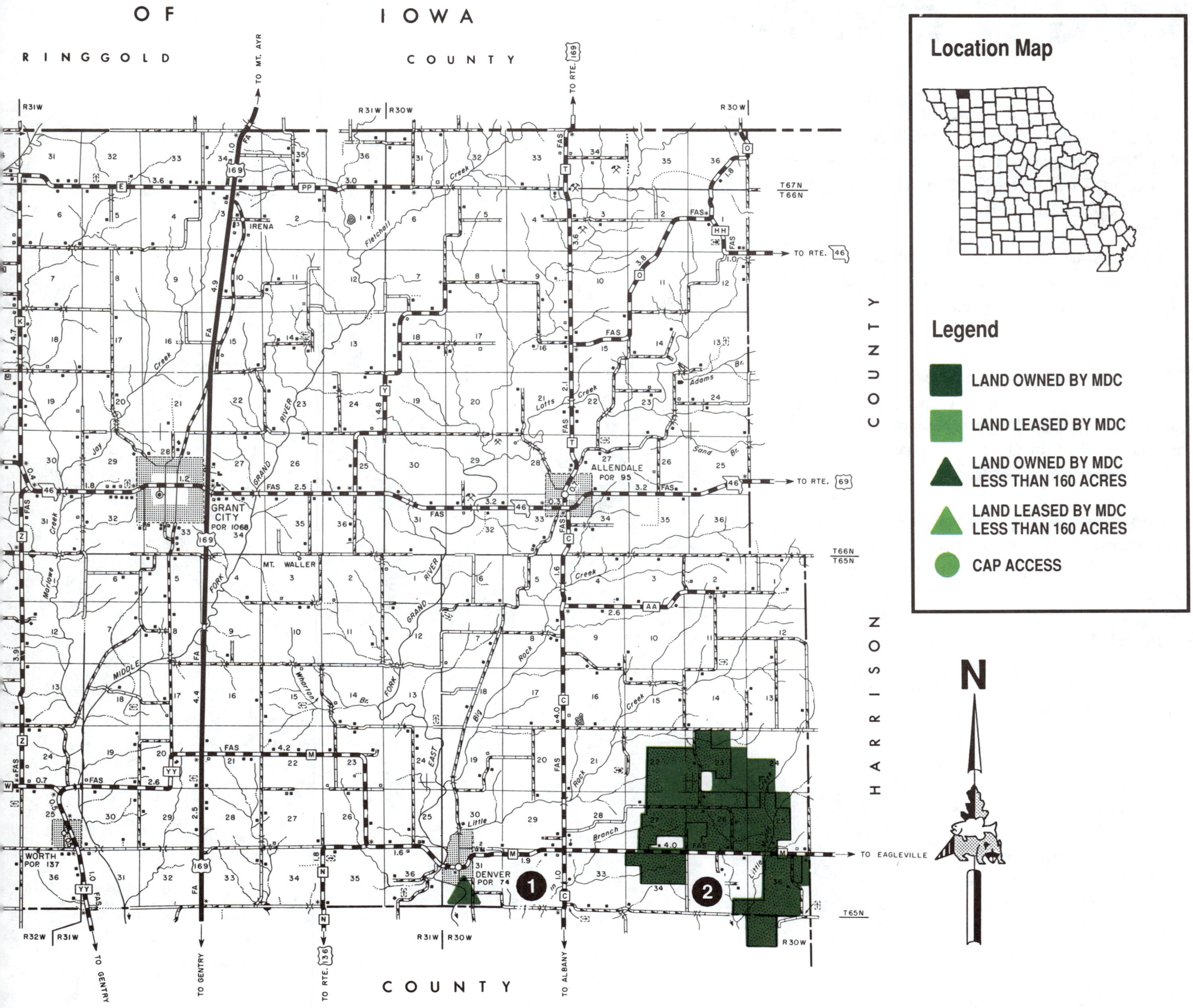
RINGGOLD OF IOWA COUNTY
HARRISON COUNTY
COUNTY
GRANT CITY POP 1068
ALLENDALE POP 95
DENVER POP 74
WORTH POP 137
IRENA
MT. WALLER
TO MT. AYR
TO RTE. 169
TO RTE. 46
TO RTE. 69
TO EAGLEVILLE
TO GENTRY
TO RTE. 136
TO ALBANY
Location Map
Legend
LAND OWNED BY MDC
LAND LEASED BY MDC
LAND OWNED BY MDC LESS THAN 160 ACRES
LAND LEASED BY MDC LESS THAN 160 ACRES
CAP ACCESS
N

WRIGHT COUNTY

1 **Wilbur Allen Memorial Conservation Area** is 1 mile north of Manes on Highway 95, and 1.5 miles west on Radford Drive. This 380-acre area contains a stretch of the Gasconade River (with a boat ramp), and there is a picnic area.

2 **Blue Slip Towersite** is on Old Highway 60, 2 miles west of Norwood. The road to this 4.6-acre area is closed to vehicles when the tower is not occupied.

3 **Buzzard Bluff Access** is 6.5 miles east of Hartville on Highway 38, and 2.25 miles north on Highway E. This is 82 acres with a boat ramp on the Gasconade River. There is also a picnic area.

4 **Camp Branch Access** is 2.5 miles east of Hartville on Highway 38. This is a 28-acre area on the Gasconade River with no boat ramp.

5 **Cheerful Hill Access** is 2.75 miles north of Manes on Highway 95, then 2.5 miles north on Highway AF. Beaver Creek flows .25 mile along this 65-acre area.

6 **John Alva Fuson Conservation Area** is 3.75 miles north of Grovespring on Highway 5, and 1.75 miles west on Smittle Cave Road. This road leads to Parks Creek Road, which runs through this 1,280-acre area. The area is known for Smittle Cave, the fifth largest cave in the state (accessible with a permit), and Little Smittle Cave and Lowell Cave (both open without restriction). Parks Creek flows through the area .75 mile, and there are three ponds providing anglers with a total of 5 acres of fishable water.

7 **Odin Access** is on Highway 38, 6 miles west of Hartville. A 1-mile stretch of the Woods Fork of the Gasconade River flows through this 126-acre area. There are two springs and a stream revetment demonstration site.

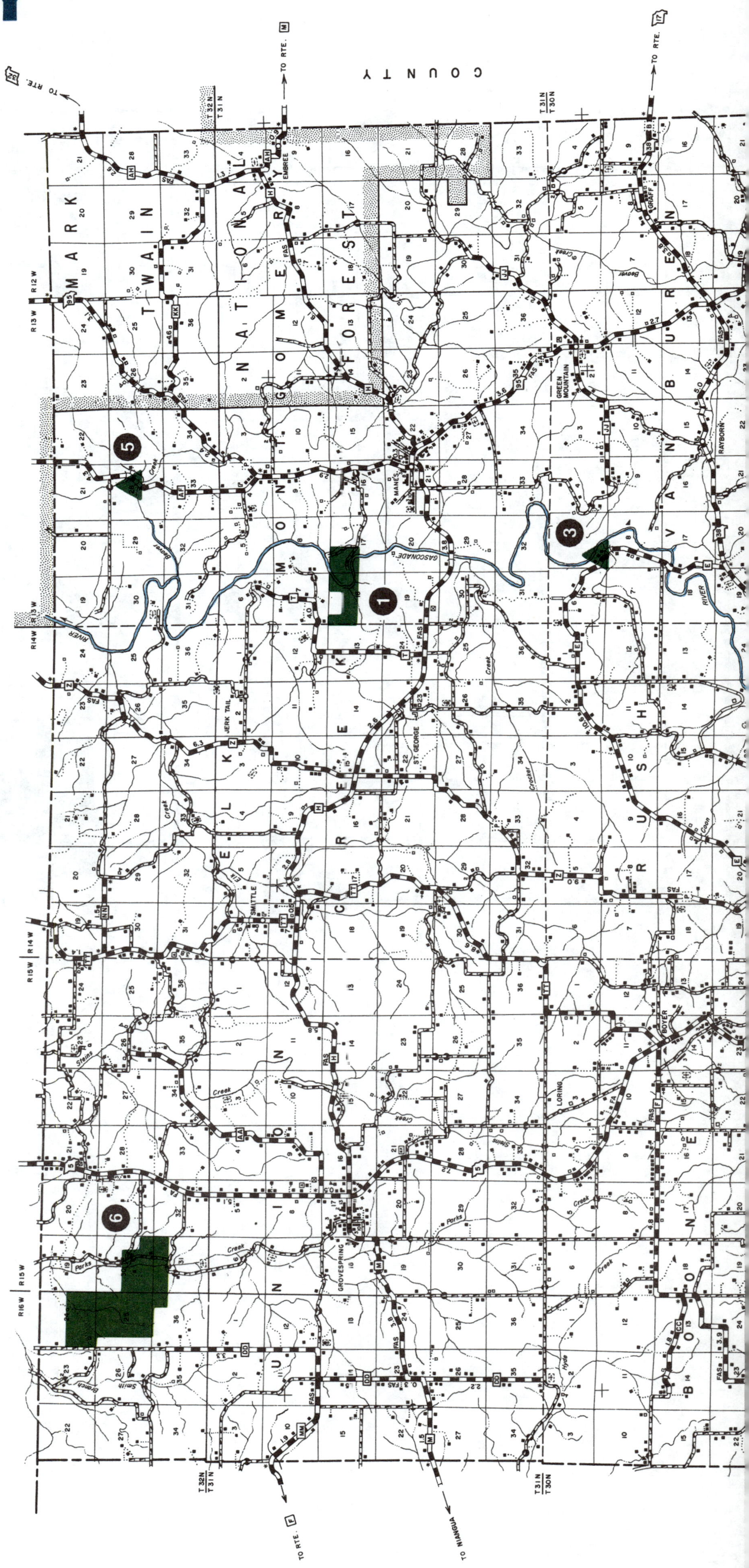

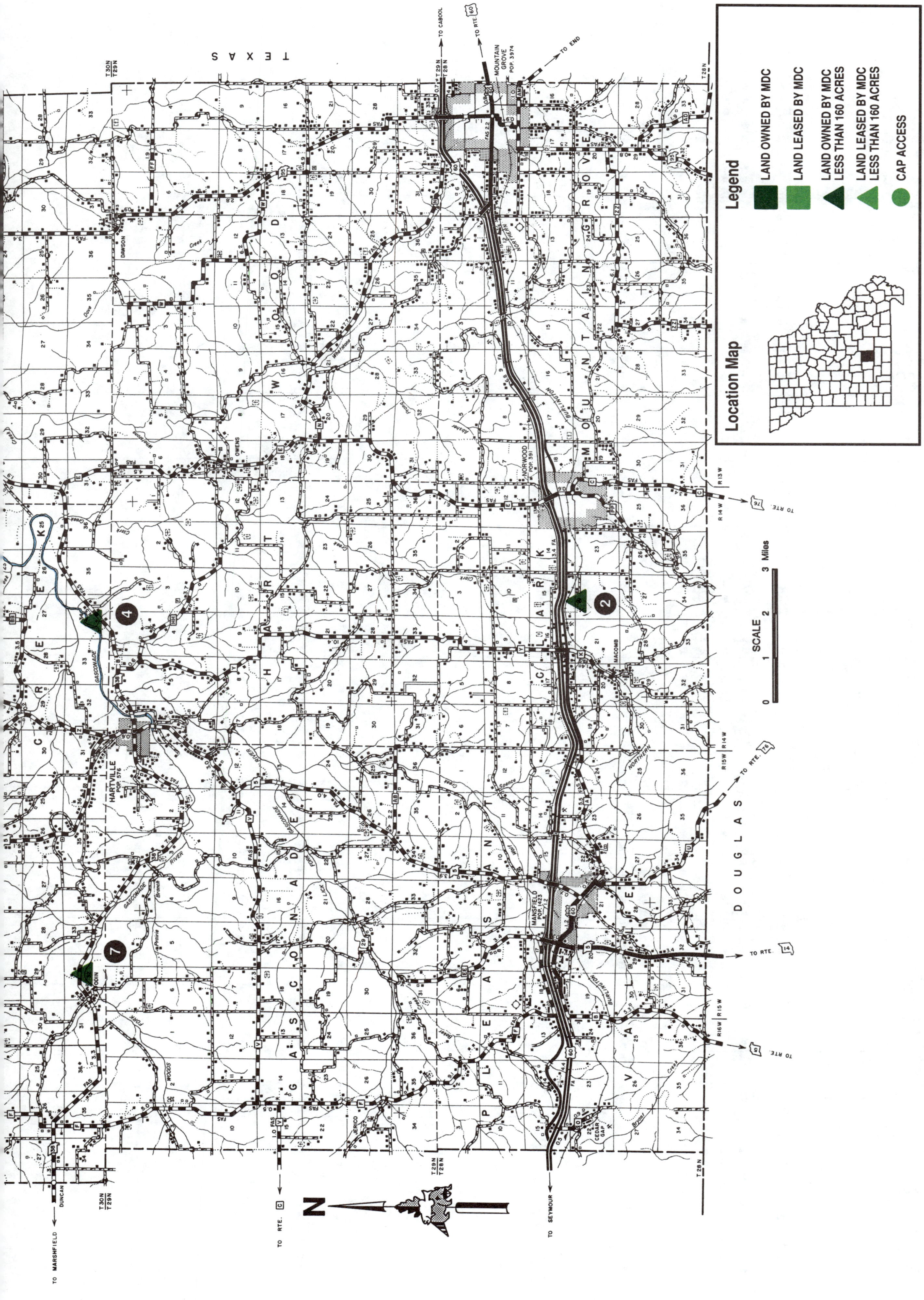

Legend
LAND OWNED BY MDC
LAND LEASED BY MDC
LAND OWNED BY MDC LESS THAN 160 ACRES
LAND LEASED BY MDC LESS THAN 160 ACRES
CAP ACCESS
Location Map
SCALE
0 1 2 3 Miles
TEXAS
DOUGLAS
HARTVILLE POP. 576
NORWOOD POP. 391
MOUNTAIN GROVE POP. 3974
MANSFIELD
MACOMB
DAWSON
OWENS
DUNCAN
CEDAR GAP
JERICO
TO MARSHFIELD
TO RTE. C
TO SEYMOUR
TO CABOOL
TO RTE. 60
TO END
TO RTE. 76
TO RTE. 14
TO RTE. 5
GASCONADE RIVER
T30N
T29N
T28N
R13W
R14W
R15W
R16W
N

	Page Number	parking lot	privy/restrooms	picnic areas	primitive camping	camper pads	river access	boat ramps	fishing docks/jetties	boat rental	viewing blinds	waterfowl blinds	interpretive programs	exhibits/displays	MDC office	hunting	waterfowl hunting	fishing	hiking trails	horse trails	canoeing	archery range	firearms range	field trials	dog training	trapping	Notes
ADAMS (ANNA M.) ACCESS	114	■					■											■			■					■	
AGENCY ACCESS	26	■					■											■			■						
AGENCY CA	26	■					■									■		■									
ALEX GEORGE LAKE	102	■	■	■														■									
ALLEN BRIDGE ACCESS	154	■					■	■										■			■						
ALLEN (WILBUR) MEM CA	238	■	■	■	■		■	■								■		■			■				■	■	
ALLENTON ACCESS	200	■	■				■	■										■			■						
ALLRED LAKE NA	28	■																	■		■					■	
AMARUGIA HIGHLANDS CA	42	♿	♿					■	♿							■	■	■							■		
AMIDON MEM CA	22	■			■		■									■		■	■						■		
ANDERSON (EDWARD) CA	172	■			■											■										■	
ANGELINE CA	210	■			■								■		■	■			■	■			■	■	■	■	
ANTIMI LAKE	24	■	■	■									■	■	■			■	■					■	■		
APPLE CREEK CA	36	♿	♿		■		■	■								■	■	■	■	■	■	■	♿	■	■	■	
AQUILLA ACCESS	214															■		■			■			■	■	■	
ARCHANGEL ACCESS	206	■					■											■			■						
ARCHIE ACCESS	42						■											■									
ARROW-WOOD CA	212	■			■		■	■								■	■	■			■			■	■	■	
ASHE JUNIPER NA	216																										
ASHLAND LAKE	24	■						■										■			■						
ATHENS ACCESS	51						■											■			■						
ATLANTA CA	130	■			■			■								■	■	■			■		■	■	■	■	
ATLANTA CITY LAKE	130	■			■											■	■	■									
AUSTIN COMMUNITY LAKE	222	♿	♿	■				■										■			■					■	
BAGNELL DAM ACCESS	140	♿	♿	■			■	■										■			■						
BALL MILL RESURGENCE NA	166	■																	■								
BALLTOWN ACCESS	226															■	■	■			■				■		
BALTIMORE BEND CA	116	■			■											■		■								■	
BAPTIST CAMP ACCESS	222	■	■	■			■											■			■						
BARN HOLLOW NA	222	■														■			■							■	
BARNES (P. F.) CA	222	■			■											■										■	
BAT CAVE CA	140															■											
BEAR CREEK CA	114	■			■											■			■	■				■	■	■	
BEAVER CREEK CA	170	■	■	■												■			■			■					
BEE HOLLOW CA	130															■	■	■			■			■	■	■	
BEE TREE LAKE	200	♿	♿	■					♿				■	■				■	■								
BELCHER BRANCH LAKE CA	26	♿	♿					■	♿							■	■	■									
BELL CHUTE ACCESS	134	■	■		■		■	■										■			■						
BELLEFONTAINE PARK LAKE	200	♿		■									■					■									
BEN BRANCH LAKE CA	160	♿	♿	■	■	■		■	♿							■	■	■			■						
BENNETT SPRING ACCESS	64	■	■				■	■										■			■						
BENNETT SPRING HATCHERY/ TROUT PARK	64	♿	♿	■									■	■	■			■	■		■						
BENNITT (RUDOLF) CA	96	■			■											■		■	■				■				
BENTON PARK LAKE	200	■		■														■									
BERGAN LAKE	102	■	■	■	■													■	■	■							
BERRY BLUFF CA	64				■											■								■	■	■	
BETHANY CITY LAKES	88	♿	♿	■	■	■		■								■	■	■			■			■	■	■	
BETHEL PRAIRIE CA	16	■	■													■		■									
BIG BUFFALO CREEK CA	20	■			■											■		■	■		■					■	
BIG CANE CA	28	■			■											■	■	■			■				■	■	
BIG CREEK CA	6	■			■											■	■	■	■		■			■	■	■	
BIG DRYWOOD CREEK CA	226															■		■									
BIG JOHN ACCESS	64	■					■											■			■					■	
BIG OAK LAKE	142	♿	♿	■				■	♿				■	■	♿			■	♿		■						
BILBY RANCH LAKE CA	156	♿	♿	■	■			■	♿							■	■	■	■		■			■	■	■	
BINDER COMMUNITY LAKE	54	♿	♿	■				■	♿	■	♿							■			■						
BIRCH CREEK CA	210				■											■								■	■	■	
BIRD'S BLUE HOLE	142	■		■				■										■									
BIRDSONG CA	194	■	■		■		■									■		■			■			■	■	■	
BISMARCK LAKE CA	100	♿	♿					■	♿						♿	■	■	■							■	■	
BITTERN BOTTOMS CA	42	■																■	■								
BLACK HAWK ACCESS	136	■			■		■	■								■	■	■			■			■	■	■	
BLACKJACK ACCESS	194	■	■		■		■	■								■	■	■			■				■	■	
BLACKWATER BRIDGE ACCESS	58	■	■				■	■										■			■						
BLACKWELL (JEREMIAH) ACCESS	108	■					■											■			■						
BLAIR BRIDGE ACCESS	162	■	■	■	■		■	■										■			■					■	
BLANCHETTE LANDING ACCESS	192	♿	♿				■	■										■			■						
BLEES LAKE	130	♿	♿	■					♿									■									
BLIND PONY LAKE CA	204	♿	♿	■				■	♿	■					■	■		■							■		
BLOCK HOLE ACCESS	36	■					■											■			■						
BLUE JAY TRAIL ACCESS	186	■					■											■			■						
BLUE LICK CA	204	■														■											
BLUE RIVER PARKWAY	102	♿	♿	■			■	■										■			■						
BLUE SLIP TOWERSITE	238																										
BLUE SPRINGS CREEK CA	60	■					■	■								■		■			■				■	■	
BLUE SPRINGS LAKE	102	■						■										■									
BLUFF SPRINGS CA	44	■														■		■							■	■	
BLUFFWOODS CA	26	♿	♿	■	■								■			■		■	♿	■					■	■	
BLUFFWOODS CA - GOODELL MEM ANNEX	26	■														■											
BOATHOUSE LAKE	200	♿		■					■				■					■									
BOECKMAN BRIDGE ACCESS	140	■					■									■	■	■			■				■	■	
BOESL (L. A.) OUTDOOR EDUC AREA	222	■							♿										■								
BOILING SPRING ACCESS	222	■	■	■			■	■										■			■						
BOIS D'ARC CA	86	■	♿													■			■			■	♿	■	■		
BONANZA CA	30	■			■											■	■	■		■					■		
BONNOTS MILL ACCESS	160	■	■		■		■	■										■			■						
BOOTLEG ACCESS	230	■			■	■	■									■		■	■		■				■	■	
BORROW PIT CA	192															■	■	■								■	
BOSTON FERRY CA	220	■		■												■			■								
BOSWORTH ACCESS	38	■	■	■			■	■										■									
BOWLIN ROAD LAKE	102	■	♿		■	■												■									
BRANCH TOWERSITE	34	■	■													■		■							■		

Key ■ = Yes ♿ = Disabled Access

beaver	bullfrogs	coyotes	deer	doves	eagles	fox	hawks	herons	otters	owls	pheasant	quail	rabbits	racoons	ruffed grouse	shore birds	snapping turtles	songbirds	squirrels	turkey	waterfowl	hybrid bass	largemouth bass	rock bass	small mouth bass	spotted bass	striped bass	white bass	blue gill	carp	catfish	crappie	muskelunge	paddlefish	suckers	sunfish	trout	walleye	mushrooms	wildflowers
⊠	⊠	⊠	⊠	⊠	□	⊠	⊠	⊠	□	⊠		⊠	⊠	⊠		⊠	⊠	⊠	⊠	⊠	□		□	■	⊠	□		□	⊠	⊠	⊠	□			■	■			⊠	⊠
□	⊠	□		□	□		□	□	□	□		□		⊠		□	⊠	□										□	□	■	■	□		□	⊠	□					
	⊠	⊠	⊠	□	⊠		□		⊠	⊠	⊠		⊠		⊠	⊠	⊠			⊠	□	□		⊠				□	□	■	■	⊠			□	□			⊠	⊠
□	⊠	□	⊠	⊠	⊠		⊠		⊠	⊠	⊠		⊠	□	⊠	⊠	□	◆		⊠		⊠	□	□					⊠	□	⊠	□				⊠	◆		□	□
⊠	⊠	⊠	⊠	□	□	⊠	⊠	⊠		⊠		□	⊠	⊠			⊠	⊠	⊠	□	□		⊠	⊠	⊠	⊠			⊠	⊠	⊠	□			⊠	⊠			□	□
⊠	■	⊠	■	⊠	◆	□	⊠	⊠	⊠	⊠		■	⊠	⊠		□	■	⊠	⊠	⊠	□		■	■	■	⊠			■	⊠	■	⊠			■	■			⊠	⊠
⊠	⊠	⊠	⊠	⊠	⊠		⊠		⊠	⊠	⊠		⊠	⊠	⊠	⊠	⊠	⊠		⊠	■	⊠	□	⊠	⊠	■		□	■	⊠	⊠	■		□	■	■		□	⊠	⊠
⊠	■	⊠	□	□	⊠	□	■		□	■	⊠		⊠	⊠	■	■	□	■		⊠	⊠	□	□	⊠					⊠	⊠	⊠	⊠				⊠			⊠	⊠
□	⊠	⊠	⊠	⊠	⊠		⊠		⊠	⊠	⊠		⊠	⊠	■	⊠	⊠	◆		⊠		⊠	□	□					⊠	⊠	⊠	⊠				⊠			⊠	■
□	⊠	⊠	⊠	⊠	⊠		⊠		⊠	⊠	⊠		□	□	■	■	⊠	□		⊠	⊠	□		⊠	⊠	□			⊠	□	⊠	□			⊠	■			⊠	■
□	□	⊠	⊠	⊠			■		□	□	⊠	⊠			■	■	⊠						□	◆	⊠																■	■
□	□	⊠	⊠	□	⊠	⊠	⊠	□	□	⊠		□	⊠	⊠	□	□	□	⊠	⊠	⊠	□	□	⊠	⊠	⊠	□	□	□	⊠	□	□	□	□	□	⊠	⊠	□	□	⊠	⊠
□	⊠	□	⊠	⊠		□	⊠	⊠		⊠		□	⊠	⊠		□	⊠	⊠	⊠	⊠	□		⊠			□			⊠	□	⊠	□			□	⊠			⊠	⊠
⊠	⊠	■	⊠	◆	□	⊠	⊠	□		■		⊠	⊠	■		□	⊠	■	■	⊠	□		⊠	□	□	□		□	⊠	■	◆	◆		□	⊠	⊠			□	■
■	⊠	⊠	⊠	⊠	□	⊠	⊠	□	□	⊠		⊠	⊠	■		□	⊠	■	⊠	⊠	□		⊠						⊠	⊠	⊠	⊠				⊠			⊠	⊠
□	□	□	□				□	□		□			□	□			□	□	□	□			□					□	□	⊠	⊠	□			□	□		□	□	⊠
□	⊠	⊠	□	⊠		□	□	□	□	□		□	□	⊠		□	⊠	⊠			□		□					□	□	■	■	□		□	⊠	□		□	□	□
□	⊠	⊠	⊠	⊠		□	⊠	□		⊠		⊠	□	⊠		□	⊠	⊠	⊠	⊠	□		□		□				□	⊠	⊠				⊠	□			⊠	⊠
		□	□	□		□	□			□		□	□					□	□	□																				□	⊠
⊠	⊠	⊠	⊠	□		□	⊠	⊠		⊠		□	⊠	⊠		□	⊠	⊠	⊠	⊠	□		⊠						⊠		⊠	⊠				⊠			⊠	□
□	□	□	⊠	□	□	□	⊠	⊠	□	⊠			⊠	⊠		⊠	□	⊠	⊠	⊠	⊠		■		□	□		⊠	■	■	■	⊠		□	■	■		⊠	◆	◆
⊠	⊠	⊠	⊠	⊠	◆	⊠	⊠	◆		⊠		⊠	⊠	⊠		◆	⊠	⊠	⊠	⊠	◆		⊠						⊠	⊠	⊠	⊠			⊠	⊠		⊠	◆	◆
⊠	⊠	⊠	⊠	⊠	□	⊠	⊠	⊠	□	⊠		⊠	⊠	⊠		□	⊠	⊠	⊠	⊠	⊠		■						■		■	■				⊠			⊠	⊠
⊠	■	□	⊠	□	□	□	□	⊠		□		⊠	⊠	⊠		□	■	⊠	⊠	□	◆		■						■		■	□				■			⊠	⊠
⊠	⊠	□	□		◆	⊠	⊠	◆	□	□			⊠	⊠		⊠	⊠	■	⊠	□	◆	⊠	■	■	■	■	⊠	■	■	■	■	■	□	■	■	■		⊠	⊠	⊠
		⊠	⊠	□	□	⊠	⊠			⊠		□	□	⊠				⊠	⊠	⊠																				⊠	⊠
⊠	◆	⊠	⊠	□	□	□	⊠	□	□	⊠		□	□	⊠		□	◆	■	⊠	⊠	◆		□			□			⊠	⊠	⊠	□		◆	⊠	⊠			◆	◆
□	■	■	■	◆	□	⊠	⊠	⊠		⊠		⊠	⊠	■			■	■	■	⊠	□		⊠		□				⊠	□	⊠	⊠				⊠			■	■
⊠	⊠	□	□	□	□	□	□	□	⊠	□		□	□	□		□	■	⊠	⊠	□	□		⊠	■	■	⊠			■	⊠	⊠	⊠			■	■			⊠	⊠
⊠	□	⊠	⊠	□	□	⊠	⊠	□		⊠		⊠	□	⊠			□	■	■	⊠																				⊠	■
			⊠	■	□		⊠	■			■		□	□	⊠				■	■	■																		⊠	■
⊠	□	⊠	⊠	⊠	◆	⊠	⊠	□	□	⊠			■	⊠		⊠		■	■	⊠																				⊠	⊠
	□	■	⊠	□	⊠	⊠	□		⊠		⊠	⊠	⊠			□	⊠	■	⊠																				⊠	⊠	
	□	□	⊠	⊠		□	⊠	□		⊠		□	⊠	⊠			□	■	⊠	□			□		□				□							□			⊠	⊠
⊠	⊠	⊠	⊠	⊠	◆	⊠	⊠	⊠	⊠	⊠	□	⊠	⊠	⊠		◆	⊠	⊠	⊠	⊠	◆		⊠						⊠	⊠	⊠	□				⊠			◆	◆
	□		□	□		□	□	⊠		□		□	■	⊠		□	□	◆	■		⊠		■						■	■	■	⊠				■			□	◆
□	⊠	⊠	⊠	□		□	□	⊠		□		⊠	⊠	⊠			□	⊠	□	□	◆		⊠						⊠	□	⊠	□				⊠			□	□
□	⊠	□	⊠	⊠	□	□	⊠	⊠	□	⊠			⊠	⊠		□	⊠	⊠	⊠	⊠	□		⊠	⊠	⊠	⊠			⊠	⊠	⊠	⊠			⊠	⊠		⊠	□	□
	⊠			⊠									⊠						⊠			⊠						⊠		⊠	□				□			
⊠	⊠	⊠	■	⊠	□	⊠	⊠	□		⊠		⊠	⊠	⊠		□	□	■	■	■	⊠		⊠						⊠	⊠	⊠	⊠				⊠			⊠	⊠
□	⊠	⊠	⊠	⊠		□	⊠	⊠	□	⊠		⊠	⊠	⊠		□	⊠	⊠	⊠	⊠			⊠	⊠	■	⊠			⊠	⊠	⊠	□			■	⊠	■		⊠	■
□	⊠	□	⊠		□		□	⊠	□	□			⊠	⊠		□	⊠	⊠	⊠						□	□				□						⊠	□	■		⊠	⊠
□	□	■	⊠	⊠		□	⊠			⊠		⊠	⊠	⊠	□		□	⊠	⊠	⊠	□								□		□						□			⊠	⊠
				⊠														⊠											⊠	⊠									
□	⊠	⊠	■	⊠	□	⊠	⊠	⊠		⊠		⊠	⊠	⊠		□	⊠	■	■	■	□		⊠						■	□	⊠	■				■	◆		⊠	■
	⊠	■	■	□	⊠	⊠	⊠	□	□	⊠		□	⊠	⊠		□	⊠	⊠	■	■				⊠	⊠				□	□	⊠				■	■	⊠		⊠	⊠
⊠	⊠	⊠	⊠	⊠	□	⊠	⊠	⊠	□	⊠	□	■	⊠	⊠		◆	⊠	⊠	■	⊠	◆		■						■	⊠	■	■			□	■			⊠	⊠
□	□	⊠	⊠	□		□	⊠	□		□		⊠	⊠	□		□	□	⊠	□	⊠	□		□						□	□	□					□			□	■
⊠	⊠	⊠	■	⊠	□	□	⊠	□		⊠		⊠	⊠	⊠			⊠	■	⊠	■	□		⊠	⊠	⊠	□			⊠	⊠	⊠	□			⊠	⊠			⊠	⊠
⊠	■	⊠	⊠	⊠	□	⊠	□	⊠	□	⊠		⊠	■	⊠		⊠	□	■	■	⊠	⊠		□	□					⊠	⊠	⊠	□				⊠			⊠	⊠
⊠	⊠	⊠	■	◆	□	□	⊠	⊠		⊠		⊠	⊠	⊠	□		⊠	◆	⊠	⊠	□		⊠						⊠	□	⊠					⊠			⊠	⊠
⊠	◆	□	□	⊠	□	□	⊠	⊠	□	⊠		□	□	⊠		⊠	◆	⊠	⊠	□	⊠							□	□	□					□			□	⊠
⊠	⊠	⊠	⊠	⊠	⊠	⊠	⊠	⊠	⊠	⊠		⊠	⊠	⊠		⊠	⊠	⊠	⊠	⊠	⊠		□	⊠	⊠	□			⊠	⊠	⊠	⊠			⊠	⊠	□		⊠	⊠
■	⊠	■	⊠	◆	◆	□	□	■		□			□	⊠		□	□	⊠	■	□	□		⊠						■		⊠	□				■			□	□
⊠	⊠	⊠	⊠	⊠	□	⊠	⊠	⊠		⊠	■	⊠	⊠	⊠		◆	⊠	⊠	□	□	◆		⊠						⊠		⊠				⊠	⊠			◆	⊠
⊠	⊠	□	⊠	⊠	◆	□	⊠	■		⊠		□	□	⊠		□	⊠	⊠	⊠	⊠	◆		■						■	■	■	■				■			⊠	⊠
□	□	⊠	⊠	□	□	⊠	⊠	□	□	⊠		□	⊠	⊠	□	□	□	⊠	⊠	⊠																					⊠	⊠
⊠	⊠	□	□	□	□	□	□	⊠		□			□	□		□	⊠	⊠	□	□	□		□	□	□		□	□	□	⊠	■	⊠			⊠	⊠			⊠	⊠
□	⊠	⊠	■	⊠		⊠	⊠	⊠	□	⊠		⊠	⊠	⊠		□	⊠	■	■	■	□		⊠	□	□	□		□	⊠	⊠	⊠	□			⊠	⊠		□	◆	■
⊠	⊠	⊠	⊠	⊠	□	⊠	⊠	⊠		⊠		⊠	⊠	⊠		□	⊠	⊠	■	⊠	⊠		⊠	□		□			⊠	■	■	⊠		□	□	■			⊠	⊠
⊠	⊠	⊠	□	□	□	□	⊠	⊠		⊠		□	□	⊠		⊠	⊠	■	□	□	◆		□						□	□	□					□			□	■
⊠	⊠	⊠	⊠	⊠	□	⊠	⊠	□		⊠		⊠	⊠	⊠	□	□	⊠	■	⊠	⊠	⊠		□		□			□	□	■	■	□			⊠	□		□	⊠	⊠
⊠	⊠	⊠	⊠	⊠	□	⊠	⊠	⊠	□	⊠		⊠	⊠	⊠		◆	⊠	⊠	⊠	⊠	⊠	⊠	⊠	⊠	⊠	□	⊠	⊠	⊠	⊠	⊠	⊠		◆	⊠	⊠		◆	◆	■
□	⊠	□	□	⊠	□	□	◆	◆	□	⊠		□	□	⊠		□	⊠	⊠	⊠	□	□		⊠					□	⊠	⊠	⊠	⊠		□	⊠	⊠		□	□	□
⊠	⊠	⊠	⊠	⊠	□	□	⊠	⊠		⊠		⊠	⊠	⊠		⊠	⊠	⊠	⊠	⊠	⊠		■	⊠	■	■			⊠	⊠	⊠	⊠			■	■			□	⊠
□	⊠	□	□	□	◆	□	⊠	⊠	□	□		□	□	□		□	⊠	⊠	□	□	◆		⊠	⊠	⊠	□			⊠	⊠	⊠				⊠	⊠	⊠	⊠	□	□
⊠	⊠	□	⊠	⊠	◆	⊠	◆	⊠		⊠			⊠	⊠		◆	⊠	◆	⊠	□	⊠	□	⊠					■	⊠	■	⊠	■		□	⊠	⊠		□	□	□
□	⊠	□	□	□	□	□	□	⊠	□	□		□	⊠	⊠		□	⊠	⊠	⊠	□	□		■						■		□	□				□			⊠	⊠
□	■	■	■	◆	□	□	⊠	◆	⊠	■		■	■	■		■	■	■	⊠	⊠	◆		■	□					■		■	⊠		□		■			⊠	⊠
□	⊠	⊠	□	□				⊠	□	□		□	□	⊠		□	⊠	⊠	⊠	□	□	□	⊠	□	□	⊠	□	◆	⊠	⊠	⊠	⊠		□	⊠	⊠		◆	□	□
⊠	⊠	⊠	⊠	⊠	□	⊠	◆	⊠	□	⊠	□	⊠	⊠	⊠		□	⊠	⊠	⊠	⊠	□		⊠						⊠	⊠	⊠	⊠		□	□	⊠			⊠	⊠
		■	⊠	⊠		□	⊠	□		⊠		⊠	⊠	⊠		□		⊠	⊠	⊠	□																				◆	◆
□	⊠	□	□	⊠	□	□	⊠	⊠		⊠		□	⊠	□		⊠	□	⊠	⊠	□	□		⊠					□	⊠	⊠	⊠	⊠			□	⊠		□	□	□
		□	□	□		□	□			□		□	□	□				⊠	□	□																					□	□
⊠	⊠	⊠	⊠	□	□	⊠	⊠	⊠	□	⊠	□	⊠	⊠	⊠	□	⊠	⊠	⊠	⊠	⊠	□	□	⊠	⊠	⊠	□	□	□	⊠	⊠	⊠	□	□	□	⊠	⊠	⊠	□	⊠	⊠
□	□	□	□	□	□	□	□	◆		□		□	□	□		□	□	⊠	□	□	◆	⊠	⊠					⊠	⊠	■	⊠	■			□	■		□	□	⊠
	□	⊠	■	□		⊠	⊠	□		⊠		□	□	⊠		□	□	■	■	■	□		⊠						⊠							⊠			◆	■
□	□	⊠	⊠	⊠	□	⊠	⊠	□		⊠		⊠	⊠	⊠	□	□	□	■	■	■																				■	■
		□	⊠	□	□	⊠	⊠			⊠		□	⊠	⊠	□			■	■	⊠																				■	⊠
□	⊠			⊠	□	□	□	□		□			□	□		□	⊠	⊠	⊠		⊠		⊠						⊠	⊠	⊠	□				⊠	⊠		□	□
⊠	■	⊠	■	◆	◆	⊠	⊠	⊠	□	⊠		⊠	■	■		□	■	■	■	■	◆	□	⊠	⊠	⊠	⊠	□	⊠	⊠	⊠	⊠	⊠		□	⊠	⊠			◆	⊠
		⊠	⊠	⊠		□	□			⊠		⊠	⊠	⊠				⊠	⊠	□																				□	□
⊠	⊠	□	⊠	□	◆	□	□	□	⊠	□		□	□	⊠		□	■	⊠	⊠	□	□		⊠	■	■	⊠			■	⊠	⊠	⊠			■	■			⊠	⊠
	⊠	⊠	◆	◆	□	⊠	⊠	□		⊠		■	■	⊠		□	⊠	■	□	◆	□		⊠						⊠		⊠					⊠			◆	⊠
□	⊠	⊠	⊠	⊠	□	⊠	◆	□	□	⊠	□	⊠	⊠	⊠		□	□	⊠	⊠	⊠	□		⊠	□	□	□			⊠	⊠	⊠	□			□	⊠			⊠	⊠
	⊠				□											⊠	⊠				⊠	⊠		⊠			⊠	⊠	⊠	⊠			⊠	⊠			
⊠	⊠	⊠	⊠	⊠	□	⊠	⊠	□		⊠	□	⊠	⊠	⊠	□	□	⊠	⊠	⊠	⊠	⊠		⊠		⊠	⊠			■	⊠	⊠				■	■			◆	◆
⊠	⊠	⊠	■	□	◆	⊠	⊠	⊠	□	⊠		□	□	■		□	⊠	■	⊠	⊠	⊠		□					□	□	■	⊠	□		□	□	□			⊠	⊠
□	□	□	■	□	□	□	⊠	□		⊠			⊠	⊠		□		⊠	⊠	⊠	□		□						□			□			⊠	□	⊠		⊠	⊠
□	⊠	⊠	⊠	⊠	□	□	⊠	□	□	□	□	⊠	⊠	⊠		□	⊠	⊠	⊠	⊠	□		□	□	□	□			⊠	■	■	⊠		□	□	⊠		□	⊠	⊠
□	⊠	□	⊠	□	□	□	□	□		□		□	⊠	⊠		□	□	⊠	⊠	⊠	□		■						■		⊠	⊠				■	⊠		□	⊠
	⊠	□	⊠	⊠		□	⊠			⊠		⊠	⊠	□				⊠	⊠	⊠	□		□						⊠		⊠					⊠			□	⊠

Key ■ = Abundant ⊠ = Common □ = Uncommon ◆ = Seasonal

	Page Number	parking lot	privy/restrooms	picnic areas	primitive camping	camper pads	river access	boat ramps	fishing docks/jetties	boat rental	viewing blinds	waterfowl blinds	interpretive programs	exhibits/displays	MDC office	hunting	waterfowl hunting	fishing	hiking trails	horse trails	canoeing	archery range	firearms range	field trials	dog training	trapping	Notes
BRAYS ACCESS	140																										
BRICKLEY HOLLOW ACCESS	20															■		■			■					■	
BRICKYARD HILL CA	10	■	■	■	■			■								■	■	■	■		■	■		■	■	■	
BRIDGER (JIM) URBAN CA	102	♿	♿	■									■	■	♿				♿								
BRIDGEWATER ACCESS	156				■		■											■			■						
BRISTLE RIDGE ACCESS	156	■					■											■			■						
BRISTOW CA	226	■														■		■							■	■	
BROOKFIELD CITY LAKE	124	■	■	■	■			■								■	■	■									
BROWN BEND ACCESS	34	♿	♿	■			■	■	♿									■			■						
BROWN (BOB) CA	94	■	■		■		■									■	■	■			■			■	■	■	
BROWN (TOM) ACCESS	8	■					■											■			■						
BROWN'S FORD ACCESS	108	♿	♿	■			■	■										■			■						
BRUNS (DR. BERNARD) ACCESS	160	■			■		■									■		■			■						
BRUNSWICK ACCESS	46	■	■	■	■		■	■										■			■						
BRUSH CREEK CA	176	■														■											
BUCK AND DOE RUN CA	51	■			■											■	■	■						■	■	■	
BUFFALO HILLS NA	128	♿														■			■							■	
BUFFALO WALLOW PRAIRIE CA	16	■														■											
BUFORD MOUNTAIN CA	100	■			■											■			■							■	
BULL SHOALS RESERVOIR	162	■	■	■	■	■	■	■	♿	■	♿				■	■	■	■			■				■	■	
BUNCH HOLLOW CA	38	■			■											■	■	■		■					■		
BURR OAK WOODS CA	102	♿	♿	■					♿		♿		■	■	♿				♿								
BURTON BRIDGE ACCESS	26	■					■											■			■						
BUSCH (AUGUST A.) MEM CA	192	♿	♿	■					♿	■	♿		■	■	♿	■		■	♿			■	♿	■	■	■	
BUSHWHACKER LAKE CA	16	♿	■		■			■	♿							■	■	■			■	■			■	■	
BUSIEK SF & WA	48	■			■											■		■	■				■				
BUTLER CITY LAKE	18	♿	♿					■	♿									■									
BUTTIN ROCK ACCESS	210	■					■	■										■			■						
BUZZARD BLUFF ACCESS	238	■		■	■		■	■								■		■			■					■	
CABOOL TOWERSITE	222															■											
CALDWELL MEM WA	76	♿	♿						■																		
CALLAGHAN MOUND ACCESS	136	■			■		■									■	■	■			■			■	■	■	
CALUMET CREEK ACCESS	172	■	■	■	■		■	■										■			■						
CAMERON (RESERVOIRS 1,2, AND 3)	68	■	■					■								■	■	■	■		■	■					
CAMP BRANCH ACCESS	238																	■			■						
CAMPBELL BRIDGE ACCESS	60	♿	♿				■	■										■			■						
CANAAN CA	80	■			■											■									■	■	
CANEY MOUNTAIN CA	162	♿	♿	■	■						■		■		♿	■			■			♿	■	■	■	■	
CANTON FERRY ACCESS	120	♿		■			■	■										■			■						
CAPE WOODS CA	36						■											■									
CAPITOL VIEW ACCESS	32	■	♿				■	■										■			■						
CARL JUNCTION ACCESS	106	■					■	■										■			■						
CARMICHAEL (MAC & ZELMA) SF	28															■										■	
CARPENTER MEM CA	150	■			■											■									■	■	
CARTER CREEK CA	40				■											■									■	■	
CARUTHERSVILLE ROOKERY CA	164	■																■									
CASH (BEN) MEM CA	74															■	■	■			■					■	
CASSVILLE CITY GREENWAY	14	■		■			■						■					■									
CASTOR RIVER CA	22	■			■	■										■			■	■							
CATAWISSA CA	76	■					■	■								■	■	■			■				■	■	
CEDAR BLUFF ACCESS	146	■			■		■									■		■			■						
CEDAR GROVE CA	70				■											■											
CEDAR HILL ACCESS	108	■	■				■											■			■						
CENTERVILLE ACCESS	188				■		■											■			■						
CEPHAS FORD ACCESS	226	■	■		■		■	■								■	■	■			■				■		
CHALK BLUFF TRAIL ACCESS	74															■	■	■			■					■	
CHAMOIS ACCESS	160	■					■	■		■								■									
CHAPEL VIEW PRAIRIE CA	90	■														■		■								■	
CHARITY ACCESS	64	■					■	■								■	■	■			■			■	■	■	
CHEERFUL HILL ACCESS	238															■		■								■	
CHERRY CORNER ACCESS	154	■					■											■			■						
CHESAPEAKE FISH HATCHERY	118	♿	♿	■											♿												
CHILLICOTHE ACCESS	126	♿	♿				■	■										■			■						
CHLOE LOWRY MARSH CA	138	■																	■								
CHOUTEAU CLAIM ACCESS	76	■	■				■	■										■			■						
CHRISTIE (JAMES D.) CA	8	■			■											■		■									
CLARK CA	51	■			■											■	■	■			■			■	■	■	
CLEAR CREEK CA	16	■														■	■	■	■								
CLEARWATER CA	188	■			■											■										■	
CLEARWATER LAKE	188	♿	♿	■	■	■	■	■	♿		♿		■	■	♿	■	■	■	■		■				■	■	
CLEMENT (R. F.) MEM FOREST & WA	70	■			■											■											
CLIFTON HEIGHTS PARK LAKE	200	■	■						♿									■									
CLIFTY CREEK CA	134	■			■											■			■								
CLIFTY CREEK NA	134															■											
CLIMAX SPRINGS TOWERSITE	34	♿																									
CLOVER DELL PARK	168	■	♿					■	♿									■	■			■					
CLUBB CREEK CA	22				■											■											
COFFIN CAVE CA	114															■										■	
COFFMAN BEACH ACCESS	150	♿	♿				■	■	♿									■			■						
COFFMAN TOWERSITE	198	■													♿	■									■		
COLDWATER ACCESS	232				■		■									■	■	■			■					■	
COLDWATER CA	232	■			■											■										■	
COLE CO JAYCEE PARK LAKE	54	♿	♿	■					♿				■					■									
COLTER'S LANDING ACCESS	76	■	■				■	■										■								■	
COMMUNITY CLUB LAKE	193	♿	♿	■					♿									■									
COMPTON HOLLOW CA	234	■			■											■			■			■					
COMSTOCK PRAIRIE CA	16	■														■	■	■									
COOK (ELMER A.) MEM ACCESS	6	■																■									
COOK (FLORENCE) ACCESS	162	■					■	■										■			■					■	
COOLEY LAKE CA	52	■	■				■	■			■					■		■	■		■				■	■	
COON ISLAND CA AND ACCESS	28	■	■				■	■	■							■	■	■			■					■	
COOPER HILL CA	160				■		■											■			■					■	

Key ■ = Yes ♿ = Disabled Access

beaver	bullfrogs	coyotes	deer	doves	eagles	fox	hawks	herons	otters	owls	pheasant	quail	rabbits	racoons	ruffed grouse	shore birds	snapping turtles	songbirds	squirrels	turkey	waterfowl	hybrid bass	largemouth bass	rock bass	small mouth bass	spotted bass	striped bass	white bass	blue gill	carp	catfish	crappie	muskelunge	paddlefish	suckers	sunfish	trout	walleye	mushrooms	wildflowers
⊠	⊠	⊠	⊠	◆	□	⊠	⊠	⊠		⊠		□	■	■		□	⊠	■	■	⊠	◆	□	⊠	⊠	⊠	⊠	□	◆	⊠	□	⊠	⊠			⊠	⊠			⊠	⊠
□	⊠	⊠	⊠	⊠	□	□	⊠	⊠	□	⊠		⊠	⊠	⊠			⊠	⊠	⊠	⊠	⊠	□	⊠	⊠	⊠	⊠	□	◆	⊠	⊠	■	■	⊠	◆	⊠	⊠		◆	⊠	⊠
□	⊠	⊠	■	◆		⊠	⊠	⊠		⊠	⊠	⊠	⊠	■		□	□	⊠	⊠	■	□		⊠						⊠	□	⊠	□			□	⊠			◆	⊠
	⊠	⊠	■	⊠		⊠	⊠			⊠		■	■	⊠			□	■	■	■	□																		⊠	⊠
⊠	⊠	⊠	⊠	⊠	□	⊠	⊠	⊠	□	⊠	□	□	□	⊠		□	⊠	⊠	□	□	□	□	⊠						⊠	⊠	⊠	□			□	⊠				⊠
□	⊠	□		□				⊠			□	□		⊠		□	⊠	⊠			□							⊠	□	■	■	⊠		□	■	⊠				
	□	⊠	□	⊠		□	⊠	□		⊠		⊠	⊠	⊠		□	⊠	⊠	⊠	□	□		⊠						⊠		□					⊠			◆	□
□	⊠	⊠	⊠	□	□	⊠	⊠	⊠	□	⊠		⊠	⊠	⊠		□	⊠	⊠	⊠	⊠	⊠		■						■	■	⊠	■				⊠			□	⊠
□		□	□	□	□	□	⊠	⊠		□	□	□	□	□	□	⊠	□	⊠	⊠	□	⊠		⊠	□	□	⊠	⊠	■	■	⊠	■	◆	□	⊠	⊠	■		⊠	□	□
⊠	⊠	⊠	■	◆	◆	⊠	⊠	⊠	□	⊠	■	⊠	⊠	■		◆	⊠	⊠	⊠	□	◆		□					□	□	■	■	⊠		□	⊠	□		□	◆	⊠
■	⊠	⊠	⊠	⊠	◆	□	⊠	⊠	⊠	□	⊠	⊠	⊠	⊠		⊠	⊠	⊠	□	□	◆		□					⊠	□	■	■	⊠		⊠	⊠	□			⊠	⊠
⊠	⊠	⊠	⊠	⊠	□	⊠	⊠	⊠	□	⊠		□	⊠	⊠		⊠	⊠	⊠	⊠	⊠	⊠		■	⊠	■	■			⊠	⊠	⊠	⊠			■	■			□	⊠
⊠	⊠	⊠	⊠	⊠	□	⊠	⊠	⊠		⊠		⊠	⊠	⊠		⊠	⊠	⊠	⊠	⊠	⊠		⊠	⊠	⊠	⊠			⊠	⊠	⊠	⊠			⊠	⊠			⊠	⊠
□	□	□	□	□	□	□	□	□	□	□		□	□	□		□	□	□	□	□	□		□	□	□	□			⊠	■	■	⊠		□	□	⊠		□	□	□
		⊠	⊠	□		□	⊠			⊠		□	□	⊠				⊠	⊠	⊠																			⊠	⊠
⊠	⊠	⊠	⊠	□	□	⊠	⊠	□		■			□	■		□	⊠	■	■	⊠	◆		□						□	□	□	□				□		□	◆	□
	□	⊠	⊠	⊠	□	⊠	⊠			⊠		⊠	⊠	⊠			□	⊠	⊠	□																			⊠	⊠
	□	⊠	⊠	□		□	⊠	□		□		⊠	⊠	□		□	□	⊠	□	□			□						□							□			□	■
		⊠	⊠	⊠	□	⊠	⊠			⊠		□	⊠	⊠				⊠	⊠	⊠																			⊠	⊠
⊠	⊠	⊠	■	⊠	◆	⊠	⊠	⊠	□	⊠		⊠	⊠	⊠	□	⊠	⊠	■	⊠	⊠	◆	□	■	□	⊠	■	□	■	■	■	■	⊠			■	■		⊠	⊠	■
□	⊠	⊠	⊠	⊠	□	⊠	◆	□		⊠	□	⊠	⊠	⊠		□	⊠	⊠	⊠	⊠	□		⊠	□		□			⊠	⊠	⊠	□			□	⊠			⊠	⊠
⊠	■	■	■	◆	□	⊠	⊠	□		⊠		⊠	■	■			⊠	■	■	⊠	□		⊠						⊠	□	⊠	□				⊠			■	■
⊠	⊠	⊠	⊠		□	□	⊠	⊠				⊠	⊠	⊠		□	□	⊠	□				⊠			□			⊠	■	■	□		□	□	⊠				
⊠	⊠	⊠	■	⊠	□	⊠	⊠	⊠		⊠	□	■	■	■	□	⊠	⊠	■	■	■	⊠	□	■		□				■	⊠	■	■			□	■			◆	■
□	⊠	⊠	■	⊠	□	□	⊠	⊠		□		⊠	■	□		⊠	⊠	■	⊠	⊠	◆		⊠						■		⊠	⊠				■			□	⊠
□	□	□	⊠	□		□	⊠	□		⊠		□	⊠	⊠				■	■	⊠				□	□										□	□			⊠	■
□	⊠	⊠	□	⊠		□		□		□		□	⊠	⊠		□	⊠	⊠	□		◆		⊠						⊠		⊠	⊠				⊠			□	□
□	⊠	□	□		⊠	□	⊠	⊠		□		□	□	□		□	⊠	⊠	□	□	□	□	□	⊠	⊠	⊠	⊠	□	⊠	□	□	□		□	⊠	⊠	□	□	⊠	⊠
⊠	■	⊠	⊠	⊠	◆	□	⊠	⊠	⊠	□		⊠	⊠	⊠		□	■	⊠	⊠	⊠	□		⊠	■	■	⊠			■	⊠	■	⊠			■	■			⊠	⊠
		⊠	⊠	□		□	□			⊠		□	□	⊠				⊠	⊠	⊠																			⊠	□
□	⊠	⊠	⊠	⊠	□	⊠	⊠	⊠		⊠	□	⊠	⊠	⊠	□		⊠	■	⊠	⊠	□		□	⊠	⊠	□			■	⊠	⊠	□				■			◆	◆
⊠	⊠	⊠	⊠	⊠	□	⊠	⊠	□		⊠		⊠	⊠	⊠		□	⊠	■	⊠	⊠	□		□		⊠				□	■	■	□			⊠	□			⊠	⊠
□	⊠	□	□	□	◆	□	⊠	□	□	□		□	□	⊠		□	⊠	⊠	□	□	□	□	⊠	□	□	□	□	⊠	⊠	⊠	⊠	⊠		□	□	⊠			□	⊠
□	⊠	□	■	⊠	□	□	⊠	⊠		□	□	⊠	⊠	⊠		□	■	⊠	⊠	⊠	◆	□	■						■	□	■	⊠				□			□	□
⊠	⊠	⊠	⊠	□	◆	□	⊠	⊠	□	□		□	⊠	⊠		□	■	⊠	⊠	□	□		⊠	■	■	⊠			■	⊠	■	⊠			■	■			⊠	⊠
⊠	⊠	⊠	⊠	⊠	□	□	⊠	⊠	□	□		⊠	⊠	⊠		⊠	⊠	⊠	⊠	⊠	⊠		⊠	⊠	⊠	□			⊠	⊠	⊠	□			⊠	⊠	□	□		□
	□	■	□	⊠		□	⊠	⊠		⊠		□	□	⊠		□	□	⊠	⊠	⊠																			◆	◆
□	⊠	⊠	■	⊠	◆	■	■	⊠		⊠		⊠	■	■		□	⊠	■	■	■	◆																		⊠	■
					⊠		□	⊠		□						⊠	□	⊠	⊠		⊠	⊠	⊠		⊠			⊠	⊠	■	■	⊠		⊠	⊠	⊠		□	□	□
⊠	⊠	□	⊠	□		□	□	□	□	□			□	⊠			⊠	⊠	□		□		⊠	□		□			⊠	□	□	□			⊠	⊠			⊠	□
⊠	■	⊠	□	□		□		⊠	□	□		□	□	■		□	■	⊠	□	□	⊠	□	⊠			⊠	□	⊠	⊠	■	■	⊠		⊠	□	⊠			⊠	⊠
⊠	⊠	⊠	⊠	⊠	□	⊠	⊠	⊠		⊠		□	□	⊠		□	⊠	⊠	⊠	⊠	□		⊠	⊠	⊠	⊠			⊠	⊠	⊠	□			⊠	⊠			□	□
			□							⊠								⊠	□	□	□																		□	⊠
	□	⊠	■	◆	□	⊠	⊠	□		⊠		⊠	⊠	■		□	□	■	■	■																			⊠	⊠
		⊠	⊠	□	□	⊠	⊠			⊠		□	⊠	⊠	□			⊠	⊠	⊠																			⊠	⊠
	⊠						◆	◆		□						◆	□	⊠			⊠								□	□					□	□				
⊠	⊠	⊠	□	□	□	□	⊠	⊠	⊠	⊠		□	⊠	⊠		◆	⊠	⊠	⊠	□	◆		□			⊠		□	⊠	⊠	⊠	⊠			⊠	■			⊠	⊠
□	⊠	□	□	□	□	□	□	□	□	□		□	⊠	⊠		□	□	⊠	⊠	□	□		□	□	□	□			□	□	□	□			⊠	⊠			□	⊠
□	□	■	⊠	□	□	⊠	⊠	□		■		□	□	■	□	□	□	■	■	⊠	□																		⊠	■
⊠	⊠	⊠	⊠	⊠	□	⊠	⊠	⊠		⊠		⊠	⊠	⊠		⊠	⊠	⊠	⊠	⊠	⊠		⊠	⊠	⊠				⊠	⊠	⊠	⊠		⊠	⊠	⊠			◆	◆
□	⊠	⊠	⊠	⊠	□	⊠	⊠	⊠	□	⊠		⊠	⊠	⊠		□	⊠	⊠	⊠	⊠	□		⊠		□	□		◆	⊠	⊠	⊠	⊠			◆	⊠		◆	⊠	⊠
□		⊠	⊠	⊠	□	⊠	⊠	⊠	□	⊠		□	⊠	⊠				⊠	⊠	⊠	□																		⊠	⊠
⊠	⊠	⊠	⊠	⊠	□	□	⊠	⊠	□	⊠		⊠	⊠	⊠		⊠	⊠	⊠	⊠	⊠	⊠		■	⊠	■	■			⊠	⊠	⊠	⊠			■	■			□	⊠
	⊠	⊠	⊠	⊠	□	⊠	⊠	⊠		⊠		□	⊠	⊠		□	□	⊠	⊠	⊠	⊠			⊠	⊠			□	⊠	⊠	□	□			⊠	⊠			⊠	⊠
⊠	◆	⊠	⊠	⊠	□	□	⊠	⊠	□	⊠		⊠	⊠	⊠		□	◆	■	⊠	⊠	◆	□	□			□		⊠	⊠	■	■	□		□	□	⊠		□	⊠	⊠
⊠	⊠	⊠	□	⊠	□	□	□	⊠	□	□		□	⊠	⊠		⊠	□	⊠	□	□	⊠		□						⊠	⊠	⊠	□				⊠			⊠	⊠
																						⊠	⊠	⊠		⊠	⊠	⊠	⊠	⊠	⊠	⊠		⊠	⊠	⊠		⊠		
	⊠	⊠	⊠	◆		⊠	⊠	□		□		⊠	⊠	⊠			□	⊠	□	⊠	□		⊠						⊠		⊠					⊠			⊠	■
⊠	⊠	⊠	⊠	⊠	□	⊠	⊠	⊠	□	⊠		⊠	⊠	⊠		□	⊠	⊠	⊠	⊠	□		□	⊠	⊠	□			⊠	⊠	⊠	⊠			⊠	⊠	□		⊠	⊠
□	■	⊠	■	□	◆	□	□	⊠	□	□		□	⊠	⊠		□	■	⊠	⊠	⊠	□		⊠	■	■	□			⊠	⊠	⊠	⊠			■	■			□	□
□	■	□	□	□		□		□	□	□		□	□	■			□	□	□	□	□		⊠	⊠	■	⊠			■	□	⊠	⊠			■	■			□	□
□	⊠	□	□	□	□	□	□	⊠		□		⊠	⊠	⊠		□	⊠	⊠	⊠		□	⊠	⊠						⊠		⊠		⊠			⊠		⊠	□	⊠
⊠	⊠	□	⊠	□	□	□	⊠	⊠	⊠	⊠	□	□	□	⊠		⊠	⊠	⊠	□	⊠	⊠		⊠					□	⊠	■	■	⊠		□	⊠	■		□	⊠	⊠
⊠	⊠	⊠	⊠	⊠	□	⊠	⊠	◆		⊠	⊠	⊠	⊠	⊠		◆	⊠	⊠	⊠		◆																		⊠	⊠
⊠	⊠	⊠	⊠	⊠	□	⊠	⊠	⊠	□	⊠		□	⊠	⊠		⊠	⊠	⊠	⊠	⊠	□		⊠	⊠	⊠				⊠	⊠	⊠				⊠	⊠			◆	◆
□	⊠	⊠	⊠	◆		⊠	⊠	□		⊠	⊠	⊠	⊠	⊠		□	□	⊠	⊠	⊠	□		⊠			□					⊠	□				□			■	■
⊠	⊠	■	■	⊠		⊠	■	■	□	■	⊠	⊠	■	■		□	⊠	■	■	■	◆		■						■	⊠	■	⊠				■			◆	◆
□	□	⊠	⊠	□		□	⊠	□		□		⊠	⊠	□		□	□	■	⊠	⊠	□		⊠						⊠	□	□					⊠			□	■
⊠	⊠	⊠	⊠	⊠	□	⊠	⊠	□		⊠		⊠	⊠	⊠	□		□	⊠	⊠	⊠	□																		⊠	⊠
⊠	⊠	⊠	⊠	⊠	⊠	⊠	⊠	⊠		⊠		⊠	⊠	⊠	□	⊠	⊠	⊠	⊠	⊠	◆		⊠	□	□	⊠		⊠	⊠	■	⊠	⊠			⊠	⊠		□	⊠	⊠
		⊠	⊠	□		⊠	⊠	□		⊠		□	□	⊠				⊠	⊠	⊠																			□	□
				⊠															⊠											⊠	⊠					□				
	□	⊠	⊠	⊠		⊠	⊠	□		⊠		□	⊠	⊠			□	⊠	⊠	⊠			□																⊠	⊠
	□	⊠	⊠	□		⊠	⊠	□		⊠		□	⊠	⊠			□	⊠	⊠	⊠			□													□			⊠	⊠
□	□	□	⊠	□	□	□	⊠	□	□	⊠	□	⊠	⊠	□	□	□	□	⊠	⊠	⊠	□																			□
	⊠	⊠	⊠	⊠		□	⊠	⊠		□		⊠	⊠	⊠		□	⊠	⊠	⊠	□	□		⊠						⊠		⊠	□				□			□	⊠
		⊠	⊠	□	□	□	⊠			⊠		□	□	⊠				⊠	■	⊠																			⊠	⊠
	□	⊠	■	□	⊠	⊠	⊠			⊠		□	⊠	⊠		□	□	⊠	■	■	□																		⊠	⊠
□	⊠	□	⊠		◆	⊠	⊠	□	□	□			□	⊠		⊠	⊠	⊠	⊠	⊠	⊠	⊠	⊠	⊠	⊠	⊠	⊠	⊠	⊠	⊠	⊠	⊠		⊠	⊠	⊠		⊠	⊠	⊠
		□	⊠			⊠	⊠			⊠			□	⊠				⊠	■	⊠																			⊠	⊠
⊠	⊠	⊠	⊠	⊠	□	⊠	⊠	⊠	□	⊠		□	□	⊠			■	⊠	⊠	⊠	⊠		⊠	⊠	⊠	⊠	□	⊠	⊠	⊠	⊠	□			⊠	⊠		□	⊠	⊠
		⊠	⊠	⊠		□	⊠			⊠		□	□	⊠				■	■	⊠																			□	■
	■	□	□	⊠		□	⊠	⊠		□		□	⊠	⊠		□	■	■	⊠	□	□		■						■	□	■	⊠				■			□	■
⊠	⊠	□	□	□	□	□	⊠	⊠	□	⊠		□	⊠	⊠		⊠	⊠	⊠	⊠	□	□	□	⊠	⊠	⊠	□	□	□	⊠	⊠	⊠	⊠	□	□	⊠	⊠	□	□	⊠	⊠
⊠	■	⊠	□	⊠		□	□	⊠		□			□	⊠			⊠	⊠	⊠	□	□		■						■		■					□			□	⊠
□	□	⊠	⊠	⊠		□	⊠	□		⊠		⊠	⊠	⊠		□	□	⊠	⊠	⊠																			⊠	⊠
□	□	⊠	⊠	□		□	⊠	□		□		⊠	⊠	□		□	□	⊠	□	□	□		□						⊠	□	⊠	□				⊠			□	■
⊠	⊠	⊠	⊠		□	⊠	⊠	□	□	⊠		□	□	⊠	□	□	□	⊠	⊠	⊠	□									⊠	⊠	□								◆
⊠	⊠	⊠	□	⊠	◆	⊠	⊠	⊠	□	⊠		□	□	⊠		⊠	⊠	⊠	⊠	□	◆	⊠	⊠	■	⊠	□	⊠	■	■	■	■	□			■	■	□	◆	◆	□
■	■	⊠	⊠	◆	◆	□	⊠	■	□	⊠		■	⊠	⊠		◆	■	⊠	⊠	□	◆		□					□	⊠	■	■	□		⊠	■	⊠		□	◆	□
⊠	⊠	⊠	⊠	⊠	□	⊠	⊠	⊠	⊠	⊠		□	⊠	⊠		◆	⊠	⊠	⊠	⊠	◆		⊠	⊠	□	⊠		⊠	⊠	⊠	⊠	⊠		□	⊠	⊠		□	⊠	⊠
□	⊠	⊠	⊠		□	⊠	⊠	□	□	⊠		□	⊠	⊠		⊠	⊠	⊠	⊠	⊠	□	□	⊠	⊠	⊠	□	□	□	⊠	⊠	⊠	⊠		□	⊠	⊠		⊠	⊠	⊠

Key ■ = Abundant ⊠ = Common □ = Uncommon ◆ = Seasonal

	Page Number	parking lot	privy/restrooms	picnic areas	primitive camping	camper pads	river access	boat ramps	fishing docks/jetties	boat rental	viewing blinds	waterfowl blinds	interpretive programs	exhibits/displays	MDC office		hunting	waterfowl hunting	fishing	hiking trails	horse trails	canoeing	archery range	firearms range	field trials	dog training	trapping	Notes
CORKWOOD CA	28	■															■	■	■								■	
COSMO-BETHEL LAKE	24	■	■	■															■									
COVER (D. & M.) MEM WA	158	■			■												■		■								■	
COWSKIN ACCESS	128	■	■				■	■									■		■			■					■	
CREVE COEUR LAKE	200	♿	♿	■				■	♿										■	■		■	■					
CRIGHTON (JOE) ACCESS	86	■					■												■			■						
CROOKED CREEK CA	60	■															■										■	
CROOKED RIVER CA	186	■			■		■										■	■	■			■				■	■	
CROSS TIMBERS ACCESS	92	■	■		■		■	■									■	■	■			■					■	
CROUCH (R. H.) ACCESS	122	■					■												■			■						
CROWLEY'S RIDGE CA	214	■			■												■	■	■						■	■		
CRUSADER'S ACCESS	87						■												■			■						
CUIVRE ISLAND CA	122	■			■		■	■									■	■	■	■		■					■	
CURRENT RIVER CA	40	■			■												■		■	■							■	
DAIRY FARM LAKE #1	24	■						■														■						
DAIRY FARM LAKE #3	24																		■			■						
DALTON BOTTOMS ACCESS	46																■		■			■					■	
DAN RIVER ACCESS	28																											
DANIEL BOONE CA	228	■		■	■												■		■		■				■	■	■	
DANVILLE CA	148	■	■			■											■	■	■	♿								
DARK CYPRESS SWAMP CA	22	■			■		■	■									■	■	■			■			■	■	■	
DARK HOLLOW NA	218				■												■											
DAVE ROCK CA	194	■															■										■	
DAVIS (DEAN) CA	98	■															■		■	■								
DAVIS (EVA N.) MEM CA	8																■											
DAVIS FORD ACCESS	114	■					■	■											■			■					■	
DAVIS MEM SF	16	■															■	■	■	■								
DAVISDALE CA	96	■			■												■	■	■	■							■	
DE BOURGMONT ACCESS	58	■	■				■	■											■			■						
DEEP FORD ACCESS	128	■					■	■									■		■			■					■	
DEER RIDGE CA	120	♿	♿	■	■	■		■									■	■	■	■	■	■		■	■	■	■	
DELANEY (ROBERT G.) CA	142	♿	♿	■				■	♿										■									
DELAWARE TOWN ACCESS	48	■		■			■	■											■			■						
DENTON (ANDY) ACCESS	82	■					■										■	■	■			■				■	■	
DENVER ACCESS	236	■	■	■	■	■	■	■									■		■			■						
DEXTER CITY LAKE	214	♿																	■									
DIAMOND GROVE PRAIRIE CA	154	■															■											
DIGGS (MARSHALL I.) CA	12	■	■	■	■			■									■	■	■			■						
DIXON TOWERSITE	178	■			■												■											
DODD ACCESS	130				■														■			■						
DOG'S BLUFF ACCESS	222	■	■	■			■	■											■			■						
DONALDSON POINT CA	152	■			■												■	■	■								■	
DONIPHAN TOWERSITE	190	■																										
DORENA ACCESS	142	■					■	■											■									
DORRIS CREEK PRAIRIE CA	16	■															■	■	■									
DORSETT HILL PRAIRIE CA	42	■																										
DOUGLAS BRANCH CA	226	■															■	■	■							■		
DRESSER ISLAND ACCESS	193	♿					■	■	■								■	■	■			■				■	■	
DRIPPING SPRINGS NA	222																		■			■						
DROVER'S PRAIRIE CA	168	■																										
DRURY-MINCY CA	220	■	■		■										♿		■	■	■						■	■		
DRYNOB ACCESS	114	■					■												■			■					■	
DRYWOOD CA	16	■															■	■	■							■		
DUCK CREEK CA	22	♿	♿		■	■		■	♿	■		♿		■	♿		■	■	■			■			■	■		
DUNN FORD ACCESS	136	■			■		■										■	■	■			■			■	■	■	
DUPONT RESERVATION CA	172	■	♿		■	■	■	■									■		■	■							■	
DUPREE (ARTHUR) MEM CA	26						■										■	■	■									
EAGLE BLUFFS CA	24	♿	♿												■		■	■	■			■				■	■	
EARTHQUAKE HOLLOW CA	32	■															■			■								
ECK (PETER A.) CA	222																■		■	■		■					■	
EDMONSON ACCESS	20	■					■										■		■			■					■	
ELAM BEND CA	82	■			■		■										■	■	■			■			■	■	■	
ELMSLIE MEM CA	136	■			■		■										■	■	■			■			■	■	■	
ELMWOOD LAKE	218	■			■			■									■	■	■								■	
ELROD MILL ACCESS	8	■	■		■		■										■		■			■						
EMMENEGGER NATURE PARK	203	■	■										■							■								
ENGELMANN WOODS NA	76	■																		■								
EUGENE TOWERSITE	140																											
EWING (ELLA) LAKE	207	■	■		■			■											■			■						
EWING LAKE	120	♿	♿	■				■											■			■						
FABIUS CHUTE ACCESS	136	■					■	■									■	■	■			■					■	
FAIRGROUNDS PARK LAKE	200	■	■	■					■										■									
FARMINGTON CITY LAKE	196	♿							♿										■									
FARRINGTON PARK LAKE	90	♿	♿	■					♿	■									■	■								
FAYETTE CITY LAKES	96	♿	♿	■	■	■		■	♿			■					■	■	■	■		■	■	■				
FELLOWS LAKE	86	♿	♿	■				■	♿	■								■	■			■						
FENWAY LANDING ACCESS	120	■	■	■	■	■	■	■											■									
FEWEL (CONNOR O.) CA	90	■															■	■	■								■	
FIDDLERS FORD ACCESS	62	■					■										■		■			■						
FIERY FORK CA	34	♿	♿	■	■		■										■	■	■	■		■				■	■	
FINLEY CREEK PARK	48	♿	♿	■			■	■	♿										■			■						
FISK ACCESS	28	■					■	■											■			■						
FLAG SPRING CA	14	■			■												■			■	■			■			■	
FLAMM CITY ACCESS	108	■					■	■											■	■		■						
FLATWOOD CHURCH CA	114	■			■												■								■	■	■	
FLATWOODS CA	232	♿			■												■						■	♿				
FLIGHT LAKE CA	226	■															■	■	■							■		
FLOYD (NANNIE B.) MEM CA	46	■			■												■	■										
FOREST 44 CA	200	♿	♿										■		♿		■			♿				♿				
FORT CROWDER CA	154	■	■		■	■										■	■			■			♿		■	■	■	
FORT PIKE ACCESS	51					■	■												■			■						
FOUNTAIN GROVE CA	124	♿	♿	■	■			■	■		■	♿		■	■		■	■	■	♿		■				■		

Key ■ = Yes ♿ = Disabled Access

beaver	bullfrogs	coyotes	deer	doves	eagles	fox	hawks	herons	otters	owls	pheasant	quail	rabbits	racoons	ruffed grouse	shore birds	snapping turtles	songbirds	squirrels	turkey	waterfowl	hybrid bass	largemouth bass	rock bass	small mouth bass	spotted bass	striped bass	white bass	blue gill	carp	catfish	crappie	muskelunge	paddlefish	suckers	sunfish	trout	walleye	mushrooms	wildflowers
□	⊠	⊠	□	⊠	□	□	□	⊠	□	⊠		⊠	■	⊠		⊠	⊠	■	■	□	⊠																		⊠	⊠
□	⊠			⊠		□	⊠	□		□		□	⊠	□		□	⊠	⊠	⊠		⊠		⊠						⊠		⊠	□				⊠			⊠	⊠
	□	⊠	⊠	⊠		⊠	⊠	◆		⊠		⊠	⊠	⊠			□	⊠	⊠	⊠	◆		⊠						⊠							⊠			⊠	⊠
⊠	⊠	⊠	⊠	□	◆	⊠	⊠	⊠		⊠		□	□	⊠		□	⊠	⊠	⊠	□	□		⊠	⊠	⊠	⊠		⊠	⊠	⊠	⊠	□			⊠	⊠			□	□
□	⊠	□	□	⊠	□	□	⊠	⊠		⊠		□	⊠	⊠		⊠	⊠	■	■	□	⊠	□	■				□	□	■	⊠	⊠	■		□	⊠	⊠			□	⊠
□	□	⊠				□	□	⊠		⊠			□	⊠		□	⊠	⊠	⊠		□		⊠	⊠	⊠	□		□	⊠	⊠	⊠	⊠			⊠	⊠			□	□
	□	⊠	⊠	⊠		⊠	□			⊠		⊠	□	⊠			□	⊠	■	⊠			⊠						⊠		⊠					⊠			⊠	⊠
⊠	⊠	⊠	⊠	⊠	□	⊠	◆	◆	□	⊠	□	⊠	⊠	⊠		◆	⊠	⊠	⊠	⊠	□		⊠		□	□			⊠	⊠	⊠	□		□	□	⊠			⊠	⊠
⊠	■	□	⊠	□		□		⊠	□	□		□	⊠	■			□	■	⊠	⊠	□		⊠	■	■	⊠		□	■	□	■	□		□	■	■		□	■	■
□	⊠	□	⊠	□	□	⊠	⊠	□	□	⊠		□	□	⊠		□	⊠	⊠	⊠	⊠	□		⊠	□	□	□		⊠	⊠	⊠	⊠	⊠		□	□	⊠		□	⊠	⊠
□	⊠	■	■	■	□	⊠	■	□		■		■	■	■		□	□	■	■	■	⊠		■						■	⊠	■					■			■	■
⊠	⊠	⊠	□	□	□	□	⊠	□		⊠			⊠	⊠		□	⊠	⊠	⊠		□		□						⊠	⊠	⊠	□			⊠	⊠		□	⊠	⊠
■	■	⊠	⊠	□	◆	□	⊠	■	⊠	⊠		□	□	■		⊠	■	■	■	⊠	◆		⊠	□	□	□		■	⊠	■	⊠	⊠		□	□	⊠		□	■	■
⊠	⊠	■	⊠	⊠	□	⊠	⊠	⊠		⊠		⊠	⊠	⊠	□	□	□	■	■	⊠	□		⊠	⊠	⊠	⊠		□	⊠	⊠	⊠	□			⊠	⊠	□	⊠	⊠	⊠
□	⊠	□	□	⊠	□	□	⊠	⊠		⊠		⊠	⊠	⊠		⊠	⊠	□	□	□	◆		⊠						⊠		⊠					⊠			□	□
	⊠	□	□	⊠		□	□	⊠		□		□	□	□		□	⊠	□	□	□	◆		⊠						⊠			⊠				⊠				
⊠	⊠	⊠	⊠	⊠	◆	⊠	⊠	□	□	⊠	□	□	⊠	⊠		□	⊠	⊠	⊠	□	◆		□						⊠	⊠	⊠	□		□					◆	◆
□	⊠	□	□	◆	□	□	⊠	□	□	□	□	□	□	□	□	□	□	⊠	□	□	◆	□	□	□	□	□	□	□	⊠	⊠	⊠	□	□	□	⊠	⊠	□	□		
	⊠	⊠	⊠	⊠	□	⊠	⊠			⊠		□	□	■	⊠		□	⊠	⊠	⊠	□		⊠						⊠	□	⊠					⊠			⊠	■
□	⊠	■	■	◆	□	⊠	⊠	⊠		■		⊠	⊠	■	⊠	□	⊠	■	■	■	□		⊠	□	□	□			⊠	⊠	⊠	□			□	⊠			■	■
■	⊠	⊠	⊠	⊠	□	⊠	⊠	⊠	⊠	⊠		⊠	■	■		□	⊠	■	■	⊠	◆		⊠	⊠	⊠	⊠			⊠	⊠	⊠	⊠			⊠	⊠			⊠	⊠
	□	⊠	⊠	□	□	⊠	□			⊠		□	□	⊠	□		□	⊠	⊠	⊠																			◆	◆
	□	⊠	⊠	⊠	□	⊠	⊠			⊠		⊠	⊠	⊠			□	■	⊠	⊠																			□	⊠
	□	⊠	⊠	⊠		⊠	⊠			⊠		⊠	⊠	⊠				■	⊠	⊠									⊠							⊠			⊠	■
		⊠	⊠	◆		⊠	⊠			⊠		□	⊠	⊠				⊠	⊠	⊠																			⊠	□
⊠	⊠	⊠	⊠	⊠	⊠	⊠	⊠	⊠	□	⊠		⊠	⊠	⊠		□	⊠	⊠	⊠	⊠	⊠		□	⊠	⊠	⊠			⊠	⊠	⊠	□			⊠	⊠			⊠	⊠
□	⊠	□	⊠	□		⊠	□	□		□		□	⊠	⊠		□	⊠	■	□	□	□		□						□	□						□			⊠	□
⊠	■	□	■	■	□	□	⊠	⊠		⊠	□	■	■	⊠	□	□	⊠	■	■	■	□		⊠						⊠	⊠	⊠				⊠	⊠			◆	◆
□	⊠	□	□	⊠	□	□	◆	⊠	□	⊠		⊠	⊠	⊠		□	⊠	⊠	⊠	□	□		⊠		□	□		□	⊠	⊠	⊠	⊠		□	⊠	⊠		□	□	□
⊠	⊠	⊠	⊠	□	□	⊠	⊠	⊠		⊠		⊠	⊠	⊠		□	⊠	⊠	⊠	□	□		⊠	⊠	⊠	⊠			⊠	⊠	⊠	□			⊠	⊠			⊠	⊠
⊠	⊠	⊠	■	⊠	□	⊠	⊠	⊠		⊠	□	⊠	⊠	■	□	⊠	⊠	■	■	■	⊠		⊠		□			□	⊠	⊠	■	⊠			□	⊠			⊠	■
⊠	⊠	□	□	□	□	□	□	⊠	□	□				⊠		⊠	⊠	⊠	□	□	⊠		⊠	□	⊠	□			⊠	□	⊠	⊠			□	⊠			□	⊠
□	□	□	□	□	□	□	□	□	□	□		□	□	□		□	□	□	□	□	□		⊠	□	⊠	□		□	⊠	⊠	⊠	□		□	⊠	⊠	□	□	□	□
□	□	⊠	⊠	□		□	□	□	□	□	□	⊠	⊠	□		□	□	◆	⊠	⊠	□		□			□			⊠	⊠	⊠				□	□			◆	◆
□	□	□	□	⊠		□	□	□		□	□	⊠	⊠	□		□	□	◆	⊠	□	□									⊠	⊠				□	□			⊠	◆
	■	□		◆				⊠				□	⊠	⊠		⊠	■	⊠			⊠		■						■	⊠	■	□				■				
		□	⊠	□		□	□			□		⊠	□	□				⊠																					□	■
⊠	⊠	⊠	⊠	□	□	□	□	□		⊠	□	⊠	■	⊠	□	□	⊠	⊠	⊠	⊠	□		⊠						⊠		⊠	⊠				⊠			⊠	⊠
		⊠	⊠	□		⊠	⊠			⊠			⊠	⊠				⊠	⊠	⊠																			□	□
□	□	⊠	⊠	⊠		⊠	⊠	□	□	⊠	⊠	⊠	⊠	⊠		□	□	⊠	⊠	⊠	□		⊠						□		⊠				⊠	⊠			◆	◆
□	⊠	□		□	□		□	□	□	□			□	□		□	■	□	□				⊠	■	■	⊠			■	⊠	⊠	⊠			■	■			□	□
⊠	⊠	□	⊠	□	⊠	□	⊠	⊠	⊠	□			⊠	⊠		⊠	⊠	⊠	⊠	⊠	□	□	□	□	□	□	□	⊠	⊠	⊠	⊠	⊠		⊠	⊠	⊠			⊠	⊠
		□	□		□	□	□			□		□	□	□				⊠	⊠	□																			□	□
	□	□			□		□	⊠	□	□			□	□		⊠	⊠	□				□	□	□	□	□	□	⊠	⊠	⊠	⊠	⊠	□	⊠	⊠	⊠	□	□		
□	□	□	□	⊠		□	⊠	⊠		□		□	□	□		◆	□	■			⊠		⊠						⊠		⊠					□			□	■
□	□	⊠	⊠	⊠	□	□	⊠	□		⊠		⊠	⊠	⊠		□	□	■	⊠	⊠	□																		⊠	■
⊠	◆	⊠	⊠	⊠	□	□	⊠	⊠		⊠		□	⊠	⊠		□	□	⊠	⊠	□	□		□						⊠	□	⊠	□				⊠			□	■
■	⊠	□	⊠	⊠	◆	⊠	⊠	■	□	⊠	□	□	□	⊠		◆	⊠	⊠	⊠	□	◆	□	⊠	□		□	□	⊠	⊠	■	⊠	⊠		□		⊠			⊠	⊠
⊠	⊠	⊠	⊠	□	□	□	□	□	⊠	⊠		□	□	⊠		□	■	⊠	⊠	⊠	□		⊠	■	■	⊠			■	⊠	⊠	⊠			■	■			□	□
	⊠	⊠	⊠	□	□	□	◆	□		□		⊠	⊠	⊠		□	⊠	□	□	□	□																		□	■
⊠	□	⊠	■	□	◆	⊠	⊠	◆	□	⊠		□	⊠	⊠		◆	⊠	⊠	■	⊠	◆		□	□	⊠				⊠						◆	⊠			⊠	⊠
⊠	⊠	⊠	⊠	◆	□	⊠	⊠	⊠	□	⊠		⊠	⊠	⊠			⊠	⊠	⊠	⊠	◆		□	⊠	⊠	⊠			⊠	⊠	⊠	□			⊠	⊠			⊠	⊠
□	⊠	⊠	⊠	⊠		□	⊠	□	□	□		⊠	⊠	⊠		□	⊠	⊠	⊠	□	□		⊠		□				⊠	⊠	⊠	□				⊠			⊠	⊠
■	■	⊠	■	◆	◆	⊠	⊠	◆	⊠	■		⊠	■	■		◆	⊠	■	■	■	◆		■		□	□			■	■	■	■				■			■	■
□	⊠	⊠	⊠	⊠		□	⊠	□		⊠		⊠	⊠	⊠		□	⊠	◆	⊠	⊠	□		□		□			□	□	■	■	□		□	⊠	□		□	⊠	■
□	□	⊠	⊠	□	◆	⊠	⊠	□	□	⊠		□	□	■	⊠	□	□	■	■	⊠	□									◆	◆								⊠	■
□	◆	□	⊠	◆	◆	□	⊠	⊠	□	⊠	⊠	□	□	⊠		□		⊠	⊠	⊠	◆		□					□	⊠	■	■	■		□	□	⊠		□	□	□
⊠	⊠	⊠	■	◆	◆	□	⊠	◆	□	⊠		□	⊠	■	□	◆	⊠	⊠	⊠	⊠	◆		⊠		□	□		⊠	⊠	⊠	⊠	⊠		□	⊠	■			⊠	■
	□	⊠	⊠	◆	□	⊠	⊠			■		□	□	⊠	□			■	⊠	⊠	□																		⊠	■
⊠	⊠	■	■	□	□	⊠	□	⊠	⊠	⊠		□	□	■		□	■	⊠	■	■	□		⊠	■	■	⊠			■	⊠	■	⊠			■	■			⊠	□
□	⊠	⊠	■	⊠	□	⊠	⊠	□		⊠		⊠	⊠	⊠			⊠	⊠	⊠	⊠	□		⊠	⊠	■	□		◆	⊠	⊠	◆	◆			⊠	■			⊠	⊠
□	⊠	⊠	■	⊠		□	□	□		□	□	⊠	■	⊠		□	⊠	◆	■	⊠	□		⊠						⊠	⊠	⊠				□	⊠			◆	◆
⊠	⊠	⊠	■	⊠	□	⊠	■	□		⊠		□	⊠	⊠		□	⊠	⊠	■	■	□		□		⊠				□	■	■				⊠	□			⊠	⊠
□	⊠	⊠	⊠	□	□	⊠	⊠	⊠	□	⊠		⊠	⊠	⊠		□	□	⊠	⊠	⊠	⊠		⊠						■		⊠	■				⊠		□	□	□
⊠	□	⊠	⊠	⊠		⊠	⊠	□	□	⊠	□	⊠	⊠	⊠		□	□	⊠	⊠	⊠	□		□					□	□	⊠	⊠	□			□	□		□	⊠	⊠
□	⊠	⊠	⊠	■	□	⊠	⊠	□	□	⊠		□	⊠	⊠		□	□	■	■	⊠	⊠		⊠		⊠			□	⊠	□	⊠	⊠	□		⊠	⊠	□	□	■	■
	□	⊠	■	□	□	⊠	⊠	□		⊠	□	⊠	⊠	⊠	□		□	⊠	⊠	⊠																			◆	◆
	□	⊠	□		□	⊠	⊠	□	□	⊠			⊠	⊠			□	⊠	⊠	□																			□	□
□	■	⊠	⊠	⊠	□	⊠	⊠	⊠		■	□	■	■	■		⊠	■	■	■	⊠	◆		■						■		■	■				■			◆	◆
□	⊠	⊠	□	⊠		□		□		□		⊠	⊠	⊠		□	⊠	⊠	□	□	⊠		■						■	□	■	⊠				■			□	⊠
⊠	⊠	⊠	⊠	⊠	⊠	⊠	⊠	⊠	□	⊠		□	□	■		⊠	⊠	⊠	⊠	⊠	⊠	⊠	⊠		⊠			⊠	⊠	■	■	⊠		⊠	⊠	⊠			⊠	⊠
				⊠															⊠				□						□	⊠	⊠	□				⊠				
	⊠							□					□	□			⊠	⊠	□		□		■						■		⊠					⊠				
	□						□	□									□	□			◆		⊠						⊠		⊠	⊠				⊠			□	□
⊠	⊠	□	⊠	□		□	⊠	⊠		⊠		□	⊠	⊠		□	⊠	⊠	⊠	⊠	◆		⊠						⊠	⊠	⊠	■				⊠			⊠	⊠
⊠	⊠	⊠	⊠	⊠	□	□	⊠	⊠		⊠		⊠	⊠	⊠		⊠	⊠	■	⊠	⊠	⊠		■	□	□	□		□	■	⊠	■	⊠	□		□	■		□	□	⊠
⊠	⊠	□	□	□	◆	□	□	⊠		□			□	⊠		⊠	⊠	⊠	⊠	□	⊠		⊠		□			⊠	■	■	■	■		⊠	⊠	■		⊠	⊠	⊠
□	□	⊠	⊠	◆		⊠	⊠	⊠		⊠		⊠	⊠	⊠		□	□	⊠	⊠	⊠	⊠		⊠						⊠	□	⊠					⊠			⊠	⊠
⊠	⊠	⊠	⊠	⊠	□	□	⊠	⊠		⊠		⊠	⊠	⊠		⊠	⊠	⊠	⊠	□	□		⊠	⊠	⊠	⊠		⊠	⊠	⊠	⊠	⊠			⊠	⊠		⊠	□	⊠
⊠	⊠	□	■	⊠	⊠	⊠	⊠	⊠	□	⊠		⊠	■	⊠	□	□	⊠	⊠	⊠	■	⊠		⊠	⊠	⊠	⊠		□	⊠	⊠	⊠	□			⊠	⊠			⊠	⊠
⊠	⊠	□		□	□		□	⊠		□		□	⊠	⊠		□	□	⊠	■	□	⊠		⊠	⊠	⊠	□			■	⊠	⊠	□			⊠	■			□	⊠
⊠	⊠	⊠	□	□	□	□	□	⊠	□	□		□	⊠	⊠		⊠	⊠	⊠	⊠		□		⊠	⊠					⊠	⊠	⊠	⊠				⊠			⊠	⊠
	□	⊠	⊠	□	□	⊠	⊠			⊠		□	□	⊠			□	⊠	⊠	⊠																			⊠	⊠
⊠	⊠	⊠	⊠	⊠	◆	⊠	⊠	⊠	□	⊠		⊠	⊠	⊠		◆	⊠	⊠	⊠	⊠	◆		⊠	■	⊠	■		□	■	⊠	⊠	■		□	■	■		□	□	⊠
	□	■	⊠	⊠	□	⊠	⊠			⊠		⊠	⊠	⊠			□	⊠	■	⊠																			⊠	⊠
		⊠	⊠	□		⊠	⊠			⊠		□	□	⊠				■	■	■																			⊠	◆
□	◆	⊠	⊠	⊠	□	□	⊠	◆	□	⊠		□	⊠	⊠		◆	□	■	⊠	□	◆							□	□	□	□	□				□			□	⊠
⊠	⊠	⊠	■	⊠	□	□	⊠	⊠	□	⊠		□	□	⊠		□	⊠	■	■	■	◆								□	⊠	⊠				⊠	⊠			⊠	□
	⊠	⊠	■	⊠		⊠	⊠			⊠		⊠	⊠	⊠		□	⊠	⊠	⊠	■	⊠		□	□	□				□		□				□	□			⊠	⊠
	□	⊠	⊠	⊠	□	⊠	⊠	□		⊠		⊠	⊠	⊠			□	⊠	⊠	□																			⊠	⊠
				□	□			□						□		□		□			□		■		□	□			■	■	■	⊠		□	□	■		⊠		⊠
■	■	⊠	■	⊠	◆	⊠	⊠	◆	⊠	⊠	□	⊠	■	■		◆	⊠	⊠	■	■	◆		⊠	□	⊠				⊠	■	■	⊠			⊠	⊠		⊠	⊠	⊠

Key ■ = Abundant ⊠ = Common □ = Uncommon ◆ = Seasonal

	Page Number	parking lot	privy/restrooms	picnic areas	primitive camping	camper pads	river access	boat ramps	fishing docks/jetties	boat rental	viewing blinds	waterfowl blinds	interpretive programs	exhibits/displays	MDC office	hunting	waterfowl hunting	fishing	hiking trails	horse trails	canoeing	archery range	firearms range	field trials	dog training	trapping	Notes
FOUR RIVERS CA	18	♿	♿		■		■	■							■	■	■	■	■		■			■	■		
FOURCHE CREEK CA	190	■			■											■										■	
FOX VALLEY LAKE CA	51	♿	♿		■		■	■	♿							■		■			■	■	■	■	■	■	
FOXGLOVE CA	186																										
FRANKLIN ISLAND CA	96	■					■	■								■	■	■			■					■	
FREDERICKSBURG FERRY ACCESS	80	■	■		■	■	■	■										■			■					■	
FREDERICKTOWN CITY LAKE	132	♿	♿	■				■				■				■	■	■	■		■					■	
FREEBURG TOWERSITE	134	■			■											■											
FRENCH BOTTOM ACCESS	26						■											■			■						
FRIENDLY PRAIRIE CA	168	■																									
FRISBEE CUTOFF ACCESS	74							■								■	■	■								■	
FROST ISLAND CA	51	■			■		■									■	■	■			■			■	■	■	
FT. LEONARD WOOD TOWERSITE	178	■			■											■											
FUNK MEM SF & WA	100				■											■		■								■	
FUSON (JOHN ALVA, MD) CA	238	■			■											■		■						■		■	
GALE (LARRY R.) ACCESS	34	♿	♿				■	■	♿							■	■	■			■				■	■	
GALLATIN CA	66	■			■											■	■	■	■				■		■		
GAMA GRASS PRAIRIE CA	226	■														■									■		
GASCONADE HILLS CA	178				■											■		■									
GASCONADE PARK ACCESS	80	■	♿		■		■	■										■			■						
GAY FEATHER PRAIRIE CA	226	■																									
GAYOSO BEND CA	164				■											■	■	■								■	
GENERAL WATKINS CA	208	■		■	■	■										■		■	■							■	
GEODE ACCESS	51	■					■											■			■						
GIPSY TOWERSITE	22				■											■											
GIRVIN (J. L.& G.) CA	152	■			■											■	■	■								■	
GOODMAN TOWERSITE	154	■														■										■	
GOOSE CREEK CA	64				■											■								■	■	■	
GRAND BLUFFS CA	148	■			■											■								■	■	■	
GRAND PASS CA	204	■	■		■		■								■	■	■	■	■						■		
GRAND TRACE CA	88	■			■											■	■	■	■						■		
GRANDFATHER PRAIRIE CA	168	■														■									■		
GRANDIN TOWERSITE	190	■			■											■									■	■	
GRASSY TOWERSITE	22	■			■											■											
GRAVES MOUNTAIN CA	100	■			■											■											
GRAVOIS CREEK CA	200																										
GRAVOIS MILLS ACCESS	150	♿	♿	■			■	■	♿									■			■						
GREEN ACCESS	66	■					■											■			■						
GREEN CITY LAKE	218	■	♿		■			■										■								■	
GREENTREE PARK ACCESS	200	♿	♿	■			■	■					■					■			■						
GREENVILLE FORD ACCESS	190	■					■	■										■			■						
GRIFFITH MEM CA	130	■			■											■	■	■						■	■	■	
GRINDSTONE RESERVOIR	68	■	♿					■	♿							■	■	■									
GROVE (CECIL G.) MEM CA	116	■														■											
GRUNDY MEM WA	72	■			■											■			■						■	■	
GRUNER FORD ACCESS	196	■					■								♿			■			■						
HADORN BRIDGE ACCESS	8	■			■		■									■	■	■			■						
HALLS MILL ACCESS	120	■	■				■											■			■						
HAMBURG FERRY ACCESS	172	♿	♿				■	■										■			■						
HAMILTON CITY LAKE	30	■	■	■	■			■										■			■						
HAMS PRAIRIE ACCESS	32	■					■											■			■						
HAPPY HOLLER CA	8	■	♿		■		■		♿							■	■	■	■		■				■		
HARDIN CA	186															■	■	■							■		
HARMONY MISSION LAKE CA	18	♿	♿					■	♿							■	■	■			■				■		
HARRIMAN HILL ACCESS	58	■	■		■		■	■										■			■						
HARTER (JAMES R.) CA	42	■																■							■		
HARTSBURG ACCESS	24	■	♿				■	■										■									
HARVIELL ACCESS	28	■					■	■								■		■			■						
HAWN ACCESS	22						■									■		■							■		
HAZEL CREEK LAKE	6	■	♿					■								■	■	■								■	
HAZEL HILL LAKE	110	♿	♿					■	♿							■	■	■						■	■	■	
HAZELGREEN ACCESS	114	■					■	■										■			■						
HEADWATERS ACCESS	36	■					■	■										■			■						
HEATH (CHARLIE) MEM CA	51	■			■		■									■	■	■			■			■	■	■	
HEBRON ACCESS	72	■	♿	■	■		■	■										■			■					■	
HELDS ISLAND ACCESS	80	■	■		■	■	■	■										■			■					■	
HELTON (WAYNE) MEM WA	88	■			■											■	■	■	■					■	■		
HEMENWAY CA	190				■											■										■	
HENNING (RUTH & PAUL) CA	216	♿	♿	■			■						■	■					■								
HENRY'S MILL ACCESS	218						■											■			■						
HENSHAW (CLAY) MEM ACCESS	86	♿	♿						♿									■			■						
HERMANN RIVERFRONT PARK	80	♿	♿	■			■	■	♿								■	■	■								
HERMITAGE ACCESS	92	■					■	■										■			■						
HI LONESOME PRAIRIE CA	20	■														■									■		
HICKORY WOODS CA	200																		■								
HIDDEN HOLLOW CA	130	■			■											■										■	
HIGBEE CITY LAKE	184	■						■										■			■						
HIGGINSVILLE CITY LAKE	116	♿	♿					■	♿								■	■			■						
HILLIARD ACCESS	28	■					■	■										■			■						
HILLTOP TOWERSITE	220	■																									
HINKSON WOODS CA	24																	■	♿		■						
HITE PRAIRIE CA	150															■										■	
HOLDEN CITY LAKE	110	♿	♿	■				■									■	■			■						
HOLLAND SF	216	■			■											■											
HOLLISTER TOWERSITE	220	■		■												■											
HOLLY RIDGE CA	214	■														■			■	■		■				■	
HOLMES BEND ACCESS	66	■			■		■									■	■	■			■						
HONEY CREEK ACCESS	54	■	■		■		■	■								■	■	■			■				■	■	
HONEY CREEK CA	8	■			■		■									■		■	■		■				■		
HOOT OWL BEND ACCESS	10	■	■		■		■											■			■				■		
HOOTEN TOWN ACCESS	216	■		■			■	■										■			■						
HORNERSVILLE ACCESS	74	■					■	■										■			■						

Key ■ = Yes ♿ = Disabled Access

beaver	bullfrogs	coyotes	deer	doves	eagles	fox	hawks	herons	otters	owls	pheasant	quail	rabbits	racoons	ruffed grouse	shore birds	snapping turtles	songbirds	squirrels	turkey	waterfowl	hybrid bass	largemouth bass	rock bass	small mouth bass	spotted bass	striped bass	white bass	blue gill	carp	catfish	crappie	muskelunge	paddlefish	suckers	sunfish	trout	walleye	mushrooms	wildflowers
⊠	◆	⊠	⊠	⊠	◆	□	⊠	■	⊠	⊠		□	⊠	⊠		◆	◆	⊠	■	⊠	◆	□	⊠					□	⊠	■	■	⊠		⊠		⊠		□	◆	⊠
□	□	⊠	⊠	⊠		⊠	□			□		⊠	⊠	⊠			□	⊠	⊠	⊠																			⊠	⊠
⊠	■	⊠	■	⊠	□	⊠	⊠	■	□	■	⊠	■	■	■		◆	■	■	■	■	◆		■						■	⊠	■	■				■			◆	◆
□	□	⊠	⊠	⊠	□	⊠	◆	□		⊠	□	⊠	⊠	⊠		□	□	⊠	⊠	□																			⊠	⊠
⊠	⊠	⊠	■	■	◆	□	⊠	◆	□	⊠	■	■	■	⊠		◆	⊠	■	⊠	⊠	◆		⊠			⊠			□	⊠	⊠	⊠		⊠	⊠	⊠			◆	◆
□	⊠	□	□	⊠	□	□	⊠	⊠		□			□	⊠		⊠	⊠	⊠	⊠	□	◆		⊠	⊠	⊠	⊠			⊠	⊠	⊠	⊠		□	⊠	⊠		□	□	⊠
⊠	⊠	⊠	⊠	⊠	□	⊠	⊠	⊠		⊠		⊠	⊠	■		⊠	⊠	■	■	⊠	◆		■						■	⊠	⊠	■			⊠	■			⊠	⊠
		⊠	⊠	⊠		⊠	⊠			⊠			⊠	⊠				⊠	⊠	⊠																				
□	□	□	□	□	□	□	□	□	□	□	□	□	□	□	□	□	□	□	□	□	□		⊠			⊠		⊠	⊠	■	■	⊠		⊠	□	⊠		□	□	□
		⊠	□	□	□	□	◆			□		⊠	⊠	⊠				⊠		□	□																		□	■
⊠	⊠	□	□		□	□	□	□	□	□			□	⊠		□	⊠	⊠	□	□	■	□	□	□	□	□	□	□	⊠	⊠	⊠	⊠			□	⊠			□	□
■	⊠	■	⊠	◆	◆	⊠	■	⊠	□	⊠	⊠	⊠	⊠	■		⊠	■	⊠	⊠	⊠	◆	□	⊠	⊠	□	□	□	⊠	⊠	■	■	⊠		⊠	■	⊠		⊠	◆	⊠
		⊠	⊠	⊠		⊠	⊠			⊠			⊠	⊠				⊠	⊠	□																			⊠	□
	□	■	⊠	⊠	□	⊠	⊠			⊠		⊠	⊠	⊠	□			⊠	■	⊠	□		⊠						⊠							⊠			⊠	⊠
⊠	⊠	■	■	⊠	□	⊠	⊠	⊠		⊠		⊠	⊠	■		□	⊠	■	■	■	□		⊠	⊠	⊠	□			■	⊠	□	□			⊠	■			⊠	⊠
□	⊠	□	⊠	⊠	□	□	⊠	□	□	⊠		⊠	⊠	⊠		⊠	⊠	⊠	⊠	⊠	⊠	⊠	■		□	■	⊠	■	■	■	■	■	□	◆	⊠	■		⊠	□	⊠
⊠	⊠	⊠	■	⊠	□	⊠	⊠	◆		⊠	□	⊠	■	⊠		◆	⊠	■	⊠	⊠	⊠		⊠							⊠	⊠				□	⊠			⊠	⊠
		⊠	⊠	⊠	□	□	⊠			⊠		□	⊠	⊠				⊠	□	□																			□	■
□	□	⊠	⊠	⊠	□	⊠	⊠	□	□	⊠		⊠	⊠	⊠		□	□	⊠	⊠	⊠	□		⊠	⊠	⊠	⊠			⊠	⊠	⊠	⊠			⊠	⊠		⊠	⊠	⊠
□	⊠				⊠		⊠	⊠		⊠			□	⊠		⊠	⊠	⊠	⊠		◆		⊠	⊠	⊠	⊠			⊠	⊠	⊠	□		□	⊠	⊠		□	□	□
		⊠	⊠	⊠		□	⊠			⊠		⊠	⊠	□				⊠	⊠	⊠																			◆	⊠
⊠	⊠	□	⊠		□	□	⊠	⊠	□	□			⊠	⊠		⊠	⊠	⊠	□	□	⊠	□	□	□	□	□	□	⊠	⊠	⊠	⊠	⊠	□	⊠	⊠	⊠			⊠	⊠
⊠	⊠	⊠	⊠	□	□	⊠	⊠			⊠		□	⊠	⊠	□		□	⊠	⊠	⊠	□		⊠	⊠	⊠				⊠	□	⊠	⊠			□	⊠			⊠	⊠
□	□			□	□		□	□		□			□	□		□	□	□	□		□		□		□				■	■	■	□				⊠		□	□	□
		⊠	⊠	⊠		⊠	⊠			⊠		□	⊠	⊠				⊠	■	⊠																			◆	◆
⊠	⊠	□	⊠	□	⊠	□	⊠	⊠	□	□			⊠	⊠		⊠	⊠	⊠	⊠	□	⊠	□	□	□	□	□	□	⊠	⊠	⊠	⊠	⊠	□	⊠	⊠	⊠			◆	◆
		⊠	⊠	□		⊠	⊠			⊠		□	□	⊠				⊠	⊠	□																			⊠	⊠
	□	⊠	⊠	⊠	□	⊠	⊠	□		⊠		⊠	⊠	⊠			□	⊠	⊠	⊠																			⊠	⊠
	□	⊠	⊠	⊠	⊠	⊠	⊠			⊠		⊠	⊠	⊠	□		□	⊠	⊠	⊠	□																		⊠	⊠
⊠	■	■	■	◆	◆	⊠	◆	◆	□	⊠	⊠	⊠	⊠	⊠		◆	■	⊠	⊠	⊠	◆		□		□	□			⊠	■	■	⊠		□	□	⊠		□	■	⊠
□	⊠	⊠	■	⊠	□	⊠	⊠	◆		⊠	⊠	⊠	■	⊠		◆	⊠	■	⊠	⊠	⊠		⊠						⊠	□	⊠					⊠			⊠	⊠
	□	⊠	⊠	⊠	□	□	⊠			⊠		⊠	⊠	⊠			□	⊠	⊠	⊠																			⊠	■
	□	⊠	⊠	□	□	⊠	□			□		□	□	⊠			□	⊠	⊠	⊠																			⊠	⊠
		⊠	⊠	□		⊠	⊠			⊠		□	□	⊠				⊠	■	⊠																			□	⊠
		⊠	⊠	□		□	⊠			⊠		□	□	⊠				■	⊠	⊠																			◆	◆
⊠	⊠	⊠	⊠	⊠		⊠	⊠	◆	□	⊠			⊠	⊠		◆	⊠	■	⊠	□	◆								□	□	□				□	□			⊠	⊠
□	⊠	⊠	⊠		◆	⊠	⊠	□	□	⊠			□	⊠		◆	⊠	⊠	⊠	⊠	◆	⊠	⊠	⊠	⊠	⊠	⊠	⊠	⊠	⊠	⊠	⊠		⊠	⊠	⊠		⊠	⊠	⊠
⊠	⊠	⊠	□	□	□	□	□	□		⊠			⊠	⊠		□	⊠	⊠	⊠		□								⊠	⊠	⊠				⊠	⊠			⊠	⊠
□	⊠	□	⊠	□	□	□	□	⊠	□	□	□	⊠	⊠	⊠		□	□	⊠	□	⊠	⊠		■						■		⊠	⊠				□			□	⊠
□	■							□	□	□			⊠	■			■	■	■		□		⊠	⊠	□	⊠		□	■	■	■	⊠		□	■	■		□	□	□
⊠	⊠	⊠	⊠	□	□	⊠	□	⊠	□	□		□	⊠	⊠		□	□	⊠	⊠	⊠	□		⊠	⊠	⊠	□			⊠	⊠		□			⊠	⊠			⊠	⊠
	□	⊠	⊠	⊠		⊠	⊠			⊠	□	⊠	⊠	⊠			□	⊠	⊠	⊠																			◆	◆
□	⊠	□	■	⊠	□	□	⊠	⊠		□	□	□	⊠	⊠		□	■	⊠	⊠	⊠	◆		■						■	⊠	■	⊠			□	⊠			□	□
□	□	⊠	⊠	□		□	⊠			⊠		⊠	⊠	⊠			□	⊠	⊠	□																			⊠	⊠
	□	⊠	⊠	⊠	□	⊠	⊠			⊠		⊠	□	⊠			□	⊠	⊠	⊠																			⊠	□
	⊠	□	□			⊠	⊠	⊠		⊠			□	⊠		⊠	⊠	⊠	□		□		⊠	⊠	⊠	⊠		□	⊠	■	⊠	⊠			■	■		□	□	⊠
⊠	□	⊠	⊠	◆		⊠	⊠	□	□	⊠	□	⊠	⊠	⊠		□	□	⊠	⊠	□	⊠		□					□	□	⊠	⊠	□			□	□		□	□	⊠
	⊠	⊠	□	⊠		□	⊠	□		⊠		□	□	⊠		□	⊠	⊠	□	□	□		□		□				□	⊠	⊠				⊠	□			□	⊠
⊠	⊠	□	■	□	◆	⊠	⊠	⊠	□	⊠		□	□	⊠		■	⊠	■	⊠	⊠	□		⊠	□	□	□	□	■	■	■	■	■		⊠	□	■		□	⊠	⊠
⊠	⊠	⊠	⊠	⊠	□	⊠	⊠	⊠		⊠		⊠	⊠	⊠		◆	⊠	⊠	⊠	□	◆		⊠						■	⊠	■	■				■			□	⊠
⊠	⊠	⊠	⊠	◆	□	⊠	⊠	□	□	⊠		□	□	⊠	□	□	⊠	■	⊠	⊠	□		⊠		⊠	□			⊠	⊠	⊠	□			□	⊠			⊠	⊠
⊠	⊠	⊠	⊠	◆	◆	⊠	⊠	⊠	⊠	⊠	⊠	⊠	⊠	⊠		⊠	⊠	⊠	⊠	⊠	◆		⊠					□	⊠	⊠	⊠	□			□	□		□	⊠	⊠
⊠	⊠	⊠	⊠	⊠	□	⊠	◆	⊠	□	⊠	□	⊠	⊠	⊠		◆	⊠	⊠	□	□	◆		□	□	□				⊠	⊠	⊠	⊠		□	□	⊠			□	⊠
□	◆	⊠	⊠	◆	□	□	⊠	◆		⊠		⊠	⊠	⊠		□	◆	■	□	□	◆		■						■		⊠	■				■			□	⊠
□	⊠	□	⊠	⊠	◆	□	◆	◆	□	⊠		⊠	⊠	⊠		□	⊠	⊠	⊠	□	□		⊠		□	□		□	⊠	⊠	⊠	⊠		□	⊠	⊠		□	□	□
□	⊠	⊠	■	⊠	□	□	■	⊠		⊠		■	⊠	⊠		◆	⊠	■	⊠	⊠	◆		⊠						⊠	⊠	⊠					⊠			◆	◆
□	□	⊠	⊠	⊠	□	□	⊠	⊠	□	⊠	□	□	□	⊠	□	□	⊠	⊠	□	□	□		□						⊠	⊠	⊠	□			⊠	⊠			⊠	□
⊠	■	⊠	⊠	⊠	□	⊠	□	⊠	□	⊠		⊠	■	⊠		⊠	⊠	■	■	⊠	⊠		⊠	⊠		⊠			⊠	□	⊠	⊠				⊠			⊠	⊠
□	⊠	⊠	⊠	⊠	□	⊠	⊠	⊠		⊠		⊠	⊠	⊠	□	□	□	⊠	□	⊠	□		⊠	⊠	⊠				□	□	□				⊠	⊠			□	⊠
⊠	⊠	⊠	⊠	□	⊠	⊠	⊠	⊠	□	⊠		⊠	□	⊠		□	□	■	⊠	⊠	◆		⊠						■	⊠	⊠	⊠	⊠			⊠			□	⊠
□	⊠	⊠	□	◆	□	□	⊠	□		⊠		⊠	⊠	⊠		⊠	□	◆	□	□	◆		■						■		■	■				■			◆	◆
□	⊠			□	□		□	□	□	□			□			□	⊠	⊠			□		□	⊠	⊠	⊠			⊠	⊠	⊠	⊠			⊠	⊠			□	□
⊠	⊠	⊠	□	⊠	□	⊠	⊠	⊠		⊠		□	⊠	⊠		□	⊠	⊠	⊠	□	□		⊠	□		⊠		⊠	■	■	■	■		⊠	⊠	⊠		□	□	■
⊠	⊠	■	■	⊠	□	■	■	◆	□	■	⊠	■	■	■		◆	⊠	■	■	■	◆		■		⊠				■	■	■	■			□	■			◆	⊠
□	⊠	⊠	⊠	□	◆	□	⊠	⊠	□	⊠		□	□	⊠		□	⊠	⊠	⊠	⊠	□		□	⊠	⊠	□			⊠	⊠	⊠				⊠	⊠	⊠		◆	⊠
□	⊠	■	⊠	⊠	□	□	⊠	⊠		⊠		□	⊠	⊠		⊠	⊠	⊠	⊠	⊠	⊠		⊠	⊠	⊠	⊠			⊠	⊠	⊠	⊠		□	⊠	⊠		□	□	⊠
⊠	⊠	⊠	⊠	⊠	□	⊠	⊠	◆		⊠	⊠	⊠	⊠	⊠		◆	⊠	⊠	⊠	⊠	◆		⊠						⊠	□	⊠					⊠			⊠	⊠
□	⊠	□	□	⊠	□	□	⊠	□	□	□		□	⊠	□		□	□	□	□		□																		□	□
□	□	⊠	■	□	□	□	⊠	□		⊠		□	⊠	⊠				■	⊠	■	□			□	□										□	□			□	■
⊠	⊠	⊠	⊠		□	□	□	□	□	□		□	□	⊠		□	⊠	□	□	□			□					□	□	⊠	⊠	□			□	□			□	⊠
⊠	⊠	□	■	⊠	◆	□	⊠	□		□		□	⊠	⊠		□	⊠	□	⊠	□	■		⊠	□	⊠	□			■	⊠	■	⊠			⊠	■			□	⊠
□	□				◆		◆	□	□	□			□	□		□	⊠	⊠	□		□		□					□	⊠	■	⊠	⊠		⊠	⊠	⊠		□	□	□
⊠	⊠			□	□		⊠	⊠					⊠	□			⊠	⊠			□	⊠	⊠	⊠	⊠	⊠	⊠	◆	⊠	⊠	⊠	⊠	□		⊠	⊠		◆		□
	⊠	⊠	⊠	⊠	□	□	◆	□		⊠		⊠	⊠	⊠		□	⊠	⊠	□	□	□		⊠						⊠		⊠					⊠			⊠	■
				□		□	□			□			□	□				⊠	⊠																				□	⊠
□	□	⊠	⊠	□		⊠	□			⊠		□	□	⊠	□		□	⊠	⊠	⊠																			◆	◆
⊠	⊠	⊠	⊠	□		□	⊠	⊠		⊠		□	⊠	⊠		□	⊠	⊠	⊠	⊠	□		⊠						⊠		⊠	⊠				⊠			⊠	□
□	⊠	⊠	⊠	◆	□	□	⊠	⊠		⊠		⊠	⊠	⊠		◆	⊠	⊠	□	□	◆		■						■	⊠	■	■			□	■			□	⊠
⊠	⊠	□		□	□	□	□	⊠	□	□		□	□	⊠		⊠	□	⊠	□		⊠		⊠	⊠	⊠	□			⊠	⊠	⊠	⊠				⊠		□	□	⊠
		□	□				□			□				□				□	□	□																			□	□
⊠	⊠	□	■	⊠	◆	□	⊠	□		□		□	⊠	⊠		□	⊠	□	⊠	□	■		⊠						⊠	⊠	⊠					⊠			⊠	■
		□	□			□	□			□		□	⊠	□				⊠		□																			⊠	■
□	⊠	□	⊠	◆	□	□	⊠	⊠		□		⊠	⊠	□		□	⊠	⊠	⊠	□	⊠		■						⊠		⊠	■				⊠			□	□
		□	⊠	□		□	⊠			⊠		□	□	⊠				⊠	⊠	⊠																			□	□
		□	⊠	□		□	□			⊠		□	□	⊠				⊠	⊠	⊠																			⊠	⊠
□	⊠	⊠	⊠	⊠	□	⊠	⊠	□		⊠		□	⊠	⊠		□	□	■	⊠	⊠	□																		◆	■
□	□	⊠	⊠	⊠	□	□	□	□		⊠		⊠	⊠	⊠		□	⊠	⊠	⊠	□	□		□						⊠	⊠	⊠	□			⊠	⊠		□	⊠	⊠
⊠	⊠	⊠	⊠	□	□	□	⊠	⊠	□	⊠		⊠	⊠	⊠		□	⊠	■	⊠	□	□		⊠	⊠	□	□			⊠	⊠	⊠	⊠			⊠	⊠			⊠	⊠
□	⊠	⊠	■	◆		⊠	⊠	□	□	⊠	□	⊠	⊠	⊠	□	□	□	■	⊠	■	□		□					□	□	■	■	□			□	□		□	⊠	⊠
⊠	⊠	⊠	⊠	⊠	◆	□	⊠	⊠	□	⊠	⊠	⊠	⊠	⊠		□	□	⊠	⊠	□	⊠		□					□	□	■	■	⊠		□	⊠	□		□	◆	⊠
□	□	□	□	□	□	□	□	□		□		□	□	□			□	⊠	□		□		⊠	□	⊠	□		□	□	⊠	⊠	□		◆	⊠	⊠	□		□	□
⊠	■	⊠	□	⊠	□	□	□	■	⊠	□		□	□	⊠		■	■	■	□		⊠		■			□		□	■	■	■	⊠			⊠	■			⊠	⊠

Key ■ = Abundant ⊠ = Common □ = Uncommon ◆ = Seasonal

	Page Number	parking lot	privy/restrooms	picnic areas	primitive camping	camper pads	river access	boat ramps	fishing docks/jetties	boat rental	viewing blinds	waterfowl blinds	interpretive programs	exhibits/displays	MDC office	hunting	waterfowl hunting	fishing	hiking trails	horse trails	canoeing	archery range	firearms range	field trials	dog training	trapping	Notes
HORNERSVILLE SWAMP CA	74	■					■	■								■	■	■			■					■	
HORSE CREEK PRAIRIE CA	62															■											
HORSESHOE BEND NA	222	■														■		■			■				■	■	
HORSESHOE LAKE	200	♿		■														■									
HOUGH PARK LAKE	54	■																■									
HOUSE SPRING ACCESS	108	■	■	■			■	■										■			■						
HOUSTON TOWERSITE	224	■														■											
HOWELL ISLAND CA	193	■					■						■			■	■	■	■		■				■	■	
HUCKLEBERRY RIDGE CA	128	■			■											■			■							■	
HUGHES MOUNTAIN NA	230	■														■			■							■	
HULL FORD ACCESS	114	■					■											■			■					■	
HUMPHREY ACCESS	174	■					■	■										■			■						
HUNGRY MOTHER CA	96	■			■											■											
HUNNEWELL ACCESS	212	■					■											■			■						
HUNNEWELL LAKE CA	212	♿	♿	■	■	■			♿	■					■	■		■						■	■	■	
HUNTER TOWERSITE	40	■			■											■										■	
HURRICANE DECK TOWERSITE	34																										
HUZZAH CA	60	■	♿		■	■	■	■								■	■	■	■		■		■		■	■	
HYDE PARK LAKE	200	■	■	■					♿									■									
HYER WOODS NA	70															■											
INDIAN CAMP ACCESS	182	■			■		■	■										■			■						
INDIAN HILLS CA	207	■			■		■									■	■	■			■			■	■	■	
INDIAN TRAIL CA	70	■		■	■										■	■		■	■						■	■	
INDIGO PRAIRIE CA	62	■														■	■	■									
IRON BRIDGE ACCESS	232				■											■		■									
IRON MOUNTAIN CITY LAKE	196	■		■				■	♿									■			■						
JACKS (MAUDE SHORES) CA	12																										
JAMES BRIDGE ACCESS	186						■											■			■						
JAMESPORT CITY LAKE	66	■	♿		■			■										■			■						
JAMESPORT COMMUNITY LAKE	66	■	■					■										■									
JANUARY-WABASH LAKE	200	♿	■	■					♿						■			■									
JEFFERSON LAKE	200	♿		■					■				■					■									
JENKINS TOWERSITE	14															■										■	
JENTELL BREES ACCESS	26				■		■											■			■						
JEROME ACCESS	170	■	■				■	■										■			■						
JESSIE HOLLOW CA	217	■			■											■											
JOPLIN TOWERSITE	154																										
JUDEN CREEK CA	36						■											■	■								
KAHRS-BOGER PARK	168	■														■	■	■							■	■	
KEARN (J. N. "TURKEY") MEM WA	110	■														■	■	■						■	■	■	
KEARN (W. R.) MEM CA	168	■														■										■	
KEEVER BRIDGE ACCESS	156	■					■											■			■						
KELLOGG LAKE	106	♿		■				■	♿									■			■						
KENDZORA (ANTHONY AND BEATRICE) CA	26															■	■	■			■				■		
KERR (H. L.) ACCESS	217	■					■	■										■			■						
KESSLER MEM WA	122	■														■		■				■			■	■	
KETCHERSIDE MOUNTAIN CA	100	♿	♿	■	■											■			■							■	
KICKAPOO PRAIRIE CA	118	■														■											
KING LAKE CA	68	■		■	■	■		■								■	■	■							■		
KINGS BLUFF ACCESS	140	■	■				■	■										■			■						
KINGS PRAIRIE ACCESS	194	■	■		■		■									■		■			■					■	
KINGSTON ACCESS	230	■					■									■		■			■					■	
KIWANIS LAKE	200	♿	♿	■					♿				■					■	♿								
KLAMBERG (ROGER) WOODS CA	200																		■								
KNEIB MEM CA	26	■														■	■		■								
KNOB LICK TOWERSITE	196	■													♿	■											
LA BELLE CITY LAKE	120	■	■	■	■			■								■	■	■								■	
LA RUSSELL ACCESS	106	■					■											■			■						
LADUE BOTTOMS CA	90	■														■	■	■								■	
LAFAYETTE PARK LAKE	200	■	■	■														■									
LAKE #7-FOREST PARK	200		■																								
LAKE #8-FOREST PARK	200			■																							
LAKE GIRARDEAU CA	36	♿	♿			■		■	♿							■		■			■						
LAKE JACOMO	102	♿	♿	■	■	■		■	♿	■			■	■	♿	■		■	■		■	■					
LAKE OF THE WOODS REC AREA	24	■	■	■											■			■									
LAKE PAHO CA	138	♿	♿	■	■	■		■	♿	■			■		■	■		■							■		
LAKE SHOWME	207	■	♿	■	■	■		■	♿							■	■	■	■			■	■		■	■	
LAKE SPRINGFIELD	86	♿	♿	■			■	■	♿									■			■						
LAKEVIEW PARK LAKE	12	♿	♿	■		■												■	■		■						
LAMAR ACCESS	16	♿	♿				■	■										■			■						
LAMINE RIVER CA	58	■			■		■	■								■	■	■			■			■	■	■	
LANAGAN TOWERSITE	128	■														■										■	
LANCASTER CITY LAKE	206	■	♿		■			■								■	■	■								■	
LANGDON BEND ACCESS	10	■	■		■		■	■										■			■				■		
LAPETITE GEMME PRAIRIE CA	176	■																									
LAPLATA CITY LAKE	130	■	♿					■	■							■	■	■								■	
LATHROP BRIDGE ACCESS	53						■											■			■						
LAUGHLIN/ROUBIDOUX PARK	178	♿	♿	■	■	■	■		♿									■			■						
LAWSON CITY LAKE	186	♿	♿	■				■	♿									■	■		■						
LEACH (B. K.) MEM CA	122	■					■	■			♿	♿				■	■	■							■	■	
LEAD MINE CA	65	■			■		■									■	■	■	■		■			■	■	■	
LEADWOOD ACCESS	196	■					■								♿			■			■						
LEBANON TOWERSITE	114																										
LENOX TOWERSITE	70	■																									
LESTERVILLE ACCESS	188				■			■								■		■			■					■	
LEWIS & CLARK FISH HATCHERY	26	■													■												
LEWIS MILL ACCESS	46	■	■		■		■	■										■			■						
LIBERTY PARK POND	168	♿	♿	■					♿						♿			■									
LICHEN GLADE CA	194	■																									
LICK CREEK CA	25	■			■			■								■		■									
LIME KILN ACCESS	154	■					■	■										■			■						
LIMPP COMMUNITY LAKE	82	■	■					■										■									

Key ■ = Yes ♿ = Disabled Access

beaver	bullfrogs	coyotes	deer	doves	eagles	fox	hawks	herons	otters	owls	pheasant	quail	rabbits	racoons	ruffed grouse	shore birds	snapping turtles	songbirds	squirrels	turkey	waterfowl	hybrid bass	largemouth bass	rock bass	small mouth bass	spotted bass	striped bass	white bass	blue gill	carp	catfish	crappie	muskelunge	paddlefish	suckers	sunfish	trout	walleye	mushrooms	wildflowers
⊠	⊠	⊠	□	⊠	□	□	⊠	⊠	⊠	⊠		□	⊠	⊠		◆	⊠	⊠	⊠	□	◆		□			□			□	⊠	⊠	□			■	⊠			⊠	⊠
		□	□	□		□	◆			□		□	□	□				⊠																					□	■
⊠	⊠	⊠	■	⊠	◆	⊠	⊠	⊠	□	⊠		⊠	⊠	⊠		□	■	⊠	■	■	□		⊠	■	■	⊠			■	⊠	⊠	□			■	■			⊠	⊠
□	⊠			⊠	□	□	□	□		□			□	□		□	⊠	⊠	⊠		⊠		⊠						⊠	⊠	⊠	□				⊠			□	□
□	⊠	□	⊠	⊠		⊠	⊠	⊠		□		□	⊠	⊠		⊠	⊠	⊠	⊠	⊠	◆		■						⊠	⊠	⊠	□				⊠			□	□
⊠	⊠	⊠	⊠	⊠	□	□	⊠	⊠		⊠		⊠	⊠	⊠		⊠	⊠	⊠	⊠	⊠	⊠		⊠	⊠	■	⊠			⊠	⊠	⊠	⊠			■	■			□	⊠
		⊠	⊠	□		□	□			□		□	□	⊠				⊠	■	⊠																			□	□
⊠	⊠	□	■	⊠	◆	□	⊠	⊠	□	■		□	⊠	⊠		◆	⊠	■	⊠	■	◆		⊠					⊠	⊠	■	■	■		□	□	⊠		□	◆	◆
	□	⊠	⊠	□		⊠	⊠			⊠		□	□	⊠			□	⊠	⊠	⊠																			⊠	⊠
		⊠	⊠	⊠	□	⊠	⊠		⊠	⊠		□	⊠	⊠				⊠	⊠	⊠																			◆	◆
⊠	⊠	⊠	⊠	⊠	□	⊠	⊠	⊠	□	⊠		□	⊠	⊠		⊠	⊠	⊠	⊠	⊠	⊠		□	⊠	⊠	□			⊠	⊠	⊠	⊠			⊠	⊠			⊠	⊠
⊠	⊠	⊠	□	□	⊠	□	⊠	⊠	□	□		□	⊠	⊠		□	□	⊠	⊠	□			⊠			□			⊠	■	■	□		□	□	⊠			□	□
	□	⊠	⊠	□		□	⊠			⊠		□	□	⊠	□			⊠	⊠	⊠																			⊠	⊠
□	■	□	□	□		□		⊠	□	□		□	⊠	■		□	■	■	□	□	□		⊠	□	⊠	⊠		■	■	■	■	■		□	⊠	■		■	⊠	⊠
⊠	⊠	■	■	■	□	⊠	■	■		⊠	□	■	■	⊠		⊠	■	■	■	■	■		⊠						■	■	⊠	■				■		□	■	■
		⊠	⊠	□	□	⊠	□			⊠		□	⊠	⊠				⊠	⊠	⊠																			⊠	⊠
		⊠	⊠	◆		⊠	⊠			⊠			⊠	⊠				⊠	⊠	⊠	□																		◆	□
□	⊠	⊠	⊠	⊠	□	⊠	⊠	⊠	□	⊠	□	⊠	⊠	⊠	□	□	⊠	⊠	⊠	⊠	□	□	⊠	⊠	⊠	⊠	□	□	⊠	⊠	⊠	□	□	□	⊠	⊠	□	□	⊠	⊠
				⊠															⊠										□	⊠	⊠					□				
	□	⊠	⊠	⊠		⊠	⊠			□		□	⊠	⊠			□	⊠	⊠	⊠																			⊠	⊠
□	⊠	□	□	□	□	□	⊠	□	□	□		□	⊠	⊠			⊠	⊠	⊠	□	□		⊠		□	□		◆	⊠	⊠	⊠	⊠		□	◆	⊠		◆	⊠	⊠
⊠	■	■	■	◆	□	■	■	■	⊠	■	■	■	■	■		◆	■	■	■	■	◆		■		□				■	⊠	⊠	⊠			□	■			◆	◆
□	⊠	⊠	⊠	⊠	□	⊠	⊠	⊠		⊠		⊠	⊠	⊠		□	⊠	⊠	■	⊠	□		□	□		□			□	□	□	□			□	□			⊠	⊠
	□	□	□	□			⊠	□		□		⊠	⊠	□		□	□	⊠			□		□						□		□					⊠			□	■
⊠	◆	⊠	⊠	◆	□	⊠	⊠	□		⊠		□	⊠	⊠		□	⊠	⊠	⊠	⊠	□			⊠	⊠	⊠			■		⊠				⊠	■			◆	◆
⊠	⊠	□	⊠	⊠	□	□	□	⊠		□		□	⊠	⊠		⊠	⊠	⊠	⊠	⊠	⊠		⊠			□			⊠	⊠	⊠	□				⊠			⊠	⊠
		□	□	□						□		□	□					□	□	□																			□	□
⊠	⊠	⊠	⊠	⊠	□	⊠	◆	⊠	□	⊠	□	⊠	⊠	⊠		□	⊠	⊠	⊠	⊠	□		□	□	□				⊠	⊠	⊠	⊠		□	□	⊠			⊠	⊠
⊠	⊠	⊠	⊠	⊠	□	⊠	⊠	⊠		⊠	□	⊠	⊠	⊠		◆	⊠	⊠	■	⊠	◆		■						■		⊠	⊠				⊠			⊠	⊠
⊠	⊠	⊠	⊠	⊠	□	⊠	⊠	◆		⊠		⊠	⊠	⊠		◆	⊠	⊠	⊠	⊠	⊠		⊠						⊠	⊠	⊠	□				⊠			⊠	⊠
				⊠															⊠				□						⊠	⊠	⊠	□				■	⊠			
	⊠			⊠	□	□	□	□		□			□	□		□	⊠	⊠	⊠		□		□						⊠	⊠	⊠	□				⊠			□	□
		⊠	⊠	⊠		⊠	⊠			⊠		□	□	⊠				⊠	⊠	□																			⊠	⊠
⊠	⊠	⊠	⊠	□		□	□	◆	□	□	□	□	⊠	⊠			⊠	◆	⊠	□	◆		□	□				□	□	■	■	⊠		□	⊠	□		□		
□	■	□				□	□	⊠	□	□			□	■		□	■	■	⊠				■	■	■	■		□	■	■	■	■			■	■		□	□	□
		□	⊠			□	□			□			□	□				□	⊠	⊠																			□	□
		⊠	⊠	⊠	□	⊠	⊠			⊠		⊠	⊠	⊠				⊠	⊠																				⊠	⊠
⊠	⊠	□	⊠	□		□	□	□	□	⊠			□	■			⊠	■	■	□	□		□			□			⊠	⊠	⊠	□			⊠	⊠			⊠	■
□	⊠	⊠	⊠	⊠	□	□	⊠	◆	□	⊠		⊠	⊠	⊠		◆	⊠	⊠	⊠	⊠	□		⊠						⊠	⊠	⊠	⊠			□	⊠			⊠	⊠
□	⊠	⊠	⊠	◆	□	□	◆	⊠		⊠		⊠	⊠	⊠		□	⊠	◆	⊠	⊠	□		⊠						⊠	□	□	□			□	⊠			◆	◆
		⊠	□	◆	□	□	□			□		⊠	⊠	□				⊠	□	□																			□	■
⊠	⊠	⊠	⊠	⊠		□	⊠	⊠		⊠	□	⊠	⊠	⊠		□	□	◆	⊠	□	□								□	⊠	⊠				□	□			◆	◆
	□			□	□	□	□	□		□			□	□		□	□	⊠	⊠		⊠		⊠			□			■	□	⊠	⊠			□	■			□	□
⊠	⊠	⊠	⊠	◆	□	□	◆	⊠	□	⊠	□	⊠	⊠	⊠		◆	⊠	⊠	⊠	⊠	⊠		⊠						⊠	⊠	⊠					⊠			◆	
⊠	□	□	□	□	□	□	□	□		□		□	□	□			□	⊠	□	□	□		⊠	□	⊠	□		□	□	⊠	⊠	□		◆	⊠	⊠	□		□	□
	□	⊠	■	◆	□	⊠	⊠	□	□	⊠		■	⊠	⊠		□	□	⊠	⊠	⊠																			⊠	■
		⊠	⊠	□	□	⊠	⊠			⊠		□	□	⊠				⊠	⊠	⊠															□	□			⊠	■
	□	⊠	⊠	⊠	□	□	■			⊠		■	⊠	⊠		□	□	■	□																				□	■
⊠	⊠	⊠	⊠	◆	□	□	⊠	⊠		⊠	⊠	⊠	⊠	⊠		⊠	⊠	⊠	⊠	⊠	⊠		□						⊠	■	⊠	⊠				□			⊠	⊠
□	⊠	□	□		□	□	⊠	◆	□	□			□	⊠		◆	⊠	⊠	⊠	□	◆	□	⊠	⊠	⊠	⊠	⊠	⊠	⊠	⊠	⊠	⊠		⊠	⊠	⊠		□	□	□
□	□	⊠	⊠	◆	□	⊠	⊠	□		⊠		⊠	⊠	⊠		□	□	⊠	⊠	⊠	□		⊠	□		□		□	⊠	⊠	⊠	⊠			⊠	⊠		□	□	□
⊠	⊠	⊠	⊠	□	□	⊠	⊠	⊠	□	⊠		□	⊠	⊠		□	⊠	■	⊠	⊠	□		⊠	⊠	⊠				■	□	⊠	□			⊠	■			◆	■
	⊠			⊠			□	□					⊠	□		□		■	⊠		⊠		■						■		□					⊠				
□	⊠	□	□	□		□	⊠			□		□	□	⊠		□	□	⊠	⊠	□	□																		⊠	⊠
	□	⊠	⊠	⊠							⊠	⊠	⊠	⊠			□		⊠	□	□																			
		□	⊠			⊠	⊠			⊠			□	⊠				⊠	■	⊠																			⊠	⊠
□	⊠	⊠	⊠	□	□	□	□	⊠	□	⊠		⊠	⊠	⊠		□	□	⊠	□	⊠	□		■						■		⊠	⊠				□			□	⊠
⊠	⊠	⊠	⊠	⊠	□	⊠	⊠	⊠		⊠		□	⊠	⊠			⊠	⊠	⊠	⊠	□		⊠	⊠	⊠	⊠			⊠	⊠	⊠	□			⊠	⊠			□	□
□	□	⊠	⊠	◆	□	□	⊠	□		⊠		⊠	⊠	⊠			□	⊠	⊠	⊠	⊠		⊠						⊠	⊠	⊠	□				⊠			⊠	⊠
				⊠															⊠				□						□	⊠	⊠					⊠				
	⊠			⊠				⊠					⊠			⊠		⊠	⊠		⊠		□						□	⊠	⊠	□				⊠				
□	⊠	□		⊠	□	□	□	□		□			□	□		□	⊠	⊠	⊠				⊠						⊠	⊠	⊠	□				⊠			□	□
⊠	⊠	⊠	⊠	□	□	⊠	⊠	⊠		⊠		□	⊠	■		⊠	⊠	■	⊠	□	⊠		■			□			■	⊠	■	⊠			□	⊠			□	■
□	⊠	⊠	■	⊠	□	⊠	⊠	⊠		⊠		⊠	⊠	⊠		◆	⊠	⊠	⊠	■	■	□	■					⊠	■	⊠	⊠	■		□	□	■		⊠	⊠	⊠
□	⊠	□	□	⊠		□	⊠	□		⊠		□	⊠	□		□	⊠	⊠	⊠	□	⊠		⊠						⊠		⊠	□				⊠			⊠	⊠
□	⊠	⊠	⊠	⊠	◆	□	◆	◆		⊠	⊠	⊠	⊠	⊠		◆	⊠	⊠	⊠	□	⊠		⊠						⊠	⊠	⊠	⊠				⊠		⊠	⊠	⊠
□	⊠	⊠	⊠	□	□	⊠	⊠	⊠	□	⊠	⊠	⊠	■	⊠		□	□	⊠	⊠	⊠	⊠		■						■		⊠	⊠				⊠		⊠	□	⊠
⊠	⊠	□	■	⊠	◆	□	⊠	⊠		□		□	⊠	⊠		⊠	⊠	⊠	⊠	□	■		⊠	⊠	⊠	⊠		□	■	⊠	⊠	⊠		◆	■	■			□	⊠
□	⊠		□	⊠	□		□	□		□		□	⊠	⊠		□	□	■	⊠	□	□		■						■		■	⊠				⊠			⊠	⊠
⊠	⊠	⊠	⊠	⊠		□	⊠	⊠	□	□		□	⊠	⊠		□	⊠	⊠	⊠		⊠		⊠	□	□	⊠		⊠	⊠	⊠	■	⊠			⊠	⊠			□	□
⊠	⊠	⊠	⊠	⊠	□	□	◆	◆	⊠	⊠		⊠	⊠	⊠		◆	⊠	⊠	⊠	⊠	◆	□	⊠	□	⊠	⊠	□	□	⊠	⊠	⊠	⊠		□	⊠	⊠		□	■	⊠
	⊠	⊠	⊠	□	⊠	⊠	⊠			⊠		□	□	⊠			⊠	⊠	⊠	□																			⊠	⊠
□	⊠	⊠	□	□	□	□	□	⊠	□	□	□	□	⊠	⊠		□	□	⊠	□	⊠	□		⊠						■		⊠	⊠				⊠			□	⊠
⊠	⊠	⊠	⊠	□	◆	□	⊠	⊠		□	⊠	□	⊠	⊠		□	□	⊠	⊠	□	⊠		□					□	□	■	■	⊠		□	⊠	□		□	◆	⊠
□	□	⊠	⊠	⊠	□	⊠	⊠	□		⊠		⊠	⊠	⊠		□	□	⊠	□	□																			◆	◆
□	⊠	⊠	⊠	□	□	⊠	⊠	⊠	□	⊠	□	⊠	⊠	⊠		□	⊠	⊠	□	⊠	⊠		■						■		⊠	⊠				⊠			□	⊠
⊠	◆	⊠	⊠	□		□	⊠		□	⊠		⊠	⊠	⊠			◆	⊠	⊠	⊠			⊠			□		⊠	⊠	⊠	⊠	⊠		□	□	⊠			⊠	⊠
□	⊠	□	□	⊠		□	□	⊠		□			□	□		□	⊠	⊠	⊠	□	□		⊠	⊠	⊠	⊠			⊠	⊠	⊠	□			⊠	⊠	⊠		⊠	⊠
□	⊠	□	⊠	■	□	⊠	⊠	⊠		■		□	■	■		◆	⊠	■	■	□	◆		■						■	□	⊠	⊠				⊠			□	□
■	■	⊠	■	◆	◆	⊠	⊠	■	□	⊠		⊠	⊠	■		■	■	■	■	⊠	◆		⊠	□	□	□		⊠	⊠	■	⊠	⊠		□	□	⊠			⊠	■
⊠	⊠	■	■	⊠	⊠	⊠	⊠	⊠	□	⊠		⊠	⊠	■	□	⊠	⊠	⊠	■	■	⊠		□	■	■	□			⊠	⊠	⊠	⊠			■	■	□		■	■
□	⊠	□	□			□	⊠	⊠		□			□	□		⊠	□	⊠	⊠	□	□		■	⊠	⊠	⊠		□	⊠	⊠	⊠	⊠			■	■			⊠	⊠
		⊠	□	⊠	□	□	⊠			⊠		⊠	⊠	⊠				◆	■	□																			⊠	⊠
		□	□	□		□	⊠		□	□			□	□				⊠	⊠	□																			□	□
⊠	⊠	■	⊠	□		⊠	⊠	⊠		⊠			⊠	■		⊠	⊠	⊠	⊠	⊠	□		⊠	⊠	⊠	⊠			⊠	⊠	⊠	⊠			⊠	⊠			⊠	⊠
◆	⊠	□	□	◆	◆	◆	◆	⊠		⊠	□	□	⊠	⊠		◆	⊠	■	⊠		◆		◆						◆		◆		◆							
□	⊠	⊠	□	⊠		□		⊠	□	⊠	□	□	□	⊠			⊠	■	⊠	□	□		□			□		⊠	⊠	■	■	⊠		□	□	⊠			⊠	⊠
	□			□			⊠	□		□			⊠	□		□	□	⊠	⊠		□		⊠						⊠	□	⊠	□				⊠			□	□
		□	□	□		□	□			□			□	□				⊠	□	□																			□	⊠
□	⊠	⊠	⊠	⊠		□	⊠	□		⊠		⊠	⊠	⊠	□	□	□	⊠	⊠	⊠	□		⊠						⊠		⊠								⊠	⊠
⊠	⊠	⊠	□	⊠	□	⊠	□	⊠		□		□	□	⊠			⊠	⊠	⊠	□	□		⊠	⊠	⊠	⊠			⊠	⊠	⊠	□			⊠	⊠			□	□
□	⊠	□	⊠	□		□	□	□		□	□	⊠	⊠	□		□	⊠	◆	□	□	⊠		⊠						⊠		⊠								□	◆

Key ■ = Abundant ⊠ = Common □ = Uncommon ◆ = Seasonal

	Page Number	parking lot	privy/restrooms	picnic areas	primitive camping	camper pads	river access	boat ramps	fishing docks/jetties	boat rental	viewing blinds	waterfowl blinds	interpretive programs	exhibits/displays	MDC office	hunting	waterfowl hunting	fishing	hiking trails	horse trails	canoeing	archery range	firearms range	field trials	dog training	trapping	Notes
LINCOLN TOWERSITE	20	■																									
LION'S LAKE	110	♿	♿	■				■	♿									■			■						
LITTLE BEAN MARSH CA	174	♿	♿					■								■		■	■							■	
LITTLE BLACK CA	190				■											■		■			■	■		■	■	■	
LITTLE COMPTON LAKE CA	38	■	■		■			■								■	■	■			■				■		
LITTLE DIXIE LAKE CA	32	♿	♿	■				■	■	■			■			■		■	♿								
LITTLE INDIAN CREEK CA	76	■			■		■									■		■	■	■			■	■	■	■	
LITTLE LOST CREEK CA	228	■			■											■		■		■				■	■	■	
LITTLE RIVER CA & JERRY P. COMBS LAKE	74	■	■	■				■	■							■	■										
LITTLE OSAGE PRAIRIE	226	■																									
LITTLE PRAIRIE COMMUNITY LAKE	170	♿	♿	■				■	♿							■	■	■			■					■	
LITTLE SAC WOODS CA	86	■			■											■			■								
LITTLE WHITEWATER CA	23	■			■											■											
LOCUST CREEK CA	218	■			■		■									■	■	■			■			■	■	■	
LOGAN (CAROLINE SHERIDAN) MEM WA	26	■														■											
LOGAN CREEK CA	188				■											■		■								■	
LOGAN (WILLIAM R.) CA	122	■	■	■	■	■										■	■	■			■		■		■	■	
LON SANDERS CANYON CA	232	■														■		■	■							■	
LONE HILL TOWERSITE	28															■											
LONE JACK LAKE CA	102	♿						■	♿							■		■							■	■	
LONG BRANCH LAKE	130	■						■								■	■	■			■			■	■	■	
LONG RIDGE CA	76	■			■	■										■			■	■					■	■	
LOTTS CREEK CA	88	■			■											■											
LOUISIANA ACCESS	172	♿	♿	■	■		■	■	♿									■									
LOUTRE LICK ACCESS	148	■			■		■	■								■	■	■			■						
LOWER FLAT CREEK ACCESS	14	■					■											■			■						
LOWER TAUM SAUK LAKE	188	■	■		■	■	■	■										■			■						
MACON CITY LAKE	130	■	♿	■	■			■										■									
MADDEN FORD ACCESS	140	♿														■	■	■							■	■	
MAGNOLIA HOLLOW CA	198	■			■										■	■		■							■	■	
MAINTZ WILDLIFE PRESERVE	36	■			■											■	■	■				■		■	■		
MAITLAND ACCESS	94	■					■											■			■						
MALTA BEND COMMUNITY LAKE	204	■																■									
MAMMOTH ACCESS	108	■					■	■										■			■						
MANITO LAKE CA	144	♿	♿	■				■	♿							■	■	■	■					■	■	■	
MANSFIELD (ALICE AHART) CA	34	■			■											■			■						■	■	
MAPLE FLATS ACCESS	23	■					■	■										■			■						
MAPLE LEAF LAKE CA	116	♿	♿					■	♿							■	■	■						■	■	■	
MAPLE WOODS NA	52	■		■															■								
MARAIS TEMPS CLAIR CA	193	■											■			■	■	■	■						■	■	
MARAMEC SPRING HATCHERY/TROUT PARK	170	♿	♿	■	■								■	■	♿			■	■		■						
MARCELINE CITY LAKE	46	■	♿		■	■		■								■	■	■	■								
MARI-OSA ACCESS	54	■	■		■		■	■										■			■						
MARION ACCESS	54	■	■		■		■	■								■	■	■			■						
MARION BOTTOMS CA	54															■	■	■									
MARQUAND ACCESS	132	■					■	■							♿	■		■			■						
MARSHALL HABILITATION CENTER LAKE	204	♿		■				■	♿									■									
MARSHALL JUNCTION CA	204	■					■									■		■			■		■	■	■	■	
MASON BRIDGE ACCESS	224	■	■				■	■										■			■						
MAYER'S LANDING ACCESS	76	■					■	■										■			■						
MCALLISTER SPRINGS ACCESS	204	■					■	■										■			■						
MCCORMACK (JAMERSON C.) CA	94	■			■											■			■							■	
MCCREDIE FARM LAKE	32	■						■										■			■						
MCKAY PARK LAKE	54	♿	♿	■														■	♿								
MCKINNY CA	38	■														■	■								■		
MCPIKE ACCESS	136	■					■																				
MERAMEC CA	76	♿	♿				■									■		■	♿						■	■	
MERAMEC CA-HEYNE MEM ANNEX	76	■														■									■	■	
MERRILL HORSE ACCESS	108	■	♿				■	■								■	■	■			■					■	
META TOWERSITE	160				■											■											
MIAMI ACCESS	204	♿	♿	■	■		■	■										■			■				■		
MIDVALE CA	224				■											■										■	
MIDWAY ACCESS	8	■					■											■			■						
MILL ROCK ACCESS	78	■					■											■								■	
MILLER COMMUNITY LAKE	40	■	■	■				■										■	■		■						
MILLSTREAM GARDENS CA	132	♿	♿	■			■	■							■	■		■	♿		■	■			■	■	
MINERAL HILLS CA	180	■			■											■		■								■	
MINERAL SPRINGS ACCESS	224	■					■	■										■			■						
MINT SPRING ACCESS	60	■					■	■										■			■					■	
MINT SPRING CA	81	■																■									
MO-KO PRAIRIE	44	■																									
MO-NO-I PRAIRIE CA	16	■														■											
MOKANE ACCESS	32	■			■		■	■										■			■						
MOLES CAVE CA	34	■			■											■									■	■	
MON-SHON PRAIRIE CA	16	■														■	■	■									
MONEGAW PRAIRIE CA	44	■														■		■									
MONKEY MOUNTAIN CA	94	■			■		■									■		■	■		■				■		
MONTAUK FISH HATCHERY/TROUT PARK	70	♿													■			■									
MONTAUK TOWERSITE	70	■			■											■											
MONTGOMERY WOODS CA	6				■											■										■	
MONTROSE CA	90	♿	■	■	■			■		■		■	■		♿	■	■	■	■		■				■		
MOON VALLEY ACCESS	65	■	■				■	■										■			■					■	
MOORE (JOSEPH HUNTER) ACCESS	142	■					■	■										■									
MOORE'S MILL ACCESS	32	■			■		■									■	■	■			■						
MORA CA	20	■														■	■	■							■	■	
MOREAU 50 ACCESS	54	■					■	■										■			■						
MORTON BRIDGE ACCESS	186	■					■											■			■				■		
MOUND VIEW ACCESS	212															■	■	■			■			■	■	■	
MOUNT HULDA TOWERSITE	20	■														■											
MOUNT SHIRA ACCESS	128	■	■				■											■			■						
MOUNT VERNON PRAIRIE	118	■																									
MOUNTAIN VIEW TOWERSITE	98	■														■											

Key ■ = Yes ♿ = Disabled Access

beaver	bullfrogs	coyotes	deer	doves	eagles	fox	hawks	herons	otters	owls	pheasant	quail	rabbits	racoons	ruffed grouse	shore birds	snapping turtles	songbirds	squirrels	turkey	waterfowl	hybrid bass	largemouth bass	rock bass	small mouth bass	spotted bass	striped bass	white bass	blue gill	carp	catfish	crappie	muskelunge	paddlefish	suckers	sunfish	trout	walleye	mushrooms	wildflowers
		⊠	□	⊠		□	⊠			⊠		⊠	⊠	□				⊠	⊠	□																			⊠	⊠
	□							□									□	□			◆		■						⊠		⊠					⊠				
⊠	⊠	⊠	⊠	□	⊠		⊠	⊠		□	⊠	⊠	⊠	⊠		⊠	□	⊠	⊠	□	◆																		⊠	⊠
□	□	⊠	⊠	□	□	⊠	□	□		□		⊠	⊠	⊠			□	⊠	⊠	⊠	□		⊠	⊠	⊠				⊠	□	□	□			⊠	⊠			⊠	⊠
⊠	⊠	⊠	⊠	⊠	◆	⊠	◆	◆	⊠	⊠		⊠	⊠	⊠		◆	⊠	⊠	⊠	⊠	⊠		⊠						⊠	■	⊠	□		□	□	⊠			⊠	⊠
■	■	⊠	■	⊠	□	⊠	⊠	⊠		⊠		⊠	■	■		□	⊠	⊠	■	⊠	⊠	□	■	□		□			■	□	■	■	□	□		■			□	■
⊠	⊠	⊠	⊠	⊠	□	⊠	⊠	□		⊠		⊠	⊠	⊠		□	⊠	⊠	⊠	⊠	□		⊠		⊠				⊠						⊠	⊠			◆	◆
	□	⊠	⊠	⊠	□	⊠	⊠	□		⊠		□	□	■	⊠		□	⊠	■	⊠									□							□			⊠	■
⊠	⊠	⊠	□	■	□	□	⊠	■	■	■	□	□	□	⊠		⊠	□	■	□		□	⊠	⊠	⊠	⊠	□	□	⊠	⊠	⊠	⊠	□		⊠	⊠		□	□	□	
	□	⊠	□	□		□	⊠	□		⊠		⊠	⊠	□		□	□	⊠	□	□									⊠							⊠			◆	■
□	■	□	⊠	⊠		□	⊠	□	□	□		⊠	⊠	■		□	⊠	■	□	⊠	⊠		■						■		■	⊠				■			□	■
□	□	⊠	⊠	⊠		□	⊠	⊠		⊠		□	⊠	⊠		□	□	■	⊠	⊠			⊠	⊠	⊠	□			⊠	⊠	⊠	□			⊠	⊠			⊠	■
		⊠	⊠	□	□	□	⊠			⊠			□	⊠	□			■	◆	⊠																			⊠	□
⊠	⊠	■	■	◆	□	⊠	⊠	◆	□	⊠	□	⊠	⊠	⊠		◆	⊠	◆	⊠	⊠	◆		⊠						⊠	■	⊠	□			□	⊠		□	⊠	⊠
		⊠	⊠	□		□	□			⊠		⊠	⊠	⊠				⊠	⊠	⊠																			⊠	⊠
□	⊠	⊠	⊠	⊠	□	⊠	⊠	⊠	□	⊠	□	⊠	⊠	⊠	□	□	⊠	⊠	⊠	⊠	□																		⊠	⊠
□	⊠	■	■	◆	□	⊠	■	□	□	■		■	■	■		□	⊠	■	■	■	□		■						■		■					■			■	■
⊠	⊠	⊠	⊠	⊠	□	⊠	⊠	⊠		⊠		□	⊠	⊠			□	⊠	⊠	⊠	□		□	⊠	⊠	⊠			⊠	□	□				⊠	⊠			⊠	⊠
□	□	□	⊠	□	□	□	⊠		□	□		□	⊠	⊠			□	⊠	⊠	⊠	□																		□	⊠
⊠	⊠	⊠	⊠	⊠	□	□	⊠	⊠		⊠		⊠	⊠	■		⊠	⊠	⊠	⊠	⊠	◆		⊠						⊠		⊠	⊠				⊠			⊠	⊠
⊠	⊠	⊠	⊠	⊠	◆	⊠	⊠	◆		⊠		⊠	⊠	⊠		◆	⊠	⊠	⊠	⊠	◆		⊠						⊠	⊠	⊠	⊠			⊠	⊠		⊠	◆	◆
□	□	⊠	⊠	⊠	□	⊠	⊠	□	□	⊠	□	⊠	⊠	⊠	□	□	□	⊠	⊠	⊠	□	□	□	□	□	□	□	□	□	□	□	□	□	□	□	□	□	□	⊠	⊠
		⊠	⊠	□		□	⊠			⊠		□	⊠	⊠				⊠	⊠	⊠																			⊠	⊠
□	■				◆		□	■		□			□	■		■	■	⊠	□		⊠		⊠		□		□	■	⊠	■	■	■		⊠	■	⊠		⊠	⊠	⊠
⊠	■	■	⊠	◆	⊠	⊠	■	⊠	□	⊠		⊠	⊠	■	□	□	■	⊠	■	⊠	⊠		⊠	□	⊠	□			⊠	■	■	□			■	⊠			⊠	⊠
⊠	⊠	⊠	□	□	□	⊠	⊠	⊠		⊠		□	□	⊠		□	⊠	⊠	⊠	□	□		⊠	⊠	⊠	⊠			⊠	⊠	⊠	□			⊠	⊠			□	□
⊠	⊠	■	⊠	⊠	⊠	□	⊠	⊠	□	⊠		□	⊠	■		⊠	⊠	⊠	⊠	⊠	⊠	□	⊠	⊠	⊠	⊠	□		■	■	⊠	■			■	■			⊠	⊠
□	⊠	⊠	⊠	□	□	⊠	⊠	⊠	□	⊠		⊠	⊠	⊠		⊠	⊠	⊠	⊠	⊠	⊠		■						■	■	⊠	■				⊠			□	⊠
⊠	⊠	⊠	■	◆	□	⊠	⊠	⊠		⊠		⊠	■	■		□	⊠	■	■	⊠	◆		⊠	⊠	⊠	⊠		□	⊠	□	□	⊠			⊠	⊠			◆	⊠
⊠	⊠	⊠	■	⊠	⊠	⊠	⊠	⊠		⊠		⊠	⊠	■		⊠	□	⊠	■	■	□		⊠	□	□	□		⊠	⊠	■	⊠	⊠		□	⊠	⊠			■	■
□	⊠	⊠	⊠	⊠	□	⊠	⊠	□		⊠		■	⊠	⊠		□	⊠	■	⊠	⊠	□		⊠						⊠		⊠					⊠			⊠	⊠
□	⊠	□		□			□	⊠			□	□	□	⊠		⊠	⊠	□			□							□	□	■	■	□		□	⊠	⊠				
□	⊠	⊠	⊠	⊠	□	⊠	⊠	⊠		⊠	□	⊠	⊠	⊠		⊠	⊠	⊠	⊠	□	□		⊠						⊠	⊠	⊠	□				⊠			⊠	⊠
⊠	⊠	□	⊠	□	□	□	⊠	⊠	□	⊠		□	⊠	⊠		⊠	⊠	⊠	⊠	⊠	⊠		⊠	⊠	■	⊠			⊠	⊠	⊠	⊠			■	■			□	⊠
⊠	⊠	⊠	⊠	⊠	□	□	◆	◆	□	⊠		⊠	⊠	⊠		◆	⊠	⊠	⊠	⊠	◆		⊠						⊠	□	⊠	⊠				⊠			⊠	⊠
			⊠			□	⊠			⊠		□	⊠	⊠				⊠	⊠	⊠																			⊠	⊠
■	⊠	⊠	⊠	⊠	□	⊠	⊠	□	⊠	⊠		⊠	⊠	⊠		□	⊠	■	⊠	⊠	□		⊠	⊠	⊠	⊠			⊠	⊠	⊠	⊠			⊠	⊠			⊠	⊠
□	⊠	□	□	◆	□	□	⊠	⊠		□		□	⊠	□		■	□	⊠	□	□	◆		⊠						⊠	□	⊠	⊠				⊠			□	■
	⊠	□	■	□		⊠	⊠			⊠			⊠	⊠			□	■	■	□																			⊠	■
⊠	■	□	⊠	◆	□	□	⊠	⊠		⊠	⊠	⊠	■	⊠		◆	■	⊠	⊠	□	◆		□					⊠	■	■	⊠	■			□	■			□	■
⊠	⊠	⊠	■	⊠	◆	⊠	⊠	■	□	⊠		□	⊠	⊠	□	◆	⊠	■	■	■	□		□	⊠	⊠	□			⊠	□	□	□			■	⊠	■		■	■
□	⊠	⊠	⊠	□	□	⊠	⊠	⊠	□	⊠		⊠	⊠	⊠		□	⊠	⊠	⊠	⊠	⊠		■						■		⊠	⊠				⊠			□	⊠
⊠	⊠	□	⊠	□	◆	□	⊠	◆	□	⊠		□	⊠	⊠		□	⊠	⊠	⊠	⊠	□	□	⊠	⊠	□	□	□	□	⊠	⊠	⊠	⊠		□	⊠	⊠		□	⊠	□
⊠	⊠	□			◆		⊠	◆	□	⊠				⊠		◆	⊠	■	⊠		◆		⊠						⊠	■	⊠	⊠		□		⊠		□	⊠	
⊠	⊠	⊠	⊠	⊠	⊠		⊠	⊠		⊠				⊠		⊠	⊠	⊠	⊠		⊠		⊠		⊠				⊠	⊠	⊠	⊠			⊠	⊠			⊠	⊠
□	⊠	⊠	□		□	⊠	⊠	⊠		⊠			⊠	□		⊠	□	⊠	⊠	□	□		⊠	■	■	⊠			⊠	⊠	⊠	⊠			■	■		□	⊠	⊠
□	⊠	⊠	⊠	⊠	□	⊠	⊠	⊠	□	⊠	□	⊠	⊠	⊠	□	⊠	⊠	⊠	⊠	□	⊠		⊠					□	⊠	⊠	⊠	■				⊠			□	⊠
□	□	⊠	⊠	◆	□	□	⊠	□	□	⊠		⊠	⊠	⊠		□	□	■	⊠	⊠	□		⊠	□		□		□	⊠	⊠	⊠	□		□	⊠	⊠		□	⊠	⊠
⊠	⊠	□	□	□	□	□	□	□	□	□		□	⊠	⊠		□	■	⊠	⊠	□	□		⊠	■	■	⊠			■	⊠	⊠	⊠			■	■			□	□
⊠	⊠	□	□	□	□	□	⊠	⊠	□	⊠		□	□	⊠		⊠	⊠	⊠	□	□	□		⊠	⊠	⊠				■	⊠	⊠				⊠	■			◆	◆
□	□	⊠	□	□	□	□	□	⊠	□	□		□	□	□		□	□	⊠	□	□	□		□						□	⊠	⊠	□		□	□	□			□	⊠
		⊠	■	◆	◆	⊠	⊠	□		⊠	□	□	⊠	⊠	□			⊠	⊠	■																			◆	⊠
□	⊠	□	□	⊠		□	⊠	⊠		⊠		⊠	⊠	⊠		□	⊠	⊠	⊠	□	◆		⊠						⊠		⊠	□				⊠			□	□
□	⊠	□	□	⊠		□	□	□		□			□	□		□	⊠	⊠	⊠	□	◆		■						■	⊠	■	⊠				■			□	□
□	⊠	⊠	⊠	⊠	◆	⊠	◆	□	□	⊠	□	⊠	⊠	⊠		◆	⊠	⊠	⊠	⊠	◆																		⊠	⊠
□	□	⊠	⊠	⊠	□	□	□	□	□	⊠		□	⊠	⊠		□	□	⊠	⊠	□		□	□	□	□		□	□	⊠	⊠	⊠	□			□	⊠				
⊠	⊠	⊠	⊠	□	□	⊠	⊠	⊠		⊠		□	□	⊠		⊠	⊠	⊠	⊠	⊠	□		⊠	⊠	⊠				⊠	⊠	⊠				⊠	⊠			◆	◆
	□	⊠	⊠	⊠	□	⊠	⊠			⊠	□	⊠	□	⊠	□		□	■	⊠	⊠																			◆	◆
⊠	⊠	□	⊠	□	□	□	□	⊠	□	⊠		⊠	⊠	⊠		⊠	⊠	⊠	⊠	⊠	□		⊠	⊠	■	⊠			⊠	⊠	⊠	⊠			■	■			□	⊠
		⊠	⊠	□		⊠			□	□			□	⊠				⊠	⊠	⊠																			□	□
⊠	⊠	⊠	⊠	⊠	◆	⊠	◆	◆	□	⊠		⊠	⊠	⊠		◆	⊠	⊠	⊠	⊠	◆		□	□	□	□			⊠	■	■	⊠		□	□	⊠		□	⊠	⊠
	□	⊠	■	□		□	□			⊠		□	□	⊠				⊠	■	■																			⊠	□
□	⊠	⊠	⊠						□	□				⊠			⊠	⊠					□					□	⊠	■	■	□			⊠	⊠				
⊠	⊠	□	⊠	□	□	□	⊠	⊠	□	⊠	□	□	⊠	⊠	□	⊠	⊠	⊠	□	□	⊠	□	⊠	⊠	⊠	□	□	□	⊠	⊠	⊠	□	□	□	⊠	⊠	□	□	□	⊠
■	⊠	□	⊠	⊠	□	⊠	⊠	⊠		⊠	□	⊠	⊠	⊠			□	⊠	⊠	⊠	□		⊠	□		⊠			⊠	⊠	⊠	□			⊠	⊠			⊠	⊠
⊠	⊠	⊠	■	□	□	⊠	⊠	⊠		⊠		⊠	⊠	⊠		□	□	■	⊠	■	□		⊠	⊠	⊠	■		⊠	⊠	⊠	⊠	⊠		□	■	■		□	⊠	⊠
⊠	⊠	■	■	□		⊠	⊠		□	⊠	□	⊠	⊠	⊠	⊠		□	⊠	⊠	■			⊠						⊠		⊠								◆	◆
■	⊠	□	□	□	□	□	□	□	□	□		□	□	⊠		□	■	⊠	⊠	□	□		⊠	■	■	⊠			■	⊠	⊠	⊠			■	■			◆	⊠
□	⊠	■	□	□		⊠	⊠	⊠		⊠		□	□	⊠		⊠	⊠	⊠	⊠	⊠	◆		□	⊠	⊠	□			⊠	⊠	⊠	⊠			⊠	⊠			⊠	⊠
□	⊠	■	□	□		□	⊠	⊠		⊠		□	□	⊠		□	⊠	⊠	⊠	□	□								□							□			◆	◆
	⊠	■	⊠	□		□	■	□		⊠		⊠	□	⊠		□	⊠	■	⊠	⊠	□								⊠							⊠			◆	■
	□	□	□	□			□	□		□		□	□	□		□	□	■																					□	■
⊠	⊠	⊠	□	⊠	□			⊠	□	□		□	□	⊠		⊠	⊠	⊠	□		⊠	□					□	⊠	□	■	■	⊠		⊠	■	□			⊠	□
□			■	□		□	⊠	◆		⊠		□	□	⊠				□	⊠	■																			⊠	⊠
	□	□	□	□			□	□		□		⊠	⊠	□			□	■	□	□	□		□						□							□			□	■
	□	⊠	⊠	□		□	■	□		⊠		⊠	⊠	□		□	□	■	□	□									⊠							⊠			◆	■
□	⊠	⊠	⊠	◆		⊠	⊠	□	□	⊠	□	⊠	⊠	⊠	□	□	□	⊠	⊠	⊠	□		□					□	□	■	■	⊠			□	□		□	■	⊠
■	⊠	□	■	□	◆	□	◆	■	□	⊠		⊠	■	⊠	□	◆	■	■	■	◆	⊠		□	□	□				⊠	□	□				■	⊠	■	□	◆	■
		⊠	⊠	□		⊠				⊠			□	⊠				⊠	⊠	⊠																			□	□
□	□	⊠	⊠			⊠	□			⊠		□	□	⊠	□		□	⊠	⊠	⊠																□			◆	◆
⊠	⊠	⊠	⊠	◆	◆	□	⊠	⊠	□	⊠		⊠	⊠	⊠		◆	⊠	■	⊠	⊠	◆	⊠	■			□			□	■	■	⊠				⊠			⊠	■
⊠	⊠	□	□	◆	□	□	⊠	⊠	□	⊠		□	□	□		⊠	⊠	⊠	⊠	□	⊠		□	⊠	⊠	□			⊠	⊠	⊠	⊠			⊠	■	□		⊠	⊠
	□				□		□	□	□	□				□		□	⊠	□			□	□	□	□	□	□	□	⊠	⊠	⊠	⊠	⊠	□	⊠	⊠	⊠		□		□
⊠	⊠	■	□	⊠	□	⊠	⊠	⊠	□	⊠		⊠	⊠	⊠	□	□	⊠	⊠	⊠	□	□		⊠	□	□	□			⊠	■	■	□			■	⊠			⊠	⊠
⊠	⊠	⊠	⊠	⊠	□	□	◆	□		⊠		⊠	⊠	⊠		□	⊠	⊠	□	□	□		⊠						⊠		⊠					⊠			⊠	⊠
⊠	⊠	□	⊠		□	□	□	□	□	□		□	□	⊠		□	⊠	■	⊠	⊠	□	□	⊠	⊠	□	⊠	□	□	⊠	■	⊠	⊠		□		⊠		□	□	□
⊠	⊠	⊠	⊠	⊠	□	⊠	◆	⊠	□	⊠	□	⊠	⊠	⊠		□	⊠	⊠	⊠	⊠	□		□						⊠	⊠	⊠	□		□	□	⊠			⊠	⊠
⊠	⊠	⊠	⊠	⊠	□	⊠	⊠	□		⊠		⊠	⊠	⊠	□	□	⊠	■	⊠	⊠	⊠		⊠		⊠			◆	⊠	⊠	⊠	⊠			⊠	⊠		◆	⊠	⊠
		⊠	⊠	⊠		□	⊠			⊠		⊠	⊠	□				⊠	⊠	⊠																			⊠	⊠
⊠	⊠	⊠	⊠	⊠	□	⊠	⊠	⊠		⊠		□	□	⊠		□	⊠	⊠	⊠	□	□		⊠	⊠	⊠	⊠			⊠	⊠	⊠	□			⊠	⊠			⊠	⊠
	□	⊠	⊠	⊠	□	⊠	⊠			⊠		■	⊠	⊠			□	■	□																				□	■
		⊠	■			⊠	⊠			⊠			□	⊠				■	⊠	⊠																			⊠	⊠

Key ■ = Abundant ⊠ = Common □ = Uncommon ◆ = Seasonal

	Page Number	parking lot	privy/restrooms	picnic areas	primitive camping	camper pads	river access	boat ramps	fishing docks/jetties	boat rental	viewing blinds	waterfowl blinds	interpretive programs	exhibits/displays	MDC office	hunting	waterfowl hunting	fishing	hiking trails	horse trails	canoeing	archery range	firearms range	field trials	dog training	trapping	Notes
MOZINGO LAKE	156	■		■				■										■			■						
MUDPUPPY CA	190	■			■											■		■			■				■	■	
MULE SHOE CA	92	■			■		■									■		■	■		■					■	
MURPHY (JOHN F.) MEM SF & WA	92	■	■													■			■								
MUSSEL BOAT LANDING ACCESS	28																	■									
MUSSEL FORK CA	124	■			■											■	■	■			■			■	■	■	
MYRTLE ACCESS	158	■	■		■		■	■										■			■						
NEEPER CA	51	■			■											■	■	■						■	■	■	
NEOSHO TOWERSITE	154															■										■	
NEW BALLWIN LAKE	200	♿	♿	■					♿				■					■	■								
NEW HAVEN ACCESS	78	■					■	■										■									
NEW MADRID BEND ACCESS	152	♿					■	■										■									
NEWMAN MEM ACCESS	66	■					■											■			■						
NIANGUA CA	234	■														■			■								
NIAWATHE PRAIRIE CA	62	■																									
NIFONG PARK POND	25	■		■														■									
NODAWAY CO COMMUNITY LAKE	156	♿	■		■			■	♿							■		■	■		■			■	■	■	
NODAWAY ISLAND ACCESS	8	■	♿		■		■	■										■			■						
NODAWAY VALLEY CA	8	■			■										■	■	■	■			■			■	■	■	
NORFORK LAKE	162	■	♿	■	■	■	■	■		■	♿					■	■	■			■					■	
NORTH LAKE – HARRISONVILLE	42	♿	♿					■	♿									■					♿				
NORTH RIVERFRONT PARK ACCESS	200	♿	♿	■			■	■										■	♿		■						
NORTH RIVERFRONT PARK LAKE	200	♿		■			■											■									
NORTHCUTT (C. L.) MEM CA	12	■			■											■											
NORTHEAST PARK LAKE	12																										
NORTON WOODS ACCESS	122	■	♿	■	■		■	■										■			■						
O'FALLON PARK LAKE	200	♿		■					♿									■									
OAK RIDGE CA	214	■														■											
ODESSA CITY LAKES	116	■	■					■	■								■	■									
ODIN ACCESS	238	■	■				■									■		■								■	
OLD MONROE ACCESS	193	■					■											■									
OLD PLANTATION ACCESS	36						■									■		■			■						
OLD TOWN ACCESS	18	■			■		■	■								■	■	■			■				■		
ONYX CAVE CA	61															■										■	
OSAGE FORK CA	114	■	♿		■											■						■		■	■	■	
OSAGE PRAIRIE CA	226	■														■		■							■	■	
OSAGE-TAVERN ACCESS	140	■	■		■	■	■	■										■			■						
OTTER SLOUGH CA	28	♿	♿	■				■	♿		♿	■			♿	■	■	■	■		■						
PA SOLE PRAIRIE CA	16	■														■	■	■									
PACIFIC PALISADES CA	108	■					■	■								■	■	■			■	■		■	■	■	
PAINT BRUSH PRAIRIE CA	168	■														■									■		
PAINTED ROCK CA	160	■	■		■											■	■	■	■							■	
PAPE (EDWIN A .) LAKE	116	♿	♿	■				■	♿								■	■			■						
PARIS ACCESS	146	■	■				■	■										■			■						
PARIS SPRINGS ACCESS	118	■					■									■		■			■					■	
PARK (GUY B.) CA	174	■														■	■	■			■				■	■	
PATRICK BRIDGE ACCESS	162	■	■	■	■	■	■									■		■	■		■					■	
PAWHUSKA PRAIRIE	16																										
PAWNEE PRAIRIE CA	88	■			■											■		■	■						■		
PAYDOWN ACCESS	134	♿	♿	■		■	■	■										■			■						
PAYNE LANDING ACCESS	94	■			■		■	■										■			■				■		
PEA RIDGE CA	230	■			■											■		■							■	■	
PEABODY CA	18	■			■			■								■	■	■			■				■		
PECK RANCH CA	40	♿	♿		■								■	■	♿	■		■	■			■	■	■	■	■	
PELICAN ISLAND NA	202																	■									
PERRY CO COMMUNITY LAKE	166	♿	♿	■				■	♿									■	■							■	
PERRY (R. & M.) MEM CA	110	■			■		■	■								■	■	■			■		■	■	■	■	
PERRY TOWERSITE	166	■														■							■				
PHENIX ACCESS	86	■														■		■									
PICKLE SPRINGS NA	198	■													♿				■								
PIGEON HILL CA	26	♿	♿		■											■			■			♿	♿				
PIGG'S LANDING ACCESS	186	■					■	■										■			■						
PIKES CAMP ACCESS	54	■	■				■	■								■	■	■			■				■	■	
PILOT KNOB TOWERSITE	170				■											■											
PIN OAK CA	212	■			■											■	■	■			■			■	■	■	
PINE RIDGE LAKE	46	■	♿	■				■										■	■								
PINEY RIVER NARROWS NA	224	■														■		■			■					■	
PINHOOK ACCESS	168	■					■											■			■						
PLAD TOWERSITE	65	■																									
PLATTE FALLS CA	174	♿	♿	■	■		■									■	■	■	■		■	■	■	■	■	■	
PLEASANT HOPE CA	176	■			■											■							■				
PLOWBOY BEND CA	144				■											■	■	■			■					■	
PLUNKETT PARK LAKE	12	♿	♿	■					♿									■	■								
POAGUE (HAYSLER A.) CA	90	■			■			■								■	■	■			■				■		
POINTERS CREEK ACCESS	160	■	■		■		■	■										■			■						
POMME DE TERRE RESERVOIR	92	■					■									■	■	■			■				■	■	
PONY EXPRESS LAKE CA	68	♿	♿	■				■	♿	■						■	■	■	■					■	■		
POOSEY CA	126	♿	♿	■	■	■		■	♿							■	■	■	■	■	■	■	■	■	■	■	
POPLAR BLUFF CA	28	■			■											■		■				■				■	
PORT HUDSON LAKE CA	78	♿	♿	■				■	♿									■	■								
POSSUM WALK ACCESS	156	■					■											■			■						
POSSUM WOODS CA	202																										
POWDER VALLEY CONS NATURE CENTER	203	♿	♿										■	■	♿				♿								
POWELL TOWERSITE	128	■														■										■	
PRAIRIE HOME CA	58	■	♿		■				♿							■	■		■	■						■	
PRAIRIE LEE LAKE	102	■						■	■									■									
PRAIRIE SLOUGH CA	122	♿	♿					■								■	■	■			■				■	■	
PRICE BRIDGE ACCESS	46															■	■	■			■					■	
PROCTOR PARK LAKE	144	■	■	■														■	■								
PROCTOR TOWERSITE	150		■		■											■											
PROSPERINE ACCESS	114	■					■											■			■					■	
PROVIDENCE ACCESS	25	■	■				■	■										■									

Key ■ = Yes ♿ = Disabled Access

beaver	bullfrogs	coyotes	deer	doves	eagles	fox	hawks	herons	otters	owls	pheasant	quail	rabbits	racoons	ruffed grouse	shore birds	snapping turtles	songbirds	squirrels	turkey	waterfowl	hybrid bass	largemouth bass	rock bass	small mouth bass	spotted bass	striped bass	white bass	blue gill	carp	catfish	crappie	muskelunge	paddlefish	suckers	sunfish	trout	walleye	mushrooms	wildflowers
□	⊠	⊠	■	⊠	□	□	■	⊠		⊠	⊠	⊠	⊠	⊠		⊠	■	⊠	□	□	◆		■						■	⊠	■	⊠				⊠		⊠	□	□
□	⊠	⊠	⊠	□	□	⊠	⊠	⊠	□	□		⊠	⊠	⊠		□	□	⊠	⊠	⊠	□		⊠	⊠	⊠	□			⊠	⊠		□			⊠	⊠			⊠	⊠
□	⊠	⊠	⊠	⊠	□	□	⊠	□		⊠		⊠	⊠	⊠			⊠	⊠	⊠	⊠		□	■	□	■	□	□	□	⊠	■	⊠	⊠			⊠	⊠		□	⊠	⊠
		⊠	⊠	□		□	⊠			⊠		□	⊠	□				⊠	⊠	⊠																			⊠	⊠
⊠	⊠	⊠	□	⊠	□	□	⊠	⊠	■	■		□	□	⊠		□	□	■	⊠	□	□		■	■	■	□			■	■	■	■			■	■			■	■
⊠	⊠	⊠	⊠	⊠	□	⊠	⊠	□	□	⊠	□	⊠	⊠	⊠		□	⊠	⊠	⊠	⊠	◆		⊠			⊠			⊠	⊠	⊠	□			⊠	⊠			◆	◆
□	⊠	⊠	⊠	◆	□	⊠	⊠	◆	□	⊠		⊠	⊠	⊠			□	⊠	⊠	⊠	◆		⊠	⊠	⊠	□			⊠	□	□	□		□	⊠	⊠		□	⊠	⊠
	⊠	⊠	■	⊠		⊠	⊠	□		■	□	⊠	⊠	⊠		□	⊠	■	■	■	□		⊠						■		⊠	⊠				■			◆	◆
		⊠	⊠	□		⊠	⊠			⊠		□	□	⊠				■	⊠	□																			⊠	⊠
	⊠			⊠									⊠	□				⊠	⊠		□		⊠						⊠	⊠	⊠					⊠				
□	□							□	□	□				⊠		□	⊠	⊠	⊠		□		□					□	□	■	⊠	⊠		⊠	⊠	□		□	□	□
					□			□								□						□	□	□	□	□	□	⊠	⊠	⊠	⊠	⊠	□	⊠	⊠	⊠				
⊠	⊠	⊠	⊠	□	□	□	⊠	□		⊠		□	⊠	⊠		□	⊠	⊠	⊠		□								⊠	⊠	⊠				⊠	⊠			⊠	⊠
		⊠	⊠	□		□	⊠			⊠		□	⊠	⊠				⊠	⊠	⊠																			⊠	⊠
	□	□	□	□		□	⊠	□		□		⊠	⊠	□		◆	□	⊠		□	□		□						□		□					□			□	■
	⊠	□		⊠		□	⊠	□		⊠		□	⊠	□		□	⊠	⊠	⊠		⊠		□						■	□	□					■				
□	⊠	⊠	⊠	⊠		⊠	⊠	⊠		⊠	⊠	⊠	⊠	⊠		□	⊠	⊠	⊠		⊠		⊠						⊠	□	⊠	⊠				⊠			⊠	⊠
		⊠		□			□	□		□	□	□	□	□		□		□	□		□		□	□				□	□	■	■	⊠		□	⊠	□		□		
⊠	⊠	⊠	⊠	⊠	◆	⊠	⊠	⊠	□	⊠	⊠	⊠	⊠	⊠		◆	⊠	⊠	⊠	⊠	◆		□						□	⊠	⊠				□	⊠			◆	⊠
⊠	⊠	⊠	⊠	⊠	◆	⊠	⊠	■	□	⊠		⊠	⊠	⊠		◆	⊠	⊠	⊠	⊠	⊠	⊠	⊠	⊠	⊠	⊠	⊠	⊠	⊠	⊠	⊠	⊠			⊠	⊠		⊠	⊠	⊠
□	⊠	□	⊠	⊠	■	□	⊠	⊠		⊠		⊠	⊠	⊠		◆	⊠	■	⊠	□	◆		■						■	□	⊠	■				■			□	⊠
⊠	□			⊠	◆	□	⊠	⊠		□		□	⊠	⊠		◆	⊠	⊠	⊠		◆	□	□		□	□	□	⊠	⊠	■	■	⊠		□	■	⊠		□	□	□
	⊠			⊠				⊠					⊠			⊠		⊠	⊠				□						□	⊠	⊠	⊠				⊠				
	□	⊠	⊠	□		□	□			⊠		□	□	□	□		□	⊠	■	⊠	□																		⊠	⊠
■	⊠	⊠	⊠	□	◆	⊠	⊠	■	□	⊠		□	□	⊠		◆	⊠	■	⊠	□	◆		⊠	□	□	□		⊠	⊠	■	⊠	⊠		□	□	⊠			⊠	⊠
	⊠			⊠									⊠			⊠			⊠		⊠		□						□	⊠	⊠	□				⊠	⊠			
□	□	⊠	⊠	⊠	□	⊠	⊠	□	□	⊠		⊠	⊠	⊠		□	□	■	□	□	□																		◆	■
□	⊠	□	⊠	□	□	□	□	□		□		□	⊠	□		□	⊠	⊠	⊠	□	◆		■						■	□	⊠	⊠				■			□	⊠
⊠	⊠	⊠	■	⊠	□	⊠	□	⊠	□	□		⊠	⊠	⊠		□	■	⊠	⊠	■	□		⊠	■	■	□			■	□	□	□			⊠	■			⊠	⊠
⊠	■	□			◆	□		⊠	□	□				■		□	■	⊠	□		□		⊠			□		⊠	⊠	■	■	■		□	⊠	⊠				
⊠	⊠	⊠	⊠	⊠	□	⊠	⊠	⊠		⊠		⊠	⊠	⊠		□	⊠	⊠	□	⊠	□		⊠	⊠	⊠	⊠		□	⊠	⊠	⊠	⊠		□	⊠	⊠		□	□	⊠
⊠	◆	⊠	⊠	⊠	□	□	⊠	⊠	□	⊠		□	□	⊠		◆	◆	■	⊠	⊠	□	◆	⊠				◆	◆	⊠	⊠	⊠	◆		◆	⊠	⊠		◆	◆	◆
□	⊠	⊠	⊠	□	□	⊠	⊠	⊠	□	⊠	□	□	⊠	⊠	□	⊠	⊠	⊠	⊠	⊠	□	□	⊠	⊠	⊠	□	□	□	⊠	⊠	⊠	□	□		⊠	⊠	□	□	⊠	⊠
	□	■	■	⊠	□	⊠	⊠			⊠		⊠	⊠	■			□	◆	■	■																			⊠	⊠
	⊠	⊠	■	⊠		⊠	■	□		⊠		⊠	⊠	⊠		□	⊠	■	⊠	□	◆		⊠						⊠		⊠					⊠			◆	■
□	⊠	□	⊠		◆	⊠	⊠	⊠	□	⊠			⊠	⊠		□	⊠	⊠	⊠	⊠	◆	□	⊠	⊠	⊠	⊠	□	⊠	⊠	⊠	⊠	⊠		⊠	⊠	⊠			◆	⊠
■	■	■	⊠	⊠	◆	⊠	⊠	⊠	⊠	⊠	□	⊠	⊠	■		◆	⊠	■	⊠	⊠	◆		□						⊠	⊠	⊠	⊠			⊠	⊠			⊠	⊠
	□	□	□	⊠		□	⊠	□		□		⊠	⊠	□		⊠	□	⊠			□		⊠						⊠		□					⊠			□	⊠
□	⊠	⊠	⊠	⊠	□	⊠	⊠	⊠	□	⊠		⊠	⊠	⊠		⊠	⊠	⊠	⊠	⊠	⊠		⊠	■	⊠	■		□	■	⊠	⊠	■		□	■	■		□	⊠	⊠
	□	⊠	⊠	⊠		□	◆	□		⊠		⊠	⊠	⊠				⊠	□	□																			□	■
□	⊠	⊠	⊠	⊠	□	⊠	⊠	⊠	⊠	⊠		⊠	⊠	⊠		□	⊠	⊠	⊠	⊠	⊠		⊠			⊠			⊠	⊠	⊠	⊠			⊠	⊠			⊠	⊠
□	⊠	⊠	⊠	◆	□	□	⊠	⊠		⊠		⊠	⊠	⊠		◆	⊠	⊠	■	⊠	◆		■						■	■	■	■			□	■			⊠	■
□	⊠	□	□	□	□	□	⊠	⊠	□	⊠		□	⊠	□		□	⊠	⊠	⊠	□	□		⊠		□	□		◆	⊠	⊠	⊠	⊠			◆	⊠		◆	⊠	⊠
⊠	⊠	⊠	■	⊠	□	⊠	⊠	■	□	■		⊠	⊠	⊠		□	□	■	⊠	□	□		■	⊠	■	□		□	■	■	■	⊠			■	■	□	□	⊠	■
⊠	⊠	⊠	⊠	⊠	□	□	⊠	⊠	□	⊠	□	⊠	⊠	⊠		□	■	⊠	⊠	□	□		⊠						⊠	□	□	□			□	⊠			⊠	⊠
⊠	⊠	⊠	⊠	⊠	◆	□	⊠	⊠	□	⊠		⊠	⊠	⊠			⊠	⊠	⊠	⊠	◆	□	⊠	⊠	⊠	□	□	□	⊠	⊠	⊠	□			⊠	⊠	⊠		◆	⊠
		⊠	⊠	□		□	⊠			□		⊠	⊠	□				⊠	□	□																			⊠	■
⊠	⊠	⊠	⊠	□		□	⊠	□	□	□	□	□	⊠	⊠		□	⊠	◆	□	□	□		⊠						⊠		⊠									
□	⊠	⊠	⊠	⊠	□	⊠	⊠	⊠	□	⊠		⊠	⊠	⊠		□	⊠	⊠	⊠	⊠	□		⊠	⊠	⊠	⊠			⊠	⊠	⊠	⊠			⊠	⊠			⊠	⊠
⊠	⊠	⊠	□	⊠	□	⊠	⊠	⊠	□	⊠	⊠	⊠	⊠	⊠		□	□	⊠	□		□		□					□	□	■	■	⊠		□	⊠	□		□	□	□
⊠	⊠	⊠	⊠	⊠	□	⊠	⊠	⊠		⊠	□	⊠	⊠	⊠	□	□	⊠	⊠	⊠	⊠	□		⊠		⊠				⊠		□				⊠	⊠			◆	◆
⊠	◆	⊠	⊠	⊠	□	□	⊠	⊠		⊠		⊠	⊠	⊠		□	◆	■	⊠	□	□		⊠						■		⊠	⊠				■			◆	⊠
⊠	⊠	■	■	⊠	⊠	⊠	⊠	⊠	□	⊠		□	⊠	⊠	□	□	⊠	■	■	⊠	□		□	□	□				⊠	□	□				⊠	⊠			⊠	⊠
⊠	⊠	⊠	■	⊠	◆	□	⊠	⊠	□	⊠		□	□	⊠		⊠	⊠	■	⊠	⊠	◆	□	⊠			□	□	⊠	⊠	⊠	■	⊠		□	⊠	⊠		□	⊠	⊠
⊠	⊠	⊠	⊠	□	□	⊠	⊠	⊠		⊠		⊠	⊠	■		⊠	⊠	■	⊠	□	◆		■	□					■	■	■	■				⊠			⊠	■
□	⊠	⊠	⊠	◆	□	□	◆	⊠	□	⊠		⊠	⊠	⊠		◆	⊠	■	⊠	⊠	◆		□						⊠	■	⊠	□		□	□	⊠			◆	■
		⊠	⊠			⊠	⊠			⊠			□	⊠				⊠	■	⊠																			□	⊠
⊠	⊠	⊠	⊠	⊠	□	□	⊠	⊠		⊠		⊠	⊠	⊠		⊠	⊠	⊠	⊠	□	□		⊠	⊠	⊠	□			⊠	⊠	⊠	□			⊠	⊠			⊠	⊠
	□	⊠	⊠			⊠	⊠			⊠			□	⊠				⊠	■	⊠									⊠							⊠			⊠	■
□	⊠	⊠	⊠	□		⊠	⊠			⊠		⊠	⊠	⊠			□	⊠	⊠	⊠																			⊠	□
⊠	⊠	⊠	⊠	⊠	□	⊠	⊠	⊠	□	⊠	□	⊠	⊠	⊠		⊠	⊠	⊠	⊠	□	□		⊠						⊠	⊠	⊠	⊠		□	□	⊠			⊠	⊠
⊠	⊠	⊠	⊠	□	◆	□	⊠	⊠	□	⊠		⊠	⊠	⊠		■	⊠	■	⊠	⊠	□	□	⊠	⊠	□	⊠	□	⊠	⊠	⊠	⊠	⊠		⊠	⊠	⊠		□	⊠	⊠
		□	⊠	□		□	⊠			□		□	□	⊠				⊠	⊠	□																			□	□
⊠	⊠	⊠	⊠	⊠	□	⊠	⊠	⊠	□	⊠	□	⊠	⊠	⊠		□	⊠	⊠	⊠	⊠	⊠		⊠						⊠	⊠	⊠	□			□	⊠			◆	◆
□	⊠	⊠	⊠	□	□	⊠	⊠	⊠	□	⊠		□	■	⊠		□	⊠	⊠	⊠	⊠	⊠		■						■		⊠					⊠			□	⊠
⊠	⊠	⊠	⊠	⊠	□	□	⊠	⊠		□		⊠	⊠	⊠		□	⊠	⊠	⊠	⊠	□		⊠	■	■	⊠			■	⊠	⊠	□			■	■			⊠	⊠
□	⊠	□	□	□	□	□	◆	◆	□	⊠		□	□	⊠		□	⊠	⊠	⊠	□	□		⊠			□			⊠	⊠	⊠	⊠		□	□	⊠		□	⊠	□
	□	□	□	□		⊠	⊠			⊠		□	□	⊠				◆	⊠	□																			□	⊠
⊠	◆	⊠	⊠	◆	◆	□	■	◆	□	⊠	□	■	⊠	■		◆	⊠	⊠	⊠	□	◆		⊠					□	■	■	■	□		□	⊠	■		□	◆	□
□	□	⊠	⊠	⊠		□	⊠	□		⊠		⊠	⊠	⊠		□	□	⊠	⊠	⊠	□																		⊠	⊠
⊠	■	⊠	⊠	⊠	◆	□	⊠	⊠	□	⊠		⊠	⊠	⊠		⊠	⊠	■	⊠	⊠	◆									■	■				■				◆	◆
	⊠		□	⊠	□		□	□		□		□	⊠	⊠		□	□	■	⊠	□	□		■						■		■	□				⊠			⊠	⊠
⊠	⊠	⊠	⊠	⊠	□	□	⊠	⊠		⊠		⊠	□	⊠		□	⊠	■	⊠	□	□		⊠			□			⊠	⊠	⊠	□				⊠			⊠	⊠
□	⊠	⊠	⊠	⊠	□	⊠	⊠	⊠	□	⊠		⊠	⊠	⊠		□	⊠	⊠	⊠	⊠	□		⊠	⊠	⊠	⊠			⊠	⊠	⊠	⊠			⊠	⊠			⊠	⊠
⊠	⊠	⊠	■	⊠	◆	⊠	⊠	⊠	□	⊠		⊠	⊠	⊠		⊠	⊠	⊠	■	■	⊠		⊠	⊠	⊠	⊠		⊠	⊠	⊠	⊠	⊠	⊠		⊠	⊠		⊠	◆	■
□	◆	⊠	⊠	◆	□	□	⊠	◆		⊠	□	⊠	⊠	⊠		◆	□	⊠	⊠	⊠	⊠		⊠						⊠	⊠	⊠	⊠	⊠			⊠			⊠	⊠
⊠	⊠	⊠	■	⊠	□	⊠	⊠	⊠	□	⊠	□	⊠	⊠	⊠		⊠	⊠	⊠	⊠	⊠	⊠		⊠						■	⊠	■	⊠			□	■			⊠	⊠
		□	⊠	□		⊠	□			⊠		□		⊠				⊠	⊠	⊠																			◆	□
□	⊠	⊠	⊠	□	□	⊠	⊠	■		⊠	□	⊠	⊠	⊠	□	■	⊠	⊠	⊠	⊠	◆		⊠	⊠	⊠				⊠		⊠					⊠			◆	■
□	⊠	⊠	⊠	□		□	⊠	⊠		⊠	⊠	□	⊠	⊠		□	⊠	⊠	⊠	□	□								□	□	⊠					□			⊠	◆
□	□	□	□	⊠	□	⊠	⊠			⊠		□	⊠	⊠				■	⊠	□																			□	■
	□	□	⊠	■	□	⊠	⊠			■		□	■	■			□	■	■	⊠	□															□			■	■
		⊠	⊠	□		⊠	⊠			⊠		□	□	⊠				⊠	⊠	□																			⊠	⊠
□	⊠	■	■	⊠	□	□	⊠	⊠		⊠		■	⊠	⊠		□	⊠	⊠	⊠	■	□		⊠	⊠	⊠	⊠			⊠	□	⊠	□			□	⊠			◆	◆
□	⊠	□	⊠	⊠	□	□	⊠	⊠		⊠		□	⊠	⊠		◆	⊠	⊠	⊠	⊠	⊠	⊠	■					⊠	■	■	⊠	■			□	■			□	⊠
■	⊠	□	■	⊠	◆	⊠	⊠	■	□	⊠		□	□	■		■	⊠	■	■	■	◆		⊠	□		□		⊠	⊠	■	■	■		□	□	■			⊠	⊠
⊠	⊠	⊠	■	⊠	□	□	⊠	⊠	□	⊠		⊠	⊠	⊠		⊠	⊠	■	⊠	⊠	■	□	□			□	□	⊠	⊠	■	■	⊠		□	⊠	⊠			⊠	⊠
□	⊠	□	□	□	□	□	□	□		□		□	□	□		□	⊠	⊠	⊠	□	□		■						■	⊠	■	⊠				■			⊠	⊠
		⊠	■			⊠	⊠			⊠			□	■				■	⊠	■																			◆	⊠
⊠	⊠	⊠	□	□	□	⊠	⊠	⊠	□	□		□	⊠	⊠		⊠	⊠	◆	⊠	□	□		□	⊠	⊠	□			⊠	⊠	⊠	⊠			⊠	⊠	⊠		⊠	⊠
□	□	□	⊠	□	□	□	□	□	□	□	□	□	□	⊠	□	□	□	□	⊠	□	□	□	□	□	□	□	□	□	⊠	⊠	⊠	⊠	□	□	⊠	⊠	□	□	□	□

Key ■ = Abundant ⊠ = Common □ = Uncommon ◆ = Seasonal

	Page Number	parking lot	privy/restrooms	picnic areas	primitive camping	camper pads	river access	boat ramps	fishing docks/jetties	boat rental	viewing blinds	waterfowl blinds	interpretive programs	exhibits/displays	MDC office	hunting	waterfowl hunting	fishing	hiking trails	horse trails	canoeing	archery range	firearms range	field trials	dog training	trapping	Notes
PROVIDENCE PRAIRIE CA	118	■														■											
PUNKIN CENTER ACCESS	126	■			■		■									■	■	■			■				■		
QUARRY POND	196	♿							♿									■									
QUEENY PARK LAKE	202	♿	■										■					■									
RADER ACCESS	234						■									■		■			■						
RANACKER CA	172	■	■		■											■		■					■			■	
RAY CO COMMUNITY LAKE	186	♿	♿	■				■	♿						♿			■									
RAY (J. THAD) MEM WA	136	■														■						■				■	
REBEL'S COVE CA	180	■			■		■	■								■	■	■			■			■	■	■	
RED ROCK LANDING CA	166	■			■		■									■	■	■									
REDHORSE ACCESS	78	■					■											■			■						
REDMAN CA	130																									■	
REDWING PRAIRIE CA	16	■														■											
REED (JAMES A.) MEM WA	102	♿	♿	■	■				♿	■			■	■	♿	■		■	■	■		■		■	■	■	
REFORM CA	32	■	■													■	■	■	■								
REIFSNIDER (FRANK) SF	228	■		■	■								■			■		■	■				■	■	■	■	
REIKER FORD ACCESS	78	■					■											■			■						
REYNOLDS (S. P.) ACCESS	164	■		■			■	■										■									
RICHTER CA	61	■			■											■								■	■	■	
RIDDLE BRIDGE ACCESS	178	■	■				■	■										■			■						
RIGGS FERRY ACCESS	122	■					■											■			■						
RINGGOLD ACCESS	174						■											■			■						
RINGO FORD ACCESS	28	■					■											■			■						
RINQUELIN TRAIL COMMUNITY LAKE	134	♿	♿		■			■	♿							■	■	■			■						
RIPGUT PRAIRIE NA	18	■														■	■								■		
RIPPEE CA	72	■	■	■	■	■	■									■		■			■				■	■	
RISCH (E. B. & M. O.) CA	16	■														■	■	■									
RIVER 'ROUND CA	78	■					■	■								■		■			■				■	■	
RIVERBREAKS CA	94	♿	♿		■											■		■	■				♿		■	■	
RIVERSIDE CA	100				■											■		■								■	
RIVERVIEW ACCESS	61	■					■	■										■			■						
RIVERWOODS CA	202	■					■											■			■						
ROARING RIVER CA	14	■														■		■								■	
ROARING RIVER HATCHERY/TROUT PARK	14	♿	■				■							■	■			■									
ROBERTS BLUFF ACCESS	58	■	■		■		■	■										■			■						
ROCHESTER FALLS ACCESS	8	■	■		■	■	■											■			■						
ROCK QUARRY ACCESS	8	■			■		■	■										■			■						
ROCKWOODS RANGE	202	■																	■								
ROCKWOODS RESERVATION	202	♿	♿	■									■	■	♿				♿								
ROCKY BARRENS CA	86	■														■									■		
ROCKY CREEK CA	210	■			■		■									■		■	■							■	
ROCKY FORD ACCESS	218	■					■											■			■						
ROCKY FORK LAKES CA	25	♿	♿		■			■								■	■	■	■		■		♿	■	■	■	
ROCKY MOUNT TOWERSITE	140															■											
ROLLINS FERRY ACCESS	160	■	■		■		■	■										■			■						
ROSATI TOWERSITE	170	■			■											■									■		
ROSE HILL TOWERSITE	158	■														■										■	
ROSE POND CA	51	■			■						■					■	■	■						■	■	■	
ROSELLE ACCESS	132	■					■	■							■	■		■			■						
ROSS ACCESS	178						■											■			■						
ROTARY PARK LAKE	36	♿	♿	■					♿									■	■								
ROTHWELL PARK LAKE	184	■	♿	■	■	■		■	♿	■								■	■		■	♿		■	■		
ROUBIDOUX CREEK CA	178						■											■									
RUNGE CONSERVATION NATURE CENTER	56	♿	♿										■	■	♿				♿								
RUSH CREEK CA	52	♿	♿	■									■	■	♿				♿								
RYDEN CAVE CA	178															■											
SAEGER WOODS CA	102																										
SAINT MARY'S CITY ACCESS	198	■					■	■										■									
SALINE VALLEY CA	140	■	■		■		■									■	■	■			■				■	■	
SAND FORD ACCESS	78	■					■	■										■			■						
SAND POND CA	190	■														■										■	
SAND POND CA - ALTVATER (N. B.)	190	■														■										■	
SANDY ISLAND CA	122															■	■	■			■					■	
SANK CA	23															■								■	■		
SANTA FE ACCESS	146	■					■	■										■			■						
SAPPINGTON BRIDGE ACCESS	61	■	♿				■	■										■			■					■	
SAVAGE ACCESS	66	■					■											■			■						
SAVANNAH CITY LAKE	8	♿	♿					■										■			■						
SAXTON ACCESS	26	■			■		■											■			■						
SCHELL-OSAGE CA	194	■	♿	■	■		■	■		■		♿	■	■	■	■	■	■	■		■				■		
SCHERER LAKE	102																	■									
SCHIFFERDECKER (W. L.) MEM CA	38	■			■											■	■	■							■		
SCHIMMEL CITY ACCESS	174	■			■		■	■										■			■						
SCHLICHT SPRINGS ACCESS	178	■	■		■		■	■										■			■					■	
SCHUMAN PARK LAKE	170	■	■	■														■									
SCIOTO LAKE	170	♿	♿	■					♿				■	■				■	■								
SCOTTS FORD ACCESS	61	■	■				■											■			■						
SCRIVNER ROAD CA	54	♿	♿				■									■	■	■	■	■	■		■		■	■	
SEARS COMMUNITY LAKE	218	■	■		■											■		■			■						
SEARS (F. O. & LEDA J.) MEM WA	12																										
SEAT (EMMETT & LEAH) MEM CA	82	■	■	■	■											■	■	■			■	■		■	■	■	
SENTINEL CA	176															■											
SETTLE'S FORD CA	18	■					■									■	■	■							■		
SEVEN ISLAND CA	143	■			■			■								■	■	■									
SEVENTY-SIX CA	166	■			■		■		■							■	■	■									
SEVER (HENRY) LAKE CA	112	■	■	■	■	■		■	■	■						■		■		■	■			■	■	■	
SEYMOUR TOWERSITE	234																										
SHANKS (TED) CA	172	♿	♿		■			■				♿	■	■	♿	■	■	■	♿		■					■	
SHANNON RANCH CA	72	■			■											■			■						■	■	
SHARPS STATION ACCESS	174	■			■		■	■								■	■	■			■				■		
SHAWNEE BEND ACCESS	34	♿	♿				■	■	♿									■									
SHAWNEE MAC LAKES CA	70	♿	♿	■	■			■	♿									■			■				■		

Key ■ = Yes ♿ = Disabled Access

beaver	bullfrogs	coyotes	deer	doves	eagles	fox	hawks	herons	otters	owls	pheasant	quail	rabbits	racoons	ruffed grouse	shore birds	snapping turtles	songbirds	squirrels	turkey	waterfowl	hybrid bass	largemouth bass	rock bass	small mouth bass	spotted bass	striped bass	white bass	blue gill	carp	catfish	crappie	muskelunge	paddlefish	suckers	sunfish	trout	walleye	mushrooms	wildflowers
□	⊠	⊠	⊠	⊠	□	⊠	■	□		⊠		⊠	⊠	⊠		□	□	■	⊠	□																				■
⊠	⊠	⊠	⊠	⊠	□	⊠	⊠	⊠	⊠	⊠	□	⊠	⊠	⊠		⊠	⊠	⊠	⊠	⊠	⊠		⊠					□	⊠	■	■	⊠		□	⊠	■		□	⊠	⊠
⊠	⊠	□	□	□	□	□	⊠	⊠		□		□	⊠	□		□	⊠	⊠	⊠	□	□		⊠						■		□					■			□	⊠
	⊠	⊠	⊠	⊠	⊠	□	⊠	⊠		⊠		⊠	⊠			⊠		⊠	⊠	⊠	⊠		□						⊠	⊠	⊠	□				⊠				
⊠	⊠	⊠	⊠	⊠	□	□	⊠	⊠	□	⊠		⊠	⊠	⊠		⊠	⊠	⊠	⊠	□	□		⊠	⊠	⊠	⊠		□	⊠	⊠	⊠	□			⊠	⊠		□	⊠	⊠
□	⊠	⊠	⊠	⊠	□	⊠	⊠	□	□	■		⊠	■	⊠	□	□	⊠	■	■	⊠	□		⊠		⊠	□			⊠	⊠	⊠	□			□	⊠			■	■
⊠	⊠	⊠	⊠	⊠	□	⊠	⊠	⊠	□	⊠	□	⊠	⊠	⊠		□	⊠	⊠	⊠	⊠	⊠		⊠						⊠	⊠	⊠	⊠			□	⊠			⊠	⊠
	□	□	⊠	⊠	□	⊠	⊠	□		⊠		⊠	⊠	■		□	□	■	⊠	⊠																			⊠	■
⊠	⊠	■	■	◆	◆	⊠	⊠	◆	⊠	⊠	⊠	⊠	⊠	■	□	◆	⊠	◆	⊠	■	◆		□						⊠	■	⊠	□			□	⊠		□	⊠	⊠
⊠	⊠	■	⊠	⊠	◆	⊠	⊠	⊠	⊠	⊠		□	□	■		⊠	⊠	■	■	⊠	◆		⊠	□	□	□		⊠	⊠	■	■	⊠		□	⊠	⊠		□	⊠	⊠
⊠	⊠	⊠	⊠	⊠	□	⊠	⊠	⊠	□	⊠		⊠	⊠	⊠		⊠	⊠	⊠	⊠	⊠	⊠		⊠	⊠	⊠				⊠	⊠	⊠	⊠			⊠	⊠			◆	◆
	⊠	⊠	⊠	⊠		⊠	⊠	□		⊠	□	⊠	⊠	⊠		□	⊠	⊠	⊠	⊠	□		□						□	⊠	⊠					⊠			◆	◆
	□	⊠	□	□		□	⊠	□		□		⊠	⊠	□			□	⊠			□															□			□	■
⊠	⊠	⊠	■	◆	□	□	■	⊠		■		■	■	■		⊠	⊠	⊠	⊠	□	◆	□	⊠						⊠	□	⊠	■		□		⊠			◆	◆
■	■	■	■	◆	□	⊠	■	⊠		■		■	⊠	■	⊠	⊠	■	■	■	⊠	⊠		■	□	⊠	□			■	■	■	⊠			⊠	■			■	■
□	⊠	⊠	⊠	⊠	□	⊠	⊠	□		⊠		□	□	■	⊠	□	□	■	■	⊠	□		⊠						⊠		⊠					⊠			⊠	■
⊠	⊠	⊠	⊠	⊠	□	⊠	⊠	⊠	□	⊠	□	⊠	⊠	⊠	□	□	⊠	⊠	⊠	⊠	□		⊠	⊠	⊠				⊠	⊠	⊠				⊠	⊠			⊠	■
																						□	□	□	□	□	□	⊠	⊠	⊠	⊠	⊠	□	⊠	⊠	⊠				
□	□	⊠	⊠	□	□	⊠	⊠	□	□	⊠	□	⊠	⊠	⊠	□	□	□	⊠	⊠	⊠	□	□	□	□	□	□	□	□	□	□	□	□	□	□	□	□	□	□	⊠	⊠
⊠	⊠	⊠	⊠	⊠	□	⊠	⊠	⊠	□	⊠		⊠	⊠	⊠		□	⊠	⊠	⊠	⊠	□		⊠	⊠	⊠	⊠			⊠	⊠	⊠	⊠			⊠	⊠		⊠	⊠	⊠
	□	□		□			□	□		□			□	□		□	□	⊠	□				⊠	□	□	□		⊠	⊠	⊠	⊠	⊠		□	□	⊠		□	⊠	⊠
⊠	⊠	⊠	⊠	□	□	⊠	⊠	⊠	□	□		⊠	⊠	⊠		□	□	⊠	⊠	□			⊠			□			⊠	■	■	□		□	□	⊠				
	⊠	□						⊠	□	□				⊠			⊠	⊠					⊠	⊠	■	⊠			■	□	⊠	□			■	■		□		
⊠	⊠	⊠	⊠	⊠	□	⊠	⊠	⊠		⊠		⊠	⊠	⊠		⊠	⊠	⊠	⊠	⊠	⊠		⊠			⊠			⊠	⊠	⊠					⊠			⊠	⊠
□	◆	⊠	⊠	⊠	□	□	⊠	□	□	⊠		⊠	⊠	⊠		□	◆	⊠	□	□	□		⊠			□		□	⊠	⊠	⊠	□		◆		⊠		◆	□	■
⊠	⊠	□	■	⊠	◆	□	⊠	⊠	⊠	⊠		⊠	⊠	⊠			□	⊠	■	■	□		□	⊠	⊠	□		□	⊠	⊠	⊠				⊠	⊠	□		◆	⊠
□	□	□	⊠	□		□	⊠	□		□		⊠	⊠	□			□	⊠	□	□	□		⊠						⊠							⊠			□	■
⊠	⊠	⊠	⊠	⊠	□	⊠	⊠	⊠	□	⊠		⊠	⊠	⊠		⊠	⊠	⊠	⊠	⊠	□		⊠	⊠	⊠				⊠	⊠	⊠				⊠	⊠			⊠	⊠
⊠	⊠	⊠	⊠	□	□	□	⊠	□		⊠	□	⊠	⊠	⊠			⊠	■	⊠	⊠	□		□						⊠	□	⊠	□			□	⊠			◆	⊠
	□	⊠	⊠	⊠	□	⊠	⊠	□		⊠		⊠	⊠	⊠			□	⊠	⊠	⊠	□		⊠	⊠	⊠	⊠		□	⊠	⊠	⊠	⊠			⊠	⊠		□	⊠	⊠
□	□	□	□	□	□	□	□	□		□		□	□	□		⊠	⊠	⊠	□	□	□		⊠	⊠	□				⊠	⊠	⊠				⊠	⊠	□	□	□	⊠
⊠	⊠	□	⊠	⊠	◆	□	⊠	◆		⊠			□	⊠		◆	⊠	⊠	⊠	□	◆		⊠			□		□	⊠	⊠	■	⊠		□	⊠	⊠		□	⊠	⊠
⊠	⊠	⊠	■	□	◆	⊠	⊠	⊠		⊠		⊠	□	⊠	□		⊠	■	⊠	⊠	□		⊠	⊠	⊠	⊠			⊠	⊠	⊠	□			⊠	⊠	⊠		⊠	⊠
⊠	■	□	⊠	□	◆	⊠	⊠	■	□	■	□	□	⊠	■	□	□	⊠	■	⊠	□	□			□					□	□					■	□	■		□	■
□	⊠	□	□	⊠	□	□	◆	◆	□	⊠		⊠	□	⊠		□	⊠	⊠	⊠	□	□		⊠		□	□		□	⊠	⊠	⊠	⊠		□	⊠	⊠		□	⊠	□
□	□	□	□	□		□	□	□	□	□	□	□	⊠	□		□	□	□	⊠	□	□		□					□	□	⊠	⊠	□			□	□		□	□	□
⊠	⊠	⊠	□	⊠		□	□	□	□	□	□	□	⊠	⊠		□	⊠	■	□	□	□		□			□		□	⊠	■	■	□			□	⊠			□	⊠
□	□	⊠	⊠	⊠		⊠	⊠			⊠		□	□	⊠			□	■	■	⊠																			■	■
⊠	□	⊠	⊠	⊠		⊠	⊠			⊠		□	⊠	■			□	■	■	⊠			□	□	□	□			□		□				□	□			■	■
		⊠	⊠	⊠		⊠	⊠			⊠		□	⊠	□				⊠	⊠	⊠	□																		⊠	⊠
	□	⊠	⊠	□	□	⊠	⊠	□		⊠		□	□	⊠	□		□	⊠	⊠	⊠	□		□	⊠	□				⊠						⊠	⊠			⊠	⊠
□	⊠	□	⊠		□	□	□	□	□	□			□	□		□	⊠	⊠	□	□		□	□	□		□	□	□	□	⊠	⊠	□			□	□		□	□	⊠
■	⊠	⊠	■	◆		□	⊠	⊠		⊠		⊠	⊠	⊠		◆	⊠	■	⊠	■	◆		■						■		⊠	□				■			□	■
		⊠	■			⊠	⊠			⊠			□	■				⊠	■	■																			◆	⊠
□	⊠	⊠	⊠	⊠	□	⊠	⊠	⊠	□	⊠		□	⊠	⊠		□	⊠	⊠	⊠	⊠	⊠		⊠	⊠	⊠	⊠			⊠	⊠	⊠	⊠			⊠	⊠		⊠	⊠	⊠
	□	⊠	⊠	⊠	□	⊠	□	□		⊠		⊠	⊠	⊠		□	□	⊠	⊠	⊠																			⊠	⊠
		□	⊠	□	□	□	□			□		□	□	□				⊠	⊠	⊠																			⊠	⊠
■	■	⊠	⊠	◆	◆	⊠	■	■	□	■	⊠	⊠	⊠	■		◆	■	■	⊠	□	◆		⊠						⊠	⊠	⊠	□				⊠			◆	◆
□	⊠	⊠	□			⊠	⊠	⊠		□		□	□	□		□	□	⊠	⊠	□	□		⊠	⊠	⊠	■		⊠	⊠	■	■	⊠			■	■		□	⊠	⊠
□	⊠	□	⊠	□	□	□	⊠	⊠	□	⊠			□	⊠		□	⊠	⊠	⊠	⊠	⊠		⊠	⊠	⊠	⊠			⊠	⊠	⊠	⊠			⊠	⊠		⊠	□	□
□	■			◆				□				□	⊠	⊠		⊠	■	⊠	⊠		⊠		■						■	⊠	■	□				⊠			□	□
□	■	□	⊠	⊠		⊠	⊠	⊠		⊠		□	⊠	⊠		□	■	⊠	⊠	□	⊠		⊠						⊠	□	⊠	⊠				⊠			⊠	⊠
□	⊠	□	⊠	⊠	□	□	⊠	□	□	□			□	⊠		□	⊠	⊠	⊠	□	□		⊠	⊠	⊠	⊠			⊠	⊠	⊠	□			⊠	⊠	⊠	□	⊠	⊠
	■	□	⊠	⊠		⊠	⊠	□		⊠		⊠	■	⊠		□	□	■	■																				⊠	■
⊠	⊠	□	■	■		⊠	⊠	⊠		⊠		⊠	■	⊠			⊠	■	■	□	□		⊠					◆	⊠	⊠	⊠	□			□	⊠			⊠	■
		□	⊠	□		□	⊠			□		□	□	□				⊠	⊠	⊠																			□	⊠
	□	□	■	⊠		⊠	⊠			⊠		□	⊠	⊠				■	■	⊠	□																		⊠	■
■	■	□	⊠	□	◆	□		■	□	□		□	□	■		⊠	■	⊠	⊠	□	◆		⊠					⊠	⊠	■	■	■		□	□	⊠			■	■
⊠	⊠	⊠	■	◆	◆	⊠	⊠	⊠		⊠		⊠	■	■		□	⊠	■	■	■	◆		⊠	⊠	⊠	⊠		◆	⊠	⊠	⊠	⊠		□	⊠	⊠			◆	⊠
⊠	⊠	⊠	□	⊠	□	⊠	⊠	⊠	□	⊠		□	□	⊠		□	⊠	⊠	⊠	□	□		⊠	⊠	⊠				⊠	⊠	⊠	□			⊠	■			⊠	⊠
□	⊠	⊠	□	⊠	□	⊠	⊠	⊠	□	⊠		⊠	⊠	⊠			⊠	⊠	⊠	□	⊠																		⊠	⊠
□	⊠	⊠	□	⊠	□	⊠	⊠	□	□	⊠		⊠	⊠	⊠			⊠	⊠	⊠	□	◆																		⊠	⊠
■	⊠	⊠	■	□	◆	⊠	⊠	■	□	⊠		□	□	■		■	⊠	■	⊠	□	◆		⊠	□		□		⊠	⊠	■	⊠	⊠		□		⊠			⊠	⊠
□	□	⊠	⊠	□		⊠	⊠			⊠		⊠	⊠	⊠				■	⊠	⊠																			⊠	⊠
□	□	□	□	□	□	□	□	□	□	□		□	□	□		□	□	⊠	□	□	□		⊠		□	□		◆	⊠	⊠	⊠	⊠			◆	⊠		◆	□	□
⊠	⊠	□	□	□	□	□	⊠	⊠	□	□	□	□	□	□	□	⊠	⊠	⊠	□	□	⊠	□	⊠	⊠	⊠	⊠	□	□	⊠	⊠	⊠	□	□	□	⊠	⊠	□	⊠	□	⊠
⊠	⊠		□		□	□	□	□		⊠			□	⊠		□	⊠	⊠	⊠	□	□								⊠	⊠	⊠				⊠	⊠			⊠	⊠
□	■	□	■	⊠	□	□	□	⊠		□	□	□	⊠	⊠		□	⊠	⊠	⊠	□	◆		■						■	⊠	■	⊠				⊠			□	⊠
⊠	⊠	⊠		⊠	□	⊠	⊠	⊠	□	□		⊠	⊠	⊠		□	□	⊠	⊠				⊠			□			⊠	■	■	□		□	□	⊠				
⊠	⊠	■	■	◆	◆	⊠	⊠	■	⊠	⊠		⊠	⊠	■		◆	⊠	■	■	■	◆	□	⊠					⊠	⊠	⊠	⊠	⊠		⊠		⊠		□	◆	⊠
□	⊠	□	⊠	⊠	□	□	⊠	⊠		⊠		⊠	⊠	⊠		⊠	□	⊠	⊠	□	◆		⊠						⊠	□	⊠	□				⊠			□	□
⊠	⊠	⊠	⊠	⊠	□	⊠	◆	◆	□	⊠	□	⊠	⊠	⊠		□	⊠	⊠	⊠	⊠	⊠		⊠	□					⊠	⊠	⊠	□			□	⊠			⊠	⊠
⊠	⊠	⊠	⊠		□	□	⊠	⊠	□	□		□	⊠	⊠		□	□	⊠	⊠	□			⊠			□			⊠	■	■	□		□	□	⊠				
□	⊠	□	⊠	⊠	□	□	⊠	□	□	⊠		□	□	⊠		□	⊠	⊠	⊠	□	□		⊠	⊠	⊠	⊠			⊠	⊠	⊠	⊠			⊠	⊠		⊠	⊠	⊠
	⊠							□					□	□			⊠	⊠	□		□		⊠						⊠		□					⊠				
⊠	⊠	⊠	⊠	⊠	□	□	⊠	⊠		⊠		□	⊠	■		⊠	⊠	■	■	□	⊠		■						■	□	⊠	□				⊠			⊠	■
⊠	⊠	□	⊠	□		□		⊠	□	□		□	□	■		□	□	■	⊠	⊠	□		⊠	⊠	■	⊠			⊠	⊠	⊠	□			■	⊠	⊠	⊠	■	■
⊠	⊠	⊠	⊠	⊠	□	⊠	⊠	⊠	□	⊠		■	⊠	■		□	⊠	⊠	■	⊠	□		⊠	□	⊠	⊠		□	⊠	⊠	⊠	⊠			⊠	⊠			⊠	⊠
□	⊠	⊠	⊠	□	□	□	⊠	⊠	□	□	□	⊠	□	□		□	⊠	⊠	□	⊠	□		■						■			□				■			⊠	⊠
		□	□	⊠			⊠			□		⊠	□					⊠	□	□																			□	⊠
⊠	⊠	⊠	⊠	⊠	□	□	□	□		□	⊠	■	■	⊠		◆	⊠	◆	■	⊠	◆		⊠						⊠		⊠					⊠			◆	◆
		⊠	⊠	□		□	⊠			⊠		□	□	⊠				⊠	⊠	⊠																			⊠	⊠
⊠	⊠	⊠	⊠	◆	□	□	⊠	⊠		⊠		⊠	⊠	⊠		◆	⊠	⊠	⊠	⊠	◆		□						□	⊠	⊠	□				□			◆	■
⊠	⊠	⊠	⊠	⊠	□	□	⊠	⊠		⊠		□	⊠	⊠		◆	⊠	⊠	⊠	⊠	◆		⊠					⊠	⊠	⊠	⊠	⊠		⊠	⊠	⊠			⊠	⊠
□	□	■	⊠	⊠	◆	⊠	⊠	⊠		⊠		⊠	⊠	■		⊠	⊠	■	■	⊠	□		■	□		⊠		⊠	⊠	■	■	■		⊠	■	⊠		□	□	■
⊠	■	■	⊠	◆	□	⊠	■	⊠	□	⊠	□	⊠	■	■		◆	■	■	■	⊠	◆		■						■	■	■	■	⊠		□	■		□	◆	◆
		□	□	□	□	□	□			□		□	□	□				⊠	⊠	□																			⊠	⊠
■	■	⊠	■	⊠	◆	⊠	■	◆	⊠	■		⊠	⊠	■		◆	■	■	■	⊠	◆		⊠			□		□	■	■	■	■		□		■			■	■
	□	⊠	■	⊠	□	□	⊠	□		⊠		⊠	⊠	⊠			□	⊠	⊠	■									□							□			◆	⊠
⊠	⊠	⊠	⊠	⊠	◆	□	⊠	⊠	□	□	□	⊠	⊠	⊠		◆	⊠	⊠	⊠	□	◆		□					□	□	■	■	□		□	⊠	□		□	◆	□
□	⊠	□	□		□	□	⊠	□		⊠			□	□		◆	⊠	⊠	⊠	□	◆	⊠	⊠	⊠	⊠	⊠	⊠	⊠	⊠	⊠	⊠	⊠		⊠	⊠	⊠		⊠	◆	□
⊠	⊠	⊠	⊠	⊠	□	⊠	⊠	⊠		⊠		⊠	⊠	⊠		⊠	⊠	⊠	⊠	⊠	⊠		⊠			⊠			⊠	⊠	⊠	⊠				⊠			⊠	⊠

Key ■ = Abundant ⊠ = Common □ = Uncommon ◆ = Seasonal

	Page Number	parking lot	privy/restrooms	picnic areas	primitive camping	camper pads	river access	boat ramps	fishing docks/jetties	boat rental	viewing blinds	waterfowl blinds	interpretive programs	exhibits/displays	MDC office	hunting	waterfowl hunting	fishing	hiking trails	horse trails	canoeing	archery range	firearms range	field trials	dog training	trapping	Notes
SHAWNEE TRAIL CA	16	■	♿		■			■	♿							■	■	■							■		
SHELVIN ROCK ACCESS	48	■					■	■										■			■						
SHEPHERD OF THE HILLS FISH HATCHERY	220	♿	♿	■			■	■			■		■	■	♿			■	■								
SHERIDAN ACCESS	236	■	■				■											■			■						
SHOEMAKER CA	6	■			■											■		■								■	
SHORT BEND ACCESS	70	■			■		■									■		■			■						
SIMMONS FORD ACCESS	224	■					■	■										■			■						
SIMPSON PARK LAKE	202	♿	♿	■				■	♿									■	♿		■						
SIMS VALLEY COMMUNITY LAKE	98	■	■	■				■										■			■						
SKY PRAIRIE CA	44	■														■									■	■	
SLOAN (DR. O. E. & ELOISE) CA	62	■														■	■	■									
SMACK-OUT ACCESS	154	■					■											■			■						
SMITH (ROGER V. & VIOLA WACHAL) CA	54	■			■											■											
SMOKY WATERS CA	54						■									■	■	■			■				■	■	
SOULARD ACCESS	136	■	■		■		■	■	■							■	■	■			■					■	
SOUTH FARM R-1 LAKE	25	■																■									
SOUTH PRONG ACCESS	224	■					■											■			■						
SOUTHWOOD ACCESS	86	♿	♿					■	♿									■			■						
SPANISH LAKE	202	♿	♿	■				■	♿				■		■			■	■		■						
SPORTSMAN'S PARK ACCESS	28	■		■			■	■										■			■						
SPRING CREEK GAP CA	134	■		■	■											■									■		
SPRINGFIELD CONS NATURE CENTER	83	♿	♿								■		■	■	♿				♿								
SQUIRES TOWERSITE	73															■										■	
ST. JOHN'S BAYOU ACCESS	152	■					■											■									
ST. STANISLAUS COUNTY PARK	202	■	■										■					■	■								
ST. THOMAS FERRY ACCESS	54	■	♿				■	■										■			■						
STAR SCHOOL HILL PRAIRIE CA	10	■			■											■			■					■	■	■	
STERLING PRICE COMMUNITY LAKE	46	■	■	■	■			■	■									■			■						
STEYERMARK (JULIAN) WOODS CA	136	■																									
STOCKTON RESERVOIR	44	■			■		■	■								■	■	■	■	■	■				■	■	
STONES CORNER ACCESS	106	■	♿				■	■										■			■					■	
STONY POINT PRAIRIE CA	62	■														■		■									
STRINGTOWN BRIDGE ACCESS	54	■					■						■			■	■	■			■					■	
STUBBLEFIELD ACCESS	14	■					■									■		■			■					■	
SUGAR CREEK CA	6	■			■											■			■	■			■			■	
SUMMERSVILLE TOWERSITE	224	■														■		■									
SUMNER ACCESS	46	■					■	■										■			■						
SUN (STEPHEN J.) CA	28	■			■											■	■	■			■				■	■	
SUNBRIDGE HILLS CA	26	■					■									■		■	■								
SUNFISH LAKE	202	♿		■				■	■									■			■						
SUNKLANDS CA	210				■											■					■			■	■	■	
SUNNYSIDE ACCESS	120						■									■		■			■			■	■	■	
SUNRISE ACCESS	136						■									■	■	■			■					■	
SUNSET PARK ACCESS	176	■					■	■										■			■						
SUSON PARK LAKES	202	♿	♿	■					♿									■									
SWAN LAKE WA	46	■	■								♿	♿	■	■	■	■	■	■	■		■						
SWEETGUM ACCESS	23	■					■	■								■		■			■			■	■	■	
SWIFTWATER BEND CA	202	■					■											■	■		■						
SWINGING BRIDGE ACCESS	58																	■			■						
SYCAMORE ACCESS	162	■					■											■			■					■	
TABERVILLE ACCESS	194	■	■		■		■	■										■									
TABERVILLE PRAIRIE CA	194	■														■		■							■	■	
TAILWATERS ACCESS	86	♿	♿				■	■										■			■						
TALBOT (ROBERT E.) CA	118	■	■		■		■									■	■	■	■		■			■	■	■	
TARKIO PRAIRIE CA	10	■			■											■		■	■					■	■	■	
TARSNEY LAKE	102							■	■									■			■						
TEA ACCESS	81	■					■											■			■					■	
TECUMSEH TOWERSITE	162															■										■	
TEN MILE POND CA	143	■						■				■			■	■	■	■						■	■		
TESZARS WOODS CA	108							■	■									■	■		■						
THIRTY-FOUR CORNER BLUE HOLE	143	■			■			■										■									
THOMAS HILL RESERVOIR	130	■	■		■	■		■	■							■	■	■			■			■	■	■	
THOMASVILLE TOWERSITE	210				■											■										■	
THOMPSON FORD ACCESS	132	■					■	■							♿	■		■			■				■		
THOMPSON (ROBERT H.) CA	182	■					■	■								■		■			■						
THREE CREEKS CA	25	■			■											■		■	■	■						■	
THURNAU (H. F.) CA	94	■	■		■		■	■								■	■	■	■		■			■	■	■	
TILLES PARK LAKE	202	■	■	■					♿				■		■			■									
TIMBER KNOB TOWERSITE	162															■										■	
TIMES BEACH ACCESS	202	■					■	■										■			■						
TINGLER LAKE CA	98	■														■	■	■	■						■	■	
TIPTON FORD ACCESS	154	■	■				■									■		■			■					■	
TOLONA ACCESS	120	■			■											■	■	■			■			■	■	■	
TORONTO SPRINGS CA	34	♿		■	■		■									■	■	■			■				■	■	
TOWER ROCK NA	166	■	■	■			■									■	■	■	■								
TOWN HOLE ACCESS	128	■					■											■			■						
TREATY LINE PRAIRIE CA	17	■														■		■									
TRI-CITY COMMUNITY LAKE	25	■	■		■	■		■										■			■	■					
TRUITT (HENRY) ACCESS	6	■					■											■			■						
TRUMAN RESERVOIR WILDLIFE LANDS	20	■	♿		■		■								■	■	■	■	■		■	■	♿			■	
TURKEY CREEK CA	44															■									■	■	
TURKEY FARM LAKE	25	■																■			■						
TUSCUMBIA ACCESS	140	■					■	■										■			■						
TWIN BORROW PITS CA	164	■																■									
TWIN LAKE REC AREA	25	■	■	■					♿									■	■		■						
TYWAPPITY COMMUNITY LAKE	208	■	■	■	■			■										■	■							■	
TYWAPPITY TOWERSITE	208				■											■										■	
ULMAN TOWERSITE	140																										
UNION ACCESS	78	■					■	■										■			■						
UNION MILL ACCESS	174	■			■		■											■			■						
UNION RIDGE CA	6	■			■											■		■								■	
UNIONVILLE CITY LAKE	180	■	♿	■				■										■	■								

Key ■ = Yes ♿ = Disabled Access

beaver	bullfrogs	coyotes	deer	doves	eagles	fox	hawks	herons	otters	owls	pheasant	quail	rabbits	racoons	ruffed grouse	shore birds	snapping turtles	songbirds	squirrels	turkey	waterfowl	hybrid bass	largemouth bass	rock bass	small mouth bass	spotted bass	striped bass	white bass	blue gill	carp	catfish	crappie	muskelunge	paddlefish	suckers	sunfish	trout	walleye	mushrooms	wildflowers
□	⊠	⊠	⊠	⊠	□	□	⊠	□		□		⊠	□	□		⊠	□	⊠	□	□	◆		⊠						⊠		□					⊠			□	□
□	□	□	□	□	□	□	□	⊠	□	□		□	□	□		□	□	⊠	⊠	□	□		⊠	□	⊠	□			□	⊠	⊠	□		◆	⊠	⊠	□	□	□	□
⊠	■	□	⊠	■	◆	⊠	⊠	■		⊠		⊠	⊠	⊠		◆	⊠	⊠	■	⊠	■									■						■		■		⊠	◆
□	□	□	□	□		□	□	□		□		□	□	□		□	□	◆	□												⊠	⊠				□	□			□	◆
⊠	⊠	⊠	⊠	□	□	⊠	⊠	□	□	⊠		□	□	⊠	□		⊠	⊠	⊠	⊠	□									□	⊠					□			◆	◆
□	⊠	⊠	⊠	⊠	□	⊠	⊠	⊠		⊠		⊠	⊠	⊠		□	⊠	⊠	⊠	⊠	□		⊠	⊠	■	⊠			⊠	⊠	⊠	□			⊠	⊠			⊠	⊠
□	⊠	□	□	⊠	□	□	⊠	⊠	□	□		□	⊠	⊠		□	⊠	⊠	⊠	□	□		⊠	■	■	⊠			■	⊠	⊠	□			■	■			□	□
⊠	■	□	□	⊠	□	□	□	⊠		□		□	□	■		⊠	■	■	⊠	□	⊠		■			Y			■	⊠	■	⊠			⊠	■			□	⊠
□	⊠	⊠	■	⊠	□	⊠	⊠	⊠		⊠		⊠	⊠	⊠		□	⊠	■	⊠	⊠	◆		⊠						⊠		⊠	⊠				⊠			⊠	⊠
		⊠	⊠	□		□	⊠			⊠		⊠	⊠	⊠				■	⊠	⊠																				◆	■
	□	⊠	⊠	⊠		□	⊠	□		□		⊠	□	□		□	□	⊠	□	□	□		⊠						⊠		□					⊠			□	□
□	■	□	□	□	□	□	□	□	□	□		□	□	■		□	□	□	□	□	□		⊠	⊠	■	⊠			■	□	⊠	⊠			■	■			□	□
□		⊠	⊠	⊠		□	⊠			⊠		⊠	□	⊠	□			⊠	⊠	⊠	□																		◆	◆
⊠	⊠	□	⊠	□	□	□	⊠	⊠	□	⊠		□	□	⊠		⊠	⊠	⊠	⊠	⊠	□	□	⊠	□		⊠	□	□	⊠	⊠	⊠	⊠		⊠	⊠	⊠		⊠	⊠	⊠
⊠	⊠	□	⊠	⊠	⊠	□	⊠	⊠		⊠		□	□	⊠		⊠	⊠	⊠	□	□	⊠		□		□			□	□	■	■	⊠		□	⊠	□		□	⊠	⊠
	⊠	□	□	⊠		□	□	⊠		□		⊠	⊠	□		⊠	⊠	□	□	□	◆		⊠						⊠			⊠				⊠			□	
□	⊠	□	□	□	□	□	⊠	⊠		□		□	□	⊠		□	⊠	⊠	□	□	□		⊠	■	■	⊠			■	⊠	⊠	□			■	■				□
⊠	⊠	□	■	⊠	◆	□	⊠	⊠		□		□	⊠	⊠		⊠	⊠	⊠	⊠	□	■		⊠	□	⊠	□			■	⊠	⊠	⊠			■	■			□	⊠
	⊠		□	⊠		□	□	⊠		□		□	⊠	□		□	⊠	■	■	□	⊠		■						■	⊠	⊠	⊠				■				□
□	⊠	□		□				⊠	□	□			□	⊠			□	⊠	⊠		□		⊠	⊠	□	⊠		□	⊠	⊠	⊠	⊠		□	⊠	⊠		⊠		
		⊠	⊠	□		⊠	⊠	□		⊠		□	⊠	⊠				⊠	⊠	⊠																				⊠	⊠
□	■	□	⊠	■	□	□	□	⊠		⊠		□	⊠	⊠		□	⊠	■	■	□	◆		■	□	□	□			■	■	⊠	⊠			⊠	⊠			◆	■
		□	⊠	⊠	□	□	⊠			□		□	□	⊠				⊠	⊠	⊠																				◆	□
																						□	⊠	⊠	⊠	⊠	⊠	□	⊠	■	■	⊠		□	■	⊠				
⊠	⊠	⊠	■	⊠	◆	□	⊠	◆	⊠	□	□	⊠	⊠	⊠		◆	■	■	⊠	⊠	◆	□	⊠			□	□	⊠	⊠	⊠	■	⊠		□	⊠	⊠		□	◆	◆
□	⊠	□	□		◆	□	□	□	□	⊠		□	□	□		□	⊠	⊠	□	□	□	□	⊠	⊠	□	⊠	□	⊠	⊠	⊠	⊠	⊠		⊠	⊠	⊠		□	□	□
		⊠	⊠	◆		⊠	⊠			⊠	□	⊠	⊠	⊠				⊠	⊠	⊠																				◆	■
⊠	⊠	⊠	⊠	⊠	□	□	⊠	⊠	□	⊠	□	⊠	⊠	⊠		◆	⊠	⊠	⊠	⊠	□		⊠	□					⊠	⊠	⊠	□			□	⊠			⊠	⊠
		⊠	⊠	⊠		⊠	⊠			⊠		⊠	⊠	■				■	⊠	⊠																				⊠	⊠
⊠	⊠	⊠	■	⊠	◆	⊠	⊠	⊠	□	⊠		⊠	■	⊠		◆	⊠	⊠	⊠	⊠	◆		■	□	⊠	■		⊠	■	■	■	■			■	■		⊠	◆	■
⊠	⊠	⊠	⊠	⊠	□	⊠	⊠	⊠		⊠		□	⊠	⊠		□	⊠	⊠	□	□	□		⊠	⊠	⊠	⊠			⊠	⊠	⊠	□			⊠	⊠			□	□
□	□	⊠	⊠	□		□	⊠	□		□		⊠	⊠	□		□	□	⊠	□	□	□		□						⊠	□	⊠					⊠			□	■
⊠	⊠	⊠	⊠	□		□	□	◆		⊠		□	□	⊠			⊠	■	⊠	⊠	□		⊠						■	⊠	⊠	⊠				■			⊠	■
⊠	⊠	⊠	⊠	⊠	□	⊠	⊠	⊠		⊠		⊠	⊠	⊠		□	⊠	⊠	⊠	□	□		⊠	⊠	⊠	⊠			⊠	⊠	⊠	□			⊠	⊠			⊠	⊠
□	⊠	⊠	■	□	□	⊠	□		□	⊠		□	□	⊠	⊠		⊠	⊠	⊠	■	□																				◆	◆
		■	■	□		⊠	□			□		□	□	■				⊠	■	■																				□	□
⊠	⊠	⊠		⊠		□		⊠	□		□	□	□	⊠			⊠	⊠			□		□			□		□	⊠	■	■	⊠		□	⊠	⊠			□	□
⊠	■	⊠	■	⊠	□	⊠	□	□	□	■		□	⊠	⊠		⊠	⊠	■	■	⊠	⊠		⊠	⊠	⊠	□			⊠	□	□	⊠				⊠			□	⊠
		□	⊠		◆	⊠	⊠			□		⊠	⊠	⊠				■	⊠	⊠			□	□		□			□	■	■	⊠		□	⊠	□			■	⊠
□	⊠	□	□	⊠	□	□	□	□		□			□	□		□	⊠	⊠	⊠	□	⊠		⊠						⊠	⊠	⊠	□				⊠			⊠	⊠
□	⊠	⊠	⊠	□	□	⊠	⊠	□	□	⊠		□	⊠	⊠	□		⊠	⊠	⊠	⊠																Y		⊠	⊠
⊠	⊠	⊠	■	◆	◆	⊠	⊠	⊠	□	⊠	□	■	■	⊠		⊠	⊠	■	■	■	◆		⊠	□	□	□		⊠	■	■	■	■	□	□	■	■		□	■	■
⊠	⊠	⊠	⊠	⊠	□	⊠	⊠	⊠		⊠		□	□	⊠		⊠	⊠	⊠	□	⊠	⊠		□		□			□	□	⊠	⊠	□			⊠	□			⊠	⊠
⊠	⊠	⊠	□	□		□	⊠	⊠	□	⊠		□	□	⊠		□	□	⊠	⊠	□	□		⊠	⊠	⊠	□		□	⊠	⊠	⊠	⊠			⊠	⊠		□	⊠	⊠
	⊠			⊠			□	⊠					⊠	□		⊠	□	⊠	⊠		⊠		□						⊠	⊠	⊠	⊠				⊠	□		
■	■	⊠	■	◆	◆	⊠	◆	◆	⊠	■	□	■	■	⊠		◆	■	■	■	■	◆		⊠						⊠	■	■	■		⊠	■	⊠			⊠	■
■	⊠	⊠	⊠	⊠	□	⊠	⊠	□	□	⊠		⊠	⊠	⊠		□	⊠	■	⊠	⊠	□		⊠	⊠	⊠	⊠			⊠	⊠	⊠	⊠			⊠	⊠			⊠	⊠
□	⊠	⊠	⊠	⊠	□	⊠	⊠	⊠	□	⊠		⊠	⊠	⊠		⊠	⊠	⊠	⊠	⊠	⊠		⊠	■	⊠	■		□	■	⊠	⊠	■		□	■	■		□	⊠	⊠
□	⊠	□	⊠	⊠	□	□	◆	◆	□	⊠		⊠	□	⊠		□	⊠	⊠	⊠	□	□		⊠		□	□		□	⊠	⊠	⊠	⊠		□	⊠	⊠		□	⊠	□
□	⊠	□	⊠	□	◆	⊠	⊠	⊠	□	□		□	□	⊠		⊠	⊠	⊠	□	⊠	◆		⊠	⊠	⊠	□			⊠	⊠	⊠				⊠	⊠	⊠	⊠	◆	⊠
□	⊠	□	⊠						□	⊠			□	⊠			⊠	⊠	□	□									□		■	⊠	⊠		⊠					□	□
	⊠	⊠	⊠	⊠		□	■	□		⊠		⊠	⊠	⊠		□	⊠	■	⊠	⊠			⊠						⊠							⊠			◆	■
⊠	⊠	□	⊠	⊠	◆	□	⊠	⊠	□	□		□	□	⊠		□	⊠	⊠	⊠	□	□		□	⊠	⊠	□		□	⊠	■	⊠	□		◆	■	⊠			□	⊠
⊠	⊠	⊠	■	◆	◆	⊠	⊠	■	□	■		■	⊠	⊠		⊠	⊠	■	⊠	⊠	⊠		■	□	□	□			■	⊠	■	□		□	■	■			⊠	■
□	⊠	⊠	■	◆		⊠	⊠	□		⊠	⊠	⊠	⊠	⊠		□	⊠	⊠	⊠	⊠	□		⊠						⊠		⊠	□			□	□			◆	■
□	⊠	□	□	□	□	□	□	□		□		□	□	□		□	⊠	⊠	□	□	□		⊠						■	□	⊠	⊠				■			□	⊠
□	⊠	■		□		□	⊠	⊠		⊠		□	□	⊠		⊠	⊠	⊠	⊠		□		⊠	⊠	⊠	⊠			⊠	⊠	⊠				⊠	⊠		□	◆	◆
		⊠	⊠	□	□	⊠	⊠			⊠		□	⊠	⊠				⊠	⊠	⊠																				⊠	⊠
⊠	⊠	⊠	□	⊠	◆	□	⊠	⊠	⊠	□		□	□	⊠		◆	⊠	■	□	□	◆							□	□	■	⊠	□			□				⊠	⊠
□	⊠	□	⊠	□	□	□	⊠	□	□	⊠		□	□	⊠		□	□	⊠	■	⊠	◆		⊠	⊠	⊠	⊠			⊠	⊠	⊠	⊠			□	⊠		□	⊠	■
	⊠		□	□	□		□	⊠	□	□			□	□		⊠	□	⊠	□		□		⊠	□	⊠	⊠			⊠	⊠	⊠	⊠			⊠	⊠			□	⊠
⊠	⊠	⊠	⊠	⊠	◆	⊠	⊠	◆	□	⊠	□	⊠	⊠	⊠		◆	⊠	⊠	⊠	⊠	◆		⊠						⊠	⊠	⊠	⊠		□	⊠	⊠			◆	◆
		⊠	⊠	□	□	⊠	□			□		□	□	⊠				⊠	⊠	⊠																				⊠	⊠
□	⊠	⊠	⊠			⊠	⊠	⊠		□		□	□	⊠		⊠	□	⊠	⊠	⊠	□		⊠	⊠	⊠	■		□	⊠	■	⊠	⊠		□	■	■		□	⊠	⊠
□	□	⊠	⊠	⊠		□	⊠	□		⊠		⊠	⊠	⊠	□		□	⊠	⊠	⊠	□								⊠							⊠			⊠	■
⊠	⊠	⊠	■	◆	◆	⊠	⊠	⊠	□	⊠	□	□	⊠	⊠		□	□	⊠	⊠	□	◆		□					□	□	■	■	⊠		□	⊠	□		□	◆	⊠
	□			⊠				□					⊠			⊠	□	⊠	⊠		⊠		□						⊠	⊠	⊠	□				⊠	⊠			
		■	■	⊠	□	⊠	⊠			⊠		⊠	□	⊠				⊠	⊠	■																				◆	□
□	⊠	⊠	⊠	⊠	□	⊠	⊠	⊠		⊠		⊠	⊠	⊠		⊠	⊠	⊠	⊠	⊠	⊠		⊠	■	⊠	■		□	■	⊠	⊠	■		□	■	■		□	⊠	⊠
⊠	⊠	⊠	⊠	□	□	□	⊠	⊠		⊠		⊠	⊠	⊠		□	⊠	⊠	⊠	□	⊠		□						⊠	□	□				⊠			⊠	⊠
⊠	⊠	⊠	⊠	⊠	□	⊠	⊠	⊠		⊠		⊠	⊠	⊠		□	⊠	⊠	⊠	□	□		⊠	⊠	⊠	⊠			⊠	⊠	⊠	□			⊠	⊠			⊠	⊠
⊠	⊠	⊠	⊠	⊠	□	⊠	⊠	□		⊠		⊠	⊠	⊠	□	□	⊠	■	⊠	⊠	⊠		□		⊠			□	⊠	⊠	⊠	⊠		□	⊠	⊠		□	⊠	⊠
⊠	⊠	⊠	■	⊠	□	⊠	⊠	⊠	□	⊠		⊠	⊠	⊠		□	⊠	⊠	⊠	■	⊠		⊠	⊠	⊠	⊠		□	⊠	⊠	⊠	□		⊠	⊠	⊠		□	⊠	⊠
□	⊠	⊠	⊠	□	◆	⊠	⊠	⊠		⊠		□	□	■		⊠	⊠	■	■	⊠	□		⊠	□		⊠		⊠	⊠	■	■	■		⊠	⊠	⊠		□	⊠	⊠
□	⊠						□	□		□				□			⊠	⊠				⊠	⊠	⊠	□			⊠	⊠	⊠	□			⊠	⊠			
	□	□	□	□		□	⊠	□		□		□	⊠	□			□	⊠		□	□		□						⊠		□				⊠			□	■
⊠	⊠	⊠	⊠	□	□	□	⊠	◆		⊠		□	□	⊠	□	□	⊠	■	⊠	□	◆		⊠			□			■		⊠	□				■			⊠	⊠
□	□	□	□				□	□		□			□	□			□	□	□	□			□					□	□	⊠	⊠	□			□	□		□	⊠	□
⊠	⊠	⊠	■	◆	◆	⊠	⊠	⊠	□	⊠		■	■	⊠		⊠	⊠	■	■	■	◆		⊠						■	■	■	⊠			■	■			■	■
		⊠	■	□		⊠	⊠			⊠		□	□	⊠				⊠	⊠	⊠	◆																				◆	◆
□	⊠	⊠	⊠	⊠		□	⊠	⊠		⊠		⊠	⊠	⊠			⊠	⊠	⊠	⊠	□		⊠						⊠		⊠				⊠			⊠	□
□	⊠				◆		⊠	□		□					◆	⊠	⊠			◆	□	⊠	⊠	⊠	⊠	□	◆	⊠	⊠	⊠	⊠		⊠	⊠	⊠				
	⊠		□		□		□	⊠		□			□				⊠	⊠			⊠		□		□	□	□		□	⊠	⊠	⊠			⊠	⊠			□	□
□	⊠	□	⊠	⊠		□	⊠	⊠		⊠		□	⊠	□		□	⊠	⊠	⊠	□	◆		⊠						⊠	□	⊠	⊠			□	⊠			⊠	⊠
⊠	⊠		⊠	□		□		□		□		□	□			□	□	⊠	⊠	⊠	□		⊠	⊠	⊠				⊠	⊠	⊠	⊠			□	⊠			⊠	⊠
		□	⊠			□	⊠			⊠				⊠				⊠	⊠	⊠																				⊠	⊠
		⊠	□			□	⊠			⊠			⊠	⊠				⊠	□	□																						□
⊠	⊠	⊠	□		□	□	⊠	⊠		⊠		□	⊠	⊠		□	⊠	⊠	⊠	□			⊠	⊠	⊠				■	⊠	⊠				⊠	■			⊠	■
□	□	⊠	□		□	□	⊠	⊠	□			□	□	⊠		□	□	⊠	□					⊠				□			⊠	■	■	□		□	□	⊠			□	□
⊠	⊠	■	⊠	◆	□	⊠	⊠	□	□	⊠	□	⊠	⊠	⊠	⊠		□	■	⊠	⊠	□		⊠						⊠	□	⊠	□				⊠			⊠	⊠
□	⊠	⊠	■	□	□	⊠	⊠	⊠	□	⊠	□	⊠	■	⊠		□	□	⊠	■	■	⊠		■						■	⊠	⊠	■				⊠			□	⊠

Key ■ = Abundant ⊠ = Common □ = Uncommon ◆ = Seasonal

	Page Number	parking lot	privy/restrooms	picnic areas	primitive camping	camper pads	river access	boat ramps	fishing docks/jetties	boat rental	viewing blinds	waterfowl blinds	interpretive programs	exhibits/displays	MDC office	hunting	waterfowl hunting	fishing	hiking trails	horse trails	canoeing	archery range	firearms range	field trials	dog training	trapping	Notes
UNIVERSITY FOREST CA	28	■														■			■	■						■	
UPPER MISSISSIPPI CA	120	■			■		■	■				■				■	■	■	■		■				■	■	
URICH ACCESS	90						■											■			■						
URICH CA	90	■														■	■	■							■		
VALLEY PARK ACCESS	202	■					■	■										■			■						
VALLEY VIEW GLADES NA	108	■														■			■								
VANDALIA COMMUNITY LAKE	12	♿	♿	■	■	■		■	■									■			■						
VANDERHOEF (ARCHIE & GRACIE) MEM SF	98	■		■												■		■	■								
VERA CRUZ ACCESS	73	■		■	■		■									■		■			■				■	■	
VETERAN'S MEM PARK LAKE	202	♿	♿						♿				■		♿			■									
VICTORIA GLADES CA	108	■														■											
VLASIS PARK LAKE	202	♿	♿	■					♿						♿			■									
VONAVENTURE MEM FOREST & WA	122	■														■		■							■	■	
WABASH CROSSING ACCESS	66	■					■	■										■			■						
WAGNER (FRANK E.) CA	186	■														■									■		
WAH-KON-TAH PRAIRIE	194	■																									
WAH-SHA-SHE PRAIRIE	106	■																									
WALKER LAKE	202	♿	♿	■					♿				■		♿			■									
WALNUT WOODS CA	52	♿	♿	■															♿								
WALTER WOODS CA	154	♿	♿	■															♿								
WAPPAPELLO RESERVOIR	232	■			■		■									■	■	■	■		■					■	
WARBLER WOODS CA	74																										
WARM FORK CA	158	■														■	■	■			■				■	■	
WARREN BRIDGE ACCESS	162						■											■			■					■	
WARRENTON TOWERSITE	228																										
WASHINGTON CITY ACCESS	78	♿	■	■			■	■	♿					■				■									
WATER WORKS LAKE	184	♿	♿	■	■	■		■	♿									■	■		■	♿		■	■		
WATERS (H. J.) & MOSS (C. B.) MEM WA	25	♿	■												♿				■								
WATSON ACCESS	10	■	■		■		■	■										■			■				■		
WELDON SPRING CA	193	■	■				■	■					■			■	■	■	■						■	■	
WELLSVILLE LAKE CA	148	♿	■					■								■	■	■			■		■				
WENKEL FORD ACCESS	78	■					■											■								■	
WEST LAKE	172	■																■			■						
WEST PLAINS TOWERSITE	98																										
WHETSTONE CREEK CA	32	■	♿	■	■		■		♿						■	■	■	■					■	■	■		
WHITE (GEORGE O.) SF NURSERY	224	♿	♿	■											♿	■		■	■						■	■	
WHITE OAK BEND ACCESS	112	■			■		■									■	■	■			■			■	■	■	
WHITE RANCH CA	98	■	♿	■	■											■		■					♿		■	■	
WHITE RIVER TRACE CA	70	■			■									■		■	■	■	■					■	■		
WHITE (ROBERT M., II) CA	12	■			■											■	■	■			■					■	
WHITE (WM. G. & E. P.) MEM WA	122	♿	♿					■	♿							■	■	■			■				■	■	
WILD ACRES PARK LAKE	202	♿		■														■									
WILDCAT ACCESS	154	■					■	■										■			■						
WILDCAT GLADE NA	154	■																									
WILHELMINA CA	28	■		■	■											■	■	■			■					■	
WILLIAMS CREEK PARK	118	♿	♿	■			■		♿									■	■								
WILLIAMS FORD ACCESS	65															■	■	■			■					■	
WILLINGHAM (RUBY CLARK) MEM WA	146				■											■											
WILMORE PARK NORTH LAKE	202	♿	■	■					♿				■					■									
WILMORE PARK SOUTH LAKE	202	■	■	■					■				■					■									
WILSON CAMP ACCESS	140	■					■									■	■	■			■					■	
WIRE ROAD CA	217	■					■									■		■									
WOLF BAYOU CA	164	■		■	■			■								■		■								■	
WOOD LAKE	102																	■			■						
WOODLAWN ACCESS	146	■					■									■		■			■						
WOODS (WOODSON K.) MEM CA	61	■	■		■		■									■		■			■						
WOOLDRIDGE ACCESS	58	■	■				■											■			■						
WORTH CO COMMUNITY LAKE	236	■	■					■										■	■		■						
WORTHWINE ISLAND CA	8	■			■		■									■	■	■	■		■				■		
WRIGHT (T. L.) MEM ACCESS	190	♿	♿				■	■										■			■						
WYACONDA CROSSING CA	120	■			■		■									■	■	■			■			■	■	■	
WYATT LAKE	102	■			■													■									
YELLOW CREEK CA	46	■														■	■	■	■		■						
YOKUM SCHOOL CA	232	■														■											
YOUNG CA	108	■														■		■	■						■	■	
YOUNGDAHL (MARK) URBAN CA	26	♿	♿	■					♿				■					■	♿								

Key ■ = Yes ♿ = Disabled Access

beaver	bullfrogs	coyotes	deer	doves	eagles	fox	hawks	herons	otters	owls	pheasant	quail	rabbits	racoons	ruffed grouse	shore birds	snapping turtles	songbirds	squirrels	turkey	waterfowl	hybrid bass	largemouth bass	rock bass	small mouth bass	spotted bass	striped bass	white bass	blue gill	carp	catfish	crappie	muskelunge	paddlefish	suckers	sunfish	trout	walleye	mushrooms	wildflowers
	⊠	⊠	⊠	□	□	□	⊠	□		■		⊠	⊠	⊠		□	□	■	■	⊠	□																		⊠	⊠
■	■	⊠	■	⊠	◆	■	■	■	□	■		□	⊠	■		◆	■	■	■	■	◆	□	■	□	□	□	□	■	■	■	■	■		⊠	□	■		⊠	■	■
⊠	□	□	□	□	□	□	⊠	□	□	□		□		⊠		□	⊠	⊠	□	□	□		□						□	⊠	⊠	□				□				□
□	⊠	⊠	⊠	◆	□	□	⊠	⊠	□	⊠		⊠	⊠	⊠		□	⊠	■	⊠	⊠	◆		□						□	⊠	⊠	□				□			⊠	■
□	■	□	□					⊠	□	□			□	■		□	■	■	⊠		□		■	⊠	⊠	⊠		⊠	■	■	■	⊠		□	■	■		□	⊠	⊠
		⊠	⊠	⊠		⊠	⊠			⊠		⊠	⊠	⊠				⊠	⊠	⊠																			■	■
□	⊠	□	□	□	□		□	□		□		⊠	□	□		□	⊠	⊠	□	□	⊠		■			□			■		■	□				■			◆	◆
⊠	⊠	⊠	■	⊠		⊠	⊠	⊠		■		⊠	⊠	⊠			⊠	■	■	⊠	□		⊠	⊠	⊠	⊠			⊠	⊠	□	□			⊠	⊠			⊠	■
⊠	■	⊠	⊠	⊠	◆	⊠	⊠	⊠	□	⊠		⊠	⊠	■			⊠	⊠	⊠	⊠	□		□	⊠	⊠	□			⊠	⊠	⊠				⊠	⊠			◆	⊠
	⊠			⊠			⊠	⊠					⊠	□		⊠	⊠	⊠	⊠		⊠		⊠						⊠	⊠	⊠	□				⊠				
		⊠	⊠	⊠		⊠	⊠			■		⊠	⊠	⊠				■	⊠	⊠																			■	■
	⊠			⊠														⊠	⊠		⊠		□						⊠		⊠	□				⊠				
□	□	⊠	■	□	□	⊠	⊠	□	□	■		□	□	⊠		□	□	■	■	■	□		□						□		□					□			■	■
⊠	⊠	⊠	□	□	□	□	⊠	□		⊠		□	⊠	⊠		□	⊠	⊠	□		□								⊠	⊠	⊠				⊠	⊠			⊠	⊠
□	□	⊠	⊠	⊠	□	⊠	◆	□	□	⊠	□	⊠	⊠	⊠	□	□	□	⊠	⊠	⊠	□																		⊠	⊠
	□	⊠	⊠	□		□	■	□		□		⊠	⊠	□		□	□	⊠	□	⊠			⊠						⊠							⊠			◆	■
	□	□	□	□		□	⊠	□		□		□	□	□		⊠	□	■			◆		□						□	□	□					□			□	■
□	⊠			⊠	□	□	□	□		□			□	□		□	⊠	⊠	⊠				⊠						⊠	⊠	⊠	□				⊠	⊠		□	□
□	⊠	□	⊠	□		⊠	⊠	□		⊠		□	□	■			□	■	■	⊠	□		⊠					◆	⊠	⊠	⊠	□			□	⊠			⊠	⊠
□	⊠	⊠	⊠	⊠		⊠	□	□		⊠		⊠	⊠	⊠		□	⊠	⊠	⊠	□	□		⊠						⊠		⊠					⊠			⊠	⊠
⊠	⊠	⊠	⊠	⊠	□	⊠	⊠	■	□	⊠		□	⊠	⊠		□	⊠	■	⊠	⊠	□		■	■	⊠	■		□	■	■	⊠	⊠	□		■	■		□	⊠	■
□	⊠	⊠	□	⊠		□	□	⊠		□	□	□	⊠	⊠		⊠	□	⊠	⊠	□																			⊠	⊠
⊠	⊠	⊠	⊠	□	□	□	⊠	⊠		⊠			□	⊠	□	□	⊠	⊠	⊠	□	□		□		⊠				⊠	□	□				⊠	⊠	⊠	□	□	⊠
⊠	⊠	⊠	□	⊠	◆	⊠	⊠	⊠	□	⊠		□	□	⊠			⊠	⊠	⊠	□	□	□	⊠	⊠	⊠	□	□	□	⊠	⊠	⊠	□			⊠	⊠	□		◆	□
	□	⊠	⊠	⊠		⊠	⊠			□		□	⊠	⊠				⊠	⊠	⊠																			□	⊠
□	□							□	□	□				□		□	⊠	⊠	□		□		□					□	□	■	■	⊠		⊠	⊠	□		□		□
□	⊠	□	⊠	⊠	□	□	⊠	⊠		⊠			⊠	□		□	⊠	⊠	⊠	□	⊠		⊠						⊠	⊠	⊠	⊠				⊠			□	□
□	⊠	□	⊠	⊠		□	□			⊠		⊠	■	⊠				■	■	⊠			□						⊠							⊠			⊠	■
⊠	⊠	⊠	⊠	⊠	◆	□	⊠	⊠		⊠	□	□	⊠	⊠		□	⊠	⊠	⊠	□	◆		□					□	□	■	■	⊠		□	⊠	□		□	◆	⊠
⊠	⊠	⊠	■	⊠	◆	□	⊠	□		■		⊠	⊠	⊠	□	□	⊠	■	■	■	⊠		■					⊠	■	■	■	■		□	□	■		□	■	■
⊠	⊠	⊠	⊠	□	□	⊠	⊠	□		⊠		□	□	⊠	□	□	⊠	⊠	⊠	□	□		■						■	■	■	⊠			□	■			⊠	⊠
⊠	⊠	⊠	⊠	□	□	□	⊠	⊠	□	□	□	□	⊠	⊠	□	⊠	⊠	⊠	⊠	□	⊠	□	⊠	⊠	⊠	□	□	□	⊠	⊠	⊠	□	□	□	⊠	⊠	□	□	□	⊠
□	■	⊠	⊠	⊠	□	⊠	⊠	□		⊠		⊠	⊠	⊠		□	⊠	⊠	⊠	⊠	⊠		■						■	■	■	■				■			⊠	⊠
		□	⊠	□		□	⊠			⊠		□	⊠	⊠				⊠	⊠	⊠																			□	⊠
■	■	■	⊠	◆	□	⊠	■	⊠		■		■	■	■	□	⊠	■	■	■	⊠	⊠		■	⊠	⊠	□			■	■	■	⊠			■	■			■	■
□	⊠	■	■	⊠	□	⊠	⊠	⊠		⊠		⊠	■	⊠		⊠	⊠	■	■	■	□		⊠						⊠	⊠	□					⊠			■	■
⊠	⊠	⊠	■	◆	□	⊠	■	■	□	■	□	⊠	■	■		□	⊠	■	■	■	□		⊠		⊠				■	⊠	⊠	⊠			□	■		□	◆	◆
⊠	⊠	⊠	■	□		⊠	⊠	⊠		■		⊠	⊠	■			⊠	■	■	■	□		⊠	⊠	⊠	⊠			⊠	□	□	□			⊠	⊠			⊠	■
□	□	⊠	■	◆		□	⊠	□		⊠		■	⊠	⊠		□	□	■	⊠	⊠	□		⊠						⊠		⊠					⊠			⊠	■
□	⊠	⊠	⊠	⊠		□	⊠	□		⊠		■	■	⊠		□	⊠	⊠	■	⊠	◆		⊠	⊠	□	⊠			⊠	□	⊠	□			□	⊠			◆	◆
□	⊠	⊠	■	◆	□	⊠	⊠	□	□	⊠		■	■	■		□	⊠	■	■	⊠	□		■						■		■	■				■			■	■
□	□			□		□	□	□					■	□			□	◆	■		⊠		■						■	□	■					■			□	◆
⊠	⊠	□			□	□	□	⊠		⊠			□	⊠		□	⊠	⊠	⊠				■	⊠	⊠	⊠			⊠	⊠	⊠	□			⊠	⊠				
		□	□	⊠		□	⊠			⊠		□	⊠	□				■	⊠																				□	⊠
⊠	■	⊠	□	□	□	□	□	⊠	□	⊠		□	⊠	⊠		⊠	⊠	■	⊠	□	⊠		□	□					⊠	⊠	⊠	□				⊠			⊠	⊠
⊠	⊠	□	□	□		□	□	⊠	□	□		□	⊠	⊠			⊠	⊠	⊠	□	□		⊠						⊠		⊠					⊠				□
⊠	⊠	⊠	⊠	⊠	⊠	⊠	⊠	⊠	⊠	⊠		⊠	⊠	⊠			⊠	◆	⊠	⊠	⊠		□	⊠	⊠	□			⊠	⊠	⊠	⊠			⊠	⊠	□		⊠	⊠
		□	⊠	⊠		□	⊠			⊠		□	□	⊠				⊠	■	⊠	□																		⊠	⊠
□	⊠		□	⊠	□	□	□	□		□			□	□		□	⊠	⊠	⊠		⊠		⊠						⊠	⊠	⊠	□				⊠			□	□
	□			⊠				□					⊠					⊠	⊠		⊠		□						⊠	⊠	⊠	□				■				
⊠	⊠	⊠	⊠		◆	⊠	⊠	□		⊠		⊠	⊠	⊠		◆	⊠	⊠	⊠	⊠	◆	□	⊠	⊠	⊠	⊠	□	⊠	⊠	⊠	⊠	⊠			⊠	⊠			◆	⊠
⊠	□	□	⊠	⊠	□	□	⊠	⊠		□		□	⊠	⊠		□		⊠	⊠	□	□			□	□					□					⊠	□	⊠		□	⊠
■	⊠	□	⊠	□	□		□	⊠		□			⊠	□		□	⊠	⊠	⊠	⊠	⊠		⊠	□	⊠	□			⊠	⊠	⊠	⊠			⊠	⊠			□	⊠
□	⊠	□	□	□	□	□	□	□		□		□	□	□		□	⊠	⊠	□	□	□		⊠	□					■	□	⊠	⊠				■			□	⊠
⊠	⊠	⊠	□	⊠	□	⊠	⊠	⊠	□	⊠		⊠	□	⊠		□	⊠	⊠	□	□	□		⊠		□	□		◆	⊠	⊠	⊠	⊠			◆	⊠			⊠	⊠
⊠	⊠	⊠	■	⊠	□	⊠	⊠	⊠		⊠		⊠	⊠	⊠	□	□	⊠	⊠	⊠	■	⊠		⊠	⊠	⊠	⊠			⊠	⊠	⊠	⊠			⊠	⊠	⊠	□	⊠	⊠
⊠	⊠	□	□					⊠					□	□		⊠		⊠	⊠	□	□		□						□	⊠	□				□	⊠			□	⊠
□	⊠	⊠	⊠	□		□		□		□	□	⊠	⊠	□		□	⊠	◆	□		⊠		⊠						⊠		⊠	□				□			◆	◆
⊠	⊠	⊠	■	◆	◆	⊠	⊠	⊠	□	⊠	□	□	⊠	⊠		□	⊠	⊠	⊠	⊠	◆		□					□	□	■	■	⊠		□	⊠	□		□	■	⊠
□	⊠	□	□	□	□	□	□	⊠	□	□			□	□		□	⊠	⊠	⊠	□	□		⊠	⊠	⊠	□		□	⊠	⊠	⊠	□		□	⊠	⊠		□	⊠	⊠
⊠	□	⊠	⊠	⊠	□	⊠	⊠	□		⊠		□	□	■	□	□	⊠	⊠	⊠	⊠	⊠		□		□				□	■	■			□	⊠	□			⊠	⊠
□	⊠	□	⊠	⊠	□	□	⊠	□		⊠		⊠	⊠	⊠		□	□	⊠	⊠	□	□		⊠						■	□	⊠	□			□	■			⊠	⊠
■	■	⊠	■	□	◆	□	◆	◆	⊠	⊠	□	□	□	■		◆	■	■	■	■	◆		□						□	■	⊠	□		□	⊠	□			⊠	■
		⊠	■	⊠		⊠	⊠			⊠		⊠	⊠	⊠				⊠	■	■																			◆	◆
□	⊠	⊠	■	⊠		⊠	⊠	□		⊠		□	⊠	⊠			□	■	■	■	□		■	⊠	⊠	⊠			■		⊠				⊠	■			⊠	⊠
	□	□	⊠	⊠		⊠	⊠	□		⊠		■	⊠	■		□	□	■	⊠				⊠						⊠		⊠								⊠	■

Key ■ = Abundant ⊠ = Common □ = Uncommon ◆ = Seasonal

GLOSSARY

access used commonly as a verb (to reach, get to, or approach), but also as a noun. An access refers to a place near a river, stream or lake where visitors may reach the water. Not all access areas have boat ramps. Some are simply streamside trails or gravel bars where people may wade and fish or launch a canoe. When searching for a boat ramp, it is best to check the index.

cantilever sign hangs from a single suspended post, shingle-style. All Conservation Department areas are identified with signs, unless an area is a recent acquisition or under construction.

CAP lake acronym for Community Assistance Program, a Department-sponsored program which provides fisheries management to local communities for developing and up-grading public use areas, usually lakes or ponds. In return, local communities maintain the area and facilities, and open their lakes to the general public.

demonstration site area used to show visitors how land can be managed for the benefit of wildlife or improvement of other natural resources. For example, certain trees may be planted near a stream to help stabilize the banks and control erosion.

Department of Natural Resources operates state parks and historic sites for preservation and interpretation, and recreation such as fishing, boating, swimming, hiking and camping. The organization is also responsible for enforcing pollution control laws, regulating solid and hazardous wastes, responding to environmental emergencies, monitoring drinking water supplies, administering soil and water conservation programs and overseeing the wise use of mineral and water resources.

Designated Natural Areas refuges for biodiversity and examples of what Missouri looked like in the time before European settlement. They are biological communities or geographical sites that preserve and are managed to perpetuate the natural character, diversity and ecological processes of Missouri's native landscapes.

ditch any artificial drainageway, tributary to a stream or body of water containing sufficient water to support fish.

fen seepy, bog-like communities that are usually fertile sites for unusual plants.

glade naturally occurring rocky openings, usually surrounded by woodlands. They are most common in hilly terrain and have shallow, rocky soil. Glades generally are good places to observe wildflowers and native grasses.

hand launch a gravel area on a lake, river or stream bank suitable for putting in a canoe or other shallow-drafting boat.

loess a buff to gray color, fine-grained, calcareous (chalky) silt or clay.

nature center houses indoor exhibits, offers educational nature programs and outdoor opportunities to visitors free of charge. The Conservation Department has nature centers in Blue Springs, Springfield, Kirkwood and Jefferson City. The centers are featured on pages 56, 83, 104 and 202.

primitive camping water, cooking fuel and other basic supplies must be brought in by the user; no special facilities provided. Camping on these areas is intended as an adjunct of hiking, hunting, fishing, nature study and similar uses.

public contact office provides for the sale of permits and publications and is staffed by MDC employees.

savannah area with sparse trees and a rich ground covering of grasses, sedges and other flowering plants.

spoil banks mounds of rock and soil found on areas previously mined.

towersite areas where MDC has (or previously had) observation towers used to help detect forest fires.

Trophy Trout Management Area specifically managed to provide trophy-sized trout. Within these designated areas, the daily limit is three trout and only trout 15 inches or longer may be kept. Anglers always should check for regulations posted at area bulletin boards.

U.S. Army Corps of Engineers manages some Missouri lakes and surrounding land for flood control, hydroelectric power, recreation, water supply, navigation and fish and wildlife conservation.

Missouri Department of Conservation

Offices and Nature Centers

NORTHWEST

Blue Springs
816/228–3766

Burr Oak Woods Conservation Nature Center
816/228–3766

Kansas City
816/356–2280

St. Joseph
816/271–3100

NORTH CENTRAL

Brookfield
816/258–7226

Chillicothe
816/646–6122

Kirksville
816/785–2420

NORTHEAST

Columbia
573/882–9880
573/884–6861

Hannibal
573/248–2530

Mexico
573/581–7220

WEST CENTRAL

Clinton
816/885–6981

El Dorado Springs
417/876–5226

Sedalia
816/530-5500

EAST CENTRAL

Powder Valley Conservation Nature Center
314/301-1500

Rolla
573/368–2225

St. Charles
314/441–4554

St. Louis
314/301–1500

Sullivan
573/468–3335

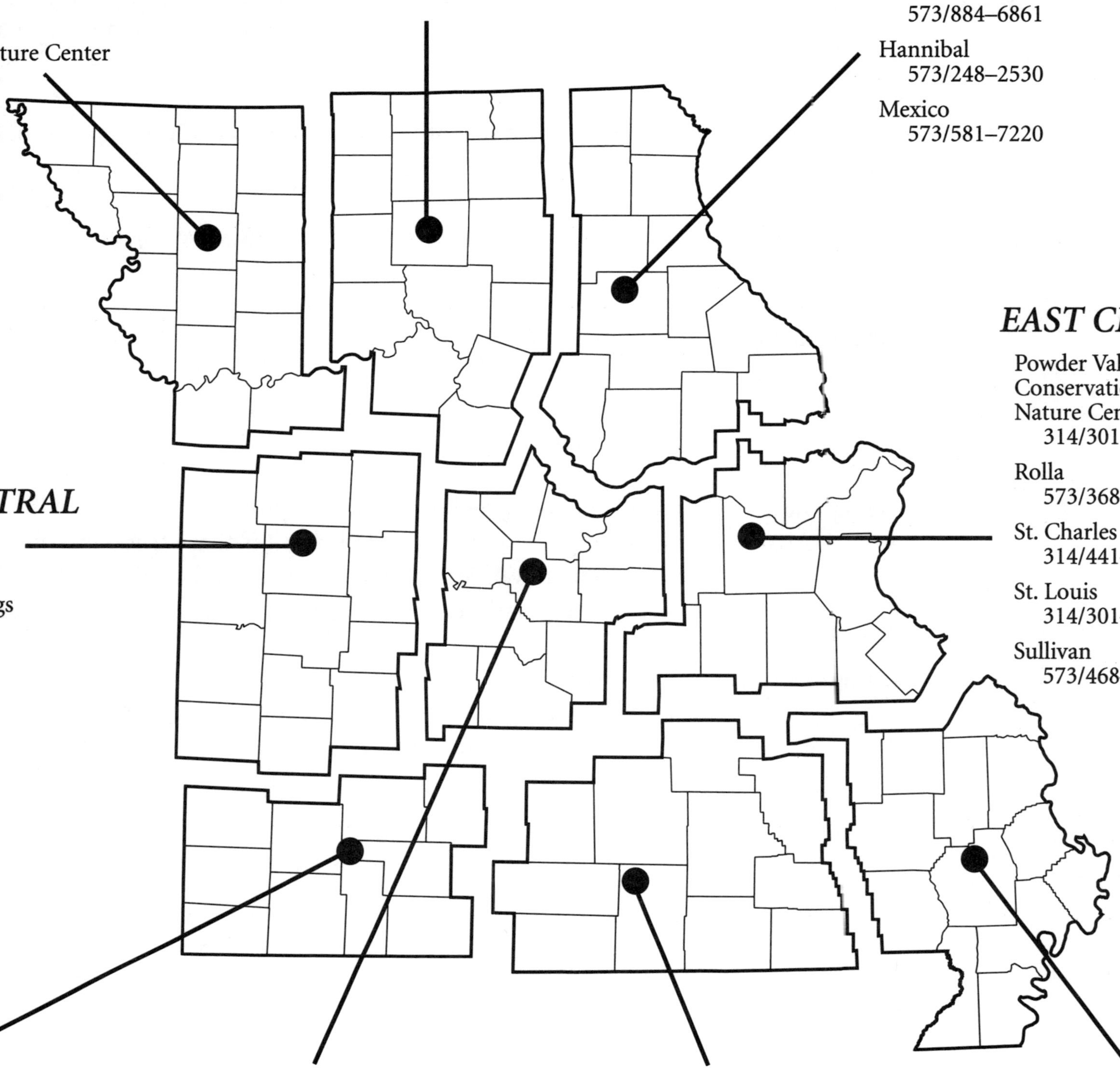

SOUTHWEST

Neosho
417/451-4158

Springfield
417/895–6880

Springfield Conservation Nature Center
417/888-4237

CENTRAL

Camdenton
573/346–2210

Jefferson City
573/751–4115

Lebanon
417/532–7612

Runge Conservation Nature Center
573/526-5544

OZARK

Eminence
573/226–3616

Salem
573/729–3182

West Plains
417/256–7161

SOUTHEAST

Cape Girardeau
573/290–5730

Dexter
573/624–7483

Perryville
573/547–4537

Piedmont
573/223–4525

Poplar Bluff
573/840-9788

Puxico
573/222–3337